Advanced Mechanics of Solids

Build on the foundations of elementary mechanics of materials texts with this modern textbook on the analysis of stresses and strains in elastic bodies.

Key features include the following.

- Presentation of advanced strength of materials through an integrated framework that focuses on four key components: computational tools, a step-by-step methodology for problem solving, treatment of the work–energy concept, and solving advanced strength of materials problems.
- A force-based finite element method alongside the conventional displacement-based (stiffness) finite element method.
- Detailed description of both uniform and nonuniform torsion problems, including the nonuniform torsion of members with general cross-sections.
- Consideration of three-dimensional stress, strain, and stress–strain relations in detail with matrix–vector relations.
- Extensive integration of MATLAB® throughout.
- A complete online teaching package that includes slides, a solutions manual, and MATLAB® code.

Based on classroom-proven material, this valuable resource provides a unified approach useful for advanced undergraduate and graduate students, practicing engineers, and researchers.

Lester W. Schmerr Jr. is Emeritus Professor of Aerospace Engineering at Iowa State University.

Advanced Mechanics of Solids
Analytical and Numerical Solutions with MATLAB®

Lester W. Schmerr, Jr.
Iowa State University

CAMBRIDGE
UNIVERSITY PRESS

University Printing House, Cambridge CB2 8BS, United Kingdom

One Liberty Plaza, 20th Floor, New York, NY 10006, USA

477 Williamstown Road, Port Melbourne, VIC 3207, Australia

314–321, 3rd Floor, Plot 3, Splendor Forum, Jasola District Centre, New Delhi – 110025, India

79 Anson Road, #06–04/06, Singapore 079906

Cambridge University Press is part of the University of Cambridge.

It furthers the University's mission by disseminating knowledge in the pursuit of education, learning, and research at the highest international levels of excellence.

www.cambridge.org
Information on this title: www.cambridge.org/9781108843317
DOI: 10.1017/9781108910132

© Lester W. Schmerr Jr. 2021

This publication is in copyright. Subject to statutory exception and to the provisions of relevant collective licensing agreements, no reproduction of any part may take place without the written permission of Cambridge University Press.

First published 2021

Printed in the United Kingdom by TJ Books Limited, Padstow Cornwall

A catalogue record for this publication is available from the British Library.

Library of Congress Cataloging-in-Publication Data

Names: Schmerr, Lester W., author.
Title: Advanced mechanics of solids : analytical and numerical solutions with Matlab® / Lester W. Schmerr, Jr., Iowa State University.
Description: New York, NY : Cambridge University Press, [2021] | Includes bibliographical references and index.
Identifiers: LCCN 2020037696 (print) | LCCN 2020037697 (ebook) | ISBN 9781108843317 (hardback) | ISBN 9781108910132 (epub)
Subjects: LCSH: Engineering mathematics.
Classification: LCC TA350 .S362 2021 (print) | LCC TA350 (ebook) | DDC 620.1/1292–dc23
LC record available at https://lccn.loc.gov/2020037696
LC ebook record available at https://lccn.loc.gov/2020037697

ISBN 978-1-108-84331-7 Hardback

Cambridge University Press has no responsibility for the persistence or accuracy of URLs for external or third-party internet websites referred to in this publication and does not guarantee that any content on such websites is, or will remain, accurate or appropriate.

Contents

Preface	page xi
1 Introduction	1
1.1 Axially Loaded Members	1
1.2 Beam Bending	6
1.2.1 Pure Bending	6
1.2.2 Engineering Beam Theory	9
1.3 Torsion of a Circular Shaft	11
1.4 Limitations and Extensions of the Theories	16
1.5 The Foundations of Deformable Body Problems	17
1.6 About This Book	19
1.7 Problems	20
Reference	27
2 Stress	28
2.1 The Stress Vector	28
2.1.1 Stress Vector on an Arbitrary Plane	30
2.2 Normal and Shear Stresses on an Oblique Plane	32
2.2.1 Stress Components on Oblique Planes	33
2.2.2 Stress Transformation Equations	34
2.3 A Relationship between the Normal Stress and Total Shear Stress	41
2.4 Principal Stresses and Principal Stress Directions	42
2.4.1 Special Case when $I_3 = 0$, Plane Stress	50
2.4.2 Special Case when Principal Stresses are Equal	51
2.5 Mohr's Circle for Three-Dimensional States of Stress	53
2.6 Stresses on the Octahedral Plane	60
2.7 Stress Notations and the Concept of Stress – A Historical Note	61
2.8 Problems	63
References	68
3 Equilibrium	69
3.1 Equations of Equilibrium – Cartesian Coordinates	69
3.1.1 Plane Stress	72

3.2 Strength of Materials Solutions	73
3.3 Force and Moment Equilibrium	76
3.4 Equations of Equilibrium – Cylindrical and Spherical Coordinates	80
3.4.1 Plane Stress	84
3.5 Problems	84

4 Strain   88

4.1 Definitions of Strains	88
4.1.1 Normal Strain	88
4.1.2 Shear Strain	89
4.2 Strain–Displacement Relations and Strain Transformations	91
4.2.1 Dilatation	98
4.2.2 Plane-Strain Principal Strains	98
4.3 Strain Compatibility Equations	99
4.3.1 Plane-Strain and Plane-Stress Compatibility Equations	102
4.3.2 Multiply Connected Bodies	103
4.4 Satisfaction of Compatibility – Strength of Materials Solutions	104
4.5 Strains in Cylindrical Coordinates	112
4.6 Problems	114
References	118

5 Stress–Strain Relations   119

5.1 Linear Elastic Materials	119
5.1.1 Linear Elastic Isotropic Materials	119
5.1.2 General Linear Elastic Stress–Strain Relations	122
5.2 Plane Stress and Plane Strain	129
5.3 Transformation of Elastic Constants	131
5.4 States of Stress and Strain on a Surface from Strain Gage Measurements	139
5.5 Problems	142

6 Governing Equations and Boundary Conditions   146

6.1 Governing Equations in Three Dimensions	146
6.2 Governing Equations in Two Dimensions	150
6.2.1 Plane Stress	150
6.2.2 Plane Strain	154
6.3 Boundary Conditions	158
6.3.1 Saint-Venant's Principle	160
6.4 Governing Equations in Matrix–Vector Form	163
6.5 Equivalent Algebraic Matrix–Vector Governing Equations for Structures	166
6.5.1 Displacement Formulation	170
6.5.2 Stress (Force) Formulation	172

6.6	Structural Analysis – A Brief Preview	174
6.7	Problems	176
	References	179

7 Analytical Solutions — 180

- 7.1 Displacement-Based Solutions — 180
  - 7.1.1 Axisymmetric Solutions in Cylindrical Geometries — 180
  - 7.1.2 Thick-Wall Pressure Vessel — 183
  - 7.1.3 Shrink-Fits — 187
- 7.2 Airy Stress Function Solutions in Cartesian Coordinates — 190
  - 7.2.1 Solutions of the Biharmonic Equation in Cartesian Coordinates — 191
  - 7.2.2 A Simply Supported Beam — 191
- 7.3 Airy Stress Function Solutions in Polar Coordinates — 197
  - 7.3.1 Michell's Solutions — 197
  - 7.3.2 Stress Concentration at a Circular Hole in a Large Plate — 198
  - 7.3.3 Pure Bending of a Curved Beam — 202
  - 7.3.4 Comparison with a Curved Beam Strength of Materials Solution — 205
  - 7.3.5 Concentrated Force on a Wedge — 208
  - 7.3.6 Concentrated Force on a Planar Surface — 209
  - 7.3.7 Elastic Bodies in Contact — 212
- 7.4 Stress Singularities — 214
- 7.5 Problems — 217
- References — 223

8 Work–Energy Concepts — 224

- 8.1 Work Concepts — 224
  - 8.1.1 Work for a Deformable Body — 224
- 8.2 Work–Strain Energy — 226
  - 8.2.1 Linear Elastic Material — 227
  - 8.2.2 Isotropic Case — 230
  - 8.2.3 Distortional Strain Energy — 232
- 8.3 Complementary Strain Energy — 233
  - 8.3.1 Linear Elastic Material — 234
- 8.4 Strain Energy for Strength of Materials Problems — 234
  - 8.4.1 Axial Loads — 235
  - 8.4.2 Bending — 236
  - 8.4.3 Torsion — 238
- 8.5 Principle of Virtual Work and Minimum Potential Energy — 239
  - 8.5.1 General Deformable Body — 239

8.6 Principle of Complementary Virtual Work and Minimum Complementary
    Potential Energy   249
    8.6.1 General Deformable Body   249
8.7 Work–Energy Principles and Discrete Forces and Moments   254
    8.7.1 Virtual Work and Potential Energy   255
    8.7.2 Complementary Virtual Work and Complementary Potential Energy   259
    8.7.3 Principle of Least Work   263
8.8 Reciprocity   267
8.9 Problems   270
References   275

9 Computational Mechanics of Deformable Bodies   276

9.1 Numerical Solutions – Axial Loads   277
    9.1.1 Principles of Virtual Work and Complementary Virtual Work   277
9.2 Stiffness-Based Finite Elements for Axial-Load Problems   281
9.3 Force-Based Finite Elements for Axial-Load Problems   295
    9.3.1 A Summary and Discussion   307
9.4 Generation of the Compatibility Equations   312
9.5 Numerical Solutions – Beam Bending   315
    9.5.1 Principles of Virtual Work and Complementary Virtual Work   316
9.6 Stiffness-Based Finite Elements for Beam Bending   320
9.7 Force-Based Finite Elements for Beam Bending   329
9.8 Some Extensions of a Force-Based Approach   337
9.9 Finite Elements for General Deformable Bodies   356
    9.9.1 Stiffness-Based Finite Elements   356
    9.9.2 Force-Based Finite Elements   358
9.10 The Boundary Element Method   361
9.11 Problems   366
References   371

10 Unsymmetrical Beam Bending   372

10.1 Multiple Axis Bending of Nonsymmetrical Beams   372
10.2 Shear Stresses in Thin, Open Cross-Section Beams   384
10.3 The Shear Center for Thin, Open Cross-Section Beams   388
10.4 Thin, Closed Cross-Section Beams   402
10.5 Problems   404

11 Uniform and Nonuniform Torsion   409

11.1 Torsion of Circular Cross-Sections – A Summary   409
11.2 Uniform Torsion of Noncircular Cross-Sections – Warping Function   411

11.3	Uniform Torsion of Noncircular Cross-Sections – Prandtl Stress Function	417
11.4	Nonuniform Torsion	426
11.5	General Solution of the Nonuniform Torsion Problem	440
11.6	Torsion of Thin, Open Cross-Sections	442
11.7	Torsion of Thin, Closed Cross-Sections	454
11.8	Problems	467
	References	474

12 Combined Deformations 475

12.1	Bending and Torsion of Thin, Open Cross-Sections	475
12.2	Bending and Torsion of Thin, Closed Cross-Sections	477
	12.2.1 Single-Cell Cross-Sections	477
	12.2.2 Multiple-Cell Cross-Sections	479
12.3	Twisting Induced by Axial Stresses	488
12.4	Problems	494
	References	495

13 Material Failure and Stability 496

13.1	Theories of Static Failure	496
	13.1.1 Maximum Normal-Stress Theory	496
	13.1.2 Maximum Shearing-Stress Theory	497
	13.1.3 Maximum Distortional Strain Energy Theory	499
13.2	Fatigue Failure	502
13.3	Fracture Mechanics	508
13.4	Stability	515
13.5	Problems	520
	References	525

Appendix A Cross-Section Properties	527
A.1 Parallel Axis Theorem	529
A.2 Area Moments in Rotated Coordinates and Principal Axes	531
A.3 Calculating Centroids and Area Moments in MATLAB®	533
Appendix B The Beltrami–Michell Compatibility Equations	537
B.1 Compatibility Equations for Stresses	537
Appendix C The Sectorial Area Function	540
Appendix D MATLAB® Files	557
Index	565

Preface

Mechanics of deformable solids is a highly developed discipline that is rooted in the desire to analyze and design safe and reliable structures and components. Engineers first encounter this subject in a strength of materials course that covers the extension, bending, and torsion of simple members, and presents the fundamental concepts of stress and strain. Traditionally, an intermediate or advanced strength of materials course builds on that elementary course and often introduces a wide variety of topics. These include bending of beams with nonsymmetrical cross-sections and curved beams, torsion of noncircular shafts, stability of columns, energy methods, thermal stresses, inelastic behavior, elements of elasticity theory, numerical methods, bending of thin plates, and the membrane theory of thin shells, elements of fatigue and fracture analysis, as well as others. Covering such a wide range of topics often makes an advanced strength text seem like the objective is simply to describe as many of the important results of the past as possible. Having taught elementary, intermediate, and advanced strength courses, as well as courses on elasticity, for over 40 years from different texts and notes, I have developed a deep appreciation of the role that these intermediate and advanced courses play in providing a sound foundation for present-day engineers. This book, *Advanced Mechanics of Solids*, is designed to communicate the course content I have developed to a wider audience and to enhance that content through a new, integrated framework. There are five key elements to the approach found in this text.

(1) A detailed description of stress and strain concepts using modern computational tools.

The treatment of stress and strain in introductory strength of materials courses is very cursory by necessity. These topics are of fundamental importance to understanding the mechanics of deformable bodies, so a higher-level course should cover these concepts in more depth. In this book the discussions of stress, strain, and stress–strain relations are conducted with matrix–vector algebra, and MATLAB® is used to obtain numerical solutions. One benefit of this approach is that it allows students to treat the stress–strain relations for anisotropic materials with little additional effort – a topic that is normally left to elasticity courses or courses on composite materials. In fact, matrices and MATLAB® are used throughout the book, in contrast to the solution-by-hand emphasis found in most other texts.

(2) The formulation and solution of problems in terms of the four fundamental principles (called the four "pillars") of equilibrium, compatibility, stress–strain relations, and boundary conditions.

In elementary strength of materials courses, the problems considered are based on physically and experimentally motivated assumptions on the deformations and/or stresses expected. While this method is very effective for the problems considered in an introductory course, second-level and advanced courses must provide a more fundamental basis that is suitable for treating the complex applications found in modern engineering. The four "pillars" of equilibrium, compatibility, stress–strain relations, and boundary conditions provide this fundamental framework so an entire chapter is devoted to examining the most common ways that stress analyses are structured around these four principles. Important canonical problems such as the stress concentration around a hole, deformation of a thick-walled cylindrical pressure vessel, pure bending of a curved beam, etc., are solved to illustrate the use of these four pillars. I should note that, unlike some other texts, I have not used the Einstein summation notation (also called indicial notation) to describe equations and relations. Although that notation makes expressions more compact and "elegant" by avoiding multiple summation signs, it adds an extra layer of abstraction that I have tried to avoid. Matrix–vector notation, however, is freely used throughout the book because of the frequent use of MATLAB®, and that notation also makes complex relations more compact.

(3) A detailed treatment of work–energy concepts and their connection to numerical solutions with either displacement-based methods or force-based methods.

Since the 1960s, computational mechanics has arisen as the primary tool for analyzing forces and deformations in deformable bodies. As a consequence, most recent advanced texts have included sections on numerical stress analysis, primarily the finite element method. Normally, a finite element discussion centers around a stiffness-based approach using the principle of virtual work. We will also examine this approach. However, we will also discuss a more recently developed force-based finite element method based on both the principle of virtual work and the principle of complementary virtual work. Unlike the stiffness-based finite element method, the force-based method will be shown to be a discrete model of the four "pillars" of stress analysis, thus tying the finite element method more directly to traditional analytical analyses. This force-based method is not described in any current advanced mechanics of materials text. We will also briefly describe a method called the boundary element method that provides an important alternative numerical approach to that of finite elements.

(4) The solution of advanced strength of materials problems such as the multiaxis bending of beams with nonsymmetrical cross-sections and the torsion of noncircular shafts.

Introductory strength of materials courses consider only simple geometries and loadings. There are, however, important extensions of those elementary analyses possible that provide a more complete description. For example, one can examine the bending of beams of nonsymmetrical cross-sections and the torsion of noncircular shafts. The torsion problems are covered in a new, unified way that emphasizes the distinction between uniform and nonuniform torsion problems. In particular, the governing equations for the nonuniform torsion of general cross-sections are developed that describe both the primary and secondary warping deformations and stresses. This formulation, like the force-based finite element method, is also not found in current texts.

(5) A discussion of theories of failure and instability.

The classical static theories for describing brittle and ductile material failure are examined as well as some of the key aspects of fatigue failure. The fracture mechanics of cracks and crack growth are also considered. The role of nondestructive evaluation (NDE) inspections in evaluating the remaining life of structures is briefly described. Finally, bifurcation instabilities (buckling) and other forms of instability – such as limit-load instabilities and snap-through buckling – are examined.

There are, of course, many other topics, as described previously, not covered in this book. The basic philosophy behind the topics that were chosen is to provide a bridge between an elementary strength of materials course and higher-level courses such as elasticity, finite elements, fracture mechanics, advanced structural analysis, and others. To build this bridge, one must emphasize both the fundamentals and the foundations for numerical analyses. The book, however, is not only a textbook, as I have included many derivations and background material that allow the book to be also used as a valuable reference and resource. In a course setting, these materials can be covered selectively. MATLAB® has been used throughout the book as a modern computational tool to replace otherwise tedious calculations. I have titled this book "Advanced Mechanics of Solids" rather than "Advanced Strength of Materials" to emphasize that it does not follow a traditional path in both content and approach. There are significant additional materials available with the book including a solutions manual, PowerPoints of the text figures, PowerPoint lectures, and MATLAB® m-files associated with both the text (see Appendix D) and the solutions. The solutions manual is available in both PDF and PowerPoint forms that give an instructor great flexibility in their use. All of these additional materials are available through the publisher at www.cambridge.org/schmerr.

MATLAB® and Simulink® are registered trademarks of The MathWorks, Inc.

1 Introduction

Mechanics of materials (also called strength of materials) is an area of engineering that describes the deformation of bodies and the distribution of forces within those bodies. Introductory courses in this discipline normally consider three types of simple deformations – the stretching or compression of axially loaded bars (Fig. 1.1a), the bending (flexure) of laterally loaded beams (Fig. 1.1b), and the torsion (twisting) of circular shafts (Fig. 1.1c). Many complex structures are composed of members that have these simple types of behavior. Advanced mechanics of deformable solids, the subject of this book, will examine the behavior of the structures of Fig. 1.1 in more detail and extend those examinations to more complex configurations. We will describe the fundamentals underlying the description of deformation and forces in general, and cover numerical methods that can be used for obtaining solutions. We will make extensive use of vectors and matrices in this book, since quantities such as stress and strain are best described in those terms, and matrix-based numerical methods, such as finite elements, are now in widespread use for solving complex engineering problems. MATLAB® will be used frequently in our problem solutions since this software package is specifically designed to efficiently make matrix–vector calculations.

This chapter will outline the three types of problems seen in Fig. 1.1 to set the stage for our later discussions and to review the basic assumptions made so that those assumptions can be later examined and, in some cases, relaxed.

1.1 Axially Loaded Members

Figure 1.2 shows a deformable bar that is being extended by an axial load, P. Although the bar is shown having a rectangular cross-section, a bar of any constant cross-sectional area can be considered in exactly the same fashion. As we will see in later chapters, this problem can be treated by applying basic principles of equilibrium and deformation. Those principles lead to partial differential equation boundary value problems, which require a fairly comprehensive background in applied mathematics. In an introductory mechanics of materials approach, however, we can learn most of the essentials of this problem by making some strong assumptions on the deformation and load distributions we expect to see. In this case, for example, we assume that the primary deformation of the body is a uniform extension of the body in the x-direction, which is along the axis of the bar. Consider a bar of length L and cross-sectional area, A, shown in Fig. 1.3a. The bar is fixed at $x = 0$ and experiences an x-displacement, Δ, at the end $x = L$ where the load, P, is

2 Introduction

Figure 1.1 (a) An axially loaded member, (b) a laterally loaded beam, and (c) torsion of a circular shaft.

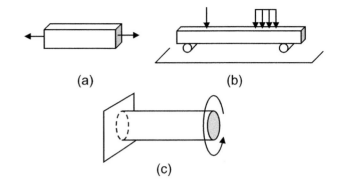

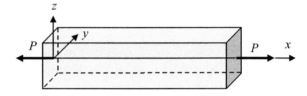

Figure 1.2 An axially loaded member.

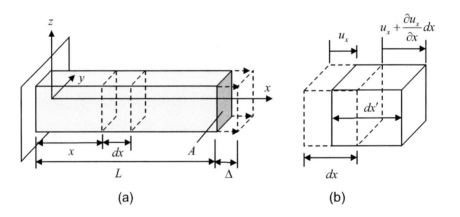

Figure 1.3 (a) The deformation of an axially loaded bar. (b) The local deformation of a small element.

applied. Since the x-displacement, u_x, is zero at $x = 0$ and Δ at $x = L$, we will assume that the displacement at any location, x, just varies linearly between those two limits, i.e.,

$$u_x(x) = \Delta\left(\frac{x}{L}\right) \tag{1.1}$$

It then follows that the displacement gradient, $\partial u_x/\partial x$, is just a constant throughout the bar given by

$$\frac{\partial u_x}{\partial x} = \frac{\Delta}{L} \tag{1.2}$$

[Note: since the displacement here is a function of only the one variable, x, strictly speaking we should write the displacement gradient as a total derivative, i.e., du_x/dx. However, in more general deformations u_x can be a function of all three coordinates (x,y,z) so we will use the partial derivative here in anticipation of that fact.] To see what this displacement gradient means in terms of the physical deformation, consider a small element of the bar of length dx at point x along the initially unloaded bar, as seen in Fig. 1.3a. When the bar is loaded, the left-hand side of this element experiences a displacement, u_x, while the right-hand side, which is located at $x + dx$, sees a slightly different displacement, $u_x + \frac{\partial u_x}{\partial x}dx$. From Fig. 1.3b we see that

$$u_x + dx' = u_x + \frac{\partial u_x}{\partial x}dx + dx \tag{1.3a}$$

or

$$\frac{\partial u_x}{\partial x} = \frac{dx' - dx}{dx} \tag{1.3b}$$

So from Eq. (1.3 b) we see that the displacement gradient is just the elongation of the element divided by its original length, a quantity that is also called the *normal strain* in the x-direction, e_{xx}, i.e.,

$$e_{xx} = \frac{dx' - dx}{dx} = \frac{\partial u_x}{\partial x} \tag{1.4}$$

From Eq. (1.2), therefore, it follows that the normal strain is also a constant in the bar.

These results, of course, are a consequence of our assumptions on the deformation in the bar. We can verify that the normal strain $e_{xx} = \Delta/L$ by measuring this strain (at least on the surfaces of the bar) with resistance strain gages. (See Chapter 5 for a discussion of strain gages.) However, such gages will reveal that there are also normal strains e_{yy}, e_{zz} in the y- and z-directions, respectively. For a material that is isotropic and linearly elastic (see generalized Hooke's law for an isotropic material in Chapter 5) we have

$$e_{yy} = e_{zz} = -\nu e_{xx} \tag{1.5}$$

Here ν is a material constant called *Poisson's ratio*. Thus, an axially loaded bar has a deformation characterized by the three normal strains (e_{xx}, e_{yy}, e_{zz}) and the picture of the deformations in Fig. 1.3 is actually incorrect and should also show the bar cross-sectional area reducing in the y- and z-directions (because of the minus sign in Eq. (1.5)) when the bar is being stretched in the x-direction.

Figure 1.4 (a) The bar separated into two parts, showing the internal normal force, F_x, acting in the bar. (b) The force/unit area (stress) acting across the cross-section, where it is assumed this stress is uniformly distributed.

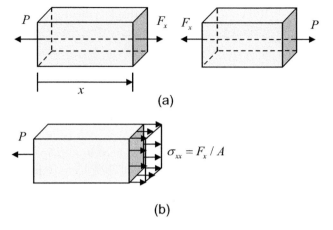

Now, let's turn our attention to the forces acting in the bar. If we imagine passing an imaginary cutting plane through the bar whose normal is along the x-axis and that separates the bar into two parts (Fig 1.4a), by equilibrium we see that there must be an internal normal force, F_x, acting on the imaginary cross-section and that we must have $F_x = P$. This result is true regardless of the x-location of the cutting plane so that the internal normal force must be a constant in the bar. However, F_x is actually the resultant of forces that are distributed over the entire cross-sectional area A at the cutting plane. Since the bar, according to our deformation assumptions, has a uniform strain deformation throughout the bar, it reasonable to assume that the internal force, F_x, is also distributed uniformly across the cross-sectional area. Under that assumption, therefore, we have a uniformly distributed force/unit area acting across A in the x-direction, called the *normal stress*, σ_{xx}, where (see Fig. 1.4b)

$$\sigma_{xx} = \frac{F_x}{A} = \frac{P}{A} \tag{1.6}$$

We can describe the stress acting within the bar by cutting out a very small element at any location and showing the normal stress, σ_{xx}, acting on this element (Fig. 1.5a). Since there is only one stress acting on the element this is called a uniaxial state of stress.

There is a restriction on how the load P is applied if our assumption of a uniform stress σ_{xx} distribution is to be valid. This can be seen by relating the stress distribution to the total internal forces and moments in the bar that σ_{xx} can produce. If we assume that the internal force, $F_x = P$, acts at a point C in the cross-section (see Fig. 1.5a), then by integrating over the cross-sectional area A of a cut, as shown in Fig. 1.5b, we must have

$$F_x = \int_A \sigma_{xx} dA = P$$

$$M_y = \int_A z\sigma_{xx} dA = 0 \tag{1.7}$$

$$M_z = -\int_A y\sigma_{xx} dA = 0$$

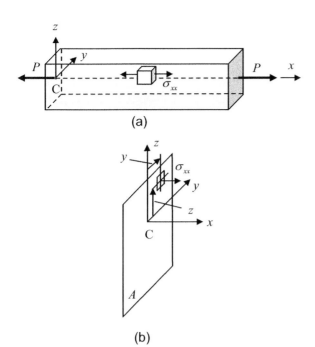

Figure 1.5 (a) The stress, σ_{xx}, acting on an element within an axially loaded bar. (b) The stress, σ_{xx}, acting on a cross-sectional cut.

where F_x is the total internal force in the x-direction and (M_y, M_z) are the internal moments about point C that F_x can produce in the y- and z-directions (the moments are zero since we have assumed F_x acts through C). We see that the force equation is satisfied by a uniform stress $\sigma_{xx} = P/A$ but the moment equations can only be satisfied if

$$\int_A y dA = \int_A z dA = 0 \tag{1.8}$$

Equation (1.8) requires that the origin of the (x, y, z)-coordinate system, point C, through which the applied load P acts, must be the centroid of the cross-sectional area, A. If P acts at a noncentroidal point then moments (M_y, M_z) may be nonzero. These moments will produce nonuniform strains across the cross-section, thus rendering our deformation assumptions invalid.

For a body that deforms in a linear elastic manner the normal stress, σ_{xx}, and the normal strain, e_{xx}, are related through *Hooke's law*:

$$\sigma_{xx} = E e_{xx} \tag{1.9}$$

where E is a material constant called *Young's modulus*. Relating the stress to the applied load P and the strain to the elongation Δ of the bar, we then have equivalently the load–elongation relation for the bar

$$P = \left(\frac{EA}{L}\right)\Delta \tag{1.10}$$

where the quantity $K = EA/L$ acts like a linear spring constant.

1.2 Beam Bending

1.2.1 Pure Bending

Lateral loads acting on a beam, as shown in Fig. 1.1b, will cause that beam to bend. To describe the deformation of the beam, let us first examine the special case of so-called pure bending, where the beam is subjected to applied moments, M, as shown in Fig. 1.6a. In this case, internally in the beam there is constant internal moment, $M_y = -M$, as can be seen from equilibrium of any section of the beam, as shown in Fig. 1.6b. Because any element of the beam is subjected to this same moment, we expect all elements to deform identically. Thus, consider any small element of the beam when the beam is undeformed, as shown in Fig. 1.7a. We will

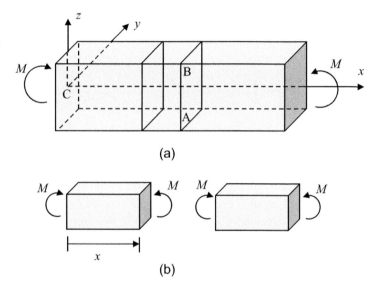

Figure 1.6 (a) Pure bending of a beam. (b) The internal moment of the beam, showing it is a constant throughout the beam and equal to the applied bending moment.

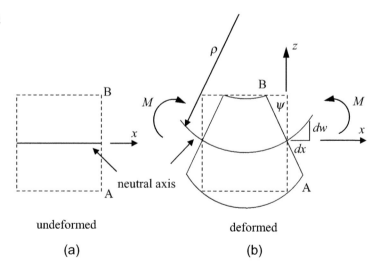

Figure 1.7 (a) An undeformed element of a beam. (b) The deformed element when a pure bending moment is applied. These deformations are greatly exaggerated in order to show them clearly.

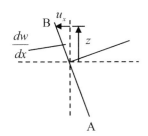

Figure 1.8 The deformed geometry of the beam, showing the x-displacement of the cross-sectional plane AB.

assume that the bending action will cause plane cross-sections of the beam such as AB to rotate through a small angle ψ about a line in the cross-section, called the *neutral axis*, without distortion, i.e., the plane section AB is assumed to remain plane after the deformation. This deformation is shown in Fig. 1.7b. [Note: all the deformations shown in Fig. 1.7 are exaggerated so that they can be more easily displayed.] We see that the neutral axis itself will rotate through a small angle dw/dx (since the tangent of a small angle is approximately equal to the angle itself), where $u_z = w(x)$ is the displacement of the neutral axis in the z-direction. If we also assume that cross-sectional planes initially at right angles to the neutral axis remain at right angles after the deformation, then we also have $\psi = dw/dx$. Under these deformation assumptions the beam will experience a displacement in the x-direction, u_x, given by (see Fig. 1.8):

$$u_x = -z \frac{dw}{dx} \tag{1.11}$$

This displacement in turn will produce an axial normal strain given by

$$e_{xx} = \frac{\partial u_x}{\partial x} = -z \frac{d^2 w}{dx^2} \tag{1.12}$$

From analytical geometry we know that if $w(x)$ describes the z-displacement of the neutral axis and the slope of the neutral axis, dw/dx, is small then d^2w/dx^2 is just the reciprocal of the radius of curvature of the neutral axis, ρ (see Fig. 1.7b), i.e.,

$$\frac{d^2 w}{dx^2} = \frac{1}{\rho} \tag{1.13}$$

In addition to our assumptions on the deformation, we will also assume that the only significant stress in the beam is the normal stress, σ_{xx}, and, like the case of the axially loaded bar, this stress will be related to the strain by Hooke's law, i.e., $\sigma_{xx} = E e_{xx}$. Thus,

$$\sigma_{xx} = -Ez \frac{d^2 w}{dx^2} = \frac{-Ez}{\rho} \tag{1.14}$$

Equations (1.12) and (1.14) show that there is no axial strain or stress at $z = 0$, which is the reason that this axis is called the *neutral axis*. Although there is only a single normal stress present in the beam, as in the axial load case there will be strains in the y- and z-directions, which we will write in terms of the axial strain and Poisson's ratio:

$$e_{yy} = e_{zz} = -\nu e_{xx} \tag{1.15}$$

Also, as in the axial load case, the stress of Eq. (1.14) must produce the appropriate internal forces and moments, which gives (see Fig. 1.5b):

$$\int_A \sigma_{xx} dA = F_x = 0$$

$$\int_A y\sigma_{xx} dA = -M_z = 0 \qquad (1.16)$$

$$\int_A z\sigma_{xx} dA = M_y = -M$$

where F_x is the internal force in the x-direction and (M_y, M_z) are internal moments acting in the y- and z-directions, respectively. Placing Eq. (1.14) into Eq. (1.16) we find

$$\int_A z dA = 0$$

$$I_{yz} = 0 \qquad (1.17)$$

$$\frac{EI_{yy}}{\rho} = -M_y = M$$

where

$$I_{yz} = \int_A yz\, dA$$

$$I_{yy} = \int_A z^2 dA \qquad (1.18)$$

The quantities I_{yz}, I_{zz} are geometrical properties of the cross-sectional area A called the mixed second area moment and second area moment, respectively (see Appendix A). The first condition in Eq. (1.17) says that the neutral axis must go through the centroid of the beam cross-sectional area, A. The second condition requires that the mixed second area moment, I_{yz}, must vanish. This can be satisfied as long as either the y- or z-axis is an axis of symmetry for the cross-section. Finally, the last condition in Eq. (1.17) is called the *moment–curvature relationship* since it relates the internal bending moment in the beam to the radius of curvature of the neutral axis of the deformed beam. Because the bending moment is a constant throughout the beam, this relationship shows that the radius of curvature is also a constant so that the shape of the deformed neutral axis is a circular arc. The moment–curvature relationship can also be written in terms of the z-displacement of the neutral axis as

$$\frac{d^2 w}{dx^2} = -\frac{M_y}{EI_{yy}} \qquad (1.19)$$

which, when placed into Eq. (1.14), leads to the bending (flexure) stress expression

$$\sigma_{xx} = \frac{M_y z}{I_{yy}} = -\frac{Mz}{I_{yy}} \qquad (1.20)$$

1.2.2 Engineering Beam Theory

When beams are used in practice, they normally support lateral loads (see Fig. 1.9a) rather than just the applied moment considered previously. Physically, this has two consequences. First, the internal bending moment in the beam is no longer a constant. This varying internal moment can be found by applying equilibrium to an arbitrary section of the beam, as shown in Fig. 1.9b. Second, from the moment–curvature relationship the curvature of the neutral axis will also not be a constant. In engineering beam theory, it is assumed that at each location x the beam acts as if it was in pure bending but where the internal moment and curvature are allowed to vary with the location x in the beam. Thus, the governing engineering beam theory equations for the displacement, strain, and stress are assumed to be the same as in the pure bending case but where now $M_y = M_y(x)$ and $1/\rho = 1/\rho(x) = d^2w(x)/dx^2$, giving

$$u_x(x,z) = -z\frac{dw(x)}{dx}$$

$$e_{xx}(x,z) = -z\frac{d^2w(x)}{dx^2}, \quad e_{yy} = e_{zz} = -\nu e_{xx}$$

$$\frac{d^2w(x)}{dx^2} = -\frac{M_y(x)}{EI_{yy}} \tag{1.21}$$

$$\sigma_{xx}(x,z) = Ee_{xx} = \frac{M_y(x)z}{I_{yy}}$$

However, as seen in Fig. 1.9b, equilibrium conditions will also require that an internal force acting in the z-direction, $V_z(x)$, must exist in addition to the internal bending moment, $M_y(x)$. This force is called a *shear force* and it will be produced by a distribution of *shear stresses*, σ_{xz}, acting tangent to a cross-sectional cut in addition to the bending normal-stress distribution, as shown in Fig. 1.10b. When we examine a general state of stress in a body in Chapter 2, we will see that these shear stresses will act on both horizontal and vertical cutting

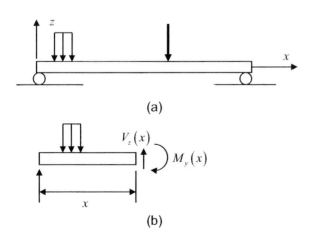

Figure 1.9 (a) A simply supported beam carrying lateral loads. (b) A free-body diagram of a section of the beam showing the internal shear force and bending moments, which are both functions of the location, x, in the beam.

Figure 1.10 (a) The internal shear force and bending moment acting on an AB cutting plane in a beam and the corresponding state of stress present in the beam. (b) The shear-stress and normal-stress distributions on plane AB that produce the shear force and bending moment.

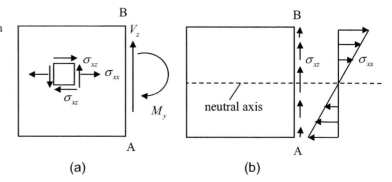

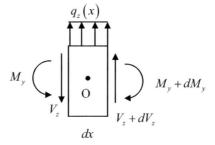

Figure 1.11 A small element of a beam showing the shear force, bending moment and applied load/length acting on it.

planes, so that a small element within the beam will have the normal and shear stresses shown in Fig. 1.10a.

The variation of the shear force in the beam is related to the bending moment. We can obtain this relationship by looking at the equilibrium of a small element of the beam as shown in Fig. 1.11. The varying shear force and bending moment are shown, as is any applied lateral load/unit length distributed along the beam. Summing moments about the center point O of the element, and noting that for a small element the load/unit length is nearly a constant so gives no net moment about O, we have

$$\sum M_{Oy} = (M_y + dM_y) - M_y - V_z \frac{dx}{2} - (V_z + dV_z)\frac{dx}{2} = 0 \quad (1.22)$$

which reduces to

$$\frac{dM_y}{dx} = V_z(x) \quad (1.23)$$

To obtain the shear-stress distribution we can imagine passing a horizontal cutting plane through a small element at a height z' above the neutral axis to obtain a section of area $A(z')$ of the beam as shown in Fig. 1.12, where the thickness of the beam at the bottom of this area is $t(z')$. In the x-direction there will be a varying normal stress acting on the front and back faces of the section as well as an average shear stress, σ_{xz}, acting on the bottom face area tdx. From equilibrium in the x-direction we find

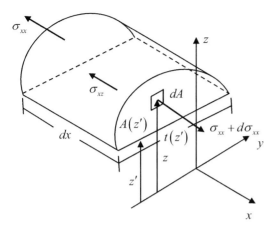

Figure 1.12 Geometry of a section of the beam for obtaining the shear stress.

$$\int_{A(z')} (\sigma_{xx} + d\sigma_{xx})dA - \int_{A(z')} \sigma_{xx}dA - \sigma_{xz}tdx = 0 \tag{1.24}$$

which gives

$$\sigma_{xz} = \frac{1}{t(z')} \int_{A(z')} \frac{d\sigma_{xx}}{dx} dA$$

$$= \frac{1}{I_{yy}t(z')} \frac{dM_y}{dx} \int_{A(z')} z dA \tag{1.25}$$

$$= \frac{V_z(x)Q(z')}{I_{yy}t(z')}$$

where $Q(z')$ is a first area moment of the area $A(z')$. This is the formula for the shear stress usually derived in elementary mechanics of materials texts. For a rectangular cross-section where the thickness, t, is a constant, it is easy to show that the shear stress has a quadratic variation in the distance z', being zero at the top and bottom of the beam and a maximum at the neutral axis.

1.3 Torsion of a Circular Shaft

Now consider a circular shaft of radius c and length L that is fixed to a rigid plane at $x = 0$ and is subjected to a torque, T_0, that produces a total twist, ϕ_0, at $x = L$ where the torque is applied (Fig. 1.13). In the deformation we will assume that cross-sectional planes whose normals are along the x-axis rotate about the bar axis. As seen in Fig. 1.14a, if two cross-sectional planes A and B, for example, both rotate through the same amount then the

Figure 1.13 A circular shaft subjected to a torque.

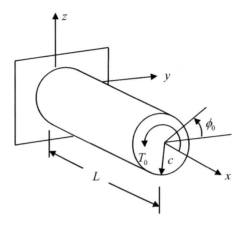

Figure 1.14 (a) The case where two cross-sectional planes A and B of a shaft have the same rotation, causing the shaft to have a rigid body rotation (no deformation). (b) The case where the plane at A is fixed and plane B rotates, causing the shaft to twist.

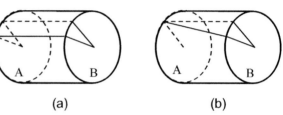

(a)   (b)

bar simply has a rigid body rotation and there is no deformation. However, if cross-section A remains fixed but plane B rotates, then the bar between A and B will experience a twist, as shown in Fig. 1.14b. Thus, we will assume that the applied torque causes the cross-sectional planes of the shaft to rotate through a small angle, $\phi(x)$, which varies from plane to plane, i.e., it is a function of x. For the bar of Fig. 1.13 ϕ varies from zero at the fixed end $x = 0$ to the angle (twist) ϕ_0 at $x = L$ where the torque T_0 is applied. We will assume this variation is linear so $\phi(x) = \phi_0 x/L$.

When a cross-sectional plane rotates through a small angle, $\phi(x)$, points in that cross-section will have a displacement in the θ-direction, u_θ, that varies linearly in the radial distance r, as shown in Fig. 1.15a. Thus, the assumed deformation of the shaft can be described by the displacements

$$\begin{aligned} u_x &= 0 \\ u_\theta &= r\phi(x) \end{aligned} \quad (1.26)$$

We can decompose the displacement u_θ into y- and z-components, as seen in Fig. 1.15b, so we have finally the assumed displacements of the shaft in Cartesian coordinates given by

$$\begin{aligned} u_x &= 0 \\ u_y &= -u_\theta \sin\theta = -r\phi(x)\sin\theta \\ &= -z\phi(x) \\ u_z &= u_\theta \cos\theta = r\phi(x)\cos\theta \\ &= y\phi(x) \end{aligned} \quad (1.27)$$

1.3 Torsion of a Circular Shaft 13

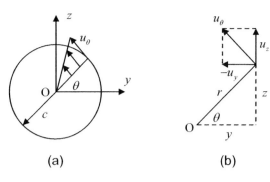

Figure 1.15 (a) The displacement of a cross-section due to a rotation about the *x*-axis. (b) The decomposition of that displacement into *y*- and *z*-components.

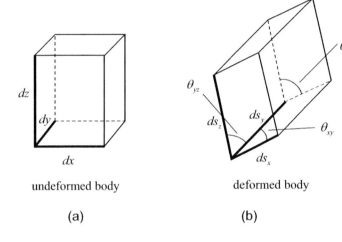

Figure 1.16 (a) A small element in an undeformed body whose sides are parallel to a set of Cartesian axes. (b) When the body is loaded this element is transformed into the deformed shape seen, where the lengths of the sides are changed as are the angles between the sides.

Before we continue with the torsion problem, we need to say a few words about strains in a more general setting. Figure 1.16a shows a small element in a body before any loads are applied so that it is undeformed. When the body is deformed this element will change to the element shown in Fig. 1.16b. The lengths of the sides of the element will have changed from (dx, dy, dz) to (ds_x, ds_y, ds_z) and the angles between the sides will have changed from $\pi/2$ (90°) to the angles $(\theta_{xy}, \theta_{xz}, \theta_{yz})$. If these changes are all small, then the element will experience small normal strains (e_{xx}, e_{yy}, e_{zz}) defined by

$$e_{xx} = \frac{ds_x - dx}{dx}$$
$$e_{yy} = \frac{ds_y - dy}{dy} \quad (1.28)$$
$$e_{zz} = \frac{ds_z - dz}{dz}$$

which can be also written in terms of displacement gradients (see Chapter 4)

$$e_{xx} = \frac{\partial u_x}{\partial x}$$

$$e_{yy} = \frac{\partial u_y}{\partial y} \qquad (1.29)$$

$$e_{zz} = \frac{\partial u_z}{\partial z}$$

These normal strains characterize completely the changes of lengths of the small element. We can also define strains that characterize the small changes in the angles between the sides of the element. These are called *engineering shear strains* ($\gamma_{xy}, \gamma_{xz}, \gamma_{yz}$) and are given by

$$\gamma_{xy} = \frac{\pi}{2} - \theta_{xy}$$

$$\gamma_{xz} = \frac{\pi}{2} - \theta_{xz} \qquad (1.30)$$

$$\gamma_{yz} = \frac{\pi}{2} - \theta_{yz}$$

In Chapter 4 we will see that these strains can also be written in terms of displacement gradients as

$$\gamma_{xy} = \frac{\partial u_x}{\partial y} + \frac{\partial u_y}{\partial x}$$

$$\gamma_{xz} = \frac{\partial u_x}{\partial z} + \frac{\partial u_z}{\partial x} \qquad (1.31)$$

$$\gamma_{yz} = \frac{\partial u_y}{\partial z} + \frac{\partial u_z}{\partial y}$$

From the assumed displacements in our torsion problem, Eq. (1.27), we obtain the strains

$$e_{xx} = e_{yy} = e_{zz} = 0$$

$$\gamma_{xy} = -\frac{d\phi}{dx}z$$

$$\gamma_{xz} = \frac{d\phi}{dx}y \qquad (1.32)$$

$$\gamma_{yz} = 0$$

We also assume that the shear stresses ($\sigma_{xy}, \sigma_{xz}, \sigma_{yz}$) are just proportional to these engineering shear stresses, so in the shaft we have

$$\sigma_{xy} = G\gamma_{xy} = -G\phi'z$$

$$\sigma_{xz} = G\gamma_{xz} = G\phi'y \qquad (1.33)$$

$$\sigma_{yz} = G\gamma_{yz} = 0$$

where $\phi' = d\phi/dx$ is the twist per unit length in the shaft and G is the shear modulus.

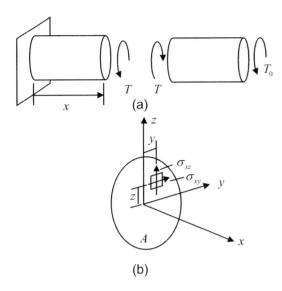

Figure 1.17 (a) Passing a cutting plane through the shaft, showing the internal torque, T, acting on a cross-section. (b) The shear stresses acting on a cross-section.

By equilibrium the internal torque, T, acting across any cross-section of the shaft, as shown in Fig. 1.17a, must be a constant equal to the applied torque, T_0. This internal torque is produced by the shear stresses acting on the cross-section (Fig. 1.17b) so that

$$T = \int_A (y\sigma_{xz} - z\sigma_{xy})dA$$
$$= G\phi' \int_A (y^2 + z^2)dA \qquad (1.34)$$
$$= G\phi' J$$

where

$$J = \int_A (y^2 + z^2)dA = \int_A r^2 dA \qquad (1.35)$$

is the *polar area moment* of the cross-sectional area, A. Since $T = T_0$ is a constant throughout the shaft and the twist/unit length, $\phi' = \phi_0/L$, it follows from Eq. (1.34) that

$$\phi_0 = T_0 L/GJ \qquad (1.36)$$

which is the moment–twist relation for the bar. Using Eq. (1.34) in Eq. (1.33) it follows that the shear stresses can also be written as

$$\sigma_{xy} = \frac{-Tz}{J}$$
$$\sigma_{xz} = \frac{Ty}{J} \qquad (1.37)$$

Figure 1.18 Combining the shear stresses acting on a cross-section to form the total shear stress, τ.

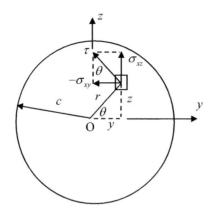

The total shear stress, τ, in the cross-section (Fig. 1.18) acts perpendicular to a radial direction whose origin is at the center of the circular cross-section and is given by

$$\begin{aligned}\tau &= -\sigma_{xy}\sin\theta + \sigma_{xz}\cos\theta \\ &= -\sigma_{xy}\frac{z}{r} + \sigma_{xz}\frac{y}{r} \\ &= \frac{Tr}{J}\end{aligned} \qquad (1.38)$$

which is a familiar result seen in elementary mechanics of materials texts and shows that the maximum total shear stress occurs on the outer boundary of the cross-section where $r = c$.

1.4 Limitations and Extensions of the Theories

The elementary theories we have presented of axial extension, bending, and torsion have varying degrees of restrictions on the geometries and loads involved. The axial-load case has some of the fewest restrictions since the bar can have any cross-sectional area and still have a uniform stress distribution $\sigma_{xx} = P/A$ acting throughout the bar (as long as the applied load acts through the centroid of the cross-section as the theory requires). The actual distribution of the applied loads at the ends of the bar, however, can locally produce highly nonuniform stresses near those ends. Thus, a bar must be long enough for the elementary theory to be valid over the most of the bar length. The distribution of the end axial loads can also have significant effects on a thin structure, producing twisting of the structure in additional to an axial extension. This will cause variations of the normal stress in the cross-section along the entire length of the structure and introduce a new load quantity call a bimoment. We will discuss this case in Chapter 12. For the elementary theory to be valid, strictly speaking the bar must have a constant cross-section, but the theory can still be applied when the cross-sectional area is slowly varying and the theory can also accommodate having axial loads distributed along the length of the bar. In both cases the stress will no longer be a constant, i.e., we will have $\sigma_{xx}(x) = F_x(x)/A(x)$. Such extensions of the theory are often found in elementary mechanics of materials texts.

The elementary theory of beam bending requires that the mixed second area moment $I_{yz} = 0$ so that theory is typically valid only for beams having symmetrical cross-sections. If a bending moment is applied about the y-axis to a beam with an unsymmetrical cross-section, bending deformations will typically be produced about both the y- and z-axes, not just about the y-axis as the elementary theory assumes. Thus, by generalizing the assumed displacements to account for these multiaxis bending effects, one can develop a beam theory that describes the normal stresses and deflections of an unsymmetrical beam (see Chapter 10). The shear-stress formula, $\sigma_{xz} = V_z Q/I_{yy} t$, is more problematic since – except for cross-sections with vertical sides, such as rectangular cross-sections – this shear – stress distribution does not properly satisfy the boundary conditions (see Chapter 6) for the cross-section. However, for general thin cross-sections (which can be symmetrical or unsymmetrical), it is possible to generalize the shear stress formula (which is often written in terms of a *shear flow* (force/length) rather than the shear stress) for those thin sections (see Chapter 10).

The theory of torsion presented in Section 1.3 is valid only for shafts having solid or hollow circular cross-sections. This strong geometry restriction is essential since, if we apply a torque to a noncircular shaft, out-of-plane (axial) warping displacements will be present, rendering our displacement assumptions invalid. The elementary theory, however, can be modified to include warping effects using either a warping function approach or a Prandtl stress function approach (Chapter 11). In either of those approaches one must solve partial differential equation boundary value problems so that computer-based solutions are often employed. For cross-sections consisting of thin closed or open sections, more analytically based formulas can be developed, as shown in Chapter 11.

1.5 The Foundations of Deformable Body Problems

As can be seen from the discussions in the previous sections, the usual approach to problems in elementary mechanics of materials is to make some strong assumptions about the displacement deformations in a particular case and to relate those displacements to the corresponding local deformations as described by the strains. The stress–strain relations of a body then give the corresponding stresses. When we relate the stresses to the applied forces or moments, we then have an explicit relationship between loads and deformations that we can use to describe how a structure will deform and to evaluate the magnitude of the stresses produced by the loads. This elementary approach requires that each new problem be based a different set of assumptions, and the approach becomes more difficult to apply as the geometries and loadings of the structure become more complex. In this book we will continue to apply this elementary approach to extensions of the problems found in elementary texts. However, we will also emphasize a set of basic principles that are based on a

Figure 1.19 The four foundational "pillars" that underlie all deformable body problems.

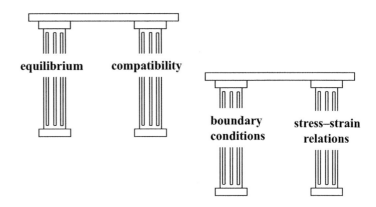

general theory of deformable bodies (also called the theory of elasticity) and provide the foundation for solving even the most difficult problems. These four principles we have called the *four pillars* of all stress analyses (Fig. 1.19). First, there is equilibrium. As we will see in Chapter 3, stresses must locally satisfy equilibrium equations at every point within a body. In Chapter 3 we will examine whether or not the stresses seen in the elementary solutions of this chapter satisfy local equilibrium and, if not, what must be modified in the analysis. Second, compatibility conditions are likewise equations that the strains must satisfy locally at every point in the body. As shown in Chapter 4, these conditions guarantee that the strains will produce physically meaningful displacements of the body, and we will examine in that chapter if the elementary strength of materials solutions satisfy the compatibility relations. In analyzing more complex problems such as the torsion of noncircular shafts, for example, we will see that compatibility plays a significant role in obtaining a solution. Stress, which is a concept discussed extensively in Chapter 2, is a quantity that characterizes how forces are distributed within a body. Similarly, strains, as described in detail in Chapter 4, are measures of the deformation distributions in the body. A third pillar, stress–strain relations, therefore, is a key element that relates these internal force and deformation distributions. In Chapter 5 we will examine stress–strain relations for both isotropic and anisotropic elastic materials. Finally, the fourth pillar involves the conditions that must be satisfied on either displacements or stresses (or combinations of stresses and displacements) at the boundaries of a body. These boundary conditions are essentially a detailed description of how a body is supported or loaded. It is difficult in practice to know precisely the boundary conditions on the displacements and/or stresses so that boundary conditions are one pillar of an analysis that may be difficult to specify. Fortunately, as discussed in Chapter 6, Saint-Venant's principle gives us a practical way to often relax the boundary conditions so that one can deal with any uncertainty that is present.

A solution that satisfies the four pillars of local stress equilibrium, strain compatibility equations, the stress–strain relations, and boundary conditions is an exact solution to a deformable body problem. In most cases one can find exact, analytical solutions to only a few relatively simple problems. In Chapter 7 we will obtain some important exact solutions.

We will also examine numerical solutions (see Chapter 9) that can allow us to solve much more complex problems using methods such as the finite element method. We will see that energy methods (Chapter 8) are often used to replace the pillars of equilibrium and compatibility when formulating such numerical solutions.

1.6 About This Book

Mechanics of materials is a highly developed area of engineering that has a long and rich history with many important developments. In an advanced-level book on this subject, therefore, one has a wide range of topics one can choose to cover. I have chosen not to follow the path of many "Advanced Strength" or "Advanced Mechanics of Materials" texts that provide a very broad coverage of many topics. Instead, I have structured this book as a bridge between elementary strength of materials and higher-level courses where the text stresses the fundamentals underlying the field. Thus, Chapters 2 through 6 cover in some detail the four foundational principles described in the previous section. Chapter 2 gives an in-depth look at the concept of stress while Chapter 4 covers the concept of strain and the important associated concept of compatibility. Both of these concepts are at the heart of how we describe the distributions of forces and deformations in a body. Chapter 3 covers the local equations of equilibrium and, as mentioned previously, reexamines the elementary strength of materials problems described in this chapter in light of those equilibrium conditions. Chapter 5 examines stress–strain relations, including the relations for anisotropic materials such as composite materials. In all these introductory chapters, MATLAB® is used extensively to solve problems since this is a modern engineering tool that is a natural successor to slide rule and calculator solutions that formed the basis of previous generations of such texts. All of these foundational principles are combined in Chapter 6 to describe the governing equations and boundary conditions that must be satisfied to solve deformable body problems. As might be expected, there are different choices one can make in these sets of governing equations. Those choices have important consequences, so in Chapter 6 we describe both displacement-based and stress-based approaches as well as the use of auxiliary functions such as the Airy stress function to formulate problems. Although there are relatively few exact solutions to the governing equations and boundary conditions of deformable bodies, there are some important "canonical" problems that are solvable, such as the thick wall pressure vessel, stress concentration around a hole, and others that can help to show the behavior of the stresses and strains explicitly. Thus, Chapter 7 is devoted to examining a selection of those important problems. In Chapter 8 we consider work–energy concepts – concepts that underlie many of the approaches used for solving problems numerically. It is shown how the principles of virtual work and complementary virtual work can serve as replacements for the equilibrium and compatibility equations, respectively, and so provide an alternative path for formulating and solving problems. In Chapter 9 these work–energy principles are shown to be the natural foundations for finite element methods, which are the computational tools most frequently used to solve complex engineering problems. Problems such as axial-load problems and beam-bending problems are described in Chapter 9 so that all of the fundamental aspects of finite elements can be described in a simple context. A unique aspect of Chapter 9 is that it covers both displacement-based (stiffness) finite elements as well as a force-based finite element method. While the stiffness-based finite element approach is the method normally described, force-based finite elements is an important alternative that has not been adequately covered in texts at any level and is in many respects closer to the fundamental principles of this field (equilibrium, compatibility, etc.) so that it deserves to have a wider audience. Another alternative to stiffness-based finite elements as a computational tool is the boundary element method, so Chapter 9 provides a brief overview of that important method as well. Three chapters of the book, Chapters 10 through 12, reexamine the problems of bending and torsion and axial loads for cases that cannot be treated with the assumptions found in elementary strength of materials. These include the bending of beams with nonsymmetrical cross-sections, the torsion of noncircular shafts, and twisting induced by axial loads. The final chapter, Chapter13, examines various types of material failure and instabilities.

1.7 PROBLEMS

P1.1 In elementary strength of materials courses, one learns how to obtain shear force distributions, $V_z(x)$, and bending moment distributions, $M_y(x)$, in a beam. Normally, this is done by using force and moment equilibrium, considering each section of the beam where the free-body diagram is different. The resulting shear force and bending moment functions are then described by functions that change from section to section in the beam. It would be nice, however, if one could express the shear force and bending moment in terms of functions that are valid for the entire beam. Such functions are called *singularity functions*. For example, consider a long beam which has a concentrated force, P, acting at $x = a$ in the beam as shown in Fig. P1.1a. If we examine equilibrium of the beam at a section taken before and after the load, as shown in Fig. P1.1b, then the shear force and bending moment can be written as

$$V_z(x) = \begin{cases} 0 & x < a \\ P & x > a \end{cases}$$
$$M_y(x) = \begin{cases} 0 & x < a \\ P(x-a) & x > a \end{cases}$$
(P1.1)

But we can express these functions in terms of singularity functions as

$$V_z(x) = P\langle x-a \rangle^0$$
$$M_y(x) = P\langle x-a \rangle^1$$
(P1.2)

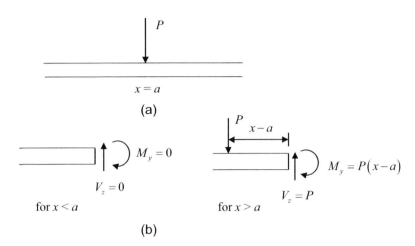

Figure P1.1 (a) A concentrated force on an infinitely long beam. (b) The shear force and bending moments in the beam both before and after where the force is applied.

where the singularity functions are defined as

$$\langle x - a \rangle^n = \begin{cases} 0 & x < a \\ (x - a)^n & x > a \end{cases} \quad n \geq 0 \tag{P1.3}$$

Note that singularity functions can be differentiated and integrated like ordinary functions since

$$\frac{d}{dx}\langle x - a \rangle^n = n\langle x - a \rangle^{n-1} \quad n \geq 1$$

$$\int \langle x - a \rangle^n dx = \frac{\langle x - a \rangle^{n+1}}{n + 1} + C \quad n \geq 0 \tag{P1.4}$$

(a) Determine the shear force $V_z(x)$ and bending moment $M_y(x)$ for the load distributions shown in Fig. P1.2 in terms of singularity functions and use the appropriate singularity functions to describe the loaded beam shown in Fig. P1.3. Sketch the shear force and bending moment distributions.

(b) Write a MATLAB® function that implements these singularity functions. The most commonly occurring first four of these distributions are shown in Fig. P1.4. Call this function s_func. The general MATLAB® function call should be

```
y = s_func(x, a, n);
```

where x is a vector of position values that span the length of the entire beam, a is the x-location where the singularity function begins, and n is an integer that is greater than or equal to zero that defines the power of the function. Note that for $n = 0$ the function has a discontinuity at $x = a$ so let this function = 1.0 for $x \geq a$ and zero for $x < 0$. The other functions are all continuous at $x = a$.

Figure P1.2 Load distributions for (a) a concentrated couple, (b) a concentrated force, (c) a constant force/unit length that starts at $x = a$, and (d) a linearly increasing force distribution that starts at $x = a$.

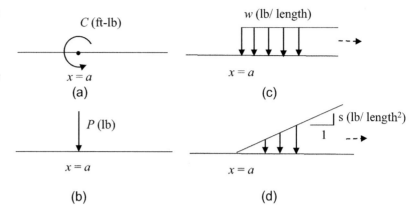

Figure P1.3 A beam in equilibrium under the given loads.

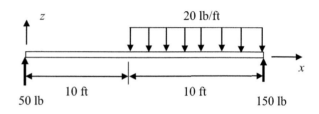

Figure P1.4 Four singularity functions, which are (a) a step function, (b) a linear function, (c) a quadratic function, and (d) a cubic function, all starting at $x = a$ and zero for $x < a$.

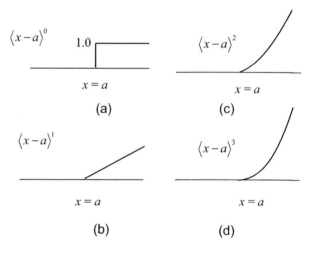

Use these singularity functions to express the shear force and bending moment in MATLAB® for the beam problem of part (a) and plot the shear force and bending moment distributions for the entire beam.

P1.2 Consider a beam with the rather complex load distribution shown in Fig. P1.5. Note that the origin of the x-coordinate is not at the left-hand end of the beam. Also note

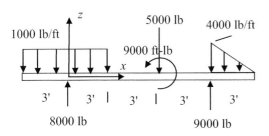

Figure P1.5 A beam in equilibrium under the given loads.

that the singularity functions of Fig. P1.2 for distributed loads start at $x = a$ but they never stop, so distributions such as the 1000 lb/ft load in Fig. P1.5 must be described by the superposition of several singularity functions.

(a) Use the singularity functions obtained in problem P1.1(a) to express the shear force and bending moment for the entire beam.
(b) Express the shear force and bending moment in terms of the MATLAB® singularity function developed in problem P1.1(b) and plot the shear force and bending moment for the entire beam.

Note that once we have the shear force, V_z, we can get the bending moment, M_y, directly from integration since

$$M_y(x) = \int V_z(x)dx + C \qquad (P1.5)$$

where C is a constant of integration. We can use Eq. (P1.5) directly since from Eq. (P1.4) we know how to integrate the singularity functions. The net constant term after the integrations of all the singularity functions are done will be zero if the bending moment is zero for x-values less than the starting point for the beam (which is $x = -3$ in the present example) and this is always the case so we can set $C = 0$.

(c) Show that your results for the bending moment of part (a) are consistent with using Eq. (P1.5) to integrate the shear force expression.

You may wonder why these functions are called singularity functions, because all the functions we have defined do not have any singular behavior. But if, for example, we differentiate the function $\langle x - a \rangle^0$ we will get a singular function called a delta function that is zero everywhere except $x = a$ where the delta function is infinite. Further differentiations will produce even more singular functions [1]. Such singular functions arise if we try to express concentrated loads or concentrated couples as special types of distributed loads but when we work directly with the shear force and bending moment expressions, as done here, there is no need to introduce truly singular functions since the shear force and bending moment are always relatively well-behaved functions (although they have "jumps" at concentrated forces or moments).

P1.3 Since we know how to integrate singularity functions (see problem P1.2), if we obtain the bending moment in terms of singularity functions, we can then integrate the moment–curvature relationship of Eq. (1.19) given by

$$\frac{d^2w}{dx^2} = -\frac{M_y}{EI_{yy}} \tag{P1.6}$$

to obtain the beam deflection. There will be constants of integration in this process, which can be found from the displacement or rotation boundary conditions. For the beam shown in Fig. P1.6:
(a) Obtain the reaction forces at the supports A and B and write the bending moment for the beam in terms of singularity functions, using the results of problem P1.1.
(b) Integrate the moment–curvature relationship of Eq. (P1.6) to obtain the deflection, $w(x)$, in terms of singularity functions. Plot the deflection of the beam, using the MATLAB® function of problem P1.1.

P1.4 A cantilever beam (Fig. P1.7a) is a simple structure often considered in elementary strength of materials texts. Consider the case when there is a small gap, Δ, between the

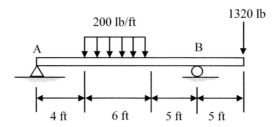

Figure P1.6 A simply supported beam.

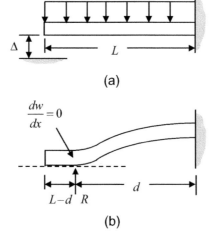

Figure P1.7 (a) A cantilever beam where there is a small gap, Δ, between the beam and a rigid floor. (b) The case where the distributed load is large enough to cause a portion of the beam to be in contact with the floor.

beam and a rigid floor, where the distributed load, q_0, is large enough to close the gap. As a consequence, a reaction force, R, exists at the left side of the beam.

(a) Determine an expression for the reaction force, R, in terms of Δ and q_0. Similarly, determine an expression for the slope of the beam at the reaction force.

(b) What is the smallest value of q_0, $q_0 \equiv q_1$, that will just close the gap? What is the value of q_0, $q_0 \equiv q_2$, that will cause the slope of the beam at the reaction force to go to zero?

(c) The slope of the beam at the floor cannot be less than zero since we have assumed that the floor is rigid. Thus, for $q_0 > q_2$ a portion of the beam must come in contact with the floor, as shown schematically in Fig. P1.7b. In this case, determine the reaction force, R, in terms of Δ and q_0. Plot the reaction force, R, versus q_0 from $q_0 = 0$ to values $q_0 > q_2$. What is the difference between the cases when $q_0 < q_2$ and $q_0 > q_2$?

Let Young's modulus of the beam be E and let the area moment $I_{yy} = I$.

P1.5 The moment–curvature relationship for beams (Eq. (1.19)) allows us to obtain the beam deflection from a knowledge of the internal bending moment. This relationship also allows us to solve statically indeterminate problems where the internal moment cannot be obtained from equilibrium alone. Consider, for example, the beam shown in Fig. P1.8 where the beam is loaded by a constant distributed load/unit length, q_0, over its entire length, L, and is rigidly fixed at both ends A and B. Use the moment–curvature relationship to obtain the reactions at A and B.

Problems P1.6 to P1.9 are axial-load problems (see Fig. P1.9) where you are asked to obtain analytical solutions for the displacement and internal force and plot your results for some specific problem values. These problems are also used in Chapter 9, where you are asked to obtain these solutions numerically via a number of different finite element methods. Thus, these analytical solutions can serve as a check on those numerical finite element results.

P1.6 The bar shown in Fig. P1.9a is fixed between two rigid walls and carries a uniform load $q_x = q_0$. The cross-sectional area of the bar is A, and its Young's modulus is E. Determine the internal axial force $F_x(x)$ acting in the bar and its displacement, $u_x(x)$. Plot the displacement and force for $q_0 = 10$ N/mm, $E = 10^4$ N/mm^2, $A = 9$ mm^2, $L = 3000$ mm.

P1.7 The bar shown in Fig. P1.9b is fixed between two rigid walls. A concentrated axial force, P, acts at its center. The cross-sectional area of the bar is A, and its Young's modulus is E. Determine the internal axial force $F_x(x)$ acting in the bar and its displacement, $u_x(x)$. Plot the displacement and force for $P = 20\,000$ N, $E = 10^4$ N/mm^2, $A = 9$ mm^2, $L = 3000$ mm.

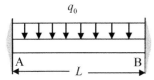

Figure P1.8 A statically indeterminate beam problem where the beam is rigidly fixed at both supports A and B.

Figure P1.9 (a) A bar fixed at both ends with a uniform distributed force $q_x = q_0$ (force/unit length) acting over its entire length. (b) A bar fixed at both ends with a concentrated force, P, at its center. (c) A bar fixed at both ends with a uniform distributed force $q_x = q_0$ (force/unit length) acting over its right half. (d) A bar fixed at both ends with a linearly increasing distributed force $q_x = q_0 x/L$ (force/unit length) acting over its entire length.

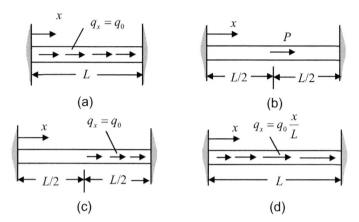

P1.8 The bar shown in Fig. P1.9c is fixed between two rigid walls. A constant distributed axial force, q_0, (force/unit length) acts over its right half. The cross-sectional area of the bar is A, and its Young's modulus is E. Determine the internal axial force $F_x(x)$ acting in the bar and its displacement, $u_x(x)$. Plot the displacement and force for $q_0 = 10 \text{ N/mm}$, $E = 10^4 \text{ N/mm}^2$, $A = 9 \text{ mm}^2$, $L = 3000 \text{ mm}$.

P1.9 The bar shown in Fig. P1.9d is fixed between two rigid walls. A linearly varying distributed force, $q_x = q_0 x/L$ (force/unit length) acts over its entire length. The cross-sectional area of the bar is A, and its Young's modulus is E. Determine the internal axial force $F_x(x)$ acting in the bar and its displacement, $u_x(x)$. Plot the displacement and force for $q_0 = 10 \text{ N/mm}$, $E = 10^4 \text{ N/mm}^2$, $A = 9 \text{ mm}^2$, $L = 3000 \text{ mm}$.

Problems P1.10 to P1.13 are beam problems (see Fig. P1.10) where you are asked to obtain analytical solutions for the displacement and internal shear force and bending moment and plot your results for some specific problem values. These problems are also used in Chapter 9, where you are asked to obtain these solutions numerically via a number of different finite element methods. Thus, these analytical solutions can serve as a check on those numerical finite element results.

P1.10 The beam shown in Fig. P1.10a is fixed at both ends. A force, P, acts at its middle. The Young's modulus of the beam is E and its area moment about the y-axis is $I_{yy} = I$. Determine the internal moment, $M(x) = -M_y(x)$, and shear force, $V_z(x)$, acting in the beam and the vertical deflection, $w(x)$, of the neutral axis. Plot the moment, shear force, and displacement in the beam for $P = 20{,}000 \text{ N}$, $E = 2 \times 10^5 \text{ N/mm}^2 (200 \text{ GPa})$, $I = 3 \times 10^6 \text{ mm}^4$, $L = 3000 \text{ mm}$.

P1.11 The beam shown in Fig. P1.10b is fixed at both ends. A constant distributed force, q_0, acts over its entire length. The Young's modulus of the beam is E and its area moment about the y-axis is $I_{yy} = I$. Determine the internal moment, $M(x) = -M_y(x)$, and shear force, $V_z(x)$, acting in the beam and the vertical deflection, $w(x)$, of the neutral axis. Plot the moment, shear force, and displacement in the beam for $q_{10} = 10 \text{ N/mm}$, $E = 2 \times 10^5 \text{ N/mm}^2 (200 \text{ GPa})$, $I = 3 \times 10^6 \text{ mm}^4$, $L = 3000 \text{ mm}$.

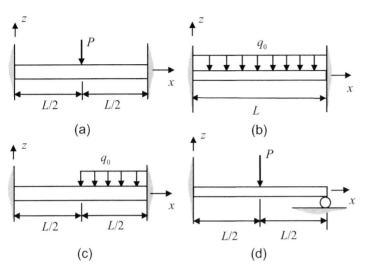

Figure P1.10 (a) A beam fixed at both ends and acted upon by a concentrated force, P, at its center. (b) A beam fixed at both ends and acted upon by a constant distributed force, q_0, over its length. (c) A beam fixed at both ends and acted upon by a constant distributed force, q_0, over its right half. (d) A concentrated force, P, acting on a beam fixed at its left end and sitting on a roller at its right end.

P1.12 The beam shown in Fig. 1.10c is fixed at both ends and carries a constant distributed force, q_0, over its right half. The Young's modulus of the beam is E and its area moment about the y-axis is $I_{yy} = I$. Determine the internal moment, $M(x) = -M_y(x)$, and shear force, $V_z(x)$, acting in the beam and the vertical deflection, $w(x)$, of the neutral axis. Plot the moment, shear force, and displacement in the beam for $q_0 = 10$ N/mm, $E = 2 \times 10^5$ N/mm^2(200 GPa), $I = 3 \times 10^6$ mm^4, $L = 3000$ mm.

P1.13 The beam shown in Fig. 1.10d is fixed at the left end and sits on a roller at the right end. It carries a concentrated force, P, at its center. The Young's modulus of the beam is E and its area moment about the y-axis is $I_{yy} = I$. Determine the internal moment, $M(x) = -M_y(x)$, and shear force, $V_z(x)$, acting in the beam and the vertical deflection, $w(x)$, of the neutral axis. Plot the moment, shear force, and displacement in the beam for $P = 20\,000$ N, $E = 2 \times 10^5$ N/mm^2(200 GPa), $I = 3 \times 10^6$ mm^4, $L = 3000$ mm.

Reference

1. E. Volterra and J. H Gaines, *Advanced Strength of Materials* (Englewood Cliffs NJ: Prentice-Hall, 1971)

2 Stress

Applied forces acting on a body affect both the internal and external deformation of that body. The internal deformations in particular depend on how the forces are distributed throughout the body. *Stress* is a key concept that gives us a way to characterize those internal force distributions. This chapter will discuss in depth the stress concept.

2.1 The Stress Vector

The basic quantity that we use to describe internal force distributions in a deformable body is called the *stress vector* or, equivalently, the *traction vector*. To obtain the stress vector, we imagine that we pass a cutting plane through a point P in a body that separates it into two pieces as seen in Fig. 2.1. The unit normal to this cutting plane (for one of the pieces) is called **n**, as shown. Let the internal force, acting over a small element of area, ΔA, at P be $\Delta \mathbf{F}$. Then the stress vector, $\mathbf{T}^{(\mathbf{n})}(\mathrm{P})$, at point P is defined as

$$\mathbf{T}^{(\mathbf{n})}(\mathrm{P}) = \lim_{\Delta A \to 0} \frac{\Delta \mathbf{F}}{\Delta A} \tag{2.1}$$

which gives a measure of the force/unit area acting at point P. The superscript **(n)** is present in the definition since the stress vector depends on the orientation of this imaginary cutting plane as well as location of the point P. For economy of notation, however, we will normally omit showing the explicit dependency of the stress vector on point P and write it simply as $\mathbf{T}^{(\mathbf{n})}$. Since the internal forces always occur in equal and opposite pairs, it follows that

$$\mathbf{T}^{(-\mathbf{n})} = -\mathbf{T}^{(\mathbf{n})} \tag{2.2}$$

Because the stress vector depends on the orientation (**n**) of the cutting plane through the point P it might appear that, in order to use this concept, we will need to know values of the stress vector for an infinite number of possible cutting planes through point P. However, as we will see shortly, this is not the case and to completely determine the stress vector on any cutting plane we need only to determine the stress vectors on three orthogonal cutting planes. Figure 2.2 shows, for example, the stress vectors $\left(\mathbf{T}^{(\mathbf{e}_x)}, \mathbf{T}^{(\mathbf{e}_y)}, \mathbf{T}^{(\mathbf{e}_z)}\right)$ acting on the three cutting planes whose unit normals are along the (x, y, z) axes, where $(\mathbf{e}_x, \mathbf{e}_y, \mathbf{e}_z)$ are unit vectors for this Cartesian coordinate system (the cutting planes are actually all taken through point P but they are shown as being the front faces of the small cube seen in

2.1 The Stress Vector

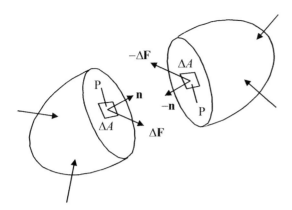

Figure 2.1 The separation of a body into two pieces by an imaginary cutting plane, showing the internal force, ΔF, acting at a point P over a small area, ΔA, of the left-hand piece, as well as the equal and opposite internal force acting on the other piece.

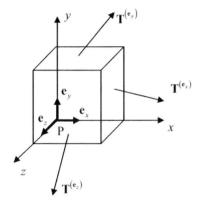

Figure 2.2 The stress vectors acting on three cutting planes at P that are along a set of Cartesian (x, y, z) axes.

Fig. 2.2 (for ease of display). The Cartesian components of these three stress vectors are what we call the stresses at point P with respect to the (x, y, z) axes. For example, $\sigma_{xx}, \sigma_{xy}, \sigma_{xz}$ are the components of the stress vector, $\mathbf{T}^{(\mathbf{e}_x)}$, so that we have

$$\mathbf{T}^{(\mathbf{e}_x)} = \sigma_{xx}\mathbf{e}_x + \sigma_{xy}\mathbf{e}_y + \sigma_{xz}\mathbf{e}_z \tag{2.3}$$

These stresses are shown in Fig. 2.3. If we similarly express the other stress vectors acting along the y- and z-planes into their components, we have altogether

$$\begin{aligned}\mathbf{T}^{(\mathbf{e}_x)} &= \sigma_{xx}\mathbf{e}_x + \sigma_{xy}\mathbf{e}_y + \sigma_{xz}\mathbf{e}_z \\ \mathbf{T}^{(\mathbf{e}_y)} &= \sigma_{yx}\mathbf{e}_x + \sigma_{yy}\mathbf{e}_y + \sigma_{yz}\mathbf{e}_z \\ \mathbf{T}^{(\mathbf{e}_z)} &= \sigma_{zx}\mathbf{e}_x + \sigma_{zy}\mathbf{e}_y + \sigma_{zz}\mathbf{e}_z\end{aligned} \tag{2.4}$$

which, letting $(x, y, z) \to (1, 2, 3)$, we can write more compactly as

$$\mathbf{T}^{(\mathbf{e}_i)} = \sum_{j=1}^{3} \sigma_{ij}\mathbf{e}_j \quad (i = 1, 2, 3) \tag{2.5}$$

These nine stresses form up what is called the *state of stress* at any point in a body (Fig. 2.4). The directions of the stresses

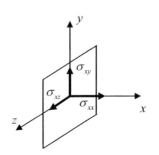

Figure 2.3 The Cartesian stress components acting on a cutting plane with a normal along the x-axis.

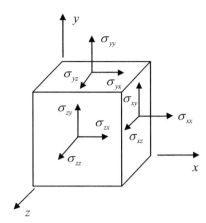

Figure 2.4 Complete state of stress at a point in a deformable body.

$(\sigma_{xx}, \sigma_{yy}, \sigma_{zz})$ are normal to the planes on which they act so they are called *normal stresses*, while the stresses $(\sigma_{xy}, \sigma_{xz}, \sigma_{yx}, \sigma_{yz}, \sigma_{zx}, \sigma_{zy})$ are called *shear stresses*. There are actually only six independent stresses since it can be shown that the shear stresses are symmetric, i.e.,

$$\begin{aligned}\sigma_{xy} &= \sigma_{yx} \\ \sigma_{yz} &= \sigma_{zy} \\ \sigma_{xz} &= \sigma_{zx}\end{aligned} \qquad (2.6)$$

This result is a direct consequence of the moment equation for a small element such as the one shown in Fig. 2.4 but we will not derive Eq. (2.6) here (see problem P2.1). An easy way to remember the meaning of the subscripts associated with these stresses is to note that the first subscript indicates the direction of the normal to the cutting plane on which the stress acts while the second subscript indicates the direction of the stress on the cutting plane. Thus, σ_{yz}, for example, is a stress that acts on a cutting plane, the normal of which is in the y-direction and the stress acts in the z-direction on this plane (see Fig. 2.4).

2.1.1 Stress Vector on an Arbitrary Plane

If we know the stress vectors acting on the (x, y, z) planes (and their corresponding stress components) at point P then we can find the stress vector at P for any other cutting plane as well. Thus, the state of stress at a point completely determines stresses on all planes at that point. To see this, consider cutting out a small tetrahedron at point P whose back faces are parallel to the x-, y-, and z-axes and whose front face is along an oblique plane whose unit normal is **n** (Fig. 2.5).

The area of the oblique face of the tetrahedron is ΔA_n and the areas of the back faces are $(\Delta A_x, \Delta A_y, \Delta A_z)$, as shown in Fig. 2.5. From Newton's second law we have

$$\mathbf{T}^{(\mathbf{n})}\Delta A_n - \mathbf{T}^{(\mathbf{e}_x)}\Delta A_x - \mathbf{T}^{(\mathbf{e}_y)}\Delta A_y - \mathbf{T}^{(\mathbf{e}_z)}\Delta A_z + \rho\mathbf{g}\Delta V = \rho\mathbf{a}\Delta V \qquad (2.7)$$

where ρ is the density (mass/unit volume) of the body, **a** is the acceleration (if the body is in motion), and ΔV is the volume of the

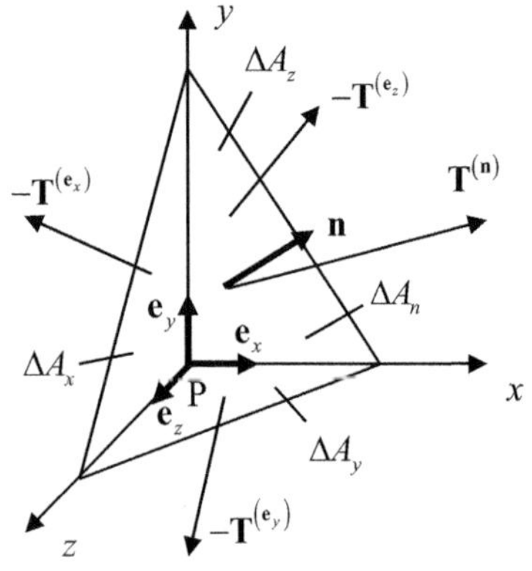

Figure 2.5 Stress vectors acting on the faces of a small tetrahedron at a point P in a body.

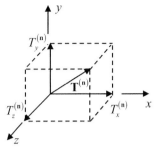

Figure 2.6 The Cartesian components $\left(T_x^{(\mathbf{n})}, T_y^{(\mathbf{n})}, T_z^{(\mathbf{n})}\right)$ of the stress vector $\mathbf{T}^{(\mathbf{n})}$.

tetrahedron. However, one can show (see problem P2.2) that the ratios of the areas of the faces of the tetrahedron are just the components of the unit normal, $\mathbf{n}$, i.e.,

$$n_x = \frac{\Delta A_x}{\Delta A_n}, n_y = \frac{\Delta A_y}{\Delta A_n}, n_z = \frac{\Delta A_z}{\Delta A_n} \tag{2.8}$$

so that we can rewrite Eq. (2.7) as

$$\mathbf{T}^{(\mathbf{n})} - \mathbf{T}^{(\mathbf{e}_x)} n_x - \mathbf{T}^{(\mathbf{e}_y)} n_y - \mathbf{T}^{(\mathbf{e}_z)} n_z + \rho \mathbf{g} \frac{\Delta V}{\Delta A_n} = \rho \mathbf{a} \frac{\Delta V}{\Delta A_n} \tag{2.9}$$

Taking the limit as we shrink the tetrahedron to the point P and using the fact that volume goes to zero faster than the area of the oblique face, we find

$$\mathbf{T}^{(\mathbf{n})} = \mathbf{T}^{(\mathbf{e}_x)} n_x + \mathbf{T}^{(\mathbf{e}_y)} n_y + \mathbf{T}^{(\mathbf{e}_z)} n_z \tag{2.10}$$

which shows that indeed a knowledge of the three stress vectors $\left(\mathbf{T}^{(\mathbf{e}_x)}, \mathbf{T}^{(\mathbf{e}_y)}, \mathbf{T}^{(\mathbf{e}_z)}\right)$ or, equivalently, the state of stress in terms of the stress components given previously allows us to determine the stress vector for any cutting plane taken through the body at point P. If we decompose the stress vector $\mathbf{T}^{(\mathbf{n})}$ into its three components along the (x, y, z) axes (see Fig. 2.6), i.e.,

$$\mathbf{T}^{(\mathbf{n})} = T_x^{(\mathbf{n})} \mathbf{e}_x, T_y^{(\mathbf{n})} \mathbf{e}_y, T_z^{(\mathbf{n})} \mathbf{e}_z \tag{2.11}$$

and use Eq. (2.4) together with Eq. (2.10), then equating components we find

$$T_x^{(\mathbf{n})} = \sigma_{xx} n_x + \sigma_{yx} n_y + \sigma_{zx} n_z$$
$$T_y^{(\mathbf{n})} = \sigma_{xy} n_x + \sigma_{yy} n_y + \sigma_{zy} n_z \tag{2.12}$$
$$T_z^{(\mathbf{n})} = \sigma_{xz} n_x + \sigma_{yz} n_y + \sigma_{zz} n_z$$

or, more compactly,

$$T_i^{(\mathbf{n})} = \sum_{j=1}^{3} \sigma_{ji} n_j \quad (i = 1, 2, 3) \tag{2.13}$$

We can also write Eq. (2.13) in vector–matrix notation by writing the stress vector and unit normal as column vectors and the state of stress as a 3×3 matrix:

$$\mathbf{T}^{(\mathbf{n})} = \left\{T^{(\mathbf{n})}\right\} = \begin{Bmatrix} T_1^{(\mathbf{n})} \\ T_2^{(\mathbf{n})} \\ T_3^{(\mathbf{n})} \end{Bmatrix}, \mathbf{n} = \{n\} = \begin{Bmatrix} n_1 \\ n_2 \\ n_3 \end{Bmatrix}, [\sigma] = \begin{bmatrix} \sigma_{11} & \sigma_{12} & \sigma_{13} \\ \sigma_{21} & \sigma_{22} & \sigma_{23} \\ \sigma_{31} & \sigma_{32} & \sigma_{33} \end{bmatrix} \tag{2.14}$$

which gives

$$\{T^{(\mathbf{n})}\} = [\sigma]^T \{n\} \tag{2.15}$$

where $[\sigma]^T$ is the transpose of the stress matrix $[\sigma]$. However, since the shear stresses are symmetric $[\sigma]^T = [\sigma]$ so that we can also write Eq. (2.15) as

$$\{T^{(\mathbf{n})}\} = [\sigma]\{n\} \tag{2.16}$$

2.2 Normal and Shear Stresses on an Oblique Plane

We can always break the stress vector on an arbitrary cutting plane into two components — one acting along the unit normal, **n**, to the plane, and the other component in the plane on which the stress vector acts. We will call these two components the *normal stress*, σ_{nn}, and the *total shear stress*, τ_s, respectively, and we will let **s** be a unit vector that defines the direction of the total shear-stress component in the plane (see Fig. 2.7). Since σ_{nn} is just the Cartesian component of $\mathbf{T}^{(\mathbf{n})}$ in the **n** direction we have

$$\begin{aligned} \sigma_{nn} &= \mathbf{T}^{(\mathbf{n})} \cdot \mathbf{n} \\ &= T_x^{(\mathbf{n})} n_x + T_y^{(\mathbf{n})} n_y + T_z^{(\mathbf{n})} n_z \end{aligned} \tag{2.17}$$

From the expressions for the stress vector components in terms of the stresses, Eq. (2.12), we find

$$\begin{aligned} \sigma_{nn} &= \sigma_{xx} n_x^2 + \sigma_{yy} n_y^2 + \sigma_{zz} n_z^2 \\ &\quad + 2\sigma_{xy} n_x n_y + 2\sigma_{xz} n_x n_z + 2\sigma_{yz} n_y n_z \end{aligned} \tag{2.18}$$

or, in matrix–vector notation

$$\sigma_{nn} = \{n\}^T [\sigma] \{n\} \tag{2.19}$$

To obtain the total shear stress we note that its magnitude can be found since

$$\begin{aligned} |\tau_s| &= \sqrt{|T^{(n)}|^2 - \sigma_{nn}^2} \\ &= \sqrt{\left(T_x^{(n)}\right)^2 + \left(T_y^{(n)}\right)^2 + \left(T_z^{(n)}\right)^2 - \sigma_{nn}^2} \end{aligned} \tag{2.20}$$

To find the direction of the total shear stress we let

$$\mathbf{s} = s_x \mathbf{e}_x + s_y \mathbf{e}_y + s_z \mathbf{e}_z \tag{2.21}$$

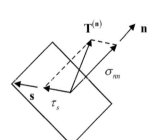

Figure 2.7 The stress vector, $\mathbf{T}^{(\mathbf{n})}$, acting on an arbitrary cutting plane and its normal-stress component, σ_{nn}, and total shear-stress component, τ_s.

and note that

$$|\tau_s|\mathbf{s} = \mathbf{T}^{(\mathbf{n})} - \sigma_{nn}\mathbf{n} \qquad (2.22)$$

which, in component form, leads to three equations that can be solved for (s_x, s_y, s_z)

$$\begin{aligned} |\tau_s|s_x &= T_x^{(\mathbf{n})} - \sigma_{nn}n_x \\ |\tau_s|s_y &= T_y^{(\mathbf{n})} - \sigma_{nn}n_y \\ |\tau_s|s_z &= T_z^{(\mathbf{n})} - \sigma_{nn}n_z \end{aligned} \qquad (2.23)$$

2.2.1 Stress Components on Oblique Planes

The results of the previous section gave us the equations needed to solve for the normal stress and total shear stress on a general oblique cutting plane. In some cases, however, we need to know instead the individual shear-stress components acting on an oblique plane along two mutually orthogonal directions in that plane that we will call t and v (see Fig. 2.8). These shear stress components are σ_{nt}, σ_{nv}. If we let $\mathbf{t}$ and $\mathbf{v}$ be unit vectors in the t and v directions, respectively, then we can write them in terms of their Cartesian components as

$$\begin{aligned} \mathbf{t} &= t_x\mathbf{e}_x + t_y\mathbf{e}_y + t_z\mathbf{e}_z \\ \mathbf{v} &= v_x\mathbf{e}_x + v_y\mathbf{e}_y + v_z\mathbf{e}_z \end{aligned} \qquad (2.24)$$

Just as we obtained the normal stress from the stress vector, Eq. (2.17), we can obtain these shear stress components:

$$\begin{aligned} \sigma_{nt} &= \mathbf{T}^{(\mathbf{n})} \cdot \mathbf{t} \\ \sigma_{nv} &= \mathbf{T}^{(\mathbf{n})} \cdot \mathbf{v} \end{aligned} \qquad (2.25)$$

or, in component form

$$\begin{aligned} \sigma_{nt} &= T_x^{(\mathbf{n})}t_x + T_y^{(\mathbf{n})}t_y + T_z^{(\mathbf{n})}t_z \\ \sigma_{nv} &= T_x^{(\mathbf{n})}v_x + T_y^{(\mathbf{n})}v_y + T_z^{(\mathbf{n})}v_z \end{aligned} \qquad (2.26)$$

which can be expressed in terms of the stresses as

$$\begin{aligned} \sigma_{nt} = {} & \sigma_{xx}n_xt_x + \sigma_{yy}n_yt_y + \sigma_{zz}n_zt_z \\ & + \sigma_{xy}(n_yt_x + n_xt_y) \\ & + \sigma_{xz}(n_zt_x + n_xt_z) \\ & + \sigma_{yz}(n_zt_y + n_yt_z) \end{aligned} \qquad (2.27a)$$

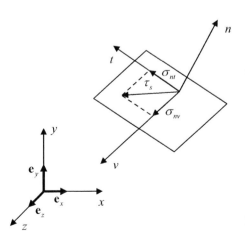

Figure 2.8 Stress components acting on a general oblique plane.

and

$$\begin{aligned}\sigma_{nv} = &\sigma_{xx}n_xv_x + \sigma_{yy}n_yv_y + \sigma_{zz}n_zv_z \\ &+ \sigma_{xy}(n_yv_x + n_xv_y) \\ &+ \sigma_{xz}(n_zv_x + n_xv_z) \\ &+ \sigma_{yz}(n_zv_y + n_yv_z)\end{aligned} \quad (2.27b)$$

Both of these equations can be written more compactly in the same form as we obtained for σ_{nn}. Summarizing all these results we have

$$\sigma_{nn} = \sum_{i=1}^{3}\sum_{j=1}^{3}\sigma_{ij}n_in_j$$

$$\sigma_{nt} = \sum_{i=1}^{3}\sum_{j=1}^{3}\sigma_{ij}n_it_j \quad (2.28)$$

$$\sigma_{nv} = \sum_{i=1}^{3}\sum_{j=1}^{3}\sigma_{ij}n_iv_j$$

or, in matrix–vector form

$$\begin{aligned}\sigma_{nn} &= \{n\}^T[\sigma]\{n\} \\ \sigma_{nt} &= \{n\}^T[\sigma]\{t\} \\ \sigma_{nv} &= \{n\}^T[\sigma]\{v\}\end{aligned} \quad (2.29)$$

2.2.2 Stress Transformation Equations

We have found the results for the stress components acting on a single plane whose unit normal was **n**. However, since **n** is arbitrary, we can use those results, appropriately modified, to obtain the stresses acting along three mutually orthogonal oblique planes in terms of the stresses acting along the (x,y,z) planes. These relationships are called the *stress transformation equations*. To obtain the stress transformation equations, we note that we can identify **n**, **t**, and **v** as unit vectors along three mutually orthogonal axes (x'_1, x'_2, x'_3) that are rotated relative to the (x,y,z) axis, as shown in Fig. 2.9. Note also that all these unit vectors can play dual roles as either unit normal vectors to various planes as well as unit tangent vectors within a given plane. Thus, for example, **n** can be a unit normal to the cutting plane whose normal is along the n-axis but it can also play the role as a unit vector in the n-direction that is tangent to the planes where the normal is **t** or **v**. If we let (e'_1, e'_2, e'_3) be unit vectors along the (x'_1, x'_2, x'_3) axes and (e_1, e_2, e_3) unit vectors along the (x_1, x_2, x_3), respectively, then we have

2.2 Normal Stress and Shear Stresses on an Oblique Plane

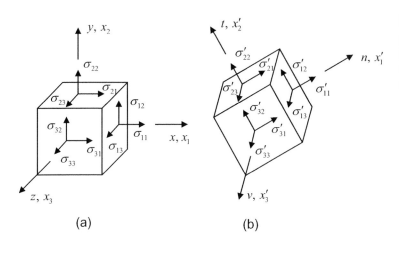

Figure 2.9 (a) The stress acting at a point along the Cartesian axes (x_1, x_2, x_3). (b) The stresses at the same point acting along the rotated Cartesian axes (x_1, x_2, x_3).

$$\begin{aligned}\mathbf{e}'_1 &= \mathbf{n} = n_1\mathbf{e}_1 + n_2\mathbf{e}_2 + n_3\mathbf{e}_3 \\ \mathbf{e}'_2 &= \mathbf{t} = t_1\mathbf{e}_1 + t_2\mathbf{e}_2 + t_3\mathbf{e}_3 \\ \mathbf{e}'_1 &= \mathbf{n} = v_1\mathbf{e}_1 + v_2\mathbf{e}_2 + v_3\mathbf{e}_3 \end{aligned} \tag{2.30}$$

It is more convenient to write these components in a more uniform way, namely

$$n_i = l_{i1}, \quad t_i = l_{i2}, \quad v_i = l_{i3} \quad (i = 1, 2, 3) \tag{2.31}$$

in terms of the nine quantities l_{ij} $(i = 1, 2, 3), (j = 1, 2, 3)$. Then Eq. (2.30) can be written as

$$\mathbf{e}'_i = \sum_{k=1}^{3} l_{ki}\mathbf{e}_k \quad (i = 1, 2, 3) \tag{2.32}$$

From Eq. (2.32) we see that

$$\mathbf{e}'_i \cdot \mathbf{e}_j = \sum_{k=1}^{3} l_{ki}(\mathbf{e}_k \cdot \mathbf{e}_j) = \sum_{k=1}^{3} l_{ki}\delta_{kj} = l_{ji} \tag{2.33}$$

where $\mathbf{e}_k \cdot \mathbf{e}_j = \delta_{kj}$ since the unit vectors are orthogonal to each other, where δ_{kj} is called the *Kronecker delta*, whose definition can be written as

$$\delta_{ij} = \begin{cases} 1 & \text{for } i = j \\ 0 & \text{otherwise} \end{cases} \tag{2.34}$$

If we interchange the i and j subscript labels in Eq. (2.33) and use the properties of the dot product it follows from Eq. (2.33) that

$$l_{ij} = (\mathbf{e}_i \cdot \mathbf{e}'_j) = \cos(x_i, x'_j) \tag{2.35}$$

i.e., l_{ij} is the cosine of the angle between the x_i axis and the x'_j axis. Thus, the $[l]$ matrix is called a *direction cosine matrix*. Note that, from Eq. (2.31), the **n**, **t**, and **v** unit vectors are just the columns of this direction cosine matrix. Thus, we can write the [*l*] matrix in the two equivalent forms

$$[l] = \begin{bmatrix} l_{11} & l_{12} & l_{13} \\ l_{21} & l_{22} & l_{23} \\ l_{31} & l_{32} & l_{33} \end{bmatrix} = \begin{bmatrix} n_1 & t_1 & v_1 \\ n_2 & t_2 & v_2 \\ n_3 & t_3 & v_3 \end{bmatrix} \qquad (2.36)$$

Now, consider the stress components $(\sigma_{nn}, \sigma_{nt}, \sigma_{nv})$ we obtained previously in Eq. (2.29). These are just the stresses on the $x'_1 = $ constant plane so we have

$$\begin{aligned} \sigma'_{11} &= \sigma_{nn} = \sum_{i=1}^{3}\sum_{j=1}^{3} l_{i1}l_{j1}\sigma_{ij} \\ \sigma'_{12} &= \sigma_{nt} = \sum_{i=1}^{3}\sum_{j=1}^{3} l_{i1}l_{j2}\sigma_{ij} \\ \sigma'_{13} &= \sigma_{nv} = \sum_{i=1}^{3}\sum_{j=1}^{3} l_{i1}l_{j3}\sigma_{ij} \end{aligned} \qquad (2.37a)$$

However, if the normal to the cutting plane was in the x'_2 direction instead we would have

$$\begin{aligned} \sigma'_{22} &= \sigma_{tt} = \sum_{i=1}^{3}\sum_{j=1}^{3} l_{i2}l_{j2}\sigma_{ij} \\ \sigma'_{21} &= \sigma_{tn} = \sum_{i=1}^{3}\sum_{j=1}^{3} l_{i2}l_{j1}\sigma_{ij} \\ \sigma'_{23} &= \sigma_{tv} = \sum_{i=1}^{3}\sum_{j=1}^{3} l_{i2}l_{j3}\sigma_{ij} \end{aligned} \qquad (2.37b)$$

and, similarly, for a cutting plane with a normal in the x'_3 direction

$$\begin{aligned} \sigma'_{33} &= \sigma_{vv} = \sum_{i=1}^{3}\sum_{j=1}^{3} l_{i3}l_{j3}\sigma_{ij} \\ \sigma'_{31} &= \sigma_{vn} = \sum_{i=1}^{3}\sum_{j=1}^{3} l_{i3}l_{j1}\sigma_{ij} \\ \sigma'_{32} &= \sigma_{vt} = \sum_{i=1}^{3}\sum_{j=1}^{3} l_{i3}l_{j2}\sigma_{ij} \end{aligned} \qquad (2.37c)$$

Because the results in Eqs. (2.37a, b, c) are in a uniform notation, we can write them all together as

$$\sigma'_{mn} = \sum_{i=1}^{3}\sum_{j=1}^{3} l_{im}l_{jn}\sigma_{ij} \quad (m=1,2,3), (n=1,2,3) \qquad (2.38)$$

or, in matrix form, as

$$[\sigma'] = [l]^T [\sigma][l] \tag{2.39}$$

where

$$[\sigma'] = \begin{bmatrix} \sigma'_{11} & \sigma'_{12} & \sigma'_{13} \\ \sigma'_{21} & \sigma'_{22} & \sigma'_{23} \\ \sigma'_{31} & \sigma'_{32} & \sigma'_{33} \end{bmatrix}, \quad [\sigma] = \begin{bmatrix} \sigma_{11} & \sigma_{12} & \sigma_{13} \\ \sigma_{21} & \sigma_{22} & \sigma_{23} \\ \sigma_{31} & \sigma_{32} & \sigma_{33} \end{bmatrix} \tag{2.40}$$

are the stress states associated with the (x'_1, x'_2, x'_3) and (x_1, x_2, x_3) axes, respectively. Equation (2.39), therefore, is the general transformation of the state of stress from the (x_1, x_2, x_3) axes to the (x'_1, x'_2, x'_3) axes. It is easy to show that we can also write the stresses with respect to the (x_1, x_2, x_3) axes in terms of the stresses for the (x'_1, x'_2, x'_3) axes as

$$[\sigma] = [l][\sigma'][l]^T \tag{2.41}$$

since the direction cosine matrix $[l]$ is an *orthogonal matrix* which satisfies

$$[l][l]^T = [l]^T[l] = [I] \tag{2.42}$$

It is called a *proper orthogonal matrix* if its determinant is equal to one, i.e.,

$$\det([l]) = 1 \tag{2.43}$$

If we dot both sides of Eq. (2.32) with a general position vector, $\mathbf{x}$, we have

$$\mathbf{x} \cdot \mathbf{e}'_i = \sum_{k=1}^{3} l_{ki} \mathbf{x} \cdot \mathbf{e}_k \quad (i = 1, 2, 3) \tag{2.44}$$

or, equivalently, in component or matrix–vector form

$$x'_i = \sum_{k=1}^{3} l_{ki} x_k \tag{2.45}$$

$$\{x'\} = [l]^T \{x\}$$

Note that some authors define the direction cosine matrix components instead as

$$Q_{ij} = l_{ji} = (\mathbf{e}'_i \cdot \mathbf{e}_j) = \cos(x'_i, x_j) \tag{2.46}$$

Thus, $[Q] = [l]^T$ and we have

$$x'_i = \sum_{k=1}^{3} Q_{ik} x_k \tag{2.47}$$

$$\{x'\} = [Q]\{x\}$$

You should be careful to determine which definition of the direction cosine matrix is being used when encountering expressions for transforming position vectors, stresses, etc. If we take the transpose of Eq. (2.36), we see that the **n**, **t**, and **v** unit vectors are the rows of the [Q] direction cosine matrix.

If the coordinates (x'_1, x'_2, x'_3) and (x_1, x_2, x_3) are both right-handed coordinates, then both [l] and its transpose will be proper orthogonal matrices. It is important to only use right-handed coordinates so that there are no reflections of coordinates present that can have unintended effects such as, for example, changing the directions of vector cross products from their usual definitions. A right-handed coordinate system (x_1, x_2, x_3) with unit vectors $(\mathbf{e}_1, \mathbf{e}_2, \mathbf{e}_3)$ along those coordinates will satisfy $\mathbf{e}_3 = \mathbf{e}_1 \times \mathbf{e}_2$.

Example 2.1 The Stress Vector, Stresses and Stress Transformations

As stated before, a knowledge of the state of stress at a point allows us to calculate the stress vector on any cutting plane and stress components in any direction at that point. As seen in the previous sections, most stress and stress-vector relations can be written in a matrix–vector form. Thus, MATLAB®, which is designed specifically to handle such matrix–vector manipulations, is an ideal software package for performing all the necessary algebra. In this example we will use MATLAB® to explore a number of the matrix–vector relationships associated with stresses and stress vectors. We will begin by forming a general three-dimensional (3-D) state of stress with respect to a set of (x, y, z) axes as a 3×3 matrix in MATLAB®:

```
stress = [ 20   40   30;  40  -10   5;  30   5   20]
stress =   20   40   30
           40  -10    5
           30    5   20
```

where the stress values are assumed to be given in MPa. Note that the stress matrix is symmetrical, as it should be since the shear stresses in the off-diagonal terms across the main diagonal of the matrix must be equal. If we are given this state of stress, we can calculate the stress vector on any cutting plane with Eq. (2.16). Thus, consider defining a cutting plane with a unit normal **n**, which we will define in MATLAB®:

```
n = [ 3;  -4;   5] /sqrt(9+16+25)
n =   0.4243
     -0.5657
      0.7071
```

Then Eq. (2.16) in MATLAB® form is

```
Tn = stress*n
Tn =   7.0711
      26.1630
      24.0416
```

2.2 Normal Stress and Shear Stresses on an Oblique Plane

which are just the stress-vector Cartesian components $\left(T_x^{(\mathbf{n})}, T_y^{(\mathbf{n})}, T_z^{(\mathbf{n})}\right)$ written as a column vector. If we want to find the normal stress on this plane, we can use Eq. (2.17) where we compute the component of the stress vector in the normal direction via a dot product:

```
snn = Tn'*n
snn = 5.2000
```

so the normal stress is 5.2 MPa. [Note: there is also a built-in MATLAB® function dot we can use instead.] Alternatively, as shown with Eq. (2.19), we can calculate this normal stress directly from the state of stress:

```
snn = n'*stress*n
snn = 5.2000
```

If we want to find the total shear stress, tots, on the plane then from Eq. (2.20) we have

```
tots = sqrt(norm(Tn)^2 -snn^2)
tots = 35.8533
```

and the total shear-stress direction, **s**, is from Eq. (2.22) given by

```
s = (Tn-snn*n)/tots
s =  0.1357
     0.8118
     0.5680
```

We can verify that in this direction on the plane we obtain the total shear stress by calculating the component of the stress vector in the **s** direction, namely

```
tots2 = Tn'*s
tots2 = 35.8533
```

which agrees with our previous total shear-stress result.

To examine the full stress transformation equations, we need to have three mutually orthogonal unit vectors $(\mathbf{n}, \mathbf{t}, \mathbf{v})$ that form up a right-handed set of coordinate axes. We already have a vector, **n**, so let's define a vector, **t**, that is orthogonal to **n** by forming the cross vector of **n** with the unit vector, $\mathbf{e}_z$, in the z-direction and then dividing by the magnitude of that cross product to get a unit vector:

```
t = cross(n, [0;0;1])
t =  -0.5657
     -0.4243
      0
t = t/norm(t)
t =  -0.8000
     -0.6000
      0
```

Then we can define a third unit vector, **v**, that is orthogonal to both **n** and **t** by again using the cross product:

```
v = cross(n, t)
v =  0.4243
    -0.5657
    -0.7071
```

We said that we needed a set of mutually orthogonal directions (**n**, **t**, **v**) that are right-handed. The unit vectors (**n**, **t**, **v**) form a right-handed system if $\mathbf{v} = \mathbf{n} \times \mathbf{t}$. However, we obtained the vector **v** via precisely this relation so we do have a right-handed system along the (n, t, v) coordinate directions.

We now are in a position to calculate many quantities associated with these coordinates. For example, we can use Eq. (2.15) to write the stress vector associated with the t = constant and v = constant planes by simply replacing the **n** vector in that equation with either **t** or **v**, i.e.,

$$\{T^{(t)}\} = [\sigma]\{t\}$$
$$\{T^{(v)}\} = [\sigma]\{v\}$$

which in MATLAB® gives

```
Tt = stress*t
Tt =  -40
      -26
      -27

Tv = stress*v
Tv = -35.3553
      19.0919
      -4.2426
```

To obtain the full state of stress with respect to the (n, t, v) coordinates, we need the direction cosine matrix formed up with **n**, **t**, **v** unit vectors in its columns (see Eq. (2.36)):

```
l = [n   t   v]
l =  0.4243   -0.8000    0.4243
    -0.5657   -0.6000   -0.5657
     0.7071    0        -0.7071
```

Then the state of stress associated with those coordinates, ntv_stresses, is given as (see Eq. (2.39))

```
ntv_stresses = l'*stress*l
ntv_stresses =   5.2000  -21.3546  -28.8000
               -21.3546   47.6000   16.8291
               -28.8000   16.8291  -22.8000
```

2.3 A Relationship between the Normal Stress and Total Shear Stress

The stress vector has some interesting properties that can be used to develop a relationship between normal stresses and total shear stresses. For example, let $\mathbf{n}^{(1)}$ and $\mathbf{n}^{(2)}$ be unit normals for two different planes passing through the same point. Then the stress vectors associated with those planes at that point satisfy

$$\mathbf{T}^{(\mathbf{n}^{(1)})} \cdot \mathbf{n}^{(2)} = \mathbf{T}^{(\mathbf{n}^{(2)})} \cdot \mathbf{n}^{(1)} \tag{2.48}$$

i.e., the component in the $\mathbf{n}^{(2)}$ direction of the stress vector acting on the plane whose normal is $\mathbf{n}^{(1)}$ is equal to the component in the $\mathbf{n}^{(1)}$ direction of the stress vector acting on the plane whose normal is $\mathbf{n}^{(2)}$. This is easily shown by writing both sides in Eq. (2.48) in terms of the stress state components associated with the (x, y, z) axes, giving the same form in both cases:

$$\begin{aligned}\mathbf{T}^{(\mathbf{n}^{(1)})} \cdot \mathbf{n}^{(2)} &= \mathbf{T}^{(\mathbf{n}^{(2)})} \cdot \mathbf{n}^{(1)} \\ &= \sigma_{xx} n_x^{(1)} n_x^{(2)} + \sigma_{yy} n_y^{(1)} n_y^{(2)} + \sigma_{zz} n_z^{(1)} n_z^{(2)} \\ &\quad + \sigma_{xy}\left(n_x^{(1)} n_y^{(2)} + n_y^{(1)} n_x^{(2)}\right) \\ &\quad + \sigma_{xz}\left(n_x^{(1)} n_z^{(2)} + n_z^{(1)} n_x^{(2)}\right) \\ &\quad + \sigma_{yz}\left(n_y^{(1)} n_z^{(2)} + n_z^{(1)} n_y^{(2)}\right)\end{aligned} \tag{2.49}$$

We can use Eq. (2.48) as follows. Let $\mathbf{n}^{(1)}$ and $\mathbf{s}^{(1)}$ be unit vectors that define the normal to a plane on which the normal stress, σ_{nn}, acts and a tangential direction in that plane along which the total shear stress, τ_s, acts (Fig. 2.10a). Now, consider rotating by a very small angle, $d\theta$, in the plane defined by $(\mathbf{n}^{(1)}, \mathbf{s}^{(1)})$ to a new plane whose normal is $\mathbf{n}^{(2)}$ and tangential direction $\mathbf{s}^{(2)}$ (Fig. 2.10b). Since the angle $d\theta$ is small the normal stress and total shear stress on the rotated plane will be to first order $\sigma_{nn} + (\partial \sigma_{nn}/\partial \theta) d\theta$ and $\tau_s + (\partial \tau_s/\partial \theta) d\theta$, as seen in Fig. 2.10b. Expressing $(\mathbf{n}^{(2)}, \mathbf{s}^{(2)})$ in terms of $(\mathbf{n}^{(1)}, \mathbf{s}^{(1)})$ to first order we have

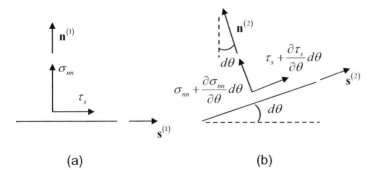

Figure 2.10 (a) The normal stress and total shear stress on an arbitrary cutting plane. (b) The normal stress and total shear stress on a cutting plane that is rotated by a small angle $d\theta$ with respect to that original plane.

42 Stress

$$\begin{aligned}\mathbf{n}^{(2)} &= -\sin(d\theta)\mathbf{s}^{(1)} + \cos(d\theta)\mathbf{n}^{(1)} \\ &\cong -d\theta\mathbf{s}^{(1)} + \mathbf{n}^{(1)} \\ \mathbf{s}^{(2)} &= \cos(d\theta)\mathbf{s}^{(1)} + \sin(d\theta)\mathbf{n}^{(1)} \\ &\cong \mathbf{s}^{(1)} + d\theta\mathbf{n}^{(1)}\end{aligned} \qquad (2.50)$$

so that, again to first order

$$\begin{aligned}\mathbf{T}^{(\mathbf{n}^{(1)})} \cdot \mathbf{n}^{(2)} &\cong \left(\tau_s \mathbf{s}^{(1)} + \sigma_{nn}\mathbf{n}^{(1)}\right) \cdot \left(-d\theta \mathbf{s}^{(1)} + \mathbf{n}^{(1)}\right) \\ &= -\tau_s d\theta + \sigma_{nn}\end{aligned} \qquad (2.51)$$

and, similarly,

$$\begin{aligned}\mathbf{T}^{(\mathbf{n}^{(2)})} \cdot \mathbf{n}^{(1)} &\cong \left[\left(\sigma_{nn} + \frac{\partial \sigma_{nn}}{\partial \theta} d\theta\right)\left(-d\theta \mathbf{s}^{(1)} + \mathbf{n}^{(1)}\right)\right] \cdot \mathbf{n}^{(1)} \\ &\quad + \left[\left(\tau_s + \frac{\partial \tau_s}{\partial \theta} d\theta\right)\left(\mathbf{s}^{(1)} + d\theta \mathbf{n}^{(1)}\right)\right] \cdot \mathbf{n}^{(1)} \\ &= \sigma_{nn} + \frac{\partial \sigma_{nn}}{\partial \theta} d\theta + \tau_s d\theta\end{aligned} \qquad (2.52)$$

Equating the expressions in Eq. (2.51) and Eq. (2.52) and dividing by $d\theta$ we find

$$\frac{\partial \sigma_{nn}}{\partial \theta} = -2\tau_s \qquad (2.53)$$

Thus, whenever the normal stress on a plane is an extremum we have $\partial \sigma_{nn}/\partial \theta = 0$ and Eq. (2.53) shows that on any such plane we must have $\tau_s = 0$ also. Planes of extreme normal stress are called *principal planes* so we have shown that principal planes must also be free of shear stress. In the next section we will see how we can use this fact to determine the explicit values of the extreme normal stresses.

2.4 Principal Stresses and Principal Stress Directions

In the previous sections we have learned how to find the stress vector or stress components acting on any cutting plane. Often, however, we are interested in the stresses acting on those particular cutting planes where the normal stresses take their extreme values since those stresses play an important role in material failure theories (see Chapter 13). These extreme normal stresses are called *principal stresses* and the planes they act on are called the *principal planes*. The unit normals to the principal planes are similarly called the *principal directions*. As shown in the previous section, on the principal planes there are no shear stresses so one of the easiest ways to obtain the principal planes is to simply find the planes on which the stress vector consists of a purely normal stress, i.e.,

$$\mathbf{T}^{(\mathbf{n})} = \sigma_p \mathbf{n} \qquad (2.54)$$

2.4 Principal Stresses and Principal Stress Directions

where σ_p is the unknown (at this stage) principal stress on a principal plane. Using the expression for the stress vector in terms of the stress components, Eq. (2.13), and the symmetry of the shear stresses $(\sigma_{ij} = \sigma_{ji})$, we have

$$T_i^{(n)} = \sum_{j=1}^{3} \sigma_{ij} n_j = \sigma_p n_i (i = 1, 2, 3) \tag{2.55}$$

Expanding these three conditions out, we find a homogeneous system of equations for the unit normal components (n_x, n_y, n_z):

$$\begin{aligned}(\sigma_{xx} - \sigma_p)n_x + \sigma_{xy}n_y + \sigma_{xz}n_z &= 0 \\ \sigma_{yx}n_x + (\sigma_{yy} - \sigma_p)n_y + \sigma_{yz}n_z &= 0 \\ \sigma_{zx}n_x + \sigma_{zy}n_y + (\sigma_{zz} - \sigma_p)n_z &= 0\end{aligned} \tag{2.56}$$

The only way that such a homogeneous system can have a nontrivial solution is for the 3×3 determinant of the coefficients to be zero, i.e.,

$$\begin{vmatrix} (\sigma_{xx} - \sigma_p) & \sigma_{xy} & \sigma_{xz} \\ \sigma_{yx} & (\sigma_{yy} - \sigma_p) & \sigma_{yz} \\ \sigma_{zx} & \sigma_{zy} & (\sigma_{zz} - \sigma_p) \end{vmatrix} = 0 \tag{2.57}$$

When expanded out this condition leads to a cubic equation which has three real roots that are the three principal stresses $(\sigma_{p1}, \sigma_{p2}, \sigma_{p3})$. This cubic equation can be written as

$$\sigma_p^3 - I_1 \sigma_p^2 + I_2 \sigma_p - I_3 = 0 \tag{2.58}$$

where (I_1, I_2, I_3) are *stress invariants* given by

$$\begin{aligned} I_1 &= \sigma_{xx} + \sigma_{yy} + \sigma_{zz} \\ I_2 &= \sigma_{xx}\sigma_{yy} + \sigma_{xx}\sigma_{zz} + \sigma_{yy}\sigma_{zz} - \sigma_{xy}^2 - \sigma_{xz}^2 - \sigma_{yz}^2 \\ I_3 &= \sigma_{xx}\sigma_{yy}\sigma_{zz} + 2\sigma_{xy}\sigma_{xz}\sigma_{yz} - \sigma_{xx}\sigma_{zy}^2 - \sigma_{yy}\sigma_{xz}^2 - \sigma_{zz}\sigma_{xy}^2 \end{aligned} \tag{2.59}$$

The last two invariants can also be written in a more compact form using determinants as

$$\begin{aligned} I_2 &= \begin{vmatrix} \sigma_{xx} & \sigma_{xy} \\ \sigma_{xy} & \sigma_{yy} \end{vmatrix} + \begin{vmatrix} \sigma_{yy} & \sigma_{yz} \\ \sigma_{yz} & \sigma_{zz} \end{vmatrix} + \begin{vmatrix} \sigma_{xx} & \sigma_{xz} \\ \sigma_{xz} & \sigma_{zz} \end{vmatrix} \\ I_3 &= \begin{vmatrix} \sigma_{xx} & \sigma_{xy} & \sigma_{xz} \\ \sigma_{xy} & \sigma_{yy} & \sigma_{yz} \\ \sigma_{xz} & \sigma_{yz} & \sigma_{zz} \end{vmatrix} \end{aligned} \tag{2.60}$$

These quantities are called invariants because they can be shown to have the same values when expressed in terms of stresses as measured in any coordinate system. Thus, in particular, in terms of the principal stresses we must have

$$I_1 = \sigma_{p1} + \sigma_{p2} + \sigma_{p3}$$
$$I_2 = \sigma_{p1}\sigma_{p2} + \sigma_{p1}\sigma_{p3} + \sigma_{p2}\sigma_{p3} \quad (2.61)$$
$$I_3 = \sigma_{p1}\sigma_{p2}\sigma_{p3}$$

Once the principal stresses are obtained, we can place one of those principal stresses into Eq. (2.56) and try to solve for the corresponding principal direction as defined by (n_x, n_y, n_z). However, because the determinant of the homogeneous system of Eq. (2.56) is zero, this means that the three equations are not all independent. Thus, the most we can do is, say, solve two of these equations for two of the unit normal components in terms of the other third component. However, we can then find this third component (to within a plus or minus sign) by using the fact that the components of the unit normal must be components of a unit vector. We can repeat this whole process for each of the other principal stresses. Thus, a procedure for finding the principal stresses and principal directions is: (1) find the three principal stresses $(\sigma_{p1}, \sigma_{p2}, \sigma_{p3})$ as roots of the cubic equation of Eq. (2.58), and (2) for each of those principal stresses, determine the corresponding principal direction vectors $(\mathbf{n}^{(1)}, \mathbf{n}^{(2)}, \mathbf{n}^{(3)})$, using the fact that these normals must be unit vectors whose components with respect to the (x, y, z) axis satisfy

$$\left(n_x^{(1)}\right)^2 + \left(n_y^{(1)}\right)^2 + \left(n_z^{(1)}\right)^2 = 1$$
$$\left(n_x^{(2)}\right)^2 + \left(n_y^{(2)}\right)^2 + \left(n_z^{(2)}\right)^2 = 1 \quad (2.62)$$
$$\left(n_x^{(3)}\right)^2 + \left(n_y^{(3)}\right)^2 + \left(n_z^{(3)}\right)^2 = 1$$

This procedure, however, is rather tedious and must be modified in some special cases, as we will see shortly. Solving the cubic equation of Eq. (2.58) can be done in a number of ways. Many calculators and various software packages have cubic equation solvers. Also, there is an explicit solution for the roots of a cubic. However, the easiest way to solve for the principal stresses and directions when using software packages such as MATLAB®, MathCad®, Mathematica®, Maple®, etc., is to recognize that this is a matrix *eigenvalue problem*. If we write Eq. (2.55) and Eq. (2.62) in matrix–vector form, then we have the problem of finding the solution of the equations

$$[\sigma]\{n\} = \sigma_p\{n\}$$
$$\{n\}^T\{n\} = 1 \quad (2.63)$$

The σ_p are called the *eigenvalues* of the matrix $[\sigma]$, which here is just the matrix defining the state of stress with respect to the (x, y, z) axes. The corresponding solutions for the normalized vectors $\{n\}$ are called the *eigenvectors*. For a real, symmetrical matrix $[\sigma]$ it can be shown that the eigenvalues are real and the eigenvectors are orthogonal to each other. Thus, the three principal directions will form three unit vectors of a coordinate system whose axes are along the principal directions. However, as we will see, the axes of these principal coordinates may not always form a right-handed system, so we may need to choose the eigenvectors so that this right-handedness is satisfied.

Example 2.2 Eigenvalue Problem Solution for Principal Stresses and Directions

Let us write a state of stress in MATLAB® by first defining a 3 × 3 state of stress matrix, stress, as

```
stress = [ 3   5   10; 5   6   4; 10   4   1]
stress =    3    5   10
            5    6    4
           10    4    1
```

where the stresses are measured in MPa. We can obtain the principal stresses in a matrix called pvals and the principal directions in a matrix pdirs by simply providing the stress matrix as an argument to the built-in MATLAB® function eig which solves the eigenvalue problem of Eq. (2.63):

```
[pdirs, pvals] = eig(stress)
pdirs =   -0.6782    0.3747    0.6322
           0.0323   -0.8442    0.5350
           0.7341    0.3833    0.5605

pvals =   -8.0625         0         0
                0    1.9648         0
                0         0   16.0977
```

The three principal stresses are approximately $\sigma_{p1} = -8.06$ MPa, $\sigma_{p2} = 1.96$ MPa, and $\sigma_{p3} = 16.10$ MPa. The matrix pvals is actually the state of stress associated with coordinates along the principal directions since the shear stresses are zero, as we have already shown. The matrix pdirs contain the three unit vectors $(\mathbf{n}^{(1)}, \mathbf{n}^{(2)}, \mathbf{n}^{(3)})$ associated with $(\sigma_{p1}, \sigma_{p2}, \sigma_{p3})$, respectively, in the columns of the pdirs matrix, i.e., it has the elements

$$\text{pdirs} = \begin{bmatrix} n_x^{(1)} & n_x^{(2)} & n_x^{(3)} \\ n_y^{(1)} & n_y^{(2)} & n_y^{(3)} \\ n_z^{(1)} & n_z^{(2)} & n_z^{(3)} \end{bmatrix} \quad (2.64)$$

These three unit vectors form the unit vectors of a coordinate system along the principal directions, which we will call (p_1, p_2, p_3) just as $(\mathbf{e}_x, \mathbf{e}_y, \mathbf{e}_z)$ are unit vectors along the (x, y, z) axes. Are the (p_1, p_2, p_3) axes a right-handed system? We can check to see by extracting the unit vectors from the pdirs matrix:

```
n1 = pdirs(:,1);
n2 = pdirs(:,2);
n3 = pdirs(:,3);
```

(where the MATLAB® vectors $nj = \mathbf{n}^{(j)}(j = 1, 2, 3)$) and perform the cross product with the built-in MATLAB® function cross, as we have done before. We should find $\mathbf{n}^{(3)} = \mathbf{n}^{(1)} \times \mathbf{n}^{(2)}$:

```
cross(n1, n2)
ans =   0.6322
        0.5350
        0.5605
```

which, when we examine the pdirs matrix shows that we do obtain the MATLAB® vector n3 ≡ $\mathbf{n}^{(3)}$.

[Note: if **e** is an eigenvector then so is − **e**. The MATLAB® eig function does not always guarantee that the signs chosen for the three eigenvectors form a right-handed system so it is important to check and, if necessary, to change the sign on an eigenvector or reorder the numbering on the eigenvectors so that the eigenvectors do form a right-handed system. A simpler check than the one given above is to compute the determinant of pdirs with the MATLAB® function det. A right-handed system must have det(pdirs) return a value of +1.]

In this case the eigenvectors do form a right-handed system so we can view the pdirs matrix as the matrix of direction cosines that transforms the stresses along a right-handed (x, y, z) axes to the stresses along a set of right-handed principal (p_1, p_2, p_3) axes. We can verify this by letting the MATLAB® matrix pdirs = [*l*] in Eq. (2.39) and computing that equation in MATLAB® as:

```
Stress p = pdirs'*stress*pdirs
Stress p =   -8.0625   -0.0000    0.0000
             -0.0000    1.9648    0.0000
              0.0000    0.0000   16.0977
```

which shows that the MATLAB® stress_p matrix contains the same stresses along the principal directions that was given by the MATLAB® matrix pvals.

Using the MATLAB® eig function makes the calculation of principal stresses and principal directions very easy. Other software packages can solve eigenvalue problems but you should check to make sure that the eigenvectors are normalized by the package so that they are unit vectors (the MATLAB® eig function does do this normalization) and that they form up a right-handed system.

Example 2.3 Direct Solution for Principal Stresses and Principal Directions

Although an eigenvalue solver such as eig is a very effective way to find principal stress and principal stresses, we do not see the details of the solution process. In this example we will examine those details with the more direct approach we outlined previously where we first

2.4 Principal Stresses and Principal Stress Directions

find the roots of a cubic equation and then solve for the principal directions explicitly. Consider, for example, the following state of stress (in MPa):

```
stress = [ 120 -55 -75; -55 55  33; -75 33 -85]
stress =  120   -55   -75
          -55    55    33
          -75    33   -85
```

To solve the cubic equation in Eq. (2.58) for this state of stress we need the coefficients of the cubic, which are in terms of the invariants (I_1, I_2, I_3). A MATLAB® function has been written, stress_invs (see Appendix D, which has listings of all the MATLAB® functions developed specifically for this book) which can be called as:

```
[ I1, I2, I3]  = stress_invs(stress)
I1 = 90
I2 = -18014
I3 = -4.7168e+05
```

where I_1 has the dimensions of MPa, I_2 has the dimensions (MPa)^2 and I_3 has the dimensions (MPa)^3. Having these invariants, we can then use the MATLAB® roots function, which takes a vector argument containing the coefficients of the cubic (the invariants together with the appropriate signs as seen in Eq. (2.58)) and returns the three roots that are the principal stresses:

```
p_stresses = roots( [ 1,-I1,I2,-I3] )
p_stresses =  176.7995
             -110.8640
               24.0644
```

We can, of course, also obtain these principal stresses with eig:

```
[ pdirs, pvals]  = eig(stress)
pdirs =   0.2872    0.4654    0.8372
         -0.0944    0.8836   -0.4587
          0.9532   -0.0527   -0.2977

pvals = -110.8640         0         0
              0      24.0644        0
              0            0   176.7995
```

which returns the same values (but in a different order).

Now, consider finding the principal directions. Let us choose first $\sigma_P^{(3)} = 176.8$ MPa and try to obtain the associated principal direction components $\left(n_x^{(3)}, n_y^{(3)}, n_z^{(3)}\right)$. From Eq. (2.56) we have

Stress

$$\left(120 - \sigma_p^{(3)}\right)n_x^{(3)} - 55n_y^{(3)} - 75n_z^{(3)} = 0$$

$$-55n_x^{(3)} + \left(55 - \sigma_p^{(3)}\right)n_y^{(3)} + 33n_z^{(3)} = 0$$

$$-75n_x^{(3)} + 33n_y^{(3)} + \left(-85 - \sigma_p^{(3)}\right)n_z^{(3)} = 0$$

Since the determinant of this homogeneous system of equations is zero, these three equations are not all independent so we cannot solve the system directly. However, we can try to find a subset of these equations that are independent and solve that system for a ratio of the principal direction components. For example, let's divide the first two equations by $n_z^{(3)}$ and try to solve that system of equations:

$$\left(120 - \sigma_p^{(3)}\right)\frac{n_x^{(3)}}{n_z^{(3)}} - 55\frac{n_y^{(3)}}{n_z^{(3)}} = 75$$

$$-55\frac{n_x^{(3)}}{n_z^{(3)}} + \left(55 - \sigma_p^{(3)}\right)\frac{n_y^{(3)}}{n_z^{(3)}} = -33$$

This is easy to do by hand but we can use MATLAB® also by forming up the coefficients of these equations and the right-hand side and then using a built-in MATLAB® solver for a linear system of equations. The steps are:

```
% choose a principal stresss
sp3 = pvals(3,3);

% form up the matrix of coefficients of a pair of equations for this
stress
C(1,1) = 120 - sp3; C(1,2) = -55; C(2,1) =-55; C(2,2)=55- sp3
C =   -56.7995  -55.0000
      -55.0000 -121.7995


% form up the right hand side as a column vector
b = [ 75;-33]
b = 75
   -33

% solve the system of equations [ C]{ r}  ={ b}  for the ratios of unit
% vector components
r = C\b
r= -2.8126
    1.5410
```

2.4 Principal Stresses and Principal Stress Directions

so that in our case we have approximately

$$\frac{n_x^{(3)}}{n_z^{(3)}} = -2.81, \quad \frac{n_y^{(3)}}{n_z^{(3)}} = 1.54$$

However, from the normalization $\left(n_x^{(3)}\right)^2 + \left(n_y^{(3)}\right)^2 + \left(n_z^{(3)}\right)^2 = 1$ we have

$$n_z^{(3)} = \frac{\pm 1}{\sqrt{1 + \left(n_x^{(3)}/n_z^{(3)}\right)^2 + \left(n_y^{(3)}/n_z^{(3)}\right)^2}}$$

so that we find from MATLAB®, choosing the plus sign

```
nz3 = 1/sqrt(1 + r(1)^2 + r(2)^2)
nz3 = 0.2977

nx3 = r(1)*nz3
nx3 = -0.8372

ny3 = r(2)*nz3
ny3 = 0.4587
```

We can repeat these calculations choosing the minus sign so we obtain, finally, the two choices

$$n_x^{(3)} = -0.8372, \quad n_x^{(3)} = +0.8372$$
$$n_y^{(3)} = +0.4586, \quad n_y^{(3)} = -0.4586$$
$$n_z^{(3)} = +0.2977, \quad n_z^{(3)} = -0.2977$$

Either of the two sign choices will give legitimate principal stress directions. If we examine the pdirs output of eig we see the eigenvalue solver made the second choice listed above, but if we compute the determinant of the the pdirs matrix we have

```
det(pdirs)
ans =   -1.0000
```

so in order for the pdirs matrix to form up the matrix of unit vectors for a right-handed system if we keep the first two columns of pdirs we should change the last column to be the unit vector

$$n_x^{(3)} = -0.8372$$
$$n_y^{(3)} = +0.4586$$
$$n_z^{(3)} = +0.2977$$

We now have to repeat this entire process twice to find the other two eigenvectors. This makes the direct process of solving the system of equations for the principal directions rather tedious. Also, there are some issues with this solution process. For example, how did we know that the two equations we chose were independent equations? We simply assumed they were independent and tried to find a solution under that assumption. If the two equations we chose were not independent, our solution would fail and we would have to make another choice. Also, what if only one of these was an independent equation? We will examine that possibility shortly.

2.4.1 Special Case when $I_3 = 0$, Plane Stress

If the stress state is such that the third invariant, I_3, is zero, then we can actually find the principal stresses in an explicit form. Thus, consider setting up the following state of stress (in MPa):

```
stress = [ 0  20  0; 20  0  10; 0  10  0]
stress =   0   20   0
          20    0  10
           0   10   0

[ I1, I2, I3] = stress_invs(stress)
I1 = 0
I2 = -500
I3 = 0
```

so this state of stress does satisfy $I_3 = 0$. (The invariant $I_1 = 0$ also in this case, but that is not important for our discussion.) We can use eig to find the principal stresses and principal directions in the normal fashion to find

```
[ pdirs, pvals] = eig(stress)
pdirs =  0.6325  -0.4472   0.6325
        -0.7071   0.0000   0.7071
         0.3162   0.8944   0.3162

pvals = -22.3607        0        0
              0  -0.0000        0
              0        0  22.3607
```

We see that one of the principal stresses is zero. If we examine the cubic equation, Eq. (2.58), when $I_3 = 0$, that equation can be rewritten as

$$\sigma_p \left(\sigma_p^2 - I_1 \sigma_p + I_2 \right) = 0 \tag{2.65}$$

Using the quadratic formula, we can explicitly find three roots of Eq. (2.65) as

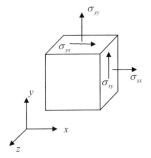

Figure 2.11 A plane state of stress (also called a biaxial state of stress).

$$\sigma_{p1}, \sigma_{p2} = \frac{I_1 \pm \sqrt{I_1^2 - 4I_2}}{2}$$
$$= \frac{I_1}{2} \pm \sqrt{I_1^2/4 - I_2} \quad (2.66)$$
$$\sigma_{p3} = 0$$

which you can verify gives the same values as found from eig. Thus, whenever $I_3 = 0$ we can find the principal stresses directly from Eq. (2.66). An important special state of stress that is often examined in elementary mechanics of materials courses is the case of a *state of plane stress* (also called a *state of biaxial stress*) where $\sigma_{zz} = \sigma_{yz} = \sigma_{xz} = 0$ (see Fig. 2.11). In this case $I_3 = 0$ and if we examine the expressions for I_1, I_2 it is easy to show that the principal stresses from Eq. (2.66) are

$$\sigma_{p1}, \sigma_{p2} = \frac{\sigma_{xx} + \sigma_{yy}}{2} \pm \sqrt{\left(\frac{\sigma_{xx} - \sigma_{yy}}{2}\right)^2 + \sigma_{xy}^2} \quad (2.67)$$
$$\sigma_{p3} = 0$$

which is a formula found in most elementary texts.

2.4.2 Special Case when Principal Stresses are Equal

Let us define a state of stress, given in MPa, as:

```
stress = [ 0    100   100; 100   0    100; 100   100  0]
stress =  0    100   100
          100   0    100
          100   100  0
```

If we use eig to find the principal stresses and principal directions, then we find

```
[pdirs, pvals] = eig(stress)
pdirs = 0.4082    0.7071    0.5774
        0.4082   -0.7071    0.5774
       -0.8165    0         0.5774

pvals = -100.0000    0          0
         0          -100.0000   0
         0           0         200.0000
```

which all seems normal, but we notice that two of the principal stresses are equal so that there are more details to this problem that the use of eig does not expose. To see what is happening, consider the equations for the principal directions (Eq. (2.56)) for the $\sigma_p^{(3)} = 200$ MPa value, which are

$$-200n_x^{(3)} + 100n_y^{(3)} + 100n_z^{(3)} = 0$$
$$100n_x^{(3)} - 200n_y^{(3)} + 100n_z^{(3)} = 0 \qquad (2.68)$$
$$100n_x^{(3)} + 100n_y^{(3)} - 200n_z^{(3)} = 0$$

We can, as done previously, solve two equations from this system and use the normalization to find

$$n_x^{(3)} = \pm 0.5774$$
$$n_y^{(3)} = \pm 0.5774 \qquad (2.69)$$
$$n_z^{(3)} = \pm 0.5774$$

which we recognize as the result found with eig if we choose the plus signs. However, if we consider instead the two repeated principal stresses $\sigma_p^{(1),(2)} = -100$ MPa, then we obtain:

$$100n_x^{(i)} + 100n_y^{(i)} + 100n_z^{(i)} = 0$$
$$100n_x^{(i)} + 100n_y^{(i)} + 100n_z^{(i)} = 0 \quad (i = 1, 2) \qquad (2.70)$$
$$100n_x^{(i)} + 100n_y^{(i)} + 100n_z^{(i)} = 0$$

These are all the same equation so even with the normalization equation, we have an infinite number of solutions. The eigenvectors associated with different eigenvalues are perpendicular to each other so this means that *every direction* in a plane P perpendicular to $\mathbf{n}^{(3)}$ must be a principal direction where the principal stress is -100 MPa. This is shown in Fig. 2.12. To find a particular direction in plane P we could, for example, set $n_z^{(2)} = 0$. Then the solution of Eq. (2.70) is $n_x^{(2)}/n_y^{(2)} = -1$ and normalizing the vector to make it a unit vector gives $n_y^{(2)} = \pm 0.7071$ so that choosing the minus sign we have

$$n_x^{(2)} = 0.7071$$
$$n_y^{(2)} = -0.7071 \qquad (2.71)$$
$$n_z^{(2)} = 0$$

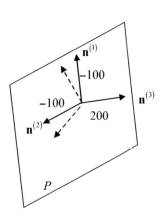

Figure 2.12 The case being considered when two principal stresses are equal. We can find a principal direction, $\mathbf{n}^{(3)}$, for the principal stress of 200 MPa (to within a $\pm$sign), and particular set of orthogonal directions in a plane P that are perpendicular to $\mathbf{n}^{(3)}$, $(\mathbf{n}^{(1)}, \mathbf{n}^{(2)})$, that define directions for the repeated principal stresses of -100 MPa, but any other set of directions in the plane, such as the dashed lines shown, for example, are also principal directions with the same principal stress.

which is identical to the result found for $\mathbf{n}^{(2)}$ with eig. To find another unit vector in the plane P so that we have a right-handed system, we can choose the plus sign in Eq. (2.69) for the components of $\mathbf{n}^{(3)}$ and then compute $\mathbf{n}^{(1)} = \mathbf{n}^{(2)} \times \mathbf{n}^{(3)}$, giving

$$n_x^{(1)} = -0.4082$$
$$n_y^{(1)} = -0.4082 \qquad (2.72)$$
$$n_z^{(1)} = 0.8165$$

which is the negative of the result obtained with eig since, as you can easily check, eig does not produce eigenvectors that form a right-handed system.

What happens if all the principal stresses are equal? In that case every direction is a principal direction so that the state of stress is only a normal-stress component, which is the same in all directions. This is similar to what we see in an ideal fluid at rest where the normal stress is negative, i.e., we have a pressure which is the same in all directions at any point in the fluid (although the pressure, of course, can vary from point to point).

2.5 Mohr's Circle for Three-Dimensional States of Stress

In elementary mechanics of materials courses, a graphical construction called Mohr's circle is often used in place of the transformation equations we have described in this chapter to analyze problems when there is a state of plane stress. Since software such as MATLAB® can effortlessly solve even complex stress relations, there is now less need to use such graphical methods to perform detailed stress calculations. However, one can gain some insight into the nature of stresses by examining the Mohr's circle construction for a general state of stress, which is what we will now do.

If we write the stresses on an arbitrary cutting plane whose normal is $\mathbf{n}$ in terms of the principal stresses and the components of $\mathbf{n}$ as measured in the principal coordinates (p_1, p_2, p_3), as shown in Fig. 2.13, then the stress-vector components along the principal directions are (see Eq. (2.12)):

$$T_1^{(\mathbf{n})} = n_1 \sigma_{p1}$$
$$T_2^{(\mathbf{n})} = n_2 \sigma_{p2} \qquad (2.73)$$
$$T_3^{(\mathbf{n})} = n_3 \sigma_{p3}$$

The normal stress and total shear stress on the arbitrary plane and the condition that the normal $\mathbf{n}$ is a unit vector are given by

$$N = \sigma_{nn} = \sigma_{p1}^2 n_1^2 + \sigma_{p2} n_2^2 + \sigma_{p3} n_3^2$$
$$S^2 = (\tau_s)^2 = \left(T_1^{(\mathbf{n})}\right)^2 + \left(T_2^{(\mathbf{n})}\right)^2 + \left(T_3^{(\mathbf{n})}\right)^2 - N^2 \qquad (2.74)$$
$$n_1^2 + n_2^2 + n_3^2 = 1$$

Figure 2.13 The normal stress, N, and total shear stress, S, on an arbitrary cutting plane with a normal, $\mathbf{n}$, as expressed in terms of the principal coordinate directions.

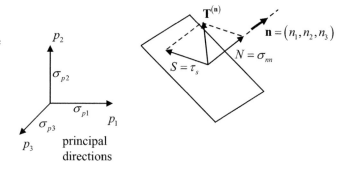

Equation (2.74) can be considered to be three equations for the unit normal components (squared) in terms of N, S, and the principal stresses. We can solve these three equations for these positive unknowns, (n_1^2, n_2^2, n_3^2), where

$$n_1^2 = \frac{S^2 + (N - \sigma_{p2})(N - \sigma_{p3})}{(\sigma_{p1} - \sigma_{p2})(\sigma_{p1} - \sigma_{p3})} \geq 0$$

$$n_2^2 = \frac{S^2 + (N - \sigma_{p1})(N - \sigma_{p3})}{(\sigma_{p2} - \sigma_{p3})(\sigma_{p2} - \sigma_{p1})} \geq 0 \qquad (2.75)$$

$$n_3^2 = \frac{S^2 + (N - \sigma_{p1})(N - \sigma_{p2})}{(\sigma_{p3} - \sigma_{p1})(\sigma_{p3} - \sigma_{p2})} \geq 0$$

If we order the stresses such that $\sigma_{p1} > \sigma_{p2} > \sigma_{p3}$ then the inequalities of Eq. (2.75) imply that

$$S^2 + (N - \sigma_{p2})(N - \sigma_{p3}) \geq 0$$
$$S^2 + (N - \sigma_{p3})(N - \sigma_{p1}) \leq 0 \qquad (2.76)$$
$$S^2 + (N - \sigma_{p1})(N - \sigma_{p2}) \geq 0$$

which can also be rewritten equivalently as

$$S^2 + \left(N - \frac{\sigma_{p2} + \sigma_{p3}}{2}\right)^2 \geq \left(\frac{\sigma_{p2} - \sigma_{p3}}{2}\right)^2$$

$$S^2 + \left(N - \frac{\sigma_{p1} + \sigma_{p3}}{2}\right)^2 \leq \left(\frac{\sigma_{p3} - \sigma_{p1}}{2}\right)^2 \qquad (2.77)$$

$$S^2 + \left(N - \frac{\sigma_{p1} + \sigma_{p2}}{2}\right)^2 \geq \left(\frac{\sigma_{p1} - \sigma_{p2}}{2}\right)^2$$

If we plot the two quantities S and N, the three inequalities in Eq. (2.77) can be interpreted as the regions exterior or interior to three circles, shown as the shaded region in Fig. 2.14, whose centers are at

2.5 Mohr's Circle for Three-Dimensional States of Stress

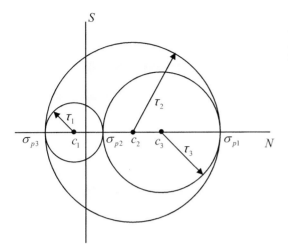

Figure 2.14 Mohr's circle construction showing all the normal stresses and total shear stresses possible at a given point in terms of the principal stresses as the shaded region between circles.

$$c_1 = \frac{\sigma_{p2} + \sigma_{p3}}{2}$$
$$c_2 = \frac{\sigma_{p1} + \sigma_{p3}}{2} \qquad (2.78)$$
$$c_3 = \frac{\sigma_{p1} + \sigma_{p2}}{2}$$

and whose radii are

$$\tau_1 = \frac{|\sigma_{p2} - \sigma_{p3}|}{2}$$
$$\tau_2 = \frac{|\sigma_{p3} - \sigma_{p1}|}{2} \qquad (2.79)$$
$$\tau_3 = \frac{|\sigma_{p1} - \sigma_{p2}|}{2}$$

which are also the magnitudes of the three extreme values of the shear stress, as seen in Fig. 2.14.

For plane-stress problems, many books use the Mohr's circle construction to analyze the stresses on cutting planes that correspond to one of the circles in Fig. 2.14. In that case, the Mohr's circle construction offers a graphical alternative to using the transformation equations discussed in this chapter. We will not discuss this type of use of Mohr's circle here but you can find the details in elementary mechanics of materials or strength of materials books. For a general 3-D state of stress, we may be interested in stresses (more precisely, N and S values) which lie anywhere within the shaded region of possible stresses shown in Fig. 2.14. In this case the Mohr's circle graphical construction is not particularly useful for finding those stresses but we can easily obtain any stresses needed with the transformation equations previously discussed. However, the Mohr's circle construction of Fig. 2.14 is useful to help locate the planes on which the extreme shear stresses act. For example, we see that there are planes of extreme shear stress when

$$N = \frac{\sigma_{p1} + \sigma_{p2}}{2}$$

$$S^2 = \left(\frac{\sigma_{p1} - \sigma_{p2}}{2}\right)^2$$

so that from Eq. (2.75) we find

$$n_1^2 = \frac{1}{2}, \quad n_2^2 = \frac{1}{2}, \quad n_3^2 = 0$$

Similarly, when

$$N = \frac{\sigma_{p1} + \sigma_{p3}}{2}$$

$$S^2 = \left(\frac{\sigma_{p1} - \sigma_{p3}}{2}\right)^2$$

we have

$$n_1^2 = \frac{1}{2}, \quad n_3^2 = \frac{1}{2}, \quad n_3^2 = 0$$

and, finally, for

$$N = \frac{\sigma_{p2} + \sigma_{p3}}{2}$$

$$S^2 = \left(\frac{\sigma_{p2} - \sigma_{p3}}{2}\right)^2$$

we have

$$n_2^2 = \frac{1}{2}, \quad n_3^2 = \frac{1}{2}, \quad n_1^2 = 0$$

These results are all summarized in Table 2.1.

Since $\sin(45°) = \cos(45°) = 1/\sqrt{2}$ and the components of the unit normal in Table 2.1 are measured with respect to the principal directions, we see that the planes of extreme shear

Table 2.1 The planes of extreme shear and the shear stress (squared) and normal stress on those planes

n_1	n_2	n_3	S^2	N
0	$\pm 1/\sqrt{2}$	$\pm 1/\sqrt{2}$	$(\sigma_{p2} - \sigma_{p3})^2/4$	$(\sigma_{p2} + \sigma_{p3})/2$
$\pm 1/\sqrt{2}$	0	$\pm 1/\sqrt{2}$	$(\sigma_{p1} - \sigma_{p3})^2/4$	$(\sigma_{p1} + \sigma_{p3})/2$
$\pm 1/\sqrt{2}$	$\pm 1/\sqrt{2}$	0	$(\sigma_{p1} - \sigma_{p2})^2/4$	$(\sigma_{p1} + \sigma_{p2})/2$

2.5 Mohr's Circle for Three-Dimensional States of Stress

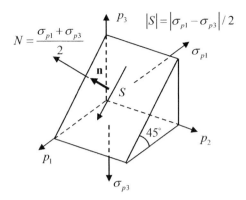

Figure 2.15 Stresses on a plane of extreme shear stress. Note that actual direction of the total shear stress may be opposite to that shown, depending on the values for the principal stresses.

all lie at $\pm 45°$ from the principal directions. One such extreme shear stress plane is shown in Fig. 2.15. Generally, however, we are interested only in the magnitude of the largest extreme shear stress.

Since we ordered the principal stresses such that $\sigma_{p1} > \sigma_{p2} > \sigma_{p3}$, the magnitude of this largest shear stress will be $|\sigma_{p1} - \sigma_{p3}|/2$ and one of the planes on which it acts is shown in Fig. 2.15. If we want to find a plane of extreme shear and the direction of the total shear stress on that plane explicitly, then this can easily be done with MATLAB® as we will now show with an example.

Example 2.4 Finding the Plane of an Extreme Shear Stress and its Direction

Consider a state of stress (in MPa) given in MATLAB® by

```
stress = [ 10   -3   0; -3   -7   2; 0   2   5]
stress = 10     -3      0
         -3     -7      2
          0      2      5
```

and use eig to calculate the principal stresses and principal directions

```
[ pdirs, pvals]  =  eig(stress)

pdirs =    0.1641      0.0884      0.9825
           0.9746      0.1390     -0.1753
          -0.1521      0.9863     -0.0633

pvals =   -7.8172           0           0
                0      5.2819           0
                0           0     10.5353
```

These principal directions do form a right-handed coordinate system since the determinant is plus one:

```
det(pdirs)
ans = 1.0000
```

(If the system were not right-handed then we would change the signs on one of the unit vectors in pdirs.) The magnitude of the largest shear stress on any cutting plane is one half the magnitude of the largest principal stress minus the smallest principal stress given here by

```
tau_max = abs(pvals(3,3) -pvals(1,1))/2
tau_max = 9.1763
```

and the normal stress on this plane is

```
norm_stress = (pvals (1,1) +pvals(3,3))/2
norm_stress = 1.3591
```

These values are easily seen in the Mohr's circle construction for this case, which is shown in Fig. 2.16. The principal stresses, ordered from the largest to the smallest, are approximately $\sigma_{p1} = 10.54$ MPa, $\sigma_{p2} = 5.28$ MPa, $\sigma_{p3} = -7.82$ MPa. Now, let the unit vectors in the columns of the pdirs matrix that are associated with these principal stresses be $\mathbf{e}_{p1}, \mathbf{e}_{p2}, \mathbf{e}_{p3}$, respectively. These unit vectors are shown schematically in Fig. 2.17a. We can extract these unit vectors from the pdirs matrix:

```
ep1 = pdirs(:,1)
ep1 =  0.1641
       0.9746
      -0.1521

ep2 = pdirs(:,2)
ep2 = 0.0884
      0.1390
      0.9863

ep3 = pdirs(:,3)
ep3 =  0.9825
      -0.1753
      -0.0633
```

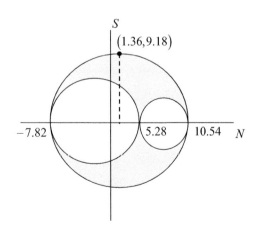

The planes of largest shear associated with the principal stresses $(\sigma_{p1}, \sigma_{p3})$ will be at $\pm 45°$ from the principal directions $(\mathbf{e}_{p1}, \mathbf{e}_{p3})$ containing either lines a–a or b–b as shown in Fig. 2.17a. Either of those planes could be chosen and we could also choose to look at the element on either side of the cutting planes through a–a or b–b so there are actually four choices we could make for planes having the largest shear stress. It is not important which of these choices we make, so let's use the cutting plane through a–a and

Figure 2.16 Mohr's circle construction for Example 2.4 showing the principal stresses and the normal stress and largest total shear values on the plane of maximum shear stress.

2.5 Mohr's Circle for Three-Dimensional States of Stress   59

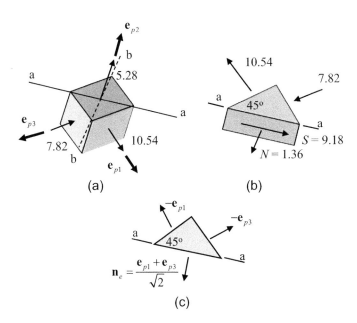

Figure 2.17 (a) The principal stresses and principal stress directions. (b) The unit normal, $\mathbf{n}_e$, acting on plane a–a, which is a plane of extreme shear. (c) Plane a–a, as seen looking in the $-\mathbf{e}_{p2}$ direction.

examine the element shown in Fig. 2.17b. If we look down the $\mathbf{e}_{p2}$ axis, this element appears as shown in Fig. 2.17c. From the geometry of Fig. 2.17c, it is easy to see that the normal to the plane containing the largest shear stress is given in terms of the principal stress directions as $\mathbf{n}_e = (\mathbf{e}_{p1} + \mathbf{e}_{p3})/\sqrt{2}$. In MATLAB® we can calculate this unit normal easily:

```
ne = (ep1+ep3)/sqrt(2)
ne = 0.8108
     0.5652
    -0.1523
```

Since we now know the normal to this cutting plane and the state of stress, it is easy to use the transformation equations discussed previously to find any information we want about the stresses on this plane. For the stress vector on the plane (see Eq. (2.16)):

```
stress_vec = stress*ne
stress_vec = 6.4119
            -6.6935
             0.3688
```

For the normal stress (Eq. (2.19)):

```
norm_stress2 = ne' * stress * ne
norm_stress2 = 1.3591
```

which agrees with our previous result. For the magnitude of total shear stress (Eq. (2.20)) we have:

```
Smax = sqrt(norm(stress_vec)^2 - norm_stress2^2)
Smax = 9.1763
```

which is obviously the largest shearing stress. The direction of this stress is just (see Eq. (2.22))

```
s = (stress_vec -norm_stress2*ne)/Smax
s = 0.5787
   -0.8131
    0.0628
```

Most books do not go into this explicit detail about determining the plane of the largest shear stress and its direction since it is only the magnitude of the maximum shearing stress that appears in the commonly discussed theories of failure [1] (see Chapter 13). However, we have given a more extensive discussion here since it also shows the importance of the transformation equations we have derived earlier.

2.6 Stresses on the Octahedral Plane

In addition to the principal planes and the planes of maximum shearing stress a particular cutting plane of interest in some material failure theories is the *octahedral plane*. The octahedral plane is by definition a plane that makes equal angles with respect to the three principal directions. Thus, if we let n_j be the component of the unit normal to the octahedral plane *as measured with respect to the to the principal directions*, then we see

$$n_1 = n_2 = n_3 = \frac{1}{\sqrt{3}} \tag{2.80}$$

defines an octahedral plane. The three components of the stress vector on the octahedral plane (as measured also with respect to the principal directions) and the normal stress on the octahedral plane are

$$\begin{aligned} T_1^{(\mathbf{n})} &= \sigma_{p1} n_1 \\ T_2^{(\mathbf{n})} &= \sigma_{p2} n_2 \\ T_3^{(\mathbf{n})} &= \sigma_{p3} n_3 \\ \sigma_{nn} &= \sigma_{p1} n_1^2 + \sigma_{p2} n_2^2 + \sigma_{p3} n_3^2 \end{aligned} \tag{2.81}$$

so that we find

$$
\begin{aligned}
(\sigma_{nn})_{oct} &= \frac{\sigma_{p1} + \sigma_{p2} + \sigma_{p3}}{3} = \frac{\sigma_{xx} + \sigma_{yy} + \sigma_{zz}}{3} = \frac{I_1}{3} \\
|\tau_s|_{oct} &= \sqrt{(\sigma_{p1}n_1)^2 + (\sigma_{p2}n_2)^2 + (\sigma_{p3}n_3)^2 - (\sigma_{p1}n_1^2 + \sigma_{p2}n_2^2 + \sigma_{p3}n_3^2)^2} \\
&= \frac{1}{3}\sqrt{(\sigma_{p1} - \sigma_{p2})^2 + (\sigma_{p2} - \sigma_{p3})^2 + (\sigma_{p1} - \sigma_{p3})^2} \\
&= \frac{1}{3}\sqrt{2I_1^2 - 6I_2}
\end{aligned}
\quad (2.82)
$$

Since the last expression for the total shear stress on the octahedral plane is in terms of the stress invariants only, we can also write this total shear stress in terms of stresses associated with the (x, y, z) axes. We find

$$|\tau_s|_{oct} = \frac{1}{3}\sqrt{(\sigma_{xx} - \sigma_{yy})^2 + (\sigma_{xx} - \sigma_{zz})^2 + (\sigma_{yy} - \sigma_{zz})^2 + 6(\sigma_{xy}^2 + \sigma_{yz}^2 + \sigma_{xz}^2)^2} \quad (2.83)$$

A closely related normal stress is called the *von Mises stress*, σ_v, given in terms of the octahedral stress as

$$\sigma_v = \frac{3}{\sqrt{2}}|\tau_s|_{oct} \quad (2.84)$$

which is a stress that appears in theories of yielding of materials [1] (also see Chapter 13). This stress is also sometimes called an *equivalent stress* or the *effective stress*. For complex 3-D states of stress, the von Mises stress is often used as the stress of choice to plot to show the internal distribution of stresses in a body.

2.7 Stress Notations and the Concept of Stress – A Historical Note

The concept of stress was developed in 1821 by Augustin-Louis Cauchy (1789–1857), so it is 200 years old. Over the years, various notations have been used to characterize the state of stress. Some of the most common of these notations are listed in Fig. 2.18 along with the names of the scientists who used them. [For a text that is notable for having pictures of many of the individuals who contributed to the field of mechanics over the years, see [2].] As you have seen, in this book we will generally write stresses as σ_{xx}, σ_{xy}, etc., or use numerical subscripts to write them as σ_{11}, σ_{12}, etc. The use of numerical subscripts is important since it facilitates writing stress relationships in matrix–vector forms. Occasionally, we will distinguish normal and shear stresses by writing normal stresses using the Greek symbol sigma (σ) and shear stresses with the Greek symbol tau (τ). The normal stress and total shear stress are examples where we wrote them as (σ_{nn}, τ_s).

The concept of stress is not an intuitive one, so before the work of Cauchy many researchers struggled with how to describe internal force distributions. Also, the concept

Stress notations

Cauchy (early work)	Poisson	Coriolis, Saint-Venant, Maxwell, Cauchy (later work)
$A\ F\ E$	$P_3\ Q_3\ R_3$	$p_{xx}\ p_{xy}\ p_{xz}$
$F\ B\ D$	$P_2\ Q_2\ R_2$	$p_{yx}\ p_{yy}\ p_{yz}$
$E\ D\ C$	$P_1\ Q_1\ R_1$	$p_{zx}\ p_{zy}\ p_{zz}$

Neuman, Kirchhoff, Rieman, Love	Kelvin	Clebsch
$X_x\ X_y\ X_z$	$P\ V\ T$	$t_{11}\ t_{12}\ t_{13}$
$Y_x\ Y_y\ Y_z$	$V\ Q\ S$	$t_{21}\ t_{22}\ t_{23}$
$Z_x\ Z_y\ Z_z$	$T\ S\ R$	$t_{31}\ t_{32}\ t_{33}$

Pearson	Current notation		
$\widehat{xx}\ \widehat{xy}\ \widehat{xz}$	$\sigma_{xx}\ \sigma_{xy}\ \sigma_{xz}$		$\sigma_{11}\ \sigma_{12}\ \sigma_{13}$
$\widehat{yx}\ \widehat{yy}\ \widehat{yz}$	$\sigma_{yx}\ \sigma_{yy}\ \sigma_{yz}$	or	$\sigma_{21}\ \sigma_{22}\ \sigma_{23}$
$\widehat{zx}\ \widehat{zy}\ \widehat{zz}$	$\sigma_{zx}\ \sigma_{zy}\ \sigma_{zz}$		$\sigma_{31}\ \sigma_{32}\ \sigma_{33}$

Figure 2.18 Stress notations used by various authors to describe the state of stress.

of stress we have presented is in a sense incomplete since we know from statics that a general distribution of forces is statically equivalent to both a force and a moment. Thus, strictly speaking, in Fig. 2.1 we should have drawn a small net force, $\Delta\mathbf{F}$, and a small net moment, $\Delta\mathbf{M}$, acting on an area ΔA of a cutting plane. In principle, therefore, we could then also define a couple stress, $\mathbf{C}^{(n)}$, as

$$\mathbf{C}^{(n)} = \lim_{\Delta A \to 0} \frac{\Delta\mathbf{M}}{\Delta A} \qquad (2.85)$$

Studies of "ordinary" structural materials have not found instances where the concept of couple stresses is needed, and in this book we do not include them. However, modern materials with complex microstructures may require such generalizations of the stress concept to describe local distributions of forces and moments in such materials.

2.8 PROBLEMS

P2.1 Consider a small element such as the one shown in Fig. 2.4. The moment equation for this element can be written as

$$\int_{\Delta V} \rho \mathbf{r} \times \mathbf{a} \, dV = \int_{\Delta V} \rho \mathbf{r} \times \mathbf{g} \, dV + \sum_{i=1}^{6} \int_{\Delta A_i} \mathbf{r} \times \mathbf{T}^{(\mathbf{n}_i)} \, dA \qquad \text{(P2.1)}$$

where ΔV is the volume of the element, ΔA_i are the six faces of the element (including the back faces, which are not shown in Fig. 2.4) on which the stress vectors $\mathbf{T}^{(\mathbf{n}_i)}$ act, ρ is the mass/volume (density) of the element, and $\mathbf{a}$ is the acceleration of the element. The quantity $\mathbf{g}$ is the acceleration of gravity. The position vector $\mathbf{r}$ is measured from the center of the element. Show that in the limit as $\Delta V \to 0$, $\Delta A_i \to 0$ that Eq. (P2.1) implies that the shear-stress components are symmetric. Note that in the limit the volume integrals go to zero faster than the surface integral so that in fact those volume integrals can be neglected.

P2.2 Prove Eq. (2.8), i.e., show that the ratio of areas of the faces of the tetrahedron shown in Fig. 2.5 are components of the unit normal of the front inclined face. (Hint: use properties of the cross product of vectors along the sides of the tetrahedron.)

P2.3 Consider the state of stress with respect to the (x, y, z) axes given in MPa by

$$[\sigma] = \begin{bmatrix} 40 & 40 & 30 \\ 40 & 20 & 0 \\ 30 & 0 & 20 \end{bmatrix}$$

(a) Determine the normal stress and total shear stress acting on a plane which is oriented at angles of 40°, 75°, and 54° *approximately* with respect to the (x, y, z) axes, respectively.

(b) Determine the direction of the total shear stress on the plane of part (a).

P2.4 For the state of stress with respect to the (x, y, z) axes given in MPa by

$$[\sigma] = \begin{bmatrix} 12 & 6 & 9 \\ 6 & 10 & 3 \\ 9 & 3 & 14 \end{bmatrix}$$

determine the normal stress and shear stress acting on the octahedral plane.

P2.5 For the state of stress with respect to the (x, y, z) axes given in MPa by

$$[\sigma] = \begin{bmatrix} 60 & 40 & -40 \\ 40 & 0 & -20 \\ -40 & -20 & 20 \end{bmatrix}$$

determine the state of stress in the (x', y', z') system shown in Fig. P2.1 obtained by making a 30° counterclockwise rotation about the x-axis as shown.

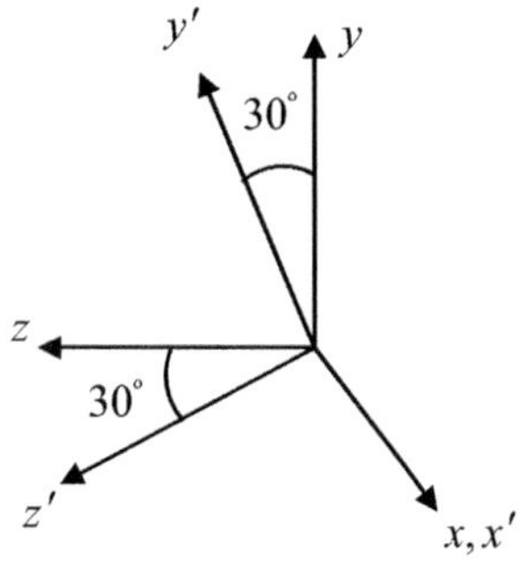

Figure P2.1 A coordinate rotation.

P2.6 Consider the state of stress with respect to the (x, y, z) axes given in MPa by

$$[\sigma] = \begin{bmatrix} 20 & -12 & 0 \\ -12 & -17 & 22 \\ 0 & 22 & 15 \end{bmatrix}$$

(a) Determine the principal stresses and principal stress directions.
(b) Determine the absolute maximum shear stress in magnitude and the plane on which it acts as well as the normal stress on this plane of extreme shear stress.

P2.7 Consider the state of stress with respect to the (x, y, z) axes given in MPa by

$$[\sigma] = \begin{bmatrix} 200 & 40 & -30 \\ 40 & 50 & -20 \\ -30 & -20 & -60 \end{bmatrix}$$

(a) Determine the (x, y, z) components of the stress vector, $\mathbf{T}^{(\mathbf{n})}$, acting on a plane P whose unit normal is

$$\mathbf{n} = \frac{3}{\sqrt{19}} \mathbf{e}_x + \frac{3}{\sqrt{19}} \mathbf{e}_y + \frac{1}{\sqrt{19}} \mathbf{e}_z$$

where $(\mathbf{e}_x, \mathbf{e}_y, \mathbf{e}_z)$ are unit vectors along the (x, y, z) axes, respectively.
(b) Determine the normal stress and total shear stress acting on plane P and the direction of the total shear stress on this plane.
(c) Two orthogonal unit vectors that lie in plane P are given by

$$\mathbf{t} = \frac{1}{\sqrt{38}} \mathbf{e}_x + \frac{1}{\sqrt{38}} \mathbf{e}_y - \frac{6}{\sqrt{38}} \mathbf{e}_z$$

$$\mathbf{v} = -\frac{1}{\sqrt{2}} \mathbf{e}_x + \frac{1}{\sqrt{2}} \mathbf{e}_y + 0 \mathbf{e}_z$$

Determine the stresses σ_{nt}, σ_{nv} acting on plane P. Show that these two shear stresses combine to produce the total shear stress (in both magnitude and direction) calculated in (b).
(d) Determine the principal stresses and principal directions and the magnitude of the maximum shearing stress.
(e) Let the (x', y', z') axes be oriented along the $(\mathbf{n}, \mathbf{t}, \mathbf{v})$ directions, respectively. Determine the complete state of stress with respect to these (x', y', z') axes. How are your results related to the results of parts (b) and (c)?

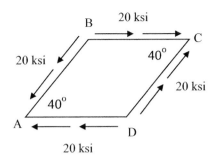

Figure P2.2 An angular plate subjected to uniform shear stresses on its edges.

P2.8 An angular plate in a state of plane stress carries stresses of 20 ksi on its sides as shown in Fig. P2.2. If the state of stress is uniform throughout the plate, determine
(a) the stress vector components $T_x^{(n)}$ and $T_y^{(n)}$ exerted on planes AB and BC;
(b) the stresses $(\sigma_{xx}, \sigma_{yy}, \sigma_{xy})$ in the plate.

P2.9 An angular plate in a state of plane stress carries stresses of 20 ksi on its sides as shown in Fig. P2.3. If the state of stress is uniform throughout the plate, determine
(a) the stress vector components $T_x^{(n)}$ and $T_y^{(n)}$ exerted on planes AB and BC;
(b) the stresses $(\sigma_{xx}, \sigma_{yy}, \sigma_{xy})$ in the plate.

P2.10 The state of stress, in MPa, with respect to a set of (x, y, z) axes, is given by

$$\sigma = \begin{bmatrix} 10 & -15 & 0 \\ -15 & 20 & 5 \\ 0 & 5 & 30 \end{bmatrix}$$

(a) Using MATLAB® determine the principal stresses and principal directions. Does the matrix of principal directions form a right-handed system?
(b) Using Eq. (2.19) show that the normal stress on the plane with a normal in the direction of the largest principal stress agrees with the results of part (a).
(c) Using Eq. (2.16) determine the (x, y, z) components of the stress vector on the plane with a normal in the direction of the largest principal stress. Show that the component of this stress vector along the largest principal stress direction agrees with the results of part (a).

P2.11 The state of stress, in MPa, with respect to a set of (x, y, z) axes, is given by

$$\sigma = \begin{bmatrix} 12 & -35 & -5 \\ -35 & 40 & 0 \\ -5 & 0 & 10 \end{bmatrix}$$

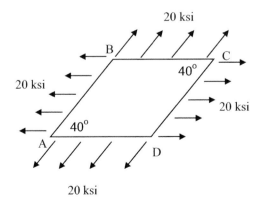

(a) Using MATLAB® determine the principal stresses and principal directions. Does the matrix of principal directions form a right-handed system? If not, make it right-handed.
(b) Determine the magnitude of the extreme shear stress, τ_{max}, and unit vectors **n**, **t** that

Figure P2.3 An angular plate subjected to uniform stresses parallel to its edges.

are normal to the τ_{max} plane and parallel to the extreme shear stress vector, respectively. Verify, using the stress transformation equations, Eq. (2.29), that you obtain the correct normal stress, σ_{nn}, and shear stress, σ_{nt}, on the plane of maximum shear stress.

P2.12 The state of stress, in MPa, with respect to a set of (x, y, z) axes, is given by

$$\sigma = \begin{bmatrix} 50 & 10 & -10 \\ 10 & 30 & 5 \\ -10 & 5 & 20 \end{bmatrix}$$

(a) Using MATLAB® determine the principal stresses and principal directions. Does the matrix of principal directions form a right-handed system? If not, make it right-handed.

(b) Determine the magnitude of the extreme shear stress, τ_{max}, and unit vectors **n**, **t** that are normal to the τ_{max} plane and parallel to the extreme shear stress vector, respectively. Verify, using the stress transformation equations, Eq. (2.29), that you obtain the correct normal stress, σ_{nn}, and shear stress, σ_{nt}, on the plane of maximum shear stress.

P2.13 At a point in a body the state of stress, as measured in ksi, with respect to a (x, y, z) coordinate system is given by

$$[\sigma] = \begin{bmatrix} 10 & 4 & -6 \\ 4 & 3 & -2 \\ -6 & -2 & 8 \end{bmatrix}$$

Consider three lines that extend from the origin of this coordinate system given by the vectors $(\mathbf{OX}, \mathbf{OY}, \mathbf{OZ})$, where

$$\mathbf{OX} = \mathbf{i} + \mathbf{j} + \mathbf{k}$$
$$\mathbf{OY} = \mathbf{i} + \mathbf{j} - 2\mathbf{k}$$
$$\mathbf{OZ} = -\mathbf{i} + \mathbf{j}$$

where the vectors $(\mathbf{i}, \mathbf{j}, \mathbf{k})$ are unit vectors along the (x, y, z) axes respectively. Note that these vectors are all perpendicular to each other and act along a right-handed (X, Y, Z) coordinate system. Determine:

(a) the (x, y, z) stress vector components for the two planes whose normals are along the **OX** and **OZ** lines, respectively;

(b) the normal stresses acting on the two planes of part (a);

(c) the shear stresses associated with the pair of lines $(\mathbf{OX}, \mathbf{OY})$ and the lines $(\mathbf{OY}, \mathbf{OZ})$.

(d) For the (X, Y, Z) axes along the $(\mathbf{OX}, \mathbf{OY}, \mathbf{OZ})$ directions, determine the state of stress with respect to these (X, Y, Z) axes. How are your results related to the stresses you calculated in parts (b) and (c)?

(e) Show that both the state of stress you found in part (d) and the original state of stress produce the same cubic equation for determining the principal stresses. Find both the principal stresses and principal stress directions.

P2.14 The state of stress, as measured in MPa, with respect to an (x, y, z) coordinate system, is given by

$$[\sigma] = \begin{bmatrix} 100 & 50 & -50 \\ 50 & 30 & -20 \\ -50 & -20 & 70 \end{bmatrix}$$

(a) Determine the (x, y, z) components of the stress vector acting on a plane P whose unit normal, **n**, makes equal angles with respect to the (x, y, z) axes. [Note: this is <u>not</u> the octahedral plane, which makes equal angles with respect to the principal axes.]

(b) Find the normal stress and total shear stress acting on plane P and the direction of the total shear stress on this plane.

(c) Two orthogonal unit vectors $(\mathbf{t}, \mathbf{v})$ that lie in the plane P are given by:

$$\mathbf{t} = \frac{1}{\sqrt{6}} \mathbf{e}_x + \frac{1}{\sqrt{6}} \mathbf{e}_y - \frac{2}{\sqrt{6}} \mathbf{e}_z$$

$$\mathbf{v} = \frac{1}{\sqrt{2}} \mathbf{e}_x + \frac{1}{\sqrt{2}} \mathbf{e}_y$$

Determine the shear stresses $(\sigma_{nt}, \sigma_{nv})$ acting on the plane P. Show that these two shear stress components combine to produce the total shear stress (in both magnitude and direction) calculated in part (b).

(d) Determine the principal stresses and principal stress directions and the magnitude of the maximum shear stress.

(e) Let the (x', y', z') axes be oriented in the $(\mathbf{n}, \mathbf{t}, \mathbf{v})$ directions, respectively. Determine the complete state of stress with respect to the (x', y', z') axes.

P2.15 The *deviatoric stresses*, s_{ij}, are related to the stresses, σ_{ij}, by subtracting off a mean normal stress component, $\sigma_m = (\sigma_{xx} + \sigma_{yy} + \sigma_{zz})/3$, from all the normal stresses, i.e.,

$$\begin{bmatrix} s_{xx} & s_{xy} & s_{xz} \\ s_{yx} & s_{yy} & s_{yz} \\ s_{zx} & s_{zy} & s_{zz} \end{bmatrix} = \begin{bmatrix} \sigma_{xx} - \sigma_m & \sigma_{xy} & \sigma_{xz} \\ \sigma_{yx} & \sigma_{yy} - \sigma_m & \sigma_{yz} \\ \sigma_{zx} & \sigma_{zy} & \sigma_{zz} - \sigma_m \end{bmatrix}$$

This deviatoric stress state is used in theories of failure with respect to slip (yielding) since it is a measure of those stresses that produce only distortional changes in a body. Consider the state of stress (in MPa):

$$[\sigma] = \begin{bmatrix} 55 & 0 & 24 \\ 0 & 46 & 0 \\ 24 & 0 & 43 \end{bmatrix}$$

(a) Determine the deviatoric stresses.
(b) Compare the cubic equation that defines the principal deviatoric stresses with the cubic equation for the principal stresses. How are they related?
(c) Compare the principal deviatoric stresses with the principal stresses. How do they differ?
(d) Compare the principal deviatoric stress directions with the principal directions of the stresses. How are they related?
(e) Draw the 3-D Mohr's circles for both the deviatoric stresses and the stresses. How is the magnitude of the maximum deviatoric shear stress related to the magnitude of the maximum shear stress?

References

1. J. R. Barber, *Intermediate Mechanics of Materials,* 2nd edn (Switzerland: Springer, 2011)
2. E. Volterra and J. H. Gaines, *Advanced Strength of Materials* (Englewood Cliffs NJ: Prentice-Hall, 1971)

3 Equilibrium

The previous chapter focused on the behavior of the stress vector and stresses at any fixed point in a body. However, stresses will also vary from point to point within a deformable body so that we need to describe those spatial variations. As we will see in this chapter, local equations of equilibrium involving the stresses must be satisfied everywhere within a body. Those governing equations of equilibrium, as described in Chapter 1, form one of the fundamental "pillars" of any analysis of stresses.

3.1 Equations of Equilibrium – Cartesian Coordinates

From statics we know that for a rigid body to be in equilibrium, there must be no net vector force, **F**, or moment, $\mathbf{M}_0$, acting about any point, O, on the body; i.e., we have

$$\begin{aligned} \mathbf{F} &= 0 \\ \mathbf{M}_0 &= 0 \end{aligned} \tag{3.1}$$

If we break a body into parts, then Eqs. (3.1) must hold for any part as well. For deformable bodies, these equilibrium conditions must also be satisfied, both for the entire body and for any part we choose. Let us consider a part of the deformable body that has a volume, V, contained within a closed surface, S (see Fig. 3.1). On S the force distribution will described by the stress vector, $\mathbf{T}^{(\mathbf{n})}$ (whose outward unit normal at any point is **n**), and within the part we will allow body forces, **f** (such as gravity), to act throughout the volume, V, where **f** has the dimensions of force/unit volume. The force and moment equilibrium of the part requires

$$\begin{aligned} \int_S \mathbf{T}^{(\mathbf{n})} dS + \int_V \mathbf{f} dV &= 0 \\ \int_S \mathbf{x}_s \times \mathbf{T}^{(\mathbf{n})} dS + \int_V \mathbf{x} \times \mathbf{f} dV &= 0 \end{aligned} \tag{3.2}$$

where dS and dV are surface and volume elements, respectively, and $(\mathbf{x}_s, \mathbf{x})$ are position vectors from point O to a surface point or point within the volume, respectively, where either $\mathbf{T}^{(\mathbf{n})}$ or **f** acts. If we use Eq. (2.10) we can write the stress vector, $\mathbf{T}^{(\mathbf{n})}$, in terms of the stress vectors, $\mathbf{T}^{(\mathbf{e}_i)}$, along the $(x, y, z) = (x_1, x_2, x_3)$ Cartesian axes (and where $\mathbf{e}_i$ are unit vectors along those axes) and the components of the unit normal to S to obtain

70 Equilibrium

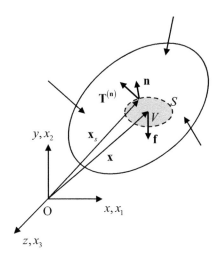

Figure 3.1 A deformed body under the action of loads (shown as concentrated forces only for illustration) and the distributed surface forces (stress vector $\mathbf{T}^{(\mathbf{n})}$) and body forces, $\mathbf{f}$, acting on an arbitrary volume V of a part of the body, where S is the closed surface surrounding V and $\mathbf{n}$ is the unit normal on S.

$$\int_S \sum_{i=1}^{3} \mathbf{T}^{(e_i)} n_i dS + \int_V \mathbf{f} dV = 0$$

$$\int_S \sum_{i=1}^{3} \mathbf{x}_s \times \mathbf{T}^{(e_i)} n_i dS + \int_V \mathbf{x} \times \mathbf{f} dV = 0$$

(3.3)

Gauss's theorem says that for any function $g(\mathbf{x}_s)$ (which could be a scalar, a vector, or a matrix) acting on S we have

$$\int_S g(\mathbf{x}_s) n_i(\mathbf{x}_s) dS = \int_V \frac{\partial g(\mathbf{x})}{\partial x_i} dV \qquad (3.4)$$

so applying this theorem to the terms in Eq. (3.3) we have

$$\int_V \left(\sum_{i=1}^{3} \frac{\partial \mathbf{T}^{(e_i)}}{\partial x_i} + \mathbf{f} \right) dV = 0$$

$$\int_V \left(\sum_{i=1}^{3} \frac{\partial}{\partial x_i} \left[\mathbf{x} \times \mathbf{T}^{(e_i)} \right] + \mathbf{x} \times \mathbf{f} \right) dV = 0$$

(3.5)

But since the part defined by the volume V is arbitrary, the integrands of Eq. (3.5) themselves must vanish so that at any point in a deformable body we must have

$$\sum_{i=1}^{3} \frac{\partial \mathbf{T}^{(e_i)}}{\partial x_i} + \mathbf{f} = 0$$

$$\sum_{i=1}^{3} \frac{\partial}{\partial x_i} \left[\mathbf{x} + \mathbf{T}^{(e_i)} \right] + \mathbf{x} \times \mathbf{f} = 0$$

(3.6)

If we expand the derivatives in the second equation of Eq. (3.6) we have

$$\sum_{i=1}^{3} \frac{\partial \mathbf{x}}{\partial x_i} \times \mathbf{T}^{(e_i)} + \mathbf{x} \times \left[\frac{\partial \mathbf{T}^{(e_i)}}{\partial x_i} + \mathbf{f} \right] = 0 \qquad (3.7)$$

3.1 Equations of Equilibrium – Cartesian Coordinates

Since $\partial \mathbf{x}/\partial x_i = \mathbf{e}_i$ and the term in square brackets in Eq. (3.7) vanishes because of the first equation in Eq. (3.6) we obtain

$$\sum_{i=1}^{3} \frac{\partial \mathbf{T}^{(\mathbf{e}_i)}}{\partial x_i} + \mathbf{f} = 0$$

$$\sum_{i=1}^{3} \mathbf{e}_i \times \mathbf{T}^{(\mathbf{e}_i)} = 0$$

(3.8)

The first equation in Eq. (3.8) are the conditions for local equilibrium in the deformable body in terms of the stress vectors acting on the Cartesian coordinate planes. We can write these stress vectors in terms of the stress components and likewise write $\mathbf{f}$ in terms of its components as

$$\mathbf{T}^{(\mathbf{e}_i)} = \sum_{j=1}^{3} \sigma_{ij} \mathbf{e}_j \quad (i = 1, 2, 3)$$

$$\mathbf{f} = \sum_{j=1}^{3} f_j \mathbf{e}_j$$

(3.9)

to obtain the equations of equilibrium for the stresses as

$$\sum_{i=1}^{3} \frac{\partial \sigma_{ij}}{\partial x_i} + f_j = 0 \quad (j = 1, 2, 3)$$

(3.10)

For a general three-dimensional (3-D) state of stress, these equilibrium equations are, therefore, explicitly, in terms of (x, y, z) components

$$\frac{\partial \sigma_{xx}}{\partial x} + \frac{\partial \sigma_{yx}}{\partial y} + \frac{\partial \sigma_{zx}}{\partial z} + f_x = 0$$

$$\frac{\partial \sigma_{xy}}{\partial x} + \frac{\partial \sigma_{yy}}{\partial y} + \frac{\partial \sigma_{zy}}{\partial z} + f_y = 0$$

$$\frac{\partial \sigma_{xz}}{\partial x} + \frac{\partial \sigma_{yz}}{\partial y} + \frac{\partial \sigma_{zz}}{\partial z} + f_z = 0$$

(3.11)

Now, let us consider the second equation in Eq. (3.8). If we again write the stress vectors in terms of the stresses this equation becomes

$$\sum_{i=1}^{3} \sum_{j=1}^{3} \sigma_{ij} \left(\mathbf{e}_i \times \mathbf{e}_j \right) = 0$$

(3.12)

We can consider the cross-product terms as the components of a vector-valued 3×3 matrix, $\mathbf{U}$, with components, $\mathbf{U}_{ij} = \mathbf{e}_i \times \mathbf{e}_j$, where

$$\mathbf{U} = \begin{bmatrix} 0 & \mathbf{e}_3 & -\mathbf{e}_2 \\ -\mathbf{e}_3 & 0 & \mathbf{e}_1 \\ \mathbf{e}_2 & -\mathbf{e}_1 & 0 \end{bmatrix} \tag{3.13}$$

which you can easily verify by carrying out the various cross products such as $\mathbf{U}_{12} = \mathbf{e}_1 \times \mathbf{e}_2 = \mathbf{e}_3$, etc. Then Eq. (3.12) can be written in matrix form as

$$\text{trace}\left([\sigma]^T[\mathbf{U}]\right) = 0 \tag{3.14}$$

We can manually carry out all the matrix operations in Eq. (3.14) but we can also let the symbolic toolkit of MATLAB® do this for us with symbolic algebra. First, we define the unit vectors and stresses as symbolic values and form up the symbolic stress and $\mathbf{U}$ matrices:

```
syms e1 e2 e3 s11 s12 s13 s21 s22 s23 s31 s32 s33
U = [ 0   e3  -e2; -e3  0   e1; e2  -e1  0]
U = [ 0,    e3,  -e2]
    [-e3,   0,    e1]
    [ e2,  -e1,   0]
S = [ s11   s12   s13; s21   s22   s23; s31   s32   s33]
S = [ s11,  s12,  s13]
    [ s21,  s22,  s23]
    [ s31,  s32,  s33]
```

where $el = \mathbf{e}_1$, $sl1 = \sigma_{11}$, etc. Then we simply carry out the matrix operations in Eq. (3.14):

```
trace(S.'*U)
ans = e3*s12 - e2*s13 + e1*s23 - e3*s21 - e1*s32 + e2*s31
```

We can rearrange this answer and set it equal to zero, obtaining

$$\mathbf{e}_1(\sigma_{23} - \sigma_{32}) + \mathbf{e}_2(\sigma_{31} - \sigma_{13}) + \mathbf{e}_3(\sigma_{12} - \sigma_{21}) = 0 \tag{3.15}$$

which gives

$$\begin{array}{ll} \sigma_{12} = \sigma_{21} & \sigma_{xy} = \sigma_{yx} \\ \sigma_{31} = \sigma_{13} \quad \text{or} & \sigma_{zx} = \sigma_{xz} \\ \sigma_{23} = \sigma_{32} & \sigma_{yz} = \sigma_{zy} \end{array} \tag{3.16}$$

i.e., the shear stresses are symmetric as a result of moment equilibrium, as we stated without proof in Chapter 2.

3.1.1 Plane Stress

Consider a thin plate with a hole as shown in Fig. 3.2 where the direction of the loading lies in the x–y plane of the plate and the loading is uniform across the plate thickness. If there are

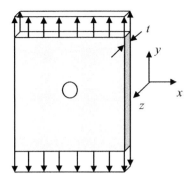

Figure 3.2 A thin plate (thickness, t, is small) with a hole under a uniaxial tensile load.

no body forces in the z-direction, and the front and back of the plate are stress-free (which means that $\sigma_{zz} = \sigma_{xz} = \sigma_{yz} = 0$ on those faces) then since the plate is thin it is reasonable to assume that $\sigma_{zz} = \sigma_{xz} = \sigma_{yz} = 0$ throughout the plate. This is an example of a state of *plane stress* where the equations of equilibrium, Eq. (3.11), reduce to

$$\frac{\partial \sigma_{xx}}{\partial x} + \frac{\partial \sigma_{yx}}{\partial y} + f_x = 0$$
$$\frac{\partial \sigma_{xy}}{\partial x} + \frac{\partial \sigma_{yy}}{\partial y} + f_y = 0 \tag{3.17}$$

and where $(\sigma_{xx}, \sigma_{xy}, f_x, f_y)$ are all functions of (x, y) only. This is an important special case since, as we will see in Chapter 7, we can solve exactly a number of useful plane-stress problems (a large plate with a hole, as seen in Fig. 3.2, is one of them).

3.2 Strength of Materials Solutions

The axial loads, bending, and torsion stresses described in Chapter 1 were developed with a number of assumptions on the deformations and the stresses. Although these stresses were shown to produce overall force and moment equilibrium, satisfaction of the stress equilibrium equations was not addressed. Do those problems satisfy the local equilibrium equations of Eq. (3.11)? For the axial-load case (Fig. 3.3a) we had only the single stress $\sigma_{xx} = P/A$ and that stress was a constant, so it is obvious that the equilibrium equations are satisfied. For the torsion of a circular rod (Fig. 3.3b) we had shear stresses (Eq. (1.37)):

$$\sigma_{xy} = \frac{-Tz}{J}$$
$$\sigma_{xz} = \frac{Ty}{J} \tag{3.18}$$

Placing these into Eq. (3.11) we see all the derivatives are zero so these torsion stresses also satisfy local equilibrium. Now, consider the case of pure bending of a beam (Fig. 3.3c), where the only stress was the bending (flexure) stress given by (Eq. (1.20)):

$$\sigma_{xx} = -\frac{Mz}{I_{yy}} \tag{3.19}$$

Placing this stress into Eq. (3.11) also shows that local equilibrium is satisfied. Finally, consider the case of engineering beam theory (Fig. 3.4a) where there were both bending and shear stresses given by (see Eq. (1.21) and Eq. (1.25)):

74   Equilibrium

Figure 3.3 Stresses present for (a) axial loading, (b) torsion, and (c) pure bending. Only the stresses acting on the front $x = $ constant face of the element are shown.

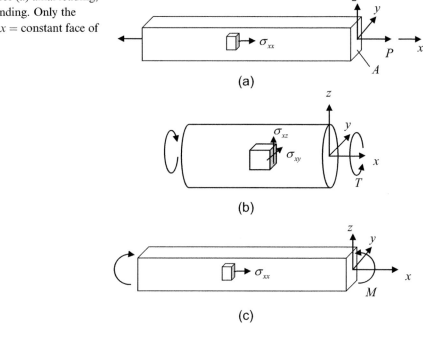

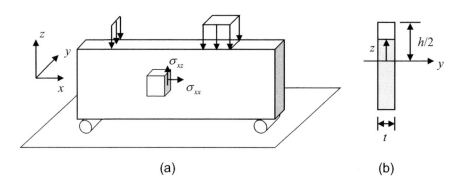

Figure 3.4 (a) Stresses in beam bending for engineering beam theory. Only the stresses acting on the $x =$ constant face of the element are shown. (b) The beam cross-section, showing the quantities needed to compute $Q(z)$.

$$\sigma_{xx}(x, z) = \frac{M_y(x)z}{I_{yy}}$$
$$\sigma_{xz}(x, z) = \frac{V_z(x)Q(z)}{I_{yy}t(z)}$$
(3.20)

All of the other stresses were ignored. Let's examine a rectangular beam where the thickness $t = $ constant. The beam is unloaded on the front and back faces where $\sigma_{yx} = \sigma_{yy} = \sigma_{yz} = 0$

so it is reasonable to assume that these stresses are zero throughout the beam. But the beam is loaded on the top face (z = constant) in the negative z-direction so that we must keep σ_{zz}. Thus, the equilibrium equations become:

$$\begin{array}{cccc}
\dfrac{\partial}{\partial x}\left[\dfrac{M_y(x)z}{I_{yy}}\right] & +0 & +\dfrac{\partial}{\partial z}\left[\dfrac{V_z(x)Q(z)}{I_{yy}t}\right] & =0 \\
0 & +0 & +0 & =0 \\
\dfrac{\partial}{\partial x}\left[\dfrac{V_z(x)Q(z)}{I_{yy}t}\right] & +0 & +\dfrac{\partial \sigma_{zz}}{\partial z} & =0
\end{array} \qquad (3.21)$$

The first equation is satisfied because we have $dM_y/dx = V_z(x)$ and for the rectangular cross-section (see Fig. 3.4b):

$$Q(z) = \int_z^{h/2} z_0 t \, dz_0 = \frac{t}{2}\left(\frac{h^2}{4} - z^2\right) \qquad (3.22)$$

$$\frac{dQ(z)}{dz} = -zt$$

Thus, we are left with the third equation, which gives

$$\frac{\partial \sigma_{zz}}{\partial z} = \frac{-dV_z/dx}{2I_{yy}}\left[\frac{h^2}{4} - z^2\right] \qquad (3.23)$$

At the top of the beam since we assume the loading is uniform across the thickness of the beam

$$\sigma_{zz}(x, z = h/2) = -q(x)/t \qquad (3.24)$$

where $q(x)$ is the applied downward force/unit length acting along the beam. Consider now the force equilibrium of a small element of the beam of length dx at a location x in the beam, as shown in Fig. 3.5. We find

$$\frac{dV_z}{dx} = q(x) \qquad (3.25)$$

Placing Eq. (3.25) into Eq. (3.23) and using the fact that the area moment for the rectangular cross-section is $I_{yy} = th^3/12$ we obtain

$$\frac{\partial \sigma_{zz}}{\partial z} = \frac{-6q(x)}{th^3}\left[\frac{h^2}{4} - z^2\right] \qquad (3.26)$$

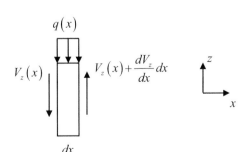

Figure 3.5 A small element of the beam showing the changing shear force and applied force/unit length acting along the beam.

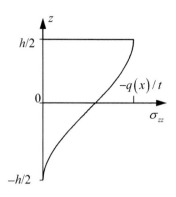

Figure 3.6 The variation of the normal stress, σ_{zz}, across the cross-section of a rectangular beam.

Integrating this equation and using Eq. (3.24) to evaluate the constant of integration, we have an explicit expression for the stress σ_{zz}:

$$\sigma_{zz}(x,z) = \frac{-6q(x)}{th^3}\left[\frac{h^2 z}{4} - \frac{z^3}{3} + \frac{h^3}{12}\right] \quad (3.27)$$

Figure 3.6 shows a plot of the variation of this stress across the cross-section. Normally, this stress is not discussed in a strength of materials solution. To see why, note that the maximum value of this stress is $|(\sigma_{zz})_{\max}| = q_{\max}/t$, where $q_{\max}$ is the largest value of the applied force/unit length. The maximum bending stress from Eq. (3.19) is $|(\sigma_{xx})_{\max}| = 6M_{\max}/th^2$, where $M_{\max}$ is the largest bending moment in the beam. Thus, the ratio of these stresses can be written as

$$\frac{|(\sigma_{zz})_{\max}|}{|(\sigma_{xx})_{\max}|} = \frac{6q_{\max}L^2}{M_{\max}}\left(\frac{h}{L}\right)^2 \quad (3.28)$$

where L is the total length of the beam. If the coefficient of $(h/L)^2$ is not large, then for long, slender beams where $h/L \ll 1$ the bending stress will be much more significant than the σ_{zz} stress and it is justified to neglect the effects of σ_{zz} in applying engineering beam theory.

3.3 Force and Moment Equilibrium

When analyzing problems with a strength of materials approach, one often must use force and moment equilibrium of a small section of the body (see Fig 1.11 or Fig. 3.5 for bending problems, for example) to describe the changes in the internal forces and moments (see Eq. (1.23) or Eq. (3.25)). The equilibrium equations of Eq. (3.11) instead involve the changes of the stresses locally at every point in a body. Clearly, there must be some relationship between these two different types of equilibrium relations. In this section we will examine that relationship for axial load, bending, and torsion problems. Let's start with the uniaxial load problem where the significant stress is σ_{xx}, which is the stress that appears in the first equilibrium equation of Eq. (3.11). If we integrate that equation across the cross-sectional area, A, of the bar (see Fig. 3.7), we find (assuming there are no body forces)

$$\frac{d}{dx}\int_A \sigma_{xx} dA + \int_A \left(\frac{\partial \sigma_{yx}}{\partial y} + \frac{\partial \sigma_{zx}}{\partial z}\right) dA = 0 \quad (3.29)$$

We recognize the first integral in Eq. (3.29) as just the internal force, $F_x(x)$, acting in the x-direction in the bar. The second integral can be converted into a closed line

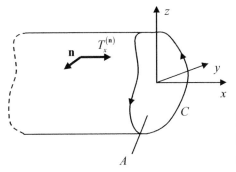

Figure 3.7 The geometry of a bar of arbitrary cross-sectional area, A. The contour, C, encloses A. Also shown is the x-component of the stress vector, $T_x^{(n)}$, acting on the surface of the bar, as well as the unit normal, **n**, to that surface.

integral around the contour, C, of the cross-sectional area using Gauss's theorem in two dimensions, which gives for any function, g,

$$\int_A \frac{\partial g}{\partial y} dA = \oint_C g n_y \, ds, \quad \int_A \frac{\partial g}{\partial z} dA = \oint_C g n_z \, ds \tag{3.30}$$

so that Eq. (3.29) becomes

$$\frac{dF_x}{dx} + \oint_C (\sigma_{yx} n_y + \sigma_{zx} n_z) ds = 0 \tag{3.31}$$

The integrand in Eq. (3.31) can be recognized as just the x-component of the stress vector, $T_x^{(n)}$, acting on the surface of the bar. Let us define a quantity, q_x, as

$$q_x = \oint_C T_x^{(n)} ds \tag{3.32}$$

Physically, $q_x = q_x(x)$ is a force/unit length acting in the x-direction generated by applied stresses acting on the bar surface. Thus, our integrated equilibrium equation is

$$\frac{dF_x}{dx} = -q_x \tag{3.33}$$

If the force/unit length is zero, then the internal axial force is just a constant, as we assumed in Chapter 1 when describing axial loads. If this force/unit length is not zero then we can use Eq. (3.33) to find the varying internal force, and, if we let $F_x = \sigma_{xx} A$, $\sigma_{xx} = E e_{xx} = E \partial u_x / \partial x$, Eq. (3.33) becomes

$$\frac{d}{dx}\left(EA \frac{\partial u_x}{\partial x}\right) = -q_x \tag{3.34}$$

which is an equation we can use to find the axial displacement, u_x, in the bar.

Now, consider the case of engineering beam theory. In that theory the two most significant stresses were a bending stress, σ_{xx}, and a shear stress, σ_{xz}. Since the bending stress again appears in the first equilibrium equation in Eq. (3.11) let us multiply that equation by z and integrate over the cross-sectional area of the beam. We obtain (again, with no body forces)

$$\frac{d}{dx}\int_A z\sigma_{xx}dA + \int_A \left(z\frac{\partial\sigma_{yx}}{\partial y} + z\frac{\partial\sigma_{zx}}{\partial z}\right)dA = 0 \tag{3.35}$$

and we can recognize the first integral in Eq. (3.35) as the internal bending moment, $M_y(x)$. We can rewrite the second integral, giving

$$\frac{dM_y}{dx} + \int_A \left[\frac{\partial}{\partial y}(z\sigma_{yx}) + \frac{\partial}{\partial z}(z\sigma_{zx})\right]dA - \int_A \sigma_{zx}dA = 0 \tag{3.36}$$

Since $\sigma_{zx} = \sigma_{xz}$ we recognize the last integral in Eq. (3.36) as the internal shear force, $V_z(x)$, acting in the z-direction in the beam. The other integral can again be transformed into an integral over the contour, C, involving the stress vector component, $T_x^{(\mathbf{n})}$, so we find

$$\frac{dM_y}{dx} + \oint_C zT_x^{(\mathbf{n})}ds - V_z = 0 \tag{3.37}$$

However, in engineering beam theory there are usually no distributed loads acting on the beam in the x-direction so $T_x^{(\mathbf{n})} = 0$ and we have, finally,

$$\frac{dM_y}{dx} = V_z \tag{3.38}$$

which is the moment–shear force relationship normally seen in beam theory. Now let us consider the shear force itself by examining the third equilibrium equation in Eq. (3.11) and integrating it over the cross-sectional area, A. In that case, we obtain

$$\frac{d}{dx}\int_A \sigma_{xz}dA + \int_A \left(\frac{\partial\sigma_{yz}}{\partial y} + \frac{\partial\sigma_{zz}}{\partial z}\right)dA = 0 \tag{3.39}$$

The first integral in Eq. (3.39) is just the shear force, $V_z(x)$, and again Gauss's theorem can be used to change the second integral into one over the contour, C, giving

$$\frac{dV_z}{dx} + \oint_C (\sigma_{yz}n_y + \sigma_{zz}n_z)ds = 0 \tag{3.40}$$

where now the integrand is just the stress vector component, $T_z^{(\mathbf{n})}$, acting in the z-direction and we can define the force/unit length in the z-direction, $q_z(x)$, acting along the beam as

$$\oint_C T_z^{(\mathbf{n})}ds = q_z \tag{3.41}$$

giving, finally,

$$\frac{dV_z}{dx} = -q_z \tag{3.42}$$

which is the same as Eq. (3.25) with $q_z = -q$ since q acts in the negative z-direction.

From the moment–curvature relationship we have $M_y = -EI_{yy}\, d^2w/dx^2$ (see Eq. (1.19)) so taking a x-derivative of Eq. (3.38) and using Eq. (3.42) we find

$$\frac{d^2}{dx^2}\left(EI_{yy}\frac{d^2w}{dx^2}\right) = q_z \tag{3.43}$$

which can be integrated to find the deflection, w, of the neutral axis. Normally, however, in elementary mechanics of materials courses the neutral axis deflection is obtained by first finding explicitly $M_y = M_y(x)$ in the beam and then integrating the moment–curvature relationship to find w. In that case only two integrations are required instead of the four integrations needed with Eq. (3.43) and one can more easily handle applied concentrated forces and couples acting on the beam since q_z cannot be described by regular functions for those types of applied loads and one must introduce generalized functions, called singularity functions, in order to use Eq. (3.43). Most elementary strength of materials texts describe these issues in the determination of beam deflections so we will not discuss them further here. In the problems of Chapter 1, however, singularity functions are used to describe the internal shear force and bending moment distributions in compact forms.

Now, consider the case of torsion. The internal twisting moment, $M_x(x)$, (called the torque, T, in Chapter 1) is given by

$$M_x = \int_A (y\sigma_{xz} - z\sigma_{xy})\, dA \tag{3.44}$$

Let's multiply the second equation in Eq. (3.11) (with no body forces) by $-z$ and the third equation by y, and integrate over the cross-sectional area, A. If we sum the two resulting equations, we obtain

$$\frac{d}{dx}\int_A (y\sigma_{xz} - z\sigma_{xy})\, dA + \int_A \left[y\left(\frac{\partial\sigma_{yz}}{\partial y} + \frac{\partial\sigma_{zz}}{\partial z}\right) - z\left(\frac{\partial\sigma_{yy}}{\partial y} + \frac{\partial\sigma_{zy}}{\partial z}\right)\right] dA = 0 \tag{3.45}$$

which can be rewritten as

$$\frac{dM_x}{dx} + \int_A \left[\frac{\partial}{\partial y}(y\sigma_{yz} - z\sigma_{yy}) + \frac{\partial}{\partial z}(y\sigma_{zz} - z\sigma_{zy})\right] dA \\ - \int_A (\sigma_{yz} - \sigma_{zy})\, dA = 0 \tag{3.46}$$

But using the symmetry of the shearing stress and Gauss's theorem, this becomes

$$\frac{dM_x}{dx} + \oint_C y(\sigma_{yz}n_y + \sigma_{zz}n_z)\, ds - \oint_C z(\sigma_{yy}n_y + \sigma_{zy}n_z)\, ds = 0 \tag{3.47}$$

or, equivalently in terms of stress vectors (Eq. (2.12))

$$\frac{dM_x}{dx} + \oint_C \left(y T_z^{(\mathbf{n})} - z T_y^{(\mathbf{n})} \right) ds = 0 \tag{3.48}$$

where the line integral represents a torque/unit length, $t_x(x)$, acting in the x-direction along the surface of the shaft, so that, finally,

$$\frac{dM_x}{dx} = -t_x \tag{3.49}$$

If t_x is zero then the internal torque is a constant in the shaft, as seen in Chapter 1. If t_x is not zero then the internal torque and the twist/unit length, $d\phi/dx$, will be functions of x. From the relationship between the torque and rate of twist, (see Eq. (1.34)), $M_x = GJ d\phi/dx$ so that

$$\frac{d}{dx}\left(GJ \frac{d\phi}{dx} \right) = -t_x \tag{3.50}$$

which is an equation that can be used to find the twist, $\phi(x)$, in the shaft.

All of the force and moment equations considered in this section were obtained by simply integrating the local equations of equilibrium appropriately. All of these equations can be obtained alternatively by examining equilibrium of small sections such as the one shown in Fig. 1.11 for bending problems. However, we have wanted to highlight here the connection between force and moment equilibrium relations and the equations of equilibrium for the stresses.

3.4 Equations of Equilibrium – Cylindrical and Spherical Coordinates

The equilibrium equations in terms of the Cartesian components of the stress vector and relationship between those stress vector components and the stresses along cutting planes whose normals are in the Cartesian coordinate directions were

$$\sum_{i=1}^{3} \frac{\partial \mathbf{T}^{(\mathbf{e}_i)}}{\partial x_i} + \mathbf{f} = 0 \tag{3.51}$$

and

$$\mathbf{T}^{(\mathbf{e}_i)} = \sum_{i=1}^{3} \sigma_{ij} \mathbf{e}_j \tag{3.52}$$

In this section we would like to generalize these equations to other coordinates such as cylindrical and spherical coordinates. We can do this if we write these equations in a coordinate-invariant form. This can be done with the aid of the vector identity

$$\nabla \cdot (\phi \mathbf{A}) = \phi \nabla \cdot \mathbf{A} + \mathbf{A} \cdot \nabla \phi \tag{3.53}$$

3.4 Equations of Equilibrium – Cylindrical and Spherical Coordinates

where ϕ is any scalar and $\mathbf{A}$ is any vector. First, consider the divergence of a vector given by

$$\sum_{i=1}^{3}\frac{\partial B_i}{\partial x_i} = \nabla \cdot \mathbf{B} = \nabla \cdot \sum_{i=1}^{3} B_i \mathbf{e}_i \qquad (3.54)$$

If we let $\phi = B_i$ and $\mathbf{A} = \mathbf{e}_i$, where the unit vectors $\mathbf{e}_i$ are not necessarily constants, then

$$\sum_{i=1}^{3}\frac{\partial B_i}{\partial x_i} = \sum_{i=1}^{3}[B_i(\nabla \cdot \mathbf{e}_i) + (\mathbf{e}_i \cdot \nabla)B_i]$$

$$= \sum_{i=1}^{3}[(\nabla \cdot \mathbf{e}_i) + (\mathbf{e}_i \cdot \nabla)]B_i \qquad (3.55)$$

which is a coordinate-invariant form of the divergence for a vector $\mathbf{B}$ written in other than Cartesian coordinates. Now, consider

$$\sum_{i=1}^{3}\frac{\partial \mathbf{B}_i}{\partial x_i} = \sum_{i=1}^{3}\frac{\partial}{\partial x_i}\left(\sum_{j=1}^{3}B_{ij}\mathbf{e}_j\right)$$

$$= \sum_{i=1}^{3}\sum_{j=1}^{3}\left(\frac{\partial B_{ij}}{\partial x_i}\mathbf{e}_j + B_{ij}\frac{\partial \mathbf{e}_j}{\partial x_i}\right) \qquad (3.56)$$

$$= \sum_{i=1}^{3}\sum_{j=1}^{3}\left[\frac{\partial B_{ij}}{\partial x_i}\mathbf{e}_j + B_{ij}(\mathbf{e}_i \cdot \nabla)\mathbf{e}_j\right]$$

But from our previous result on the divergence we can write

$$\sum_{i=1}^{3}\frac{\partial B_{ij}}{\partial x_i} = \sum_{i=1}^{3}[(\nabla \cdot \mathbf{e}_i) + (\mathbf{e}_i \cdot \nabla)]B_{ij} \qquad (3.57)$$

so that we have

$$\sum_{i=1}^{3}\frac{\partial \mathbf{B}_i}{\partial x_i} = \sum_{i=1}^{3}\sum_{j=1}^{3}\left[B_{ij}\mathbf{e}_j(\nabla \cdot \mathbf{e}_i) + \mathbf{e}_j(\mathbf{e}_i \cdot \nabla)B_{ij} + B_{ij}(\mathbf{e}_i \cdot \nabla)\mathbf{e}_j\right]$$

$$= \sum_{i=1}^{3}[\mathbf{B}_i(\nabla \cdot \mathbf{e}_i) + (\mathbf{e}_i \cdot \nabla)\mathbf{B}_i] \qquad (3.58)$$

$$= \sum_{i=1}^{3}(\nabla \cdot \mathbf{e}_i + \mathbf{e}_i \cdot \nabla)\mathbf{B}_i$$

which is a coordinate-invariant form. If we let $\mathbf{B}_i = \mathbf{T}^{(\mathbf{e}_i)}$ then the equilibrium equations in coordinate-invariant form are

$$\sum_{i=1}^{3}(\nabla \cdot \mathbf{e}_i + \mathbf{e}_i \cdot \nabla)\mathbf{T}^{(\mathbf{e}_i)} + \mathbf{f} = 0 \qquad (3.59)$$

Figure 3.8 (a) Cylindrical coordinates. (b) Spherical coordinates.

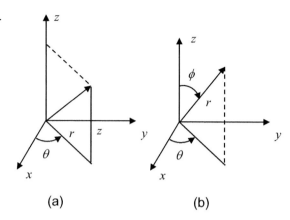

We can use Eq. (3.59) for cylindrical coordinates (r, θ, z) (see Fig. 3.8a) by letting

$$\mathbf{e}_1 \equiv \mathbf{e}_r = \cos\theta \mathbf{e}_x + \sin\theta \mathbf{e}_y$$
$$\mathbf{e}_2 \equiv \mathbf{e}_\theta = -\sin\theta \mathbf{e}_x + \cos\theta \mathbf{e}_y \tag{3.60}$$
$$\mathbf{e}_3 \equiv \mathbf{e}_z$$

and

$$\mathbf{T}^{(\mathbf{e}_1)} = \sigma_{rr}\mathbf{e}_r + \sigma_{r\theta}\mathbf{e}_\theta + \sigma_{rz}\mathbf{e}_z$$
$$\mathbf{T}^{(\mathbf{e}_2)} = \sigma_{\theta r}\mathbf{e}_r + \sigma_{\theta\theta}\mathbf{e}_\theta + \sigma_{\theta z}\mathbf{e}_z$$
$$\mathbf{T}^{(\mathbf{e}_3)} = \sigma_{zr}\mathbf{e}_r + \sigma_{z\theta}\mathbf{e}_\theta + \sigma_{zz}\mathbf{e}_z \tag{3.61}$$
$$\mathbf{f} = f_r\mathbf{e}_r + f_\theta\mathbf{e}_\theta + f_z\mathbf{e}_z$$

Using these results and the expression for the gradient in cylindrical coordinates given by

$$\nabla = \mathbf{e}_r \frac{\partial}{\partial r} + \mathbf{e}_\theta \frac{1}{r}\frac{\partial}{\partial \theta} + \mathbf{e}_z \frac{\partial}{\partial z} \tag{3.62}$$

we find

$$\nabla \cdot \mathbf{e}_1 + \mathbf{e}_1 \cdot \nabla = \left(\frac{1}{r} + \frac{\partial}{\partial r}\right)$$
$$\nabla \cdot \mathbf{e}_2 + \mathbf{e}_2 \cdot \nabla = \frac{1}{r} + \frac{\partial}{\partial \theta} \tag{3.63}$$
$$\nabla \cdot \mathbf{e}_3 + \mathbf{e}_3 \cdot \nabla = \frac{\partial}{\partial z}$$

so that the equilibrium equations in terms of the stress vectors are

$$\frac{\partial \mathbf{T}^{(\mathbf{e}_r)}}{\partial r} + \frac{\mathbf{T}^{(\mathbf{e}_r)}}{r} + \frac{1}{r}\frac{\partial \mathbf{T}^{(\mathbf{e}_\theta)}}{\partial \theta} + \frac{\partial \mathbf{T}^{(\mathbf{e}_z)}}{\partial z} + \mathbf{f} = 0 \tag{3.64}$$

3.4 Equations of Equilibrium – Cylindrical and Spherical Coordinates

or, in terms of the stresses

$$\left(\frac{\partial \sigma_{rr}}{\partial r}\mathbf{e}_r + \frac{\partial \sigma_{r\theta}}{\partial r}\mathbf{e}_\theta + \frac{\partial \sigma_{rz}}{\partial r}\mathbf{e}_z\right) + \left(\frac{\sigma_{rr}}{r}\mathbf{e}_r + \frac{\sigma_{r\theta}}{r}\mathbf{e}_\theta + \frac{\sigma_{rz}}{r}\mathbf{e}_z\right)$$
$$+\frac{1}{r}\left(\frac{\partial \sigma_{\theta r}}{\partial \theta}\mathbf{e}_r + \frac{\partial \sigma_{\theta\theta}}{\partial \theta}\mathbf{e}_\theta + \frac{\partial \sigma_{\theta z}}{\partial \theta}\mathbf{e}_z + \sigma_{\theta r}\frac{\partial \mathbf{e}_r}{\partial \theta} + \sigma_{\theta\theta}\frac{\partial \mathbf{e}_\theta}{\partial \theta}\right) \quad (3.65)$$
$$+\left(\frac{\partial \sigma_{zr}}{\partial z}\mathbf{e}_r + \frac{\partial \sigma_{z\theta}}{\partial z}\mathbf{e}_\theta + \frac{\partial \sigma_{zz}}{\partial z}\mathbf{e}_z\right) + (f_r \mathbf{e}_r + f_\theta \mathbf{e}_\theta + f_z \mathbf{e}_z) = 0$$

which gives

$$\frac{\partial \sigma_{rr}}{\partial r} + \frac{\sigma_{rr} - \sigma_{\theta\theta}}{r} + \frac{1}{r}\frac{\partial \sigma_{\theta r}}{\partial \theta} + \frac{\partial \sigma_{zr}}{\partial z} + f_r = 0$$

$$\frac{\partial \sigma_{r\theta}}{\partial r} + \frac{2\sigma_{r\theta}}{r} + \frac{1}{r}\frac{\partial \sigma_{\theta\theta}}{\partial \theta} + \frac{\partial \sigma_{z\theta}}{\partial z} + f_\theta = 0 \quad (3.66)$$

$$\frac{\partial \sigma_{rz}}{\partial r} + \frac{\sigma_{rz}}{r} + \frac{1}{r}\frac{\partial \sigma_{\theta z}}{\partial \theta} + \frac{\partial \sigma_{zz}}{\partial z} + f_z = 0$$

and, as with the Cartesian components of the stress, from moment equilibrium the shear stresses are symmetric, i.e.,

$$\sigma_{r\theta} = \sigma_{\theta r}, \quad \sigma_{z\theta} = \sigma_{\theta z}, \quad \sigma_{zr} = \sigma_{rz} \quad (3.67)$$

We can consider spherical coordinates in exactly the same fashion (see Fig. 3.8b). In that case we have

$$\mathbf{e}_1 \equiv \mathbf{e}_r = \sin\phi\cos\theta\mathbf{e}_x + \sin\phi\sin\theta\mathbf{e}_y + \cos\phi\mathbf{e}_z$$
$$\mathbf{e}_2 \equiv \mathbf{e}_\phi = \cos\phi\cos\theta\mathbf{e}_x + \cos\phi\sin\theta\mathbf{e}_y - \sin\phi\mathbf{e}_z \quad (3.68)$$
$$\mathbf{e}_3 \equiv \mathbf{e}_\theta = -\sin\theta\mathbf{e}_x + \cos\theta\mathbf{e}_y$$

with

$$\nabla = \mathbf{e}_r\frac{\partial}{\partial r} + \mathbf{e}_\phi\frac{1}{r}\frac{\partial}{\partial \phi} + \mathbf{e}_\theta\frac{1}{r\sin\phi}\frac{\partial}{\partial \theta} \quad (3.69)$$

and

$$\mathbf{T}^{(\mathbf{e}_1)} = \sigma_{rr}\mathbf{e}_r + \sigma_{r\theta}\mathbf{e}_\theta + \sigma_{r\phi}\mathbf{e}_\phi$$

$$\mathbf{T}^{(\mathbf{e}_2)} = \sigma_{\phi r}\mathbf{e}_r + \sigma_{\phi\theta}\mathbf{e}_\theta + \sigma_{\phi\phi}\mathbf{e}_\phi$$

$$\mathbf{T}^{(\mathbf{e}_3)} = \sigma_{\theta r}\mathbf{e}_r + \sigma_{\theta\theta}\mathbf{e}_\theta + \sigma_{\theta\phi}\mathbf{e}_\phi \quad (3.70)$$

$$\mathbf{f} = f_r\mathbf{e}_r + f_\theta\mathbf{e}_\theta + f_\phi\mathbf{e}_\phi$$

so that the equilibrium equations are of the form

84 Equilibrium

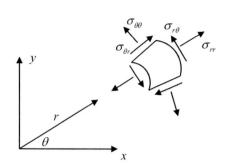

Figure 3.9 The state of stress in polar coordinates for plane stress.

$$\frac{\partial T^{(e_r)}}{\partial r} + \frac{2T^{(e_r)}}{r} + \frac{1}{r\sin\phi}\frac{\partial T^{(e_\theta)}}{\partial \theta} + \frac{1}{r}\frac{\partial T^{(e_\phi)}}{\partial \phi} + \frac{\cos\phi}{r\sin\phi}T^{(e_\phi)} + \mathbf{f} = 0 \quad (3.71)$$

Placing Eq. (3.70) into Eq. (3.71) and using Eq. (3.68), after some algebra we find

$$\frac{\partial \sigma_{rr}}{\partial r} + \frac{1}{r}\frac{\partial \sigma_{r\phi}}{\partial \phi} + \frac{1}{r\sin\phi}\frac{\partial \sigma_{r\theta}}{\partial \theta} + \frac{2\sigma_{rr} - \sigma_{\phi\phi} - \sigma_{\theta\theta} + \sigma_{r\phi}\cot\phi}{r} + f_r = 0$$

$$\frac{\partial \sigma_{r\theta}}{\partial r} + \frac{1}{r}\frac{\partial \sigma_{\phi\theta}}{\partial \phi} + \frac{1}{r\sin\phi}\frac{\partial \sigma_{\theta\theta}}{\partial \theta} + \frac{3\sigma_{r\theta} + 2\sigma_{\theta\phi}\cot\phi}{r} + f_\theta = 0 \quad (3.72)$$

$$\frac{\partial \sigma_{r\phi}}{\partial r} + \frac{1}{r}\frac{\partial \sigma_{\phi\phi}}{\partial \phi} + \frac{1}{r\sin\phi}\frac{\partial \sigma_{\theta\phi}}{\partial \theta} + \frac{3\sigma_{r\phi} + (\sigma_{\phi\phi} - \sigma_{\theta\theta})\cot\phi}{r} + f_\phi = 0$$

3.4.1 Plane Stress

A special case of the cylindrical coordinate expressions is when we have a state of plane stress where $\sigma_{rz} = \sigma_{\theta z} = \sigma_{zz} = 0$ and $f_z = 0$. Then the equations of equilibrium reduce to

$$\frac{\partial \sigma_{rr}}{\partial r} + \frac{\sigma_{rr} - \sigma_{\theta\theta}}{r} + \frac{1}{r}\frac{\partial \sigma_{\theta r}}{\partial \theta} + f_r = 0$$

$$\frac{\partial \sigma_{r\theta}}{\partial r} + \frac{2\sigma_{r\theta}}{r} + \frac{1}{r}\frac{\partial \sigma_{\theta\theta}}{\partial \theta} + f_\theta = 0 \quad (3.73)$$

A state of plane stress is shown in Fig. 3.9. We see that these stresses are described in this case in terms of the polar coordinates (r, θ) only.

3.5 PROBLEMS

P3.1 Consider a long, thick cylinder that is subjected to a constant internal pressure, p_0, and where there is no axial stress developed, i.e., $\sigma_{zz} = 0$ (Fig. P3.1). Under these conditions, because of the symmetry of the loading and the geometry it is reasonable to assume that there are no shear stresses developed and that the radial and hoop stresses $(\sigma_{rr}, \sigma_{\theta\theta})$ are functions of r only.

(a) If we let

$$\sigma_{rr} = A - \frac{B}{r^2}$$

$$\sigma_{\theta\theta} = A + \frac{B}{r^2}$$

$$\sigma_{zz} = 0$$

$$\sigma_{rz} = \sigma_{r\theta} = \sigma_{\theta z} = 0$$

where (A, B) are constants, show that the equations of equilibrium in cylindrical coordinates are satisfied identically.

(b) Determine the constants (A, B) in terms of p_0 and the radii (a, b) from the known values of the radial stress at $r = a$ and $r = b$ and the corresponding values of the radial and hoop stresses.

(c) Where do the largest stresses (in magnitude) occur for the pressurized thick cylinder? Obtain the largest stresses to first order when the thickness of the cylinder, $t = (b - a)$ is small. What can you conclude about the relative size of σ_{rr} and $\sigma_{\theta\theta}$ for a thin cylinder?

P3.2 Consider a closed thin cylindrical pressure vessel that carries a constant internal pressure, p_0, as shown in Fig. P3.2a.

(a) Assume the hoop stress, $\sigma_{\theta\theta}$, and axial stress, σ_{zz}, are the only two significant stresses in the cylinder and because the cylinder is thin further assume these two stresses are constants across the thickness. Use force equilibrium and the free body diagrams of Fig. P3.2b, c to determine those stresses. Give an argument for why these are the only two significant stresses in the cylinder.

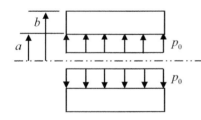

Figure P3.1 A thick cylinder carrying a uniform internal pressure, p_0.

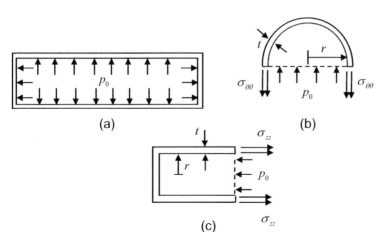

Figure P3.2 (a) A closed thin-wall cylindrical pressure vessel carrying an internal pressure, p_0. (b, c) Free body diagrams for determining the hoop stress, $\sigma_{\theta\theta}$, and the axial stress, σ_{zz}.

(b) How does the hoop stress expression here compare with the result of part (c) in problem P3.1?

P3.3 Consider a thick wall spherical pressure vessel carrying an internal pressure, p_0, as shown in Fig. P3.3. Based on the symmetry of the problem and the loading, we expect that there are no shear stresses developed in the vessel and that $\sigma_{\theta\theta} = \sigma_{\phi\phi}$ (why?). Because of the symmetry we also expect that $(\sigma_{rr}, \sigma_{\theta\theta}, \sigma_{\phi\phi})$ are all functions of the radial distance, r, only in spherical coordinates.

(a) If we let

$$\sigma_{rr} = A - \frac{B}{r^3}$$

$$\sigma_{\theta\theta} = \sigma_{\phi\phi} = A + \frac{B}{2r^3}$$

$$\sigma_{r\theta} = \sigma_{r\phi} = \sigma_{\theta\phi} = 0$$

where (A, B) are constants, show that the equations of equilibrium in spherical coordinates, Eq. (3.72), are satisfied identically.

(b) Determine the constants (A, B) in terms of p_0 and the radii (a, b) from the known values of the radial stress at $r = a$ and $r = b$ and the corresponding stresses.

(c) Where do the largest stresses (in magnitude) occur in the sphere? Obtain these stresses to first order when the thickness of the sphere, $t = (b - a)$ is small. What can you conclude about the relative size of σ_{rr}, $\sigma_{\theta\theta} = \sigma_{\phi\phi}$ for a thin sphere?

P3.4 Consider a thin spherical pressure vessel that carries a constant internal pressure, p_0, as shown in Fig. P3.4a.

(a) Assume the hoop stress, $\sigma_h = \sigma_{\theta\theta} = \sigma_{\phi\phi}$, is the only significant stress in the sphere and because the sphere is thin further assume this stress is a constant across the thickness. Use force equilibrium and the free body diagram of Fig. P3.4b to determine that stress.

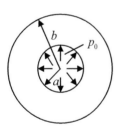

Figure P3.3 A thick-wall spherical pressure vessel.

Figure P3.4 (a) A thin-wall spherical pressure vessel. (b) A free body diagram for determining the hoop stress, σ_h.

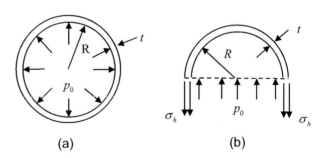

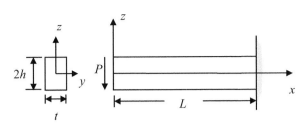

Figure P3.5 A cantilever beam.

(b) How does the hoop stress expression here compare with the result of part (c) in problem P3.3?

P3.5 Consider the cantilever beam with a rectangular cross-section as shown in Fig. P3.5. Assume the stresses in the beam are given by

$$\sigma_{xx} = a\,xz$$
$$\sigma_{xz} = b + cz^2$$
$$\sigma_{xy} = \sigma_{yz} = \sigma_{yy} = \sigma_{zz} = 0$$

(a) Determine the relationship between the constants (a, b, c) so that the equations of equilibrium are satisfied.
(b) Determine the relationship between the constants such that the top and bottom and front and back faces of the beam are stress-free.
(c) If the shear stress acting on the left face of the beam produces the load P, determine all the stresses in terms of this load and show that these stresses are consistent with engineering beam theory.

4 Strain

Just as the concept of stress gives us a measure of force distributions in a deformable body, the concept of strain describes the deformations locally at every point within the body. In this chapter we will define strains and describe how strains change with directions and with the choice of coordinates, as we have done with stresses. Strains will also be related to the displacements of the deformable body. Since the strains often found in practice are quite small, we will also consider the behavior of small strains and their relationship to displacements in this important case. In fact, this book will consider problems only for small strains. It will be shown that strains must satisfy a set of compatibility equations to ensure that they represent a well-behaved deformation of a body. As discussed in Chapter 1, these compatibility equations form one of the four "pillars" of analyzing deformable bodies. Finally, we will examine strains and strain–displacement relations in cylindrical coordinates.

4.1 Definitions of Strains

As briefly discussed in Chapter 1, locally we can describe the deformations of a body in terms of changes of length and changes of angle for a small element (see Fig. 1.16). Here, we want to analyze this local deformation in some detail and define the strains that characterize those changes.

4.1.1 Normal Strain

When a body is loaded and deforms, elements of that body elongate or contract and are distorted. The quantity that we use to measure the changes of length occurring in the body is called the *normal strain*. Consider, for example, a small directed line segment, $d\mathbf{r}_n$, between two very close points P and Q in an undeformed body (Fig. 4.1a). Let the length of this line segment be ds_n and its direction, $\mathbf{n}$, where $\mathbf{n}$ is a unit vector. When the body deforms, point P will move to a new point, P^*, and Q will move to Q^* and the original line segment will change to a new segment, $d\mathbf{R}_n$, of length dS_n (see Fig. 4.1b). The normal strain, E_{nn}, at point P is then defined as

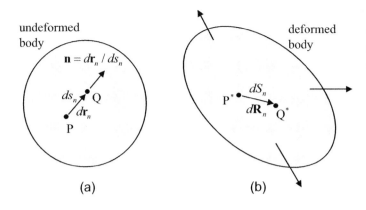

Figure 4.1 (a) A line element PQ in the direction **n** in an undeformed body. (b) The corresponding line element PQ in the deformed body.

$$\begin{aligned} E_{nn} &= \frac{1}{2}\frac{dS_n^2 - ds_n^2}{ds_n^2} \\ &= \frac{1}{2}\frac{d\mathbf{R}_n \cdot d\mathbf{R}_n - d\mathbf{r}_n \cdot d\mathbf{r}_n}{ds_n^2} \\ &= \frac{1}{2}\left(\left|\frac{d\mathbf{R}_n}{ds}\right|^2 - 1\right) \end{aligned} \qquad (4.1)$$

This definition can be used regardless of the size of the strain so that it is a general definition. However, if the changes of length occurring in the body are quite small, which often occurs in practice, then this normal strain reduces to the normal small strain, e_{nn}, defined as the relative elongation of the element, i.e.,

$$\begin{aligned} E_{nn} &= \frac{1}{2}\frac{(dS_n - ds_n)(dS_n + ds_n)}{ds_n^2} \\ &\cong e_{nn} = \frac{dS_n - ds_n}{ds_n} \end{aligned} \qquad (4.2)$$

which is the usual definition of strain seen in elementary strength of materials texts.

4.1.2 Shear Strain

In addition to changes of length occurring in a deformable body, there are distortions (corresponding to changes in angles) which also must be characterized. The *shear strain* is the quantity we use to define these changes. To obtain the shear strain, we consider two small line segments $(d\mathbf{r}_n, d\mathbf{r}_t)$ in the undeformed body along the **n** and **t** unit vector directions which are initially at right angles to each other (Fig. 4.2a). In the deformed body, these line segments, $(d\mathbf{R}_n, d\mathbf{R}_t)$, will no longer be orthogonal. We will let θ_{nt} be the new angle between the two distorted line segments. The shear strain, E_{nt}, will then be defined as

Figure 4.2 (a) Two small line PQ and PR segments that are orthogonal in an undeformed body. (b) The corresponding line segments P*Q* and P*R* in a deformed body that are no longer orthogonal.

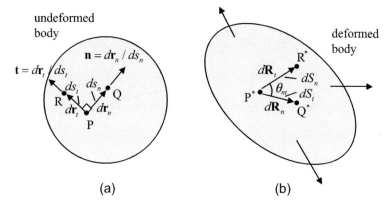

$$E_{nt} = \frac{1}{2} \frac{d\mathbf{R}_n}{ds_n} \cdot \frac{d\mathbf{R}_t}{ds_t} \tag{4.3}$$

This shear strain is related to the angular distortions occurring in the body since the cosine of the angle between any two unit vectors is just the dot product of those unit vectors. Thus,

$$\cos\theta_{nt} = \sin\left(\frac{\pi}{2} - \theta_{nt}\right) = \frac{\dfrac{d\mathbf{R}_n}{ds_n} \cdot \dfrac{d\mathbf{R}_t}{ds_t}}{\left|\dfrac{d\mathbf{R}_n}{ds_n}\right|\left|\dfrac{d\mathbf{R}_t}{ds_t}\right|} \tag{4.4}$$

Using the definitions of the normal and shear strains for the **n** and **t** directions, then

$$\sin\left(\frac{\pi}{2} - \theta_{nt}\right) = \frac{2E_{nt}}{\sqrt{1+2E_{nn}}\sqrt{1+2E_{tt}}} \tag{4.5}$$

which shows that if $\theta_{nt} = \pi/2$ (so that there is no angular distortion of these two lines) then $E_{nt} = 0$. When the changes of lengths and changes of angles are small then

$$\sin\left(\frac{\pi}{2} - \theta_{nt}\right) \cong \left(\frac{\pi}{2} - \theta_{nt}\right) \tag{4.6}$$
$$E_{nn} \ll 1, \quad E_{tt} \ll 1$$

so that

$$E_{nt} \cong e_{nt} = \frac{1}{2}\left(\frac{\pi}{2} - \theta_{nt}\right) \tag{4.7}$$

where e_{nt} is the small shear strain. It is common in engineering studies to use the *engineering shear strain*, γ_{nt}, in place of e_{nt}, where

$$\gamma_{nt} = 2e_{nt} = \frac{\pi}{2} - \theta_{nt} \tag{4.8}$$

The engineering shear strain is just equal to the change of angle so that it has a direct physical meaning. However, we will see that in strain transformations it is more convenient to use e_{nt}, which is also called the *tensor shear strain*.

4.2 Strain–Displacement Relations and Strain Transformations

Although the normal strains and shear strains define the deformation locally everywhere in a body, even knowing these strains it is not easy to directly determine what shape the deformed body actually takes. A more convenient way to define the deformation is to describe the displacement vector, $\mathbf{u}(x_1, x_2, x_3)$, for every point $P = (x_1, x_2, x_3)$ in the undeformed body. Obviously, the strains are related to these displacements so we need to determine these strain–displacement relations. From the geometry (Fig. 4.3) we see that

$$\mathbf{R}(x_1, x_2, x_3) = \mathbf{r} + \mathbf{u}(x_1, x_2, x_3)$$
$$= \sum_{i=1}^{3} x_i \mathbf{e}_i + \mathbf{u}(x_1, x_2, x_3) \tag{4.9}$$

so that if $\mathbf{r}$ changes by a small amount $d\mathbf{r}_n$ in the $\mathbf{n}$ direction, we have

$$d\mathbf{R}_n = d\mathbf{r}_n + d\mathbf{u} \tag{4.10}$$

or, dividing by the length of $d\mathbf{r}_n$, ds_n, we find

$$\frac{d\mathbf{R}_n}{ds_n} = \frac{d\mathbf{r}_n}{ds_n} + \frac{d\mathbf{u}}{ds_n}$$
$$= \mathbf{n} + \frac{d\mathbf{u}}{ds_n} \tag{4.11}$$

Using this result in the definition of the normal strain, Eq. (4.1), gives

$$E_{nn} = \frac{1}{2}\left(\frac{d\mathbf{R}}{ds_n} \cdot \frac{d\mathbf{R}}{ds_n} - 1\right)$$
$$= \mathbf{n} \cdot \frac{d\mathbf{u}}{ds_n} + \frac{1}{2}\frac{d\mathbf{u}}{ds_n} \cdot \frac{d\mathbf{u}}{ds_n} \tag{4.12}$$

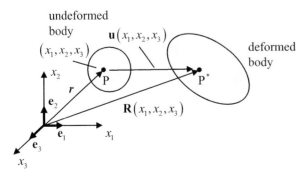

Figure 4.3 Deformations of a body defined by the displacement vector, $\mathbf{u}$, at any point in the body.

Similarly, if we let **r** change by a small amount $d\mathbf{r}_t$ in the **t** direction

$$\frac{d\mathbf{R}_t}{ds_t} = \frac{d\mathbf{r}_t}{ds_t} + \frac{d\mathbf{u}}{ds_t}$$
$$= \mathbf{t} + \frac{d\mathbf{u}}{ds_t} \qquad (4.13)$$

so, from Eq. (4.3),

$$E_{nt} = \frac{1}{2}\frac{d\mathbf{R}_n}{ds_n} \cdot \frac{d\mathbf{R}_t}{ds_t}$$
$$= \frac{1}{2}\left(\mathbf{t} \cdot \frac{d\mathbf{u}}{ds_n} + \mathbf{n} \cdot \frac{d\mathbf{u}}{ds_t} + \frac{d\mathbf{u}}{ds_n} \cdot \frac{d\mathbf{u}}{ds_t}\right) \qquad (4.14)$$

If the strains are small enough so that we can neglect the products of the displacement gradients in these definitions, then we see that for the small normal and shear strains we have

$$e_{nn} = \mathbf{n} \cdot \frac{d\mathbf{u}}{ds_n}$$
$$e_{nt} = \frac{1}{2}\left(\mathbf{n} \cdot \frac{d\mathbf{u}}{ds_t} + \mathbf{t} \cdot \frac{d\mathbf{u}}{ds_n}\right) \qquad (4.15)$$

Since Eq. (4.14) and Eq. (4.15) are valid for any directions **n** and **t**, we can compute both the finite and small strains for line segments along the Cartesian $(x_1, x_2, x_3) = (x, y, z)$ axes by simply making particular choices for **n** and **t**. For example, consider letting **n** be along the x_1-axis direction and let **t** be along the x_2-axis direction. Then $\mathbf{n} = \mathbf{e}_1, \mathbf{t} = \mathbf{e}_2$ so that

$$E_{xx} = E_{11} = \mathbf{e}_1 \cdot \frac{\partial \mathbf{u}}{\partial x_1} + \frac{1}{2}\frac{\partial \mathbf{u}}{\partial x_1} \cdot \frac{\partial \mathbf{u}}{\partial x_1}$$

$$e_{xx} = e_{11} = \mathbf{e}_1 \cdot \frac{\partial \mathbf{u}}{\partial x_1}$$

$$E_{xy} = E_{12} = \frac{1}{2}\left(\mathbf{e}_1 \cdot \frac{\partial \mathbf{u}}{\partial x_2} + \mathbf{e}_2 \cdot \frac{\partial \mathbf{u}}{\partial x_1} + \frac{\partial \mathbf{u}}{\partial x_1} \cdot \frac{\partial \mathbf{u}}{\partial x_2}\right) \qquad (4.16)$$

$$e_{xy} = e_{12} = \frac{1}{2}\left(\mathbf{e}_1 \cdot \frac{\partial \mathbf{u}}{\partial x_2} + \mathbf{e}_2 \cdot \frac{\partial \mathbf{u}}{\partial x_1}\right)$$

In exactly the same fashion, we can obtain the other normal strains and shear strains. We can, in fact, write all the normal and shear strains in the same forms as

$$E_{ij} = \frac{1}{2}\left(\mathbf{e}_i \cdot \frac{\partial \mathbf{u}}{\partial x_j} + \mathbf{e}_j \cdot \frac{\partial \mathbf{u}}{\partial x_i} + \frac{\partial \mathbf{u}}{\partial x_i} \cdot \frac{\partial \mathbf{u}}{\partial x_j}\right) \quad (i,j = 1, 2, 3)$$

$$e_{ij} = \frac{1}{2}\left(\mathbf{e}_i \cdot \frac{\partial \mathbf{u}}{\partial x_j} + \mathbf{e}_j \cdot \frac{\partial \mathbf{u}}{\partial x_i}\right) \quad (i,j = 1, 2, 3) \qquad (4.17)$$

4.2 Strain–Displacement Relations and Strain Transformations

These results are still in vector notation. However, if we write the displacement vector **u** and its gradients in terms of scalar components along (x_1, x_2, x_3) axes, i.e.,

$$\mathbf{u} = \sum_{i=1}^{3} u_i \mathbf{e}_i$$

$$\frac{\partial \mathbf{u}}{\partial x_j} = \sum_{i=1}^{3} \frac{\partial u_i}{\partial x_j} \mathbf{e}_i \quad (j = 1, 2, 3)$$

(4.18)

then the strains can be written directly in terms of these components as

$$E_{ij} = \frac{1}{2} \left(\frac{\partial u_i}{\partial x_j} + \frac{\partial u_j}{\partial x_i} + \sum_{k=1}^{3} \frac{\partial u_k}{\partial x_i} \cdot \frac{\partial u_k}{\partial x_j} \right) \quad (i,j = 1, 2, 3)$$

$$e_{ij} = \frac{1}{2} \left(\frac{\partial u_i}{\partial x_j} + \frac{\partial u_j}{\partial x_i} \right) \quad (i,j = 1, 2, 3)$$

(4.19)

These nine strains associated with the Cartesian coordinate directions are actually only six independent strains since the strains are inherently symmetric by their definitions, i.e., $E_{12} = E_{21}$, $e_{12} = e_{21}$, etc. These strains define the *state of strain* at any point in the body since they can be used to obtain the normal and shear strains in any directions at a point, just as the Cartesian components of stress allow us to obtain any normal or shear stresses at a point. We can write the states of finite or small strains in matrix form as

$$[E] = \begin{bmatrix} E_{11} & E_{12} & E_{13} \\ E_{21} & E_{22} & E_{23} \\ E_{31} & E_{32} & E_{33} \end{bmatrix} \quad [e] = \begin{bmatrix} e_{11} & e_{12} & e_{13} \\ e_{21} & e_{22} & e_{23} \\ e_{31} & e_{32} & e_{33} \end{bmatrix}$$

(4.20)

Consider finding a finite normal strain E_{nn} from the state of strain $[E]$. Since

$$\mathbf{n} = \sum_{i=1}^{3} n_i \mathbf{e}_i = \frac{d\mathbf{r}_n}{ds_n} = \sum_{i=1}^{3} \frac{dx_i}{ds_n} \mathbf{e}_i$$

$$\frac{d\mathbf{u}}{ds_n} = \sum_{i=1}^{3} \frac{du_i}{ds_n} \mathbf{e}_i = \sum_{i=1}^{3} \left(\sum_{j=1}^{3} \frac{\partial u_i}{\partial x_j} \frac{dx_j}{ds_n} \right) \mathbf{e}_i = \sum_{i=1}^{3} \sum_{j=1}^{3} \frac{\partial u_i}{\partial x_j} n_j \mathbf{e}_i$$

(4.21)

we find (see Eq. (4.12))

$$E_{nn} = \mathbf{n} \cdot \frac{d\mathbf{u}}{ds_n} + \frac{1}{2} \frac{d\mathbf{u}}{ds_n} \cdot \frac{d\mathbf{u}}{ds_n}$$

$$= \sum_{i=1}^{3} \sum_{j=1}^{3} \left(\frac{\partial u_i}{\partial x_j} + \frac{1}{2} \sum_{k=1}^{3} \frac{\partial u_k}{\partial x_i} \frac{\partial u_k}{\partial x_j} \right) n_i n_j$$

(4.22)

Note that we can write

$$\frac{\partial u_i}{\partial x_j} = \frac{1}{2}\left(\frac{\partial u_i}{\partial x_j} + \frac{\partial u_j}{\partial x_i}\right) + \frac{1}{2}\left(\frac{\partial u_i}{\partial x_j} - \frac{\partial u_j}{\partial x_i}\right) \quad (4.23)$$
$$= e_{ij} + \omega_{ij}$$

where e_{ij} are the small strain components and ω_{ij} can be shown to be local average rotations in the body [1]. But the matrix of ω_{ij} terms is antisymmetric, i.e., $\omega_{ij} = -\omega_{ji}$ so that

$$\sum_{i=1}^{3}\sum_{j=1}^{3} \omega_{ij} n_i n_j = 0 \quad (4.24)$$

giving

$$\sum_{i=1}^{3}\sum_{j=1}^{3} \frac{\partial u_i}{\partial x_j} n_i n_j = \sum_{i=1}^{3}\sum_{j=1}^{3} e_{ij} n_i n_j \quad (4.25)$$

This means that we can rewrite Eq. (4.22) as

$$E_{nn} = \sum_{i=1}^{3}\sum_{j=1}^{3} \frac{1}{2}\left(\frac{\partial u_i}{\partial x_j} + \frac{\partial u_j}{\partial x_i} + \sum_{k=1}^{3}\frac{\partial u_k}{\partial x_i}\frac{\partial u_k}{\partial x_j}\right) n_i n_j$$
$$= \sum_{i=1}^{3}\sum_{j=1}^{3} E_{ij} n_i n_j \quad (4.26)$$

which shows that we can find the normal strain in any direction from a knowledge of the state of strain. In matrix–vector form we have, equivalently,

$$E_{nn} = \{n\}^T [E]\{n\} \quad (4.27)$$

We can follow the same steps for the finite shear strain, E_{nt}, to find

$$E_{nt} = \sum_{i=1}^{3}\sum_{j=1}^{3} \frac{1}{2}\left(\frac{\partial u_i}{\partial x_j} + \frac{\partial u_j}{\partial x_i} + \sum_{k=1}^{3}\frac{\partial u_k}{\partial x_i}\frac{\partial u_k}{\partial x_j}\right) n_i n_j$$
$$= \sum_{i=1}^{3}\sum_{j=1}^{3} E_{ij} n_i t_j \quad (4.28)$$

or

$$E_{nt} = \{n\}^T [E]\{t\} \quad (4.29)$$

When the strains are small, we can simply neglect the product of the displacement derivatives and find, analogously

$$e_{nn} = \sum_{i=1}^{3}\sum_{j=1}^{3} \frac{1}{2}\left(\frac{\partial u_i}{\partial x_j} + \frac{\partial u_j}{\partial x_i}\right)n_i n_j$$

$$= \sum_{i=1}^{3}\sum_{j=1}^{3} e_{ij} n_i n_j$$

$$e_{nt} = \sum_{i=1}^{3}\sum_{j=1}^{3} \frac{1}{2}\left(\frac{\partial u_i}{\partial x_j} + \frac{\partial u_j}{\partial x_i}\right)n_i t_j$$

$$= \sum_{i=1}^{3}\sum_{j=1}^{3} e_{ij} n_i t_j$$

(4.30)

or, in matrix–vector notation again,

$$e_{nn} = \{n\}^T [e]\{n\}$$
$$e_{nt} = \{n\}^T [e]\{t\}$$

(4.31)

If we compare Eq. (4.31) with our previous results for stresses (Eq. (2.29)) we see that they are identical in form if we simply make the replacements $\sigma_{nn}, \sigma_{nt} \to e_{nn}, e_{nt}$, and $\sigma_{ij} \to e_{ij}$ (or equivalently for finite strains $\sigma_{nn}, \sigma_{nt} \to E_{nn}, E_{nt}$ and $\sigma_{ij} \to E_{ij}$). Thus, our previous equations involving stress transformations, principal stresses, Mohr's circle, etc., remain valid for strains as well. For example, using these replacements we can write the strain transformation equations relating the small strains e'_{ij} along a set of (x'_1, x'_2, x'_3) axes to the small strains e_{ij} along a set of (x_1, x_2, x_3) axes through the direction cosines as

$$e'_{mn} = \sum_{i=1}^{3}\sum_{j=1}^{3} l_{im} l_{jn} e_{ij}$$

(4.32)

or, in matrix–vector notation,

$$[e'] = [l]^T [e][l]$$

(4.33)

Similarly, the principal strains, e_p, satisfy the cubic equation (see Eq. (2.58) for the stresses)

$$e_p^3 - J_1 e_p^2 + J_2 e_p - J_3 = 0$$

(4.34)

in terms of the strain invariants

$$J_1 = e_{xx} + e_{yy} + e_{zz}$$
$$J_2 = e_{xx}e_{yy} + e_{xx}e_{zz} + e_{yy}e_{zz} - e_{xy}^2 - e_{xz}^2 - e_{yz}^2$$
$$J_3 = e_{xx}e_{yy}e_{zz} + 2e_{xy}e_{xz}e_{yz} - e_{xx}e_{zy}^2 - e_{yy}e_{xz}^2 - e_{zz}e_{xy}^2$$

(4.35)

which can be expressed in terms of the principal strains as

$$J_1 = e_{p1} + e_{p2} + e_{zp3}$$
$$J_2 = e_{p1}e_{p2} + e_{p1}e_{p3} + e_{p2}e_{p3}$$
$$J_3 = e_{p1}e_{p2}e_{p3}$$

(4.36)

Example 4.1 Strain Calculations

Consider the following state of small strains with respect to the (x, y, z) axes as defined in MATLAB®:

```
strain = [ 2670 2670 -330; 2670 2000 -1500; -330 -1500  -2000]
strain = 2670          2670           -330
        2670          2000          -1500
        -330         -1500          -2000
```

The strains here are given in microstrain (μstrain, or μ), i.e., $e_{11} = 2670\,\mu = 2760 \times 10^{-6}$, etc. If this state of strain is uniform throughout a region of a body such as the parallelepiped shown in Fig. 4.4, then we can analyze the strains in different directions by examining the behavior of finite lines in the body rather than the infinitesimally small lines considered previously in defining the strains since every portion of the finite lines will experience the same strains. First, calculate the normal strain experienced by the line EA in Fig. 4.4. Thus, we need to define a unit normal in the EA direction. In MATLAB® we can form up this unit normal as a *column* vector, **n**, in two steps:

```
EA = [ 0 -1 -2];
n = EA'/norm(EA)
n =         0
       -0.4472
       -0.8944
```

Then we can calculate the normal strain, enn, in this normal direction with Eq. (4.31) as

```
enn = n'*strain*n
enn = -2400
```

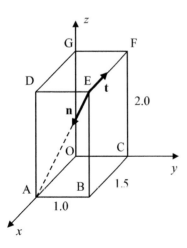

which we can write as $e_{nn} = -2400\,\mu$. Now suppose we want to find the engineering shear strain experienced by lines EA and EF in Fig. 4.4. These two lines are orthogonal to each other, as required by our definition of shear strain. Define a column unit vector **t** along EF, which is very simple in this case, and then use it in conjunction with the unit vector **n** in Eq. (4.31):

```
t = [ -1  0   0]';
ent = n'*strain*t
ent =   898.8993
```

Figure 4.4 A parallelepiped in a region of a body where there is a uniform (constant) state of strain.

4.2 Strain–Displacement Relations and Strain Transformations

The engineering shear strain for this pair of lines, therefore, is

```
engstrain = 2*ent
engstrain = 1.7978e+03
```

which we can write as $\gamma_{nt} = 1798\,\mu$. If we want to find the principal strains and principal directions for the state of strain we have been considering, then we can simply use the MATLAB® function eig as done with the stresses:

```
[pdirs, pvals] = eig(strain)

pdirs =   0.1356    0.6807    0.7199
         -0.3794   -0.6356    0.6724
         -0.9152    0.3643   -0.1721

pvals = 1.0e+03 *
         -2.5729    0         0
          0         0.0003    0
          0         0         5.2426
```

where the principal direction unit vectors are in the columns of the matrix pdirs and the principal strains $e_{p1} \cong -2573\,\mu$, $e_{p2} \cong 0.3\,\mu$, $e_{p3} \cong 5243\,\mu$ are the diagonal values in pvals. Recall that the principal directions must be checked to see if they form a right-handed system by evaluating the determinant of the pdirs matrix. (In this case the determinant is –1 so the system is not right-handed.) One of the principal strains is almost zero so Mohr's circles for this case is as shown in Fig. 4.5. The largest tensor shear strain is just equal to the radius of the largest circle so the maximum engineering shear strain is just the *diameter* of this circle, which is the difference of the largest and the smallest principal strains:

```
maxshear = pvals(3,3) - pvals(1,1)
maxshear = 7.8155e+03
```

so we have $|\gamma_{max}| \cong 7816\,\mu$.

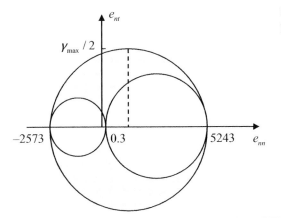

Figure 4.5 The Mohr's circle construction for the strains in Example 4.1.

Figure 4.6 Changes of a small volumetric element in terms of the small strains present.

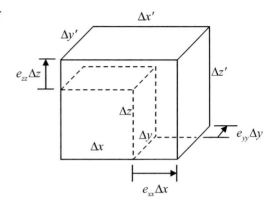

4.2.1 Dilatation

Since small normal strains are measures of changes of length, they are also related to the change of volume occurring in a deformable body. We have, from the definitions of the strains (see Fig. 4.6):

$$\Delta x' \cong \Delta x + e_{xx}\Delta x$$
$$\Delta y' \cong \Delta y + e_{yy}\Delta y \qquad (4.37)$$
$$\Delta z' \cong \Delta z + e_{zz}\Delta z$$

so that to first order (neglecting products of the small strains)

$$\Delta x' \Delta y' \Delta z' = \Delta x \Delta y \Delta z + (e_{xx} + e_{yy} + e_{zz})\Delta x \Delta y \Delta z \qquad (4.38)$$

For a small element $V = \Delta x \Delta y \Delta z$ the relative change of volume of that element is defined as the *dilatation*, Δ, given by

$$\Delta \equiv \frac{\Delta V}{V} = e_{xx} + e_{yy} + e_{zz}$$
$$= \frac{\partial u_x}{\partial x} + \frac{\partial u_y}{\partial y} + \frac{\partial u_z}{\partial z} \qquad (4.39)$$

Since the sum of the normal strains in Eq. (4.39) is an invariant, the dilatation is independent of the orientation of the Cartesian coordinate system used to describe the deformation.

4.2.2 Plane-Strain Principal Strains

The special case of plane strain is where the displacements are given as $u_x = u_x(x,y)$, $u_y = u_y(x,y)$, $u_z = 0$, which produces a state of strain where $e_{xx} = e_{xx}(x,y)$, $e_{yy} = e_{yy}(x,y)$, $e_{xy} = e_{xy}(x,y)$, $e_{xz} = e_{yz} = e_{zz} = 0$. In this case the third strain invariant $J_3 = 0$ and the principal strains are given explicitly from Eq. (4.34) as

$$e_{p1}, e_{p2} = \frac{e_{xx} + e_{yy}}{2} \pm \sqrt{\left(\frac{e_{xx} - e_{yy}}{2}\right) + e_{xy}^2} \qquad (4.40)$$

$$e_{p3} = 0$$

which are relations for the strains directly analogous to those for the stresses in the case of plane stress, as shown in Chapter 2.

4.3 Strain Compatibility Equations

We know that stresses in a deformable body must satisfy local equilibrium equations everywhere in the body. Likewise, we will find in this section that strains also must satisfy a set of compatibility equations within the body. To see this, consider a body with displacements at points P_1 and P_2 given by $\mathbf{u}_1$ and $\mathbf{u}_2$, respectively (see Fig. 4.7). Let $\Delta \mathbf{u} = \mathbf{u}_2 - \mathbf{u}_1$, which is the relative displacement of P_2 with respect to P_1. Breaking that relative displacement into components, we have

$$\Delta \mathbf{u} = \sum_{i=1}^{3} \Delta u_i \mathbf{e}_i \qquad (4.41)$$

Then these components can be written in terms of integrals over some path C from P_1 to P_2 (Fig. 4.7) in terms of the local small strains and rotations as

$$\Delta u_i = \int_{P_1}^{P_2} du_i = \sum_{j=1}^{3} \int_{P_1}^{P_2} \frac{\partial u_i}{\partial x_j} dx_j = \sum_{j=1}^{3} \int_{P_1}^{P_2} (e_{ij} + \omega_{ij}) dx_j \qquad (4.42)$$

But we have

$$\omega_{ij} dx_j = d(\omega_{ij} x_j) - \sum_{k=1}^{3} x_j \frac{\partial \omega_{ij}}{\partial x_k} dx_k \qquad (4.43)$$

so that Eq. (4.42) becomes

$$\Delta u_i = \sum_{j=1}^{3} \omega_{ij} x_j \bigg|_{P_1}^{P_2} + \sum_{j=1}^{3} \int_{P_1}^{P_2} \left(e_{ik} - \sum_{j=1}^{3} x_j \frac{\partial \omega_{ij}}{\partial x_k} \right) dx_k \qquad (4.44)$$

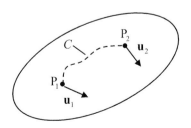

Figure 4.7 Two points, P_1 and P_2, in a deformable body where the displacement vector at those points are $\mathbf{u}_1$ and $\mathbf{u}_2$, respectively, and an integration path, C, between those two points.

Through differentiation of the strain–displacement relations, it can be verified that these derivatives of the local rotation can also be written in terms of derivatives of the strains, i.e.,

$$\frac{\partial e_{ik}}{\partial x_j} - \frac{\partial e_{jk}}{\partial x_i} = \frac{1}{2}\frac{\partial}{\partial x_k}\left(\frac{\partial u_i}{\partial x_j} - \frac{\partial u_j}{\partial x_i}\right) = \frac{\partial \omega_{ij}}{\partial x_k} \qquad (4.45)$$

giving

$$\Delta u_i = \sum_{j=1}^{3} \omega_{ij} x_j \bigg|_{P_1}^{P_2} + \sum_{j=1}^{3} \int_{P_1}^{P_2} \zeta_{ik} dx_k \qquad (4.46)$$

$$\text{where} \quad \zeta_{ik} = e_{ik} - \sum_{j=1}^{3} x_j \left(\frac{\partial e_{ik}}{\partial x_j} - \frac{\partial e_{jk}}{\partial x_i}\right)$$

The first term in the above Δu_i expression only depends on the points P_1 and P_2. The second term is independent of the path C between these two points (and hence also only depends on P_1 and P_2) if, for a simply connected region (a region with no holes), we have

$$\frac{\partial \zeta_{ik}}{\partial x_l} = \frac{\partial \zeta_{il}}{\partial x_k} \qquad (4.47)$$

(see, for example, Wylie, C. R., *Advanced Engineering Mathematics*, 4th edn, McGraw-Hill, p. 684; see also[2]). Thus, if Eq. (4.47) is satisfied everywhere in a simply connected region, the displacement will have a single value everywhere in that region. (The displacement is single-valued but not unique since we can always add a rigid body displacement that does not change the strains.) The derivatives contained in Eq. (4.47) can be written as

$$\frac{\partial \zeta_{ik}}{\partial x_l} = \frac{\partial e_{ik}}{\partial x_l} - \left(\frac{\partial e_{ik}}{\partial x_l} - \frac{\partial e_{lk}}{\partial x_i}\right) - \sum_{j=1}^{3} x_j \left(\frac{\partial^2 e_{ik}}{\partial x_j \partial x_l} - \frac{\partial^2 e_{jk}}{\partial x_i \partial x_l}\right)$$

$$\frac{\partial \zeta_{il}}{\partial x_k} = \frac{\partial e_{il}}{\partial x_k} - \left(\frac{\partial e_{il}}{\partial x_k} - \frac{\partial e_{kl}}{\partial x_i}\right) - \sum_{j=1}^{3} x_j \left(\frac{\partial^2 e_{il}}{\partial x_j \partial x_k} - \frac{\partial^2 e_{jl}}{\partial x_i \partial x_k}\right) \qquad (4.48)$$

However, the first two terms on the right-hand side of Eq. (4.48) cancel and the third terms, when placed back into Eq. (4.47), also cancel, so Eq. (4.47) reduces to

$$\sum_{j=1}^{3} x_j \left(\frac{\partial^2 e_{ik}}{\partial x_j \partial x_l} + \frac{\partial^2 e_{jl}}{\partial x_i \partial x_k} - \frac{\partial^2 e_{jk}}{\partial x_i \partial x_l} - \frac{\partial^2 e_{il}}{\partial x_j \partial x_k}\right) = 0 \qquad (4.49)$$

Since the x_j are independent, the quantity in brackets in Eq. (4.49) must vanish, and

$$\frac{\partial^2 e_{ik}}{\partial x_j \partial x_l} + \frac{\partial^2 e_{jl}}{\partial x_i \partial x_k} - \frac{\partial^2 e_{jk}}{\partial x_i \partial x_l} - \frac{\partial^2 e_{il}}{\partial x_j \partial x_k} = 0 \qquad (4.50)$$

We will write Eq. (4.50) symbolically as
$$R_{ijkl} = 0 \tag{4.51}$$

Since i, j, k, l can all have values ranging from 1 to 3, Eq. (4.51) looks like a total of 81 equations. However, because of the following symmetries and antisymmetries:
$$\begin{aligned} R_{ijkl} &= R_{klij} \\ R_{ijkl} &= -R_{jikl} = -R_{ijlk} \end{aligned} \tag{4.52}$$

we can choose only six distinct terms in Eq. (4.51) given by
$$\begin{aligned} S_{11} &\equiv R_{2323} = 0 \\ S_{22} &\equiv R_{3131} = 0 \\ S_{33} &\equiv R_{1212} = 0 \\ S_{21} &= S_{12} \equiv R_{2331} = 0 \\ S_{23} &= S_{32} \equiv R_{3112} = 0 \\ S_{13} &= S_{31} \equiv R_{1223} = 0 \end{aligned} \tag{4.53}$$

which we have written in terms of the components of a symmetric S matrix:
$$\begin{bmatrix} S_{11} & S_{12} & S_{13} \\ S_{21} & S_{22} & S_{23} \\ S_{31} & S_{32} & S_{33} \end{bmatrix} \tag{4.54}$$

All of the equations in Eq. (4.53) are not independent since the components of the S matrix can be shown to satisfy the three equations:
$$\sum_{j=1}^{3} \frac{\partial S_{ij}}{\partial x_j} = 0 \quad (i = 1, 2, 3) \tag{4.55}$$

The six equations in Eq. (4.53) are called the strain *compatibility equations*. Because of Eq. (4.55) these compatibility equations represent only three independent conditions that the strains must satisfy. Explicitly, these six equations are

$$\begin{aligned} 2\frac{\partial^2 e_{23}}{\partial x_2 \partial x_3} - \frac{\partial^2 e_{22}}{\partial x_3^2} - \frac{\partial^2 e_{33}}{\partial x_2^2} &= 0 \\ 2\frac{\partial^2 e_{31}}{\partial x_3 \partial x_1} - \frac{\partial^2 e_{33}}{\partial x_1^2} - \frac{\partial^2 e_{11}}{\partial x_3^2} &= 0 \\ 2\frac{\partial^2 e_{12}}{\partial x_1 \partial x_2} - \frac{\partial^2 e_{11}}{\partial x_2^2} - \frac{\partial^2 e_{22}}{\partial x_1^2} &= 0 \\ \frac{\partial^2 e_{33}}{\partial x_1 \partial x_2} + \frac{\partial^2 e_{12}}{\partial x_3^2} - \frac{\partial^2 e_{23}}{\partial x_3 \partial x_1} - \frac{\partial^2 e_{31}}{\partial x_3 \partial x_2} &= 0 \\ \frac{\partial^2 e_{11}}{\partial x_2 \partial x_3} + \frac{\partial^2 e_{23}}{\partial x_1^2} - \frac{\partial^2 e_{31}}{\partial x_1 \partial x_2} - \frac{\partial^2 e_{12}}{\partial x_1 \partial x_3} &= 0 \\ \frac{\partial^2 e_{22}}{\partial x_3 \partial x_1} + \frac{\partial^2 e_{31}}{\partial x_2^2} - \frac{\partial^2 e_{12}}{\partial x_2 \partial x_3} - \frac{\partial^2 e_{23}}{\partial x_2 \partial x_1} &= 0 \end{aligned} \tag{4.56}$$

4.3.1 Plane-Strain and Plane-Stress Compatibility Equations

Because the six compatibility equations are not independent, it is difficult in general 3-D problems to use these equations directly. However, for plane-strain problems, recall we have only two nonzero displacements because plane-strain conditions require

$$u_1 = u_1(x_1, x_2)$$
$$u_2 = u_2(x_1, x_2) \qquad (4.57)$$
$$u_3 = 0$$

which implies that

$$e_{33} = e_{13} = e_{23} = 0$$
$$e_{11} = e_{11}(x_1, x_2)$$
$$e_{22} = e_{22}(x_1, x_2) \qquad (4.58)$$
$$e_{12} = e_{12}(x_1, x_2)$$

so there is only one compatibility equation that is not identically zero, given by

$$2\frac{\partial^2 e_{12}}{\partial x_1 \partial x_2} - \frac{\partial^2 e_{11}}{\partial x_2^2} - \frac{\partial^2 e_{22}}{\partial x_1^2} = 0 \qquad (4.59)$$

Recall that, for plane-stress problems, we assume

$$\sigma_{13} = \sigma_{23} = \sigma_{33} = 0$$
$$\sigma_{11} = \sigma_{11}(x_1, x_2)$$
$$\sigma_{22} = \sigma_{22}(x_1, x_2) \qquad (4.60)$$
$$\sigma_{12} = \sigma_{12}(x_1, x_2)$$

In the next chapter when we discuss stress–strain relations, we will see that for a linearly elastic isotropic material the condition of plane stress implies that the corresponding conditions on the strains are:

$$e_{13} = e_{23} = 0 \quad \text{but} \quad e_{33} \neq 0$$
$$e_{11} = e_{11}(x_1, x_2)$$
$$e_{22} = e_{22}(x_1, x_2) \qquad (4.61)$$
$$e_{12} = e_{12}(x_1, x_2)$$

so that the compatibility equations reduce to

$$\frac{\partial^2 e_{33}}{\partial x_2^2} = 0$$
$$\frac{\partial^2 e_{33}}{\partial x_1^2} = 0$$
$$\frac{\partial^2 e_{33}}{\partial x_1 \partial x_3} = 0 \qquad (4.62)$$
$$2\frac{\partial^2 e_{12}}{\partial x_1 \partial x_2} - \frac{\partial^2 e_{11}}{\partial x_2^2} - \frac{\partial^2 e_{22}}{\partial x_1^2} = 0$$

Plane-stress problems usually involve thin bodies such as plates where the x_3-direction is normal to the plate and where the e_{33} strains across the thickness of the plate are controlled by the in-plane normal strains, (e_{11}, e_{22}), as we will see in the next chapter. Thus, the first three equations in Eq. (4.62) typically cannot be satisfied and plane-stress conditions can only be approximately true in such a body. However, the behavior of the out-of-plane displacement and strain of a thin body such as a plate are usually not of great interest anyway so that in solving plane-stress problems, one usually ignores the first three equations in Eq. (4.62), leading to the single plane-stress compatibility equation for the strains:

$$2\frac{\partial^2 e_{12}}{\partial x_1 \partial x_2} - \frac{\partial^2 e_{11}}{\partial x_2^2} - \frac{\partial^2 e_{22}}{\partial x_1^2} = 0 \qquad (4.63)$$

which is identical to the plane-strain compatibility equation, Eq. (4.59). If this single compatibility equation is satisfied, we expect that in both plane-stress and plane-strain problems we should be able to integrate the in-plane strains (e_{11}, e_{22}, e_{12}) to find the in-plane displacements (u_1, u_2).

4.3.2 Multiply Connected Bodies

For bodies with holes the compatibility equations are not sufficient by themselves to guarantee that the strains can be integrated to find a single-valued displacement field. In fact, in nonsimply connected bodies there may be some cases where we want displacements that are not single-valued. The split ring shown in Fig. 4.8, for example, is an example where the point P on one side of the split is fixed (displacement = 0) while the displacement of the same point P on the other side of the slit has a value of D. In this case we see that integration paths like C_1 that do not enclose the hole yield single-valued displacements while paths such as C_2 do not, i.e.,

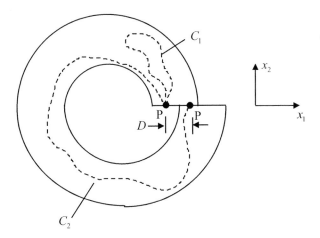

Figure 4.8 A multiply connected body where the displacement at point P has multiple values.

$$\int_{C_1} du_j = 0 \quad (j=1,2,3)$$

$$\int_{C_2} du_j = D\delta_{j1} \quad (j=1,2,3) \tag{4.64}$$

in terms of the Kronecker delta symbol, δ_{ji}, (see Eq. (2.34)). To ensure that a multiply connected body such as the ring cannot split in this manner, we must supplement the compatibility equations by additional conditions. For a body with m holes such as that shown in Fig. 4.9 if, in addition to satisfying the compatibility equations, we also satisfy the m subsidiary conditions

$$\int_{C_i} du_j = 0 \quad (j=1,2,3) \ (i=1,2,\ldots,m) \tag{4.65}$$

where the integrals are taken around each hole, then the displacements will be single-valued.

Note that, in solving any problem, if we end up obtaining a displacement field that represents the deformation of the body, then compatibility is not an issue since we can always obtain a set of strains that are compatible with those displacements simply by taking the appropriate displacement derivatives. However, when we directly solve a problem only for the strains (or corresponding stresses), it is not automatically guaranteed that a well-behaved single-valued displacement field can be obtained by integrating those strains. The compatibility equations (and any subsidiary conditions needed for multiply connected bodies) provide the guarantee that a well-behaved, single-valued displacement field can be obtained from the strain field. Later in this book we see examples of both types of problems.

4.4 Satisfaction of Compatibility – Strength of Materials Solutions

In Chapter 3 we examined if the elementary strength of materials problems of extension, bending, and torsion satisfied the local equilibrium equations. In this section we want to examine the strains in those problems and see if the strain compatibility equations are satisfied.

Axially Loaded Members

The strain field described in Chapter 1 for axially loaded members subjected to an end load P acting through the centroid of the cross-sectional area had the constant strain components

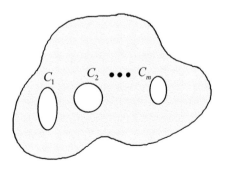

Figure 4.9 A multiply connected body with m holes.

$$e_{xx} = \frac{P}{AE}$$
$$e_{yy} = e_{zz} = -\nu e_{xx} \qquad (4.66)$$
$$e_{xy} = e_{xz} = e_{yz} = 0$$

so that compatibility is satisfied for these strains. If we use this theory and apply it to bars of varying geometry or Young's modulus and/or distributed axial loads acting on the bar surface, then we have instead

$$e_{xx} = \frac{F_x(x)}{A(x)E(x)}$$
$$e_{yy} = e_{zz} = -\nu e_{xx} \qquad (4.67)$$
$$e_{xy} = e_{xz} = e_{yz} = 0$$

where $F_x(x)$ is the varying internal axial force in the bar. In this case the compatibility equations give

$$\frac{d^2 e_{yy}}{dx^2} = 0$$
$$\frac{d^2 e_{zz}}{dx^2} = 0 \qquad (4.68)$$

which generally are not satisfied exactly except in very special cases. However, if the derivative of e_{xx} is only a slowly varying function of x then the derivative of the lateral strains will also be slowly varying in x and these compatibility equations can be approximately satisfied.

Torsion of Circular Shafts

In the torsion problem discussed in Chapter 1, where a circular shaft is subjected to an end-twisting moment, T_0, the strains present, based on the deformation assumptions, were (see Eq. (1.32))

$$e_{xx} = e_{yy} = e_{zz} = 0$$
$$e_{xy} = -\frac{1}{2}\phi' z$$
$$e_{xz} = \frac{1}{2}\phi' y \qquad (4.69)$$
$$e_{yz} = 0$$

in terms of the constant twist/unit length $d\phi/dx = \phi' = T_0/GJ$. Since these strains are linear functions of the coordinates and the compatibility equations involve only second derivatives, compatibility is satisfied. However, if we attempt to use this theory for twisting of a circular bar with varying cross-section or shear modulus and/or distributed torques acting along the length of the shaft, we will have $\phi' = T(x)/G(x)J(x)$, where $T(x)$ is the internal torque. Then the compatibility equations yield

$$\frac{\partial^2 e_{xy}}{\partial x \partial z} = -\frac{1}{2}\phi'' = 0$$

$$\frac{\partial^2 e_{xz}}{\partial x \partial y} = \frac{1}{2}\phi'' = 0$$

(4.70)

where $\phi'' = d^2\phi/dx^2$. Equation (4.70) can be satisfied approximately if the twist/unit length is slowly varying along the length of the shaft.

Beam Bending

Now, consider engineering beam theory, which includes pure bending as a special case. In this theory the strains obtained in Chapter 1 were

$$e_{xx} = \frac{M_y(x)z}{EI_{yy}}$$

$$e_{yy} = e_{zz} = -\nu e_{xx}$$

$$e_{xy} = e_{yz} = 0$$

$$e_{xz} = \frac{1}{2}\frac{V_z(x)Q(z)}{GI_{yy}t(z)}$$

(4.71)

In discussing such problems, however, we want to change the coordinates (from those used in Chapter 1) so that the vertical axis is the y-axis instead of the z-axis (see Fig 4.10a) and the internal moment on a right-hand cut is taken along the positive z-axis while the internal shear force is positive on a right-hand cut in the positive y-direction, as shown in Fig. 4.10b. This coordinate change will facilitate our later comparisons with more detailed plane-stress solutions to such problems. With these changes the strains can be written as

Figure 4.10 (a) Coordinates used for beam problems in this section. (b) The positive directions taken for the internal shear force and bending moment on a right-hand cut taken in the beam.

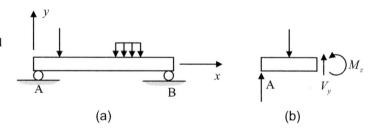

(a)   (b)

4.4 Satisfaction of Compatibility – Strength of Materials Solutions

$$e_{xx} = -\frac{M_z(x)y}{EI_{zz}}$$
$$e_{yy} = e_{zz} = -\nu e_{xx}$$
$$e_{xz} = e_{yz} = 0 \tag{4.72}$$
$$e_{xy} = \frac{1}{2}\frac{V_y(x)Q(y)}{GI_{zz}t(y)}$$

In this case the compatibility equations reduce to

$$\frac{\partial^2 e_{zz}}{\partial x^2} = 0$$
$$2\frac{\partial^2 e_{xy}}{\partial x \partial y} - \frac{\partial^2 e_{yy}}{\partial x^2} = 0 \tag{4.73}$$
$$\frac{\partial^2 e_{zz}}{\partial x \partial y} = 0$$

or, equivalently,

$$\frac{\nu M_z''(x)y}{EI_{zz}} = 0$$
$$\frac{V_y'(x)}{GI_{zz}}\frac{d}{dy}\left(\frac{Q}{t}\right) - \frac{\nu M_z''(x)y}{EI_{zz}} = 0 \tag{4.74}$$
$$\frac{\nu M_z'(x)}{EI_{zz}} = 0$$

where $()' = d()/dx$ and $()'' = d^2()/dx^2$. If the internal bending moment, M_z, is a constant and $V_y = 0$ (which is true for the pure bending case) then the compatibility equations will be satisfied exactly. Otherwise, we see that in general compatibility will not be satisfied. In engineering beam theory this incompatibility is normally not addressed since one considers only the y-displacement of the neutral axis, which we will call $v_a(x)$ here, which is obtained directly by integrating the moment–curvature relationship $d^2 v_a/dx^2 = M_z(x)/EI_{zz}$, not by integrating the strains.

Let us examine a case where we can have compatibility satisfied at least approximately so that we can integrate the strains to find the displacements throughout the beam. Note that if $dM_z/dx = -V_y$ is a constant then the first two equations in Eq. (4.74) will be satisfied, but not the third. Consider a cantilever beam (Fig. 4.11a) where the beam has a rectangular cross-section (Fig. 4.11b). For this beam, the shear force $V_y = P$ throughout the beam. Thus, the compatibility equation for the strains $e_{xx}(x,y), e_{yy}(x,y), e_{xy}(x,y)$ in the x–y plane, which is the second equation in Eq. (4.74), will be satisfied exactly, but the third equation in Eq. (4.74) for the out-of-plane strain, e_{zz}, namely $\partial^2 e_{zz}/\partial x \partial y = 0$, cannot be satisfied. This is similar to the plane stress case so we expect that we should be able to at least integrate the strains in the x–y plane to find the in-plane displacements $u_x(x,y), u_y(x,y)$, which we will now show is indeed true.

Figure 4.11 (a) A cantilever beam of length L. (b) The rectangular cross-section of width b and height, $2h$.

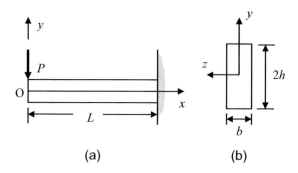

First, let us obtain the y-displacement of the neutral axis, v_a using the moment–curvature relationship. In the beam the internal force and moment are $V_y = P$ and $M_z = -Px$. For the rectangular cross-section, we have $Q(y) = b(h^2 - y^2)/2$ and $I_{zz} = 2bh^3/3$. Integrating the moment–curvature relationship twice gives

$$\frac{d^2 v_a}{dx^2} = -\frac{Px}{EI_{zz}}$$

$$\frac{dv_a}{dx} = -\frac{Px^2}{2EI_{zz}} + C_1 \qquad (4.75)$$

$$v_a = -\frac{Px^3}{6EI_{zz}} + C_1 x + C_2$$

in terms of the constants of integrations C_1, C_2, which can be found from the boundary conditions of zero deflection and slope at the fixed wall:

$$\frac{dv_a(L)}{dx} = 0 \quad \rightarrow \quad C_1 = \frac{PL^2}{2EI_{zz}}$$

$$v_a(L) = 0 \quad \rightarrow \quad C_2 = -\frac{PL^3}{3EI_{zz}} \qquad (4.76)$$

which gives the displacement of the neutral axis:

$$v_a = \frac{P}{EI_{zz}}\left[-\frac{x^3}{6} + \frac{L^2 x}{2} - \frac{L^3}{3}\right] \qquad (4.77)$$

Now, consider integrating both the normal strains e_{xx}, e_{yy}. From Eq. (4.72) these integrals are

$$\frac{\partial u_x}{\partial x} = e_{xx} = \frac{Pxy}{EI_{zz}}$$

$$u_x(x,y) = \frac{Px^2 y}{2EI_{zz}} + g(y)$$

$$\frac{\partial u_y}{\partial y} = e_{yy} = -\nu e_{xx} = -\frac{\nu Pxy}{EI_{zz}} \qquad (4.78)$$

$$u_y(x,y) = -\frac{\nu Pxy^2}{2EI_{zz}} + h(x)$$

4.4 Satisfaction of Compatibility – Strength of Materials Solutions

in terms of two arbitrary functions $g(y)$, $h(x)$. But from the shear strain we find

$$\frac{\partial u_x}{\partial y} + \frac{\partial u_y}{\partial x} = 2e_{xy} = \frac{3}{4}\frac{P}{bh^3 G}(h^2 - y^2) \tag{4.79}$$

which yields, using Eq. (4.78) and rearranging the terms,

$$\frac{Px^2}{2EI_{zz}} + \frac{dh}{dx} = \frac{3}{4}\frac{P}{bh^3 G}(h^2 - y^2) + \frac{vPy^2}{2EI_{zz}} - \frac{dg}{dy} \tag{4.80}$$

Equation (4.80) is of the form $H(x) = G(y)$ and since x and y are independent variables, this equation can only be true for all x and y if $H(x) = G(y) = C$, where C is a constant. Thus, integrating the two sides of Eq. (4.80) on x or y, respectively, and using the value of I_{zz} we find

$$h(x) = Cx - \frac{Px^3}{6EI_{zz}} + v_0$$

$$g(y) = -Cy + \frac{P}{2GI_{zz}}\left(h^2 y - \frac{y^3}{3}\right) + \frac{vPy^3}{6EI_{zz}} + u_0 \tag{4.81}$$

in terms of two constants of integration u_0, v_0. Placing these values into the displacement expressions in Eq. (4.78) gives

$$u_x(x, y) = \frac{Px^2 y}{2EI_{zz}} + \frac{P}{2GI_{zz}}\left(h^2 y - \frac{y^3}{3}\right) + \frac{vPy^3}{6EI_{zz}} - Cy + u_0$$

$$u_y(x, y) = -\frac{Px^3}{6EI_{zz}} - \frac{vPxy^2}{2EI_{zz}} + Cx + v_0 \tag{4.82}$$

The constants all have a physical meaning. In particular,

$$u_x(0, 0) = u_0$$
$$u_y(0, 0) = v_0$$
$$\left.\frac{\partial u_y}{\partial x}\right|_{x=0, y=0} = C \tag{4.83}$$

so that (u_0, v_0) are the x- and y-displacements at the load P and C is the slope of the neutral axis at the load. If, at the wall, we set

$$u_x(L, 0) = u_y(L, 0) = 0$$
$$\frac{\partial u_y(L, 0)}{\partial x} = 0 \tag{4.84}$$

we find

$$u_0 = 0, \quad v_0 = -\frac{PL^3}{3EI_{zz}}$$
$$C = \frac{PL^2}{2EI_{zz}} \tag{4.85}$$

Using the relationship between the shear modulus and Young's modulus, namely, $G = E/2(1+\nu)$, the displacements then become finally

$$u_x(x,y) = \frac{Px^2 y}{2EI_{zz}} + \frac{P(1+\nu)}{EI_{zz}}\left(h^2 y - \frac{y^3}{3}\right) + \frac{\nu P y^3}{6EI_{zz}} - \frac{PL^2 y}{2EI_{zz}}$$

$$u_y(x,y) = -\frac{Px^3}{6EI_{zz}} - \frac{\nu P x y^2}{2EI_{zz}} + \frac{PL^2 x}{2EI_{zz}} - \frac{PL^3}{3EI_{zz}}$$

(4.86)

If we examine the displacements of the neutral axis from these expressions, we have

$$u_a = u_x(x, 0) = 0$$

$$v_a = u_y(x, 0) = -\frac{Px^3}{6EI_{zz}} + \frac{PL^2 x}{2EI_{zz}} - \frac{PL^3}{3EI_{zz}}$$

(4.87)

which agrees with the strength of materials solution for the vertical displacement, Eq. (4.78). This displacement is shown (greatly exaggerated) in Fig. 4.12a. According to the assumptions of the beam theory presented in Chapter 1, plane sections initially at right angles to the neutral axis remain plane and at right angles after the loads are applied which would imply that a small pair of horizontal and vertical lines at the wall would remain horizontal and vertical, as shown in Fig. 4.12a. However, the displacements of Eq. (4.86) require that a line in the beam that is originally vertical at the wall becomes sloped, while a horizontal line will remain horizontal, as shown in Fig. 4.12b. Thus, there is an inconsistency between the original bending deformation assumptions and the displacements predicted by the theory. This is not surprising since in engineering beam theory we know that there are shear stresses generated when the bending normal stress, σ_{xx}, is not constant along the beam so that there must be associated shear strains developed as well and these shear stresses will lead to additional shearing displacements. One can remove this inconsistency and generate a modified beam theory that accounts for beam deflections due to shear that is called Timoshenko beam theory [3]. However, for long, slender beams (where $h/L \ll 1$), these shear deflections are not significant and engineering beam theory is adequate. We can see

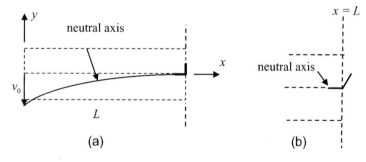

Figure 4.12 (a) The neutral axis deflections of a cantilever beam (greatly exaggerated). According to the elementary engineering beam theory deformation assumptions a small pair of horizontal and vertical lines at the neutral axis and at the wall location should remain horizontal and vertical. (b) The displacements generated by the theory, however, predict that the two lines are no longer at a right angle.

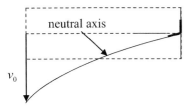

Figure 4.13 Displacement of the neutral axis of a cantilever beam in elementary engineering beam theory (greatly exaggerated) when a small vertical line along the neutral axis at the wall is held vertical. This requires the neutral axis to have a slope at the wall.

this type of behavior within the context of the present theory if we simply change the boundary conditions at the wall so that Eq. (4.84) is replaced by

$$u_x(L,0) = u_y(L,0) = 0$$
$$\frac{\partial u_y(L,0)}{\partial x} = 0 \tag{4.88}$$

This means we are requiring a small vertical line at $x = L$, $y = 0$ to remain vertical but a small line at the neutral axis can now have a slope at the wall as shown in Fig. 4.13. These boundary conditions give

$$u_0 = 0$$
$$v_0 = -\frac{PL^3}{3EI_{zz}}\left[1 + 3(1+v)\frac{h^2}{L^2}\right] \tag{4.89}$$
$$C = \frac{PL^2}{2EI_{zz}}\left[1 + 2(1+v)\frac{h^2}{L^2}\right]$$

so that there is a larger deflection and slope predicted at the load (see Eq. (4.85)) with correction terms that are of order h^2/L^2 (so that they are small for long, slender beams).

Since we have obtained the displacements throughout the beam from engineering beam theory, we can plot these displacements to examine this displacement field. It is convenient to write these equations in normalized form as

$$\frac{EI_{zz}u_x}{PL^3} = \frac{\bar{x}^2\bar{y}}{2} + (1+v)\left(\frac{h^2}{L^2}\bar{y} - \frac{\bar{y}^3}{3}\right) + \frac{v\bar{y}^3}{6} - \frac{\bar{y}}{2}$$
$$\frac{EI_{zz}u_y}{PL^3} = -\frac{\bar{x}^3}{6} - \frac{v\bar{x}\bar{y}^2}{2} + \frac{\bar{x}}{2} - \frac{1}{3} \tag{4.90}$$

where $\bar{x} = x/L$, $\bar{y} = y/L$. A MATLAB® script cantilever_d (see Appendix D) was written that computes these normalized displacements for $h/L = 0.1$, $v = 1/3$, and plots them as a "quiver" plot of arrows representing the displacements. The result is shown in Fig. 4.14. It is seen that, while the neutral axis only undergoes a vertical displacement, this is not true of other points in the beam.

Figure 4.14 Displacements of a cantilever beam.

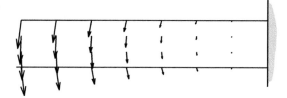

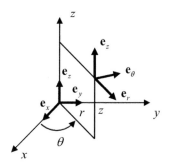

Figure 4.15 Cylindrical coordinates.

4.5 Strains in Cylindrical Coordinates

Consider the small strains along an orthogonal set of coordinates (n, t, s). From Eq. (4.15) the normal strains and shear strains associated with these coordinates can be written as

$$e_{nn} = \mathbf{n} \cdot \frac{\partial \mathbf{u}}{\partial s_n}, \quad e_{tt} = \mathbf{t} \cdot \frac{\partial \mathbf{u}}{\partial s_t}, \quad e_{ss} = \mathbf{s} \cdot \frac{\partial \mathbf{u}}{\partial s_s}$$

$$e_{nt} = \frac{1}{2}\left(\mathbf{n} \cdot \frac{\partial \mathbf{u}}{\partial s_t} + \mathbf{t} \cdot \frac{\partial \mathbf{u}}{\partial s_n}\right), \quad e_{ns} = \frac{1}{2}\left(\mathbf{n} \cdot \frac{\partial \mathbf{u}}{\partial s_s} + \mathbf{s} \cdot \frac{\partial \mathbf{u}}{\partial s_n}\right), \quad e_{ts} = \frac{1}{2}\left(\mathbf{t} \cdot \frac{\partial \mathbf{u}}{\partial s_s} + \mathbf{s} \cdot \frac{\partial \mathbf{u}}{\partial s_t}\right) \quad (4.91)$$

Before, we let (n, t, s) be along a set of (x, y, z) axes to obtain the strains in Cartesian coordinates. Similarly, we can use these expressions to find the strains in any other set of orthogonal coordinates. Consider a set of cylindrical coordinates, for example, as shown in Fig. 4.15. In that case we can take the lengths of small elements in the (r, θ, z) directions as

$$\partial s_n = \partial r$$
$$\partial s_t = r \partial \theta \quad (4.92)$$
$$\partial s_z = \partial z$$

and the unit vectors $(\mathbf{n}, \mathbf{t}, \mathbf{s})$ as

$$\mathbf{n} = \mathbf{e}_r = \cos\theta \mathbf{e}_x + \sin\theta \mathbf{e}_y$$
$$\mathbf{t} = \mathbf{e}_\theta = -\sin\theta \mathbf{e}_x + \cos\theta \mathbf{e}_y \quad (4.93)$$
$$\mathbf{s} = \mathbf{e}_z$$

4.5 Strains in Cylindrical Coordinates

The displacement vector, **u**, is given by

$$\mathbf{u} = u_r \mathbf{e}_r + u_\theta \mathbf{e}_\theta + u_z \mathbf{e}_z \tag{4.94}$$

Consider now the normal strain in the r-direction. We have

$$\begin{aligned} e_{rr} &= \mathbf{e}_r \cdot \frac{\partial \mathbf{u}}{\partial r} \\ &= \mathbf{e}_r \cdot \left(\frac{\partial u_r}{\partial r} \mathbf{e}_r + \frac{\partial u_\theta}{\partial r} \mathbf{e}_\theta + \frac{\partial u_z}{\partial r} \mathbf{e}_z \right) \\ &= \frac{\partial u_r}{\partial r} \end{aligned} \tag{4.95}$$

In the θ-direction, however,

$$\begin{aligned} e_{\theta\theta} &= \mathbf{e}_\theta \cdot \frac{1}{r} \frac{\partial \mathbf{u}}{\partial \theta} = \mathbf{e}_\theta \cdot \left(\frac{\partial u_r}{\partial \theta} \mathbf{e}_r + \frac{\partial u_\theta}{\partial \theta} \mathbf{e}_\theta + \frac{\partial u_z}{\partial \theta} \mathbf{e}_z + u_r \frac{\partial \mathbf{e}_r}{\partial \theta} + u_\theta \frac{\partial \mathbf{e}_\theta}{\partial \theta} \right) \\ &= \frac{1}{r} \frac{\partial u_\theta}{\partial \theta} + \frac{u_r}{r} \end{aligned} \tag{4.96}$$

where we use the fact that

$$\frac{\partial \mathbf{e}_r}{\partial \theta} = \mathbf{e}_\theta, \quad \frac{\partial \mathbf{e}_\theta}{\partial \theta} = -\mathbf{e}_r \tag{4.97}$$

The additional term coming from the radial displacement can be explained if we consider how such a displacement affects the elongation of an element initially in the θ-direction, as shown in Fig. 4.16. We see from that figure that a radial displacement causes a strain in the θ-direction for an element initially in that direction given by

$$(e_{\theta\theta})_{u_r} = \frac{(r+u_r)d\theta - rd\theta}{rd\theta} = \frac{u_r}{r} \tag{4.98}$$

Now consider the strain in the z-direction. We find

$$\begin{aligned} e_{zz} &= \mathbf{e}_z \cdot \frac{\partial \mathbf{u}}{\partial z} = \mathbf{e}_z \cdot \left(\frac{\partial u_r}{\partial z} \mathbf{e}_r + \frac{\partial u_\theta}{\partial z} \mathbf{e}_\theta + \frac{\partial u_z}{\partial z} \mathbf{e}_z \right) \\ &= \frac{\partial u_z}{\partial z} \end{aligned} \tag{4.99}$$

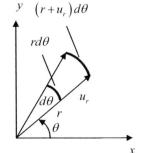

Figure 4.16 Effects of a radial displacement, u_r, on the length of an element (bold line) in the θ-direction.

Figure 4.17 (a) Two small lines along the r- and θ- directions in an undeformed body. (b) Those same two lines in the deformed body when there is a constant displacement, u_θ, of the radial line.

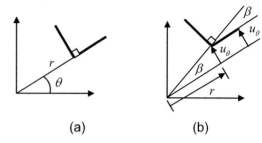

Next, consider one of the shear strains, namely

$$e_{r\theta} = \frac{1}{2}\left(\mathbf{e}_r \cdot \frac{1}{r}\frac{\partial \mathbf{u}}{\partial \theta} + \mathbf{e}_\theta \cdot \frac{\partial \mathbf{u}}{\partial r}\right)$$

$$= \frac{1}{2}\mathbf{e}_r \cdot \frac{1}{r}\left(\frac{\partial u_r}{\partial \theta}\mathbf{e}_r + \frac{\partial u_\theta}{\partial \theta}\mathbf{e}_\theta + u_r\frac{\partial \mathbf{e}_r}{\partial \theta} + u_\theta\frac{\partial \mathbf{e}_\theta}{\partial \theta}\right) + \frac{1}{2}\mathbf{e}_\theta \cdot \left(\frac{\partial u_r}{\partial r}\mathbf{e}_r + \frac{\partial u_\theta}{\partial r}\mathbf{e}_\theta + \frac{\partial u_z}{\partial r}\mathbf{e}_z\right) \quad (4.100)$$

$$= \frac{1}{2}\left(\frac{1}{r}\frac{\partial u_r}{\partial \theta} + \frac{\partial u_\theta}{\partial r} - \frac{u_\theta}{r}\right)$$

The first two terms in Eq. (4.100) are analogous to the terms that appear in Cartesian coordinates. The last terms can be explained by the fact that, even if the displacement in the θ-direction is not changing in r, it can still produce a shear strain as shown in Fig. 4.17, where

$$(e_{r\theta})_{u_\theta} = -\frac{\beta}{2} = -\frac{1}{2}\frac{u_\theta}{r} \quad (4.101)$$

(the minus sign exists because a constant displacement in the θ-direction causes the angle between two lines initially along the r- and θ-directions to be greater than 90°, as shown in Fig. 4.17b).

Finally, in exactly the same fashion we can obtain the other two shear strains, which are given as

$$e_{rz} = \frac{1}{2}\left(\frac{\partial u_z}{\partial r} + \frac{\partial u_r}{\partial z}\right)$$

$$e_{\theta z} = \frac{1}{2}\left(\frac{\partial u_\theta}{\partial z} + \frac{1}{r}\frac{\partial u_z}{\partial \theta}\right) \quad (4.102)$$

4.6 PROBLEMS

P4.1 Consider the state of strain of Example 4.1.
  (a) If we let $(\mathbf{e}_{p1}, \mathbf{e}_{p2}, \mathbf{e}_{p3})$ be unit vectors along the three principal strain directions obtained in that example, show for the lines at ±45° from the largest and smallest principal strains defined by the unit vectors

$$\mathbf{d}_1 = (\mathbf{e}_{p1} + \mathbf{e}_{p3})/\sqrt{2}$$
$$\mathbf{d}_2 = (\mathbf{e}_{p1} - \mathbf{e}_{p3})/\sqrt{2}$$

we obtain from the strain transformation equations the largest tensor shear strain (in magnitude) along these directions and that twice this largest strain is the largest engineering shear strain.

(b) Define the unit vector $\mathbf{v} = \mathbf{n} \times \mathbf{t}$ and use the strain transformation equation from the (x, y, z) coordinates to the (n, t, v) coordinates to obtain the state of strain with respect to the (n, t, v) coordinates. Identify the strains (e_{nn}, e_{nt}) previously calculated in this example for this state of strain.

P4.2 Show that the compatibility equation for a state of plane strain, Eq. (4.59), can be found directly from the displacements obtained by integrating the strain–displacement relations.

P4.3 The state of strain (in µstrain) at a point P in a deformed body is given for a set of (x, y, z) coordinates by

$$[e] = \begin{bmatrix} 200 & -300 & 0 \\ -300 & 100 & 200 \\ 0 & 200 & 100 \end{bmatrix}$$

(a) Determine the principal strains at point P.
(b) For the largest positive principal strain, determine the corresponding principal strain direction explicitly, i.e., DO NOT USE MATLAB® or any other software package, but instead obtain the principal direction components by hand. Show all your work. (You can use MATLAB® or any other package to verify your answer.)

P4.4 Consider the same state of strain as given in problem P4.3 and assume these strains are constant throughout the deformed body. Figure P4.1 shows a set of lines in this region of constant strain that have the lengths and directions as shown in the undeformed body ($|PA| = |PB| = |PC| = a$).
(a) Determine the length of line AC in the deformed body in terms of a.
(b) Determine the angle, in degrees, between lines DB and DA in the deformed body (point D is at the center of AC).

P4.5 The state of strain in a region about a point P in a stressed body is given by

$$[e] = \begin{bmatrix} 0.003 & 0 & -0.003 \\ 0 & 0.004 & 0.005 \\ -0.003 & 0.005 & 0.006 \end{bmatrix}$$

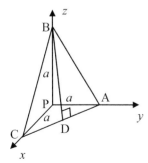

Figure P4.1 A set of lines (as seen in a body when it is undeformed) that are in a region of uniform strain in the deformed body.

and these strains are constant in this region. Consider the lines in this region shown in Fig. P4.2.

(a) Determine the change in angle (in degrees) between lines PA and PB, which are initially have a length, a, and are at a right angle to each other in the deformed body, as shown.

(b) Determine the new length of line PC in the deformed body, which originally has a length of $a\sqrt{2}$.

(c) Determine the state of strain with respect to the (x', y', z') axes shown in Fig. P4.2. How is this state of strain related to any answers in parts (a) and (b)?

P4.6 The state of strain (measured in µstrain) with respect to an (x, y, z) coordinate system is:

$$[e] = \begin{bmatrix} 100 & -30 & 10 \\ -30 & 50 & 0 \\ 10 & 0 & 250 \end{bmatrix}$$

(a) Determine the principal strains and the principal strain directions.

(b) Determine the magnitude of the largest engineering shear stress and the directions (unit vectors) of the lines in the body along which this largest shear stress occurs (see problem P4.1). Use the strain transformation equations to verify that one obtains this largest shear stress along the directions you obtained.

P4.7 The state of strain (measured in µstrain) with respect to an (x, y, z) coordinate system is:

$$[e] = \begin{bmatrix} -200 & 25 & 0 \\ 25 & 100 & 0 \\ 0 & 0 & 200 \end{bmatrix}$$

(a) Determine the principal strains and the principal strain directions.

(b) Determine the magnitude of the largest engineering shear stress and the directions (unit vectors) of the lines in the body along which this largest shear stress occurs (see problem P4.1).

Use the strain transformation equations to verify that one obtains this largest shear stress along the directions you obtained.

P4.8 The state of strain (measured in µstrain) with respect to an (x, y, z) coordinate system is:

$$[e] = \begin{bmatrix} -300 & 50 & 30 \\ 50 & -100 & 0 \\ 30 & 0 & 100 \end{bmatrix}$$

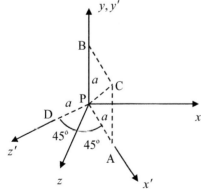

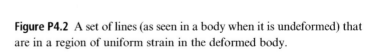

Figure P4.2 A set of lines (as seen in a body when it is undeformed) that are in a region of uniform strain in the deformed body.

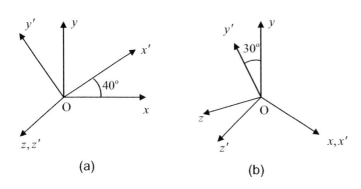

Figure P4.3 (a) A coordinate rotation of 40° about the z-axis. (b) A coordinate rotation of 30° about the x-axis.

Determine the state of strain with respect to the (x', y', z') axes that are rotated 40° about the z-axis (see Fig. P4.3a).

P4.9 The state of strain (measured in µstrain) with respect to an (x, y, z) coordinate system is:

$$[e] = \begin{bmatrix} 400 & 25 & 30 \\ 25 & 200 & 50 \\ 30 & 50 & 100 \end{bmatrix}$$

Determine the state of strain with respect to the (x', y', z') axes that are rotated 30° about the x-axis (see Fig. P4.3b).

P4.10 When the strains are uniform in a region, we can analyze the changes of lengths and angles of finite lines in that region, as considered in a number of previous problems. However, if the strains vary from point to point then this is no longer possible. Consider, for example a finite line drawn from point A to point B in an undeformed body, as shown in Fig. P4.4a. Suppose this line becomes the line drawn from A′ to B′ in the deformed body (whose deformation, assumed to be small, is greatly exaggerated). If at a point P on the line AB we draw a small line segment of length ds (which acts in the **n** direction), and if that point moves to point P′ where the line segment now has a length ds', then the small normal strain at P along the line is defined as $e_{nn} = (ds' - ds)/ds$. It follows that

$$\int_{A'}^{B'} ds' - \int_{A}^{B} ds = \int_{A}^{B} e_{nn} ds$$

$$\rightarrow L' - L = \int_{A}^{B} e_{nn} ds$$

where (L', L) are the lengths of the lines A′B′ and AB in the deformed and undeformed bodies, respectively. Another way to write this relation is

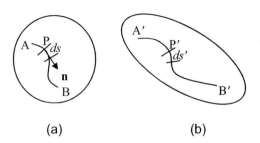

Figure P4.4 (a) A finite line in a undeformed body. (b) The line after a small deformation (which is greatly exaggerated).

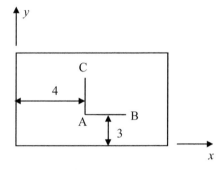

Figure P4.5 A plate geometry where two lines AB and AC are shown when the plate is undeformed. Each line is 2 in. long.

$$\bar{e}_{nn} = \frac{L' - L}{L} = \frac{1}{L}\int_A^B e_{nn}\,ds$$

where $\bar{e}_{nn}$ is the average normal strain for the finite line AB.

Consider a pair of lines AB and AC in a plate as shown in Fig. P4.5. Each of these lines is two inches long. If the plate has a 2-D displacement field given by $u_x = c(2x + y)$, $u_y = b(x - 3y^2)$, where x and y are measured in inches and b and c are both very small constants, determine the lengths of these lines in the deformed body in terms of b and c. What can you say about the change in angles?

References

1. I. S. Sokolnikoff, *Mathematical Theory of Elasticity* (New York, NY: McGraw-Hill, 1956)
2. C. R. Wylie and L. C. Barrettt, *Advanced Engineering Mathematics*, 6th edn (New York, NY: McGraw-Hill, 1995)
3. H. Reismann and P. S. Pawlik, *Elasticity-Theory and Applications* (New York, NY: John Wiley and Sons, 1980)

5   Stress–Strain Relations

Stresses describe the local distributions of forces within a deformable body, and strains describe the local deformations. In this chapter we want to describe the relations between stresses and strains as these are the key relationships that allow us to connect the loads applied to a body to its changes in shape. In this book, we will consider only linear elastic materials, where the stresses are proportional to the small strains present. Both isotropic and anisotropic linear elastic materials will be discussed. As described in Chapter 1 these stress–strain relations are one of the key "pillars" of stress analyses.

5.1   Linear Elastic Materials

5.1.1   Linear Elastic Isotropic Materials

The relationships between stresses and strains seen in the mechanics of materials problems discussed in Chapter 1 were simple since there were very few significant stresses and strains present in those elementary problems. In the axial loading and pure bending cases, for example, there was only one significant axial stress, σ, and we related this stress to the small axial strain, e, through Hooke's law $\sigma = Ee$, but we noted that there were also small strains, $e_l = e_{yy}$ or e_{zz}, generated by this axial stress in those lateral directions where $e_l = ve$. In the case of torsion we had shear stresses, $\tau = \sigma_{xy}$ or σ_{xz}, generated that we again related to shear strains, $\gamma = \gamma_{xy}$ or γ_{xz}, through a similar Hooke's law $\tau = G\gamma$. These three material constants, the elastic constant E (Young's modulus), the constant G (the shear modulus), and the constant v (Poisson's ratio), while appearing separately in those problems, are in fact the constants that help to define the general relationship between stresses and strains for small deformations of an isotropic, linearly elastic solid. The term "isotropic" means that the stress–strain relations for the material are the same in all directions at any point in a body at that point (although the stress–strain relations may vary from point to point if the body is inhomogeneous). The term "linearly elastic" means that that there is a linear relationship between the stresses and the strains. Hooke's law is obviously such a linear relationship. Remarkably, we can generate the general stress–strain equations for an isotropic, linear elastic material (also called *generalized Hooke's law*) by using the simple results of the elementary problems discussed in Chapter 1 and the principle of superposition. We start by applying to an element a uniaxial stress σ_{xx}, as seen in Fig. 5.1a, which produces both a direct small strain, e_{xx}, and lateral small strains (e_{yy}, e_{zz}). For an isotropic material, these lateral

Figure 5.1 (a) Applying a normal stress to an element acting in the x-direction, followed by adding (b) a normal stress in the y-direction, (c) a normal stress in the z-direction, and (d) a shear stresses σ_{xy} and $\sigma_{yx} = \sigma_{xy}$. (e) A complete state of stress where there are three normal stresses $(\sigma_{xx}, \sigma_{yy}, \sigma_{zz})$ and three shear stresses $(\sigma_{xy}, \sigma_{yz}, \sigma_{zx})$ and their symmetric counterparts. Note that only the stresses acting on the front faces of the element are shown.

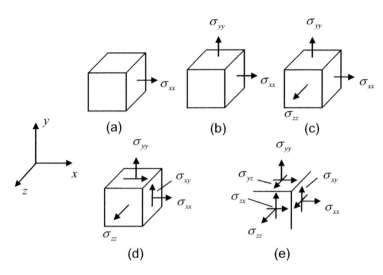

strains are equal since the behavior of the material is the same in both the y- and z-directions, giving

$$e_{xx} = \frac{\sigma_{xx}}{E}$$
$$e_{yy} = -\frac{\nu\sigma_{xx}}{E} \quad (5.1)$$
$$e_{zz} = -\frac{\nu\sigma_{xx}}{E}$$

Now, we hold the stress σ_{xx} constant and apply a normal stress in the y-direction, as seen in Fig. 5.1b. This causes both a direct strain, e_{yy}, and lateral strains (e_{xx}, e_{zz}). For an isotropic material these strains are related to the stresses with the same elastic constants used in Eq. (5.1) so for the state of stress in Fig. 5.1b we have

$$e_{xx} = \frac{\sigma_{xx}}{E} - \frac{\nu\sigma_{yy}}{E}$$
$$e_{yy} = \frac{\sigma_{yy}}{E} - \frac{\nu\sigma_{xx}}{E} \quad (5.2)$$
$$e_{zz} = -\frac{\nu\sigma_{xx}}{E} - \frac{\nu\sigma_{yy}}{E}$$

Keeping the σ_{xx} and σ_{yy} stresses constant, we then add the normal stress, σ_{zz}, as shown in Fig, 5.1c, which produces an additional direct strain, e_{zz}, as well as lateral strains, (e_{xx}, e_{yy}), giving the state of stress shown in Fig. 5.1c where:

$$e_{xx} = \frac{\sigma_{xx}}{E} - \frac{\nu\sigma_{yy}}{E} - \frac{\nu\sigma_{zz}}{E}$$
$$e_{yy} = \frac{\sigma_{yy}}{E} - \frac{\nu\sigma_{xx}}{E} - \frac{\nu\sigma_{zz}}{E} \quad (5.3)$$
$$e_{zz} = \frac{\sigma_{zz}}{E} - \frac{\nu\sigma_{xx}}{E} - \frac{\nu\sigma_{yy}}{E}$$

Now, we add the shear stress, σ_{xy} (and its equal counterpart σ_{yx}), as shown in Fig. 5.1d. This generates an engineering shear strain, γ_{xy}, given by $\gamma_{xy} = \sigma_{xy}/G$ in terms of the shear modulus. Similarly, when we apply the shear stresses σ_{yz}, σ_{zx} we will generate additional engineering shear strains $\gamma_{yz} = \sigma_{yz}/G$, $\gamma_{zx} = \sigma_{zx}/G$, producing the stress–strain relations for the complete three-dimensional state of stress shown in Fig. 5.1e where

$$e_{xx} = \frac{\sigma_{xx}}{E} - \frac{v}{E}(\sigma_{yy} + \sigma_{zz})$$
$$e_{yy} = \frac{\sigma_{yy}}{E} - \frac{v}{E}(\sigma_{xx} + \sigma_{zz})$$
$$e_{zz} = \frac{\sigma_{zz}}{E} - \frac{v}{E}(\sigma_{xx} + \sigma_{yy})$$
$$\gamma_{xy} = \frac{\sigma_{xy}}{G}, \quad \gamma_{yz} = \frac{\sigma_{yz}}{G}, \quad \gamma_{zx} = \frac{\sigma_{zx}}{G}$$

(5.4)

In obtaining this result the fact that the material was assumed to be isotropic allowed us to use the same elastic constants (E, G, v) in relating stresses to strains in the different directions. Because we also assumed linear relations between the stresses and strains, we could add the strains together through superposition. Equation (5.4) gives us the strains in terms of the stresses. We can invert these relations and instead write the stresses in terms of the strains as

$$\sigma_{xx} = \frac{E}{(1+v)(1-2v)}\left[(1-v)e_{xx} + v(e_{yy} + e_{zz})\right]$$
$$\sigma_{yy} = \frac{E}{(1+v)(1-2v)}\left[(1-v)e_{yy} + v(e_{xx} + e_{zz})\right]$$
$$\sigma_{zz} = \frac{E}{(1+v)(1-2v)}\left[(1-v)e_{zz} + v(e_{xx} + e_{yy})\right]$$
$$\sigma_{xy} = G\gamma_{xy}, \quad \sigma_{yz} = G\gamma_{yz}, \quad \sigma_{zy} = G\gamma_{zx}$$

(5.5)

While these general stress–strain relations are written in terms of the three material constants, (E, G, v), an isotropic, linearly elastic solid is actually characterized by only two independent constants since the condition of isotropy can be shown to require that $G = E/2(1+v)$.

Equations (5.4) and (5.5) are generalized Hooke's law for general 3-D states of stress and strain in a linear elastic, isotropic material. Many common structural materials such as steel and aluminum alloys, for example, can be characterized by these relations. Because Eq. (5.5) is so commonly used, we have written a MATLAB® function stress_strain that has the calling sequence

```
stress = stress_strain(E, nu, strain);
```

which takes a 3 × 3 strain matrix, strain, where the strains are measured in μstrain, and returns the 3 × 3 stress matrix, stress, with the stresses measured in MPa. Here, Young's modulus, E, is given in GPa, and nu is Poisson's ratio. In the same fashion we have a MATLAB® function strain_stress that has the calling sequence

```
strain = strain_stress(E, nu, stress);
```

which takes a 3 × 3 stress matrix, stress, where the stresses are in MPa, and returns the 3 × 3 matrix, strain, with the strains measured in µstrain. Again, Young's modulus, E, is given in GPa, and nu is Poisson's ratio.

5.1.2 General Linear Elastic Stress–Strain Relations

In the most general case when the stresses are linearly related to the small strains, the stress–strain relationship can be written in matrix form as

$$\sigma_{ij} = \sum_{k=1}^{3}\sum_{l=1}^{3} C_{ijkl} e_{kl} \tag{5.6}$$

in terms of the elastic constants C_{ijkl}. Since there are nine stresses and nine strains, there are a total of 81 constants. However, if we use the fact that the stress matrix and strain matrix are symmetric, we have

$$\begin{aligned} C_{ijkl} &= C_{jikl} \\ C_{ijkl} &= C_{ijlk} \end{aligned} \tag{5.7}$$

so that we are only really relating six stresses to six strains and there is a total of 36 constants. We can write these stress–strain relations out explicitly as

$$\begin{Bmatrix} \sigma_{11} \\ \sigma_{22} \\ \sigma_{33} \\ \sigma_{23} \\ \sigma_{13} \\ \sigma_{12} \end{Bmatrix} = \begin{bmatrix} C_{1111} & C_{1122} & C_{1133} & C_{1123}+C_{1132} & C_{1113}+C_{1131} & C_{1112}+C_{1121} \\ C_{2211} & C_{2222} & C_{2233} & C_{2233}+C_{2232} & C_{2213}+C_{2231} & C_{2212}+C_{2221} \\ C_{3311} & C_{3322} & C_{3333} & C_{3323}+C_{3332} & C_{3313}+C_{3331} & C_{3312}+C_{3321} \\ C_{2311} & C_{2322} & C_{2333} & C_{2323}+C_{2332} & C_{2313}+C_{2331} & C_{2312}+C_{2331} \\ C_{1311} & C_{1322} & C_{1333} & C_{1323}+C_{1332} & C_{1313}+C_{1331} & C_{1312}+C_{1321} \\ C_{1211} & C_{1222} & C_{1233} & C_{1223}+C_{1232} & C_{1213}+C_{1231} & C_{1212}+C_{1221} \end{bmatrix} \begin{Bmatrix} e_{11} \\ e_{22} \\ e_{33} \\ e_{23} \\ e_{13} \\ e_{12} \end{Bmatrix} \tag{5.8a}$$

and using the symmetry properties of the elastic constants

$$\begin{Bmatrix} \sigma_{11} \\ \sigma_{22} \\ \sigma_{33} \\ \sigma_{23} \\ \sigma_{13} \\ \sigma_{12} \end{Bmatrix} = \begin{bmatrix} C_{1111} & C_{1122} & C_{1133} & 2C_{1123} & 2C_{1113} & 2C_{1112} \\ C_{2211} & C_{2222} & C_{2233} & 2C_{2223} & 2C_{2213} & 2C_{2212} \\ C_{3311} & C_{3322} & C_{3333} & 2C_{3323} & 2C_{3313} & 2C_{3312} \\ C_{2311} & C_{2322} & C_{2333} & 2C_{2323} & 2C_{2313} & 2C_{2312} \\ C_{1311} & C_{1322} & C_{1333} & 2C_{1323} & 2C_{1313} & 2C_{1312} \\ C_{1211} & C_{1222} & C_{1233} & 2C_{1223} & 2C_{1213} & 2C_{1212} \end{bmatrix} \begin{Bmatrix} e_{11} \\ e_{22} \\ e_{33} \\ e_{23} \\ e_{13} \\ e_{12} \end{Bmatrix} \tag{5.8b}$$

where the factors of two comes from combining terms for the symmetric shear strains. We can eliminate those factors by writing the stress–strain relations in terms of the engineering shear strains, giving

$$\begin{Bmatrix} \sigma_{11} \\ \sigma_{22} \\ \sigma_{33} \\ \sigma_{23} \\ \sigma_{13} \\ \sigma_{12} \end{Bmatrix} = \begin{bmatrix} C_{1111} & C_{1122} & C_{1133} & C_{1123} & C_{1113} & C_{1112} \\ C_{2211} & C_{2222} & C_{2233} & C_{2223} & C_{2213} & C_{2212} \\ C_{3311} & C_{3322} & C_{3333} & C_{3323} & C_{3313} & C_{3312} \\ C_{2311} & C_{2322} & C_{2333} & C_{2323} & C_{2313} & C_{2312} \\ C_{1311} & C_{1322} & C_{1333} & C_{1323} & C_{1313} & C_{1312} \\ C_{1211} & C_{1222} & C_{1233} & C_{1223} & C_{1213} & C_{1212} \end{bmatrix} \begin{Bmatrix} e_{11} \\ e_{22} \\ e_{33} \\ \gamma_{23} \\ \gamma_{13} \\ \gamma_{12} \end{Bmatrix} \quad (5.9)$$

We will see in Chapter 8 that, if we also assume these stress–strain relations come from an internal energy function, the elastic constants have the additional symmetry relations

$$C_{ijkl} = C_{klij} \quad (5.10)$$

If you examine the relations of Eq. (5.10) for the coefficient matrix in Eq. (5.9), you will see that Eq. (5.10) means that the coefficient matrix is symmetric, which reduces the total number of constants to 21 (six diagonal terms and 15 off-diagonal terms). For computational purposes, it is advantageous to rewrite Eq. (5.9), in terms of a vector of six stresses $\sigma_I (I = 1, 2 \ldots, 6)$ and six strains $e_J (J = 1, 2 \ldots, 6)$ related by a 6×6 symmetric matrix C_{IJ} where

$$I = \begin{cases} i & \text{for } i = j \\ 9 - (i + j) & \text{for } i \neq j \end{cases}$$
$$J = \begin{cases} k & \text{for } k = l \\ 9 - (k + l) & \text{for } k \neq l \end{cases} \quad (5.11)$$

so that we have in this contracted notation

$$\sigma_I = \sum_{J=1}^{6} C_{IJ} e_J \quad (I = 1, 2 \ldots, 6) \quad (5.12)$$

or, in matrix–vector notation,

$$\{\sigma\} = [C]\{e\} \quad (5.13)$$

where

$$\{\sigma\} = [\sigma_{11}, \sigma_{22}, \sigma_{33}, \sigma_{23}, \sigma_{13}, \sigma_{12}]^T$$
$$\{e\} = [e_{11}, e_{22}, e_{33}, \gamma_{23}, \gamma_{13}, \gamma_{12}]^T \quad (5.14)$$

or, equivalently, in terms of the (x, y, z) Cartesian stress and strain components

$$\begin{Bmatrix} \sigma_{xx} \\ \sigma_{yy} \\ \sigma_{zz} \\ \sigma_{yz} \\ \sigma_{xz} \\ \sigma_{xy} \end{Bmatrix} = \begin{bmatrix} C_{11} & C_{12} & C_{13} & C_{14} & C_{15} & C_{16} \\ C_{21} & C_{22} & C_{23} & C_{24} & C_{25} & C_{26} \\ C_{31} & C_{32} & C_{33} & C_{34} & C_{35} & C_{36} \\ C_{41} & C_{42} & C_{43} & C_{44} & C_{45} & C_{46} \\ C_{51} & C_{52} & C_{53} & C_{54} & C_{55} & C_{56} \\ C_{61} & C_{62} & C_{63} & C_{64} & C_{65} & C_{66} \end{bmatrix} \begin{Bmatrix} e_{xx} \\ e_{yy} \\ e_{zz} \\ \gamma_{yz} \\ \gamma_{xz} \\ \gamma_{xy} \end{Bmatrix} \quad (5.15)$$

where the 6 × 6 *elastic constants matrix*, $[C]$, is symmetric, i.e., $C_{12} = C_{21}$, $C_{23} = C_{32}$, etc. Equation (5.15) is the general linear stress–strain relationship for an anisotropic material, written in what is called *Voigt notation*. Most anisotropic materials used in engineering practice, however, have certain directional symmetries that considerably reduce the number of independent constants. We will not derive in detail how such symmetries affect the constants matrix (for such details see theory of elasticity texts) but will simply list below some of the important special cases.

Orthotropic Material (Nine Constants)

Materials such as wood or some composite materials have an orthotropic behavior (see Fig. 5.2) where there are different constants along three orthogonal directions. Although these materials are inhomogeneous on a microscopic level, engineers usually treat them in terms of an equivalent homogeneous continuum model where these local properties are "smeared out" and averaged. Along the (x, y, z) material directions (see Fig. 5.2) the stress–strain relations reduce to

$$\begin{Bmatrix} \sigma_{xx} \\ \sigma_{yy} \\ \sigma_{zz} \\ \sigma_{yz} \\ \sigma_{xz} \\ \sigma_{xy} \end{Bmatrix} = \begin{bmatrix} C_{11} & C_{12} & C_{13} & 0 & 0 & 0 \\ C_{12} & C_{22} & C_{23} & 0 & 0 & 0 \\ C_{13} & C_{23} & C_{33} & 0 & 0 & 0 \\ 0 & 0 & 0 & C_{44} & 0 & 0 \\ 0 & 0 & 0 & 0 & C_{55} & 0 \\ 0 & 0 & 0 & 0 & 0 & C_{66} \end{bmatrix} \begin{Bmatrix} e_{xx} \\ e_{yy} \\ e_{zz} \\ \gamma_{yz} \\ \gamma_{xz} \\ \gamma_{xy} \end{Bmatrix} \quad (5.16)$$

which can also be inverted to write the strains in terms of the stresses in terms of a compliance matrix, $[D]$, where

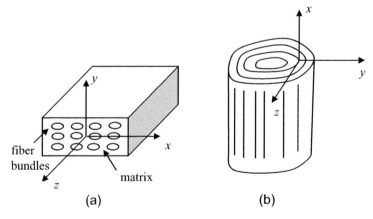

Figure 5.2 (a) A composite material consisting of elliptical-shaped bundles of fibers embedded in a matrix material. This material has different constants values along the (x, y, z) directions and can be treated as orthotropic. (b) Wood also has a fibrous structure, which produces different constants values along the wood fibers, in the radial ring growth direction, and tangential to the growth rings, leading to orthotropic behavior.

$$\begin{Bmatrix} e_{xx} \\ e_{yy} \\ e_{zz} \\ \gamma_{yz} \\ \gamma_{xz} \\ \gamma_{xy} \end{Bmatrix} = \begin{bmatrix} D_{11} & D_{12} & D_{13} & 0 & 0 & 0 \\ D_{12} & D_{22} & D_{23} & 0 & 0 & 0 \\ D_{13} & D_{23} & D_{33} & 0 & 0 & 0 \\ 0 & 0 & 0 & D_{44} & 0 & 0 \\ 0 & 0 & 0 & 0 & D_{55} & 0 \\ 0 & 0 & 0 & 0 & 0 & D_{66} \end{bmatrix} \begin{Bmatrix} \sigma_{xx} \\ \sigma_{yy} \\ \sigma_{zz} \\ \sigma_{yz} \\ \sigma_{xz} \\ \sigma_{xy} \end{Bmatrix} \quad (5.17)$$

Note that the choice of these constants or compliance values depends on our definition of the directions of the (x, y, z) axes. For the fiber-reinforced composite of Fig. 5.2a, for example, we have taken the z-axis to be along the fiber bundle direction, while for the wood case we have taken the wood fibers to be along the x-axis instead.

In engineering practice, the compliance is often written in a form similar to that of generalized Hooke's law for an isotropic material, namely

$$\begin{aligned} e_{xx} &= \frac{1}{E_x}\sigma_{xx} - \frac{\nu_{yx}}{E_y}\sigma_{yy} - \frac{\nu_{zx}}{E_z}\sigma_{zz} \\ e_{yy} &= \frac{1}{E_y}\sigma_{yy} - \frac{\nu_{xy}}{E_x}\sigma_{xx} - \frac{\nu_{zy}}{E_z}\sigma_{zz} \\ e_{zz} &= \frac{1}{E_z}\sigma_{zz} - \frac{\nu_{xz}}{E_x}\sigma_{xx} - \frac{\nu_{yz}}{E_y}\sigma_{yy} \\ \gamma_{yz} &= \frac{1}{G_{yz}}\sigma_{yz}, \quad \gamma_{xz} = \frac{1}{G_{xz}}\sigma_{xz}, \quad \gamma_{xy} = \frac{1}{G_{xy}}\sigma_{xy} \end{aligned} \quad (5.18)$$

where

$$\frac{\nu_{yz}}{E_y} = \frac{\nu_{zy}}{E_z}, \quad \frac{\nu_{xz}}{E_x} = \frac{\nu_{zx}}{E_z}, \quad \frac{\nu_{xy}}{E_x} = \frac{\nu_{yx}}{E_y} \quad (5.19)$$

so that effectively we have three different Young's moduli, three shear moduli, and three Poisson's ratios for this material. In terms of the full compliance matrix:

$$\begin{Bmatrix} e_{xx} \\ e_{yy} \\ e_{zz} \\ \gamma_{yz} \\ \gamma_{xz} \\ \gamma_{xy} \end{Bmatrix} = \begin{bmatrix} 1/E_x & -\nu_{yx}/E_y & -\nu_{zx}/E_z & 0 & 0 & 0 \\ -\nu_{xy}/E_x & 1/E_y & -\nu_{zy}/E_z & 0 & 0 & 0 \\ -\nu_{xz}/E_x & -\nu_{yz}/E_y & 1/E_z & 0 & 0 & 0 \\ 0 & 0 & 0 & 1/G_{yz} & 0 & 0 \\ 0 & 0 & 0 & 0 & 1/G_{xz} & 0 \\ 0 & 0 & 0 & 0 & 0 & 1/G_{xy} \end{bmatrix} \begin{Bmatrix} \sigma_{xx} \\ \sigma_{yy} \\ \sigma_{zz} \\ \sigma_{yz} \\ \sigma_{xz} \\ \sigma_{xy} \end{Bmatrix} \quad (5.20)$$

Note that unlike the general anisotropic materials, where all the stresses and strains are related to each other, in an orthotropic material normal strains only produce normal stresses and shear strains only produce shear stresses along the material directions (although, unlike the isotropic case, this is not true in general for other directions).

For plane stress $(\sigma_{zz} = \sigma_{zx} = \sigma_{zy} = 0)$ the stress–strain relations for an orthotropic solid reduce to

$$\sigma_{xx} = \frac{E_x}{1 - v_{xy}v_{yx}}(e_{xx} + v_{yx}e_{yy})$$

$$\sigma_{yy} = \frac{E_x}{1 - v_{xy}v_{yx}}(e_{yy} + v_{xy}e_{yy}) \quad (5.21)$$

$$\sigma_{xy} = G_{xy}\gamma_{xy}$$

Transversely Isotropic Material (Five Constants)

A transversely isotropic material gets its name from the fact that, in a plane orthogonal to a particular material axis, the material behaves as if it were isotropic. An example is the case where circular glass or graphite fibers are embedded in a polymer matrix (Fig. 5.3). In this case, assuming the z-axis is along the fiber direction, the material has isotropic behavior in the x–y "transverse" plane and the constants matrix is:

$$\begin{Bmatrix} \sigma_{xx} \\ \sigma_{yy} \\ \sigma_{zz} \\ \sigma_{yz} \\ \sigma_{xz} \\ \sigma_{xy} \end{Bmatrix} = \begin{bmatrix} C_{11} & C_{12} & C_{13} & 0 & 0 & 0 \\ C_{12} & C_{11} & C_{13} & 0 & 0 & 0 \\ C_{13} & C_{13} & C_{33} & 0 & 0 & 0 \\ 0 & 0 & 0 & C_{44} & 0 & 0 \\ 0 & 0 & 0 & 0 & C_{55} & 0 \\ 0 & 0 & 0 & 0 & 0 & (C_{11}-C_{12})/2 \end{bmatrix} \begin{Bmatrix} e_{xx} \\ e_{yy} \\ e_{zz} \\ \gamma_{yz} \\ \gamma_{xz} \\ \gamma_{xy} \end{Bmatrix} \quad (5.22)$$

and the corresponding compliance matrix can be written as

$$\begin{Bmatrix} e_{xx} \\ e_{yy} \\ e_{zz} \\ \gamma_{yz} \\ \gamma_{xz} \\ \gamma_{xy} \end{Bmatrix} = \begin{bmatrix} 1/E_x & -v_{xy}/E_x & -v_{zx}/E_z & 0 & 0 & 0 \\ -v_{xy}/E_x & 1/E_x & -v_{zx}/E_z & 0 & 0 & 0 \\ -v_{xz}/E_x & -v_{xz}/E_x & 1/E_z & 0 & 0 & 0 \\ 0 & 0 & 0 & 1/G_{xz} & 0 & 0 \\ 0 & 0 & 0 & 0 & 1/G_{xz} & 0 \\ 0 & 0 & 0 & 0 & 0 & 1/G_{xy} \end{bmatrix} \begin{Bmatrix} \sigma_{xx} \\ \sigma_{yy} \\ \sigma_{zz} \\ \sigma_{yz} \\ \sigma_{xz} \\ \sigma_{xy} \end{Bmatrix} \quad (5.23)$$

where $v_{xz}/E_x = v_{zx}/E_z$ and $G_{xy} = E_x/2(1 + v_{xy})$. To make it easier to connect these compliance values to the fiber and transverse plane directions, an f subscript is often used to denote "fiber" and a t subscript used to denote "transverse", giving

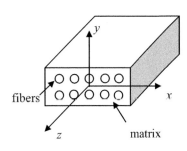

Figure 5.3 A material where fibers of circular cross-section are embedded in a material matrix. This material acts macroscopically like a transversely isotropic material.

$$\begin{Bmatrix} e_{xx} \\ e_{yy} \\ e_{zz} \\ \gamma_{yz} \\ \gamma_{xz} \\ \gamma_{xy} \end{Bmatrix} = \begin{bmatrix} 1/E_t & -v_t/E_t & -v_f/E_f & 0 & 0 & 0 \\ -v_t/E_t & 1/E_t & -v_f/E_f & 0 & 0 & 0 \\ -v_f/E_f & -v_f/E_f & 1/E_f & 0 & 0 & 0 \\ 0 & 0 & 0 & 1/G_f & 0 & 0 \\ 0 & 0 & 0 & 0 & 1/G_f & 0 \\ 0 & 0 & 0 & 0 & 0 & 1/G_t \end{bmatrix} \begin{Bmatrix} \sigma_{xx} \\ \sigma_{yy} \\ \sigma_{zz} \\ \sigma_{yz} \\ \sigma_{xz} \\ \sigma_{xy} \end{Bmatrix} \quad (5.24)$$

Cubic Material (Three Constants)

$$\begin{Bmatrix} \sigma_{xx} \\ \sigma_{yy} \\ \sigma_{zz} \\ \sigma_{yz} \\ \sigma_{xz} \\ \sigma_{xy} \end{Bmatrix} = \begin{bmatrix} C_{11} & C_{12} & C_{12} & 0 & 0 & 0 \\ C_{12} & C_{11} & C_{12} & 0 & 0 & 0 \\ C_{12} & C_{12} & C_{11} & 0 & 0 & 0 \\ 0 & 0 & 0 & C_{44} & 0 & 0 \\ 0 & 0 & 0 & 0 & C_{44} & 0 \\ 0 & 0 & 0 & 0 & 0 & C_{44} \end{bmatrix} \begin{Bmatrix} e_{xx} \\ e_{yy} \\ e_{zz} \\ \gamma_{yz} \\ \gamma_{xz} \\ \gamma_{xy} \end{Bmatrix} \quad (5.25)$$

Materials of pure aluminum crystals (as opposed to aluminum alloys) have this cubic material behavior.

Isotropic Material (Two Constants)

We have already given the stress–strain relations in the isotropic case. In terms of the elastic constants matrix we have

$$\begin{Bmatrix} \sigma_{xx} \\ \sigma_{yy} \\ \sigma_{zz} \\ \sigma_{yz} \\ \sigma_{xz} \\ \sigma_{xy} \end{Bmatrix} = \begin{bmatrix} C_{11} & C_{12} & C_{12} & 0 & 0 & 0 \\ C_{12} & C_{11} & C_{12} & 0 & 0 & 0 \\ C_{12} & C_{12} & C_{11} & 0 & 0 & 0 \\ 0 & 0 & 0 & (C_{11}-C_{12})/2 & 0 & 0 \\ 0 & 0 & 0 & 0 & (C_{11}-C_{12})/2 & 0 \\ 0 & 0 & 0 & 0 & 0 & (C_{11}-C_{12})/2 \end{bmatrix} \begin{Bmatrix} e_{xx} \\ e_{yy} \\ e_{zz} \\ \gamma_{yz} \\ \gamma_{xz} \\ \gamma_{xy} \end{Bmatrix}$$

$$(5.26)$$

where

$$C_{11} = \frac{E(1-v)}{(1+v)(1-2v)}, \quad C_{12} = \frac{Ev}{(1+v)(1-2v)}$$
$$\frac{(C_{11}-C_{12})}{2} = \frac{E}{2(1+v)} = G \quad (5.27)$$

These stress–strain relations are identical to Eq. (5.5), namely

$$\sigma_{xx} = \frac{E}{(1+v)(1-2v)}\left[(1-v)e_{xx} + v(e_{yy}+e_{zz})\right]$$

$$\sigma_{yy} = \frac{E}{(1+v)(1-2v)}\left[(1-v)e_{yy} + v(e_{xx}+e_{zz})\right] \quad (5.28)$$

$$\sigma_{zz} = \frac{E}{(1+v)(1-2v)}\left[(1-v)e_{zz} + v(e_{xx}+e_{yy})\right]$$

$$\sigma_{xy} = G\gamma_{xy}, \quad \sigma_{yz} = G\gamma_{yz}, \quad \sigma_{zx} = G\gamma_{zx}$$

By rearranging these terms, we can write the stress–strain relationship in matrix terms. We have, for example, equivalently

$$\sigma_{xx} = \frac{E}{(1+v)(1-2v)}\left[(1-2v)e_{xx} + v(e_{xx}+e_{yy}+e_{zz})\right]$$

$$\sigma_{yy} = \frac{E}{(1+v)(1-2v)}\left[(1-2v)e_{yy} + v(e_{xx}+e_{yy}+e_{zz})\right] \quad (5.29)$$

$$\sigma_{zz} = \frac{E}{(1+v)(1-2v)}\left[(1-2v)e_{zz} + v(e_{xx}+e_{yy}+e_{zz})\right]$$

$$\sigma_{xy} = \frac{E}{(1+v)}e_{xy}, \quad \sigma_{yz} = \frac{E}{(1+v)}\gamma_{yz}, \quad \sigma_{zx} = \frac{E}{(1+v)}e_{zx}$$

which now can be written, finally in the compact form

$$\sigma_{ij} = \frac{E}{(1+v)}\left(e_{ij} + \frac{v}{(1-2v)}\left(\sum_{i=1}^{3}e_{kk}\right)\delta_{ij}\right)(i,j=1,2,3) \quad (5.30)$$

where δ_{ij} is the Kronecker delta (see Eq. (2.34)).

Now, consider the compliance matrix where we have

$$\begin{Bmatrix} e_{xx} \\ e_{yy} \\ e_{zz} \\ \gamma_{yz} \\ \gamma_{xz} \\ \gamma_{xy} \end{Bmatrix} = \begin{bmatrix} 1/E & -v/E & -v/E & 0 & 0 & 0 \\ -v/E & 1/E & -v/E & 0 & 0 & 0 \\ -v/E & -v/E & 1/E & 0 & 0 & 0 \\ 0 & 0 & 0 & 1/G & 0 & 0 \\ 0 & 0 & 0 & 0 & 1/G & 0 \\ 0 & 0 & 0 & 0 & 0 & 1/G \end{bmatrix} \begin{Bmatrix} \sigma_{xx} \\ \sigma_{yy} \\ \sigma_{zz} \\ \sigma_{yz} \\ \sigma_{xz} \\ \sigma_{xy} \end{Bmatrix} \quad (5.31)$$

and $G = E/2(1+v)$. This gives Eq. (5.4):

$$e_{xx} = \frac{\sigma_{xx}}{E} - \frac{v}{E}(\sigma_{yy}+\sigma_{zz})$$

$$e_{yy} = \frac{\sigma_{yy}}{E} - \frac{v}{E}(\sigma_{xx}+\sigma_{zz})$$

$$e_{zz} = \frac{\sigma_{zz}}{E} - \frac{v}{E}(\sigma_{xx}+\sigma_{yy}) \quad (5.32)$$

$$\gamma_{xy} = \frac{\sigma_{xy}}{G}, \quad \gamma_{yz} = \frac{\sigma_{yz}}{G}, \quad \gamma_{zx} = \frac{\sigma_{zx}}{G}$$

Again, we can rearrange these equations to write them in matrix terms. Equivalently, we have

$$e_{xx} = \frac{(1+v)\sigma_{xx}}{E} - \frac{v}{E}\left(\sigma_{xx} + \sigma_{yy} + \sigma_{zz}\right)$$

$$e_{yy} = \frac{(1+v)\sigma_{yy}}{E} - \frac{v}{E}\left(\sigma_{xx} + \sigma_{yy} + \sigma_{zz}\right)$$

$$e_{zz} = \frac{(1+v)\sigma_{zz}}{E} - \frac{v}{E}\left(\sigma_{xx} + \sigma_{yy} + \sigma_{zz}\right) \tag{5.33}$$

$$e_{xy} = \frac{(1+v)\sigma_{xy}}{E}, \quad e_{yz} = \frac{(1+v)\sigma_{yz}}{E}, \quad e_{zx} = \frac{(1+v)\sigma_{zx}}{E}$$

which can be written, finally, as

$$e_{ij} = \frac{1}{E}\left((1+v)\sigma_{ij} - v\left(\sum_{i=1}^{3}\sigma_{ij}\right)\delta_{ij}\right)(i,j=1,2,3) \tag{5.34}$$

One important difference between the isotropic case and the others listed above is that while the matrix of constants for the isotropic case is the same for *any* orientation of the (x, y, z) axes associated with the stresses and strains, this is not true in general for other materials and the expressions given previously *are valid only for a particular set of materials axes* associated with the symmetries present in the materials. We will give the transformation equations shortly that will allow us to obtain the matrix of constants as measured in any coordinate system. These transformations, like the stress and strain transformations, involve the direction cosines relating a set of (x', y', z') axes to the (x, y, z) axes in which the above stress–strain relations are written. Another important difference between isotropic and anisotropic media is that the principal stress and principal strain directions *do not coincide* in general for anisotropic materials so that we need to calculate those directions (and the corresponding principal stress and principal strain values) separately for the stress and strain.

5.2 Plane Stress and Plane Strain

Two important special cases for isotropic materials we have mentioned previously are the cases of plane stress and plane strain. In the plane-stress case, it is assumed that there are no normal and shear stresses in, say, the z-direction, while the stresses acting in the x- and y-directions are functions only of (x, y), i.e.,

$$\sigma_{zx} = \sigma_{zy} = \sigma_{zz} = 0$$
$$\sigma_{xx} = \sigma_{xx}(x, y)$$
$$\sigma_{yy} = \sigma_{yy}(x, y) \tag{5.35}$$
$$\sigma_{xy} = \sigma_{xy}(x, y)$$

Figure 5.4 (a) A plane-stress problem where a thin plate lying in the x–y plane is pressurized at an interior hole. (b) A plane-strain problem where a long cylinder is pressurized at an interior hole and is constrained by rigid end containing walls (not shown) that prevent deformation in the z-direction.

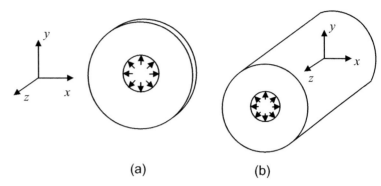

The case where a thin plate is acted upon by pressure at an interior hole (see Fig. 5.4a) is an example where we might expect to see a state of plane stress. In this case, generalized Hooke's law for an isotropic material reduces to equations only for the in-plane stresses $(\sigma_{xx}, \sigma_{yy}, \sigma_{zz})$ given by

$$\sigma_{xx} = \frac{E}{1-v^2}\left[e_{xx} + v e_{yy}\right]$$
$$\sigma_{yy} = \frac{E}{1-v^2}\left[e_{yy} + v e_{xx}\right] \qquad (5.36)$$
$$\sigma_{xy} = G\gamma_{xy}$$

but note that, in addition to the in-plane strains $(e_{xx}, e_{xy}, \gamma_{xy})$, there is also a normal strain, e_{zz}, since

$$e_{zz} = -\frac{v}{E}\left[\sigma_{xx} + \sigma_{yy}\right]$$
$$= -\frac{v}{(1-v)}\left[e_{xx} + e_{yy}\right] \qquad (5.37)$$

Physically this makes sense since in-plane loads such as the pressure in Fig. 5.4b can cause changes in the plate thickness, resulting in strains in the thickness direction.

The case of plane strain, in contrast, occurs when there is no displacement in, say, the z-direction, and the displacements in the x- and y-directions are only functions of (x, y), i.e.,

$$u_z = 0$$
$$u_x = u_x(x, y) \qquad (5.38)$$
$$u_y = u_y(x, y)$$

or, equivalently, for the strains

$$e_{zz} = \gamma_{zx} = \gamma_{zy} = 0$$
$$e_{xx} = e_{xx}(x, y)$$
$$e_{yy} = e_{yy}(x, y) \qquad (5.39)$$
$$\gamma_{xy} = \gamma_{xy}(x, y)$$

which gives the stress–strain relations for an isotropic material as

$$\sigma_{xx} = \frac{E}{(1+v)(1-2v)}\left[(1-v)e_{xx} + ve_{yy}\right]$$

$$\sigma_{yy} = \frac{E}{(1+v)(1-2v)}\left[(1-v)e_{yy} + ve_{xx}\right] \quad (5.40)$$

$$\sigma_{xy} = G\gamma_{xy}$$

But in this case, there is also an out-of-plane normal stress, σ_{zz}, given by

$$\sigma_{zz} = \frac{Ev}{(1+v)(1-2v)}\left[e_{xx} + e_{yy}\right] \quad (5.41)$$

This normal stress is present in order to enforce the condition that there can be no displacement in the z-direction. An example of a plane-strain condition is where a hollow cylinder is pressurized at its inner radius and is restrained from moving in the z-direction at its ends by rigid supports (see Fig. 5.4b). One should note, however, that the rigid supports are necessary to enforce the plane-strain conditions since, as we will show in Chapter 7 (where we consider an exact solution for a pressurized thick cylinder), if the ends of the cylinder are unloaded and free to expand the cylinder is in a state of plane stress even though it is not thin in the z-direction.

Comparing the plane-stress and plane-strain cases, we see that the stress–strain relations are very similar. In the next chapter, when we use these stress–strain relations as part of complete sets of governing equations for solving problems, we will see that through a simple transformation of the elastic constants we can turn a plane-stress problem solution for a case such as the one shown in Fig. 5.4a into a the plane-strain problem solution for a case such as shown in Fig. 5.4b, and vice versa.

5.3 Transformation of Elastic Constants

In two different sets of coordinates $\{x_i\}$ and $\{x'_i\}$ the stress–strain relations in terms of the elastic constants are (see Eq. (5.6))

$$\sigma_{ij} = \sum_{k=1}^{3}\sum_{l=1}^{3} C_{ijkl} e_{kl}$$

$$\sigma'_{ij} = \sum_{p=1}^{3}\sum_{q=1}^{3} C'_{ijpq} e'_{pq} \quad (5.42)$$

But we have the stress and strain in these coordinates also related through

$$\sigma'_{mn} = \sum_{i=1}^{3}\sum_{j=1}^{3} l_{im} l_{jn} \sigma_{ij}$$

$$= \sum_{i=1}^{3}\sum_{j=1}^{3}\sum_{k=1}^{3}\sum_{l=1}^{3} l_{im} l_{jn} C_{ijkl} e_{kl} \quad (5.43)$$

and

$$\sigma'_{mn} = \sum_{p=1}^{3}\sum_{q=1}^{3}\sum_{k=1}^{3}\sum_{l=1}^{3} C'_{mnpq} l_{kp} l_{lq} e_{kl} \qquad (5.44)$$

Equating Eq. (5.43) and Eq. (5.44) and using the fact that this equality must be true for all of the common strain components, e_{kl}, in this expression we find

$$\sum_{p=1}^{3}\sum_{q=1}^{3} l_{kp} l_{lq} C'_{mnpq} = \sum_{i=1}^{3}\sum_{j=1}^{3} l_{im} l_{jn} C_{ijkl} \qquad (5.45)$$

If we multiply Eq. (5.45) by $l_{kr} l_{ls}$ and sum over (k, l) we find

$$C'_{mnrs} = \sum_{k=1}^{3}\sum_{l=1}^{3}\sum_{i=1}^{3}\sum_{j=1}^{3} l_{kr} l_{ls} l_{im} l_{jn} C_{ijkl} \qquad (5.46)$$

Equation (5.46) is the transformation relationship for the elastic constants when we change our coordinate system from the $\{x_i\}$ to the $\{x'_i\}$ coordinates. However, in practice it is usually more convenient to use the Voigt notation where the stress–strain relations are written in terms of a reduced 6×6 constants matrix, C_{IJ}, ($I, J = 1, 2, \ldots, 6$). Thus, we need to find this transformation for the reduced constants matrix. To do this let us define

$$M_{ijkl} = l_{ki} l_{lj} \qquad (5.47)$$

Then we can write Eq. (5.46) as

$$C'_{mnrs} = \sum_{k=1}^{3}\sum_{l=1}^{3}\sum_{i=1}^{3}\sum_{j=1}^{3} M_{rskl} M_{mnij} C_{ijkl} \qquad (5.48)$$

To obtain the reduced form of this equation we can follow the same procedure used in obtaining the reduced constants matrix. First, we consider the terms in Eq. (5.48) involving M_{mnij} only. Omitting the subscripts and sums that are not involved with this matrix, we have

$$C'_{mn..} = \sum_{i=1}^{3}\sum_{j=1}^{3} M_{mnij} C_{ij..} M_{....} \qquad (5.49)$$

which is now a relationship between two symmetrical matrices C'_{mn} and C_{ij} analogous to the relationship between the stress and strain matrices. Thus, like the stress–strain relations, we can express Eq. (5.49) in terms of a 6×6 matrix M_{MI}, where

$$C'_{M..} = \sum_{I=1}^{6} M_{MI} C_{I..} M_{....} \quad (M = 1, 2 \ldots, 6) \qquad (5.50)$$

In the same fashion we can write Eq. (5.48) as

$$C'_{..rs} = \sum_{k=1}^{3}\sum_{l=1}^{3} M_{rskl} C_{..kl} M_{....} \qquad (5.51)$$

which is again a relationship between two symmetric matrices that can be written in terms of the reduced matrix M_{RK}, where

$$C'_{..R} = \sum_{K=1}^{6} M_{RK} C_{..K} M_{....} \quad (R = 1, 2 \ldots, 6) \tag{5.52}$$

Combining Eq. (5.50) and Eq. (5.52) we obtain

$$C'_{MR} = \sum_{I=1}^{6} \sum_{K=1}^{6} M_{MI} C_{IK} M_{RK} \quad (M, R = 1, 2 \ldots, 6) \tag{5.53}$$

which is the transformation relationship for the reduced constants matrix. The form of the 6×6 matrix $[M]$ in terms of the M_{ijkl} components is the same as found earlier for the form of the elastic constants matrix (see Eq. (5.8a)), i.e.,

$$[M] = \begin{bmatrix} M_{1111} & M_{1122} & M_{1133} & M_{1123} + M_{1132} & M_{1113} + M_{1131} & M_{1112} + M_{1121} \\ M_{2211} & M_{2222} & M_{2233} & M_{2223} + M_{2232} & M_{2213} + M_{2231} & M_{2212} + M_{2221} \\ M_{3311} & M_{3322} & M_{3333} & M_{3323} + M_{3332} & M_{3313} + M_{3331} & M_{3312} + M_{3321} \\ M_{2311} & M_{2322} & M_{2333} & M_{2323} + M_{2332} & M_{2313} + M_{2331} & M_{2312} + M_{3321} \\ M_{1311} & M_{1322} & M_{1333} & M_{1323} + M_{1332} & M_{1313} + M_{1331} & M_{1312} + M_{1321} \\ M_{1211} & M_{1222} & M_{1233} & M_{1223} + M_{1232} & M_{1213} + M_{1231} & M_{1212} + M_{1221} \end{bmatrix} \tag{5.54}$$

which can be written in terms of the direction cosines as

$$[M] = \begin{bmatrix} l_{11}^2 & l_{21}^2 & l_{31}^2 & 2l_{21}l_{31} & 2l_{31}l_{11} & 2l_{21}l_{11} \\ l_{12}^2 & l_{22}^2 & l_{32}^2 & 2l_{22}l_{32} & 2l_{12}l_{31} & 2l_{12}l_{22} \\ l_{13}^2 & l_{23}^2 & l_{33}^2 & 2l_{23}l_{33} & 2l_{13}l_{33} & 2l_{13}l_{23} \\ l_{12}l_{13} & l_{22}l_{23} & l_{32}l_{33} & l_{22}l_{33} + l_{32}l_{23} & l_{12}l_{33} + l_{32}l_{13} & l_{12}l_{23} + l_{22}l_{13} \\ l_{11}l_{13} & l_{21}l_{23} & l_{31}l_{33} & l_{21}l_{33} + l_{31}l_{23} & l_{11}l_{33} + l_{31}l_{13} & l_{11}l_{23} + l_{21}l_{13} \\ l_{11}l_{12} & l_{21}l_{22} & l_{31}l_{32} & l_{21}l_{32} + l_{31}l_{22} & l_{11}l_{32} + l_{31}l_{12} & l_{11}l_{22} + l_{21}l_{12} \end{bmatrix} \tag{5.55}$$

Note that in terms of matrix multiplication we can write the transformation of Eq. (5.53) more compactly as

$$[C'] = [M][C][M]^T \tag{5.56}$$

Since the 6×6 matrix $[M]$ is rather complex, we have written a MATLAB® function M_Matrix that has the calling sequence

```
M = M_Matrix(l);
```

where l is the matrix of direction cosines and the function returns the 6×6 matrix $[M]$. Note that the direction cosine matrix $[l]$ is associated with the coordinate transformation from a set of $\{x\}$ coordinates to a set of $\{x'\}$ coordinates given by

$$\{x'\} = [l]^T \{x\} \tag{5.57}$$

where (see Eq. (2.35))

$$l_{ij} = \left(\mathbf{e}_i \cdot \mathbf{e}'_j\right) = \cos\left(x_i, x'_j\right) \tag{5.58}$$

and some texts define the direction cosines matrix instead as the transpose of $[l]$. Thus, one must be careful to use the direction cosines defined by Eq. (5.58) as the input argument in the M_Matrix function. In the use of the reduced set of 6×6 elastic constants we often need to transform from 3×3 stress or strain matrices to 6×1 stress or strain vectors, and vice versa. Thus, we have also written a pair of MATLAB® functions m_to_v and v_to_m that are called as:

```
vector = m_to_v(matrix, 'type');
matrix = v_to_m(vector, 'type');
```

where matrix can be either a 3×3 matrix of stresses or tensor strains and vector can be either a 6×1 stress or strain vector. The 'type' input argument is a string having the value of 'stress' or 'strain' to indicate the type of matrix and vector quantities we are considering. The m_to_v function takes a 3×3 stress or strain matrix and outputs the corresponding stress or strain vector. If 'type' is equal to 'strain' then the function transforms the tensor strains in the input matrix to the engineering shear strains in the output strain vector. Similarly, the function v_to_m takes a 6×1 stress or strain vector and transforms it to the corresponding 3×3 stress or strain matrix. If 'type' is equal to 'strain', then the engineering shear strains present in the input strain vector are transformed to the tensor strain components in the output strain matrix.

Example 5.1 Stress, Strain, and Stress–Strain Transformations

This example will consider transformations of stress, strain, and stress–strain relations using MATLAB® to perform all the calculations. Specifically, we will consider an orthotropic material where the only nonzero elastic constants (in GPa) in a set of material coordinates are

$$C_{11} = 103; C_{22} = 50; C_{33} = 75;$$
$$C_{12} = C_{21} = 55; C_{13} = C_{31} = 25;, C_{23} = C_{32} = 40;$$
$$C_{44} = 45; C_{55} = 10; C_{66} = 27.6$$

Normally the strains are given in terms of μstrain (multiples of 10^{-6}) and the stresses given in MPa. In that case we must multiply the elastic constants in GPa by a factor of 10^{-3} so that these constants are in units of MPa/μstrain. Thus, in MATLAB® we have

```
C = zeros(6,6);              % Set up empty 6x6 matrix
C(1,1) = 103; C(2,2) = 50; C(3,3) = 75;    % define constants
C(1,2) = 55; C(1,3) = 25; C(2,3) = 40;
C(4,4) = 45; C(5,5) = 10; C(6,6) = 27.6;
```

```
C(2,1) = C(1,2); C(3,1) = C(1,3); C(3,2) = C(2,3);
C = C*10^-3               % put C in units of
% MPa/microstrain

C =

    0.1030    0.0550    0.0250         0         0         0
    0.0550    0.0500    0.0400         0         0         0
    0.0250    0.0400    0.0750         0         0         0
         0         0         0    0.0450         0         0
         0         0         0         0    0.0100         0
         0         0         0         0         0    0.0276
```

Now, let us consider the following state of strain (in μstrain) as given in the material coordinates:

```
strain = [ 300   50    20;  50   200   30;  20   30   100]  % state of
% strain matrix
strain =   300    50    20
            50   200    30
            20    30   100
```

We can change this strain matrix into a strain vector, e:

```
e = m_to_v(strain, 'strain')   % change strain matrix to strain
% vector

e =   300
      200
      100
       60
       40
      100
```

where we see that the tensor shear strains have been transformed to engineering shear strains. We can multiply these strains by the reduced elastic constants matrix to obtain the stress in vector form and change it into a stress matrix:

```
s = C*e  % calculate the stress vector
s =  44.4000
     30.5000
     23.0000
      2.7000
      0.4000
      2.7600
```

```
stress = v_to_m(s, 'stress')    % change stress vector to stress
% matrix
stress =   44.4000    2.7600    0.4000
            2.7600   30.5000    2.7000
            0.4000    2.7000   23.0000
```

Having the state of stress in the material coordinates, we can calculate the principal stresses and principal stress directions with the MATLAB® eig function

```
[pdirs, pstress] = eig(stress)   % calculate principal stress
% directions and principal stresses

pdirs =  0.0217   -0.1980   -0.9800
        -0.3130    0.9296   -0.1948
         0.9495    0.3110   -0.0418

pstress =  22.1190         0         0
                0   30.8154         0
                0         0   44.9656
```

If we check the determinant of pdirs, we see it is plus one:

```
det(pdirs)
ans = 1.0000
```

so the principal directions matrix, pdirs, is just the matrix of direction cosines, $[l]$, that relate the principal coordinates to the material coordinates (otherwise we need to change the sign on one of the columns of pdirs, as discussed Chapter 2). Thus, we can calculate the strains in the principal stress coordinates, strain_new, by using the transformation of the state of strain from material to principal stress coordinates (see Eq. (4.33)):

```
l = pdirs;                       % rename direction cosines matrix
strain_new = l'*strain*l         % calculate strains in principal
% stress coordinates

strain_new =   92.2052    -5.9178    -6.8183
               -5.9178   190.7245   -31.8258
               -6.8183   -31.8258   317.0703
```

However, these are not principal strains (obviously, since the shear strains are not zero) because the principal strain directions are not the same as the principal stress directions for this orthotropic material. We can calculate the principal stresses from these strains in the principal stress coordinates directly from the stress–strain relations if we know the reduced constants matrix in the principal stress coordinates. But we know how to transform this constants matrix. We need the M-transformation matrix, which we can get from the M_Matrix function

```
M = M_Matrix(l)    % calculate M matrix needed for
% transforming the constants matrix
```

5.3 Transformation of Elastic Constants   137

```
% to the principal stress coordinates
M =  0.0005    0.0980    0.9015   -0.5945    0.0413   -0.0136
     0.0392    0.8641    0.0967    0.5782   -0.1232   -0.3681
     0.9603    0.0379    0.0017    0.0163    0.0819    0.3817
     0.1940   -0.1811   -0.0130   -0.0994   -0.2965   -0.8724
    -0.0213    0.0610   -0.0397   -0.1719   -0.9314    0.3025
    -0.0043   -0.2910    0.2953    0.7852   -0.1813    0.0822
```

and then the new constants matrix is just

```
C_new = M*C*M'     % find the constants matrix in principal
% stress coordinates

C_new =  0.0845    0.0281    0.0280   -0.0001    0.0032   -0.0120
         0.0281    0.0677    0.0502    0.0076   -0.0062    0.0179
         0.0280    0.0502    0.1033   -0.0002    0.0025   -0.0075
        -0.0001    0.0076   -0.0002    0.0241   -0.0038   -0.0063
         0.0032   -0.0062    0.0025   -0.0038    0.0126   -0.0041
        -0.0120    0.0179   -0.0075   -0.0063   -0.0041    0.0322
```

which we see looks nothing like our original constants matrix. Changing the strain matrix strain_new in these principal stress coordinates to a corresponding strain vector, e_new, we have

```
e_new = m_to_v(strain_new, 'strain')    % change strain matrix to
% strain vector

e_new =  92.2052
        190.7245
        317.0703
        -63.6516
        -13.6365
        -11.8356
```

which we then can use to calculate the stress in vector form in these principal stress coordinates as

```
s_new = C_new*e_new    % use stress-strain relations in
% principal stress coordinates
% to calculate stress in vector form
S_new =  22.1190
         30.8154
         44.9656
         -0.0000
         -0.0000
          0.0000
```

and changing this vector to a stress matrix we have, finally

```
stress_new = v_to_m(S_new, 'stress')  % obtain stress matrix
% in principal stress coordinates
stress_new =   22.1190    0.0000   -0.0000
                0.0000   30.8154   -0.0000
               -0.0000   -0.0000   44.9656
```

which is identical with the principal stress matrix, pstress, calculated previously from eig. This example should give you a good feel for how we can use the stress and strain transformations and the transformation of the elastic constants matrix to solve problems. Now, let us go back to the question of the principal strains and the principal strain directions. We had the state of strain matrix in the original material coordinates given by the matrix strain so let's use eig to calculate the principal strains and principal strain directions:

```
[strain_dirs, pstrain] = eig(strain)    % calculate principal
% strain directions and principal strains
strain_dirs =   -0.0322   -0.4169   -0.9084
                -0.2525    0.8827   -0.3962
                 0.9670    0.2166   -0.1337
pstrain = 91.4995        0           0
                0   183.7468         0
                0         0    324.7537
```

and we see the matrix of principal strain directions, strain_dirs, is very different from the principal stress directions matrix, pdirs, calculated previously.

Before leaving this example let us consider the case of an isotropic material where, in material coordinates (in GPa), we have

$$C_{11} = C_{22} = C_{33} = 50;$$
$$C_{12} = C_{21} = C_{13} = C_{31} = C_{23} = C_{32} = 20;$$
$$C_{44} = C_{55} = C_{66} = (C_{11} - C_{12})/2 = 15$$

Thus, the elastic constants matrix, C2, in MPa/µstrain can be calculated in MATLAB® as

```
C2 = zeros(6,6);       % form up constants matrix for
% isotropic material
C2(1,1) = 50; C2(2,2) = C2(1,1); C2(3,3) = C2(1,1);
C2(1,2) = 20; C2(2,1) = C2(1,2); C2(1,3) = C2(1,2); C2(3,1) = C2(1,2);
C2(2,3) = C2(1,2); C2(3,2) = C2(1,2);
C2(4,4) = 15; C2(5,5) = C2(4,4); C2(6,6) = C2(4,4);
C2 = C2*10^-3
```

```
C2 =  0.0500    0.0200    0.0200    0         0         0
      0.0200    0.0500    0.0200    0         0         0
      0.0200    0.0200    0.0500    0         0         0
      0         0         0         0.0150    0         0
      0         0         0         0         0.0150    0
      0         0         0         0         0         0.0150
```

If we use the M matrix for the principal stress directions calculated previously to examine the elastic constants matrix in those coordinates, we find

```
C2_new = M*C2*M'       % constants matrix in principal stress
% coordinates

C2_new =  0.0500    0.0200    0.0200    0.0000   -0.0000    0.0000
          0.0200    0.0500    0.0200    0.0000   -0.0000    0.0000
          0.0200    0.0200    0.0500    0.0000   -0.0000    0.0000
          0.0000    0.0000    0.0000    0.0150    0.0000   -0.0000
         -0.0000   -0.0000   -0.0000    0.0000    0.0150    0.0000
          0.0000    0.0000    0.0000   -0.0000    0.0000    0.0150
```

which has not changed from our original constants matrix, showing that the stress–strain relations are indeed the same in both of these coordinate systems (and, in fact, in any coordinate system).

5.4 States of Stress and Strain on a Surface from Strain Gage Measurements

It is normally not possible to measure the states of stress and strain inside a deformable body. However, if one can perform a series of normal strain measurements on the free surface of a body one can deduce the states of stress and strain on that surface using the strain transformations and the stress–strain relations. In this section we will demonstrate this process for an isotropic material. Consider, for example, a free surface, S, of an isotropic elastic body, as shown in Fig. 5.5. The x- and y-coordinates are tangent to the surface at P and the z-axis is normal to the surface. The problem is to find the complete 3-D states of stress and strain at point P given by

$$[e] = \begin{bmatrix} e_{xx} & e_{xy} & e_{xz} \\ e_{yx} & e_{yy} & e_{yz} \\ e_{zx} & e_{zy} & e_{zz} \end{bmatrix}, \quad [\sigma] = \begin{bmatrix} \sigma_{xx} & \sigma_{xy} & \sigma_{xz} \\ \sigma_{yx} & \sigma_{yy} & \sigma_{yz} \\ \sigma_{zx} & \sigma_{zy} & \sigma_{zz} \end{bmatrix} \quad (5.59)$$

Since the surface is assumed to be free, as shown in Fig. 5.5, we have $\sigma_{zz} = \sigma_{zx} = \sigma_{zy} = 0$. It follows from the stress–strain law for an isotropic material that

Figure 5.5 A free surface of an elastic body.

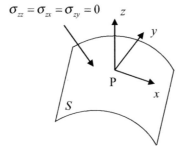

$\sigma_{zz} = \sigma_{zx} = \sigma_{zy} = 0$

Figure 5.6 (a) A resistance strain gage. (b) A strain gage rosette.

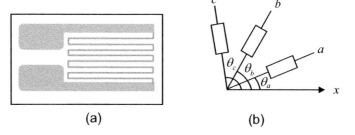

(a)   (b)

$$\sigma_{xx} = \frac{E}{(1+v)(1-2v)}\left[(1-v)e_{xx} + v(e_{yy} + e_{zz})\right]$$

$$\sigma_{yy} = \frac{E}{(1+v)(1-2v)}\left[(1-v)e_{yy} + v(e_{xx} + e_{zz})\right]$$

$$\sigma_{zz} = \frac{E}{(1+v)(1-2v)}\left[(1-v)e_{zz} + v(e_{xx} + e_{yy})\right] = 0 \qquad (5.60)$$

$$\sigma_{xy} = 2Ge_{xy} = 0$$

$$\sigma_{xz} = 2Ge_{xz} = 0$$

$$\sigma_{yz} = 2Ge_{yz} = 0$$

and from the condition $\sigma_{zz} = 0$ one finds

$$e_{zz} = -\frac{v}{(1-v)}(e_{xx} + e_{yy}) \qquad (5.61)$$

so, using the symmetry of the shear strains and shear stresses, the states of strain and stress at P are

$$[e] = \begin{bmatrix} e_{xx} & e_{xy} & 0 \\ e_{xy} & e_{yy} & 0 \\ 0 & 0 & e_{zz} \end{bmatrix}, \quad [\sigma] = \begin{bmatrix} \sigma_{xx} & \sigma_{xy} & 0 \\ \sigma_{xy} & \sigma_{xy} & 0 \\ 0 & 0 & 0 \end{bmatrix} \qquad (5.62)$$

Normal strains on the free surface of a body can be measured with resistance strain gages such as the one shown in Fig. 5.6a. A gage consists of a grid of wires or thin foil that are placed on a thin paper or plastic backing, which can then be cemented to the surface of a body. The change of length of the wires or foil causes the electrical resistance of the gage to change and this resistance change can be related to the normal strain along the gage. A set of three gages placed at a location on the surface is called a *strain gage rosette*. Figure 5.6b shows a general rosette configuration where the three gages are oriented at angles $(\theta_a, \theta_b, \theta_c)$ with respect to an *x*-axis.

Now, suppose we place a strain gage rosette on the surface where the gages are along the *x*-, *b*-, and *y*-axes, as shown in Fig. 5.7, and that these gages measure the normal strains e_a, e_b, e_c, respectively. The *a* and *c* gages of the rosette, therefore, give us the three normal strains, $e_{xx} = e_a, e_{yy} = e_c$, and $e_{zz} = -\nu(e_a + e_c)/(1-\nu)$, leaving only the shear strain e_{xy} as unknown. However, from the strain transformation equation (see Eq. (4.32)) we have that the normal strain in a x_1' direction is

$$e_{11}' = l_{11}l_{11}e_{11} + l_{21}l_{21}e_{22} + l_{31}l_{31}e_{33} + 2l_{11}l_{21}e_{12} + 2l_{11}l_{31}e_{13} + 2l_{21}l_{31}e_{23} \qquad (5.63)$$

which, if we take the *b*-axis along this x_1' axis, and the (x, y) axes as the (x_1, x_2) axes, we can rewrite as

$$e_b = e_a \cos^2\theta + e_c \sin^2\theta + 2e_{xy} \sin\theta \cos\theta \qquad (5.64)$$

since (see Eq. (2.35)) $l_{i1} = \cos(x_i, x_1') (i = 1, 2, 3)$. Equation (5.64) shows that a knowledge of the strains (e_a, e_b, e_c) and the angle θ are all that is needed to calculate the remaining unknown shear strain, e_{xy}, thus giving the complete state of strain on the surface in Eq. (5.62). The stress–strain law, therefore, also gives us the stresses in the state of stress in Eq. (5.62). Note that while the state of stress on the surface is a two-dimensional (2-D) state of stress it is not a state of plane stress since the stresses may be functions of all three coordinates (x, y, z) and not just (x, y), as assumed for plane stress. To make it easier to do analysis with a rosette, we have written the MATLAB® function rosette that has the calling sequence

```
[exx eyy exy] = rosette(anga, angb, angc, ea, eb, ec);
```

which takes the measured strains (e_a, e_b, e_c) and the angles $(\theta_a, \theta_b, \theta_c) = (\text{anga}, \text{angb}, \text{angc})$, as measured in degrees from the *x*-axis for gages (a, b, c), respectively, and returns the strains (e_{xx}, e_{yy}, e_{xy}) by solving the normal strain equations for the three gages, Eq. (5.63), written as:

$$\begin{aligned} e_a &= e_{xx}\cos^2\theta_a + e_{yy}\sin^2\theta_a + 2e_{xy}\sin\theta_a\cos\theta_a \\ e_b &= e_{xx}\cos^2\theta_b + e_{yy}\sin^2\theta_b + 2e_{xy}\sin\theta_b\cos\theta_b \\ e_c &= e_{xx}\cos^2\theta_c + e_{yy}\sin^2\theta_c + 2e_{xy}\sin\theta_c\cos\theta_c \end{aligned} \qquad (5.65)$$

Accurate measurements with strain gages require that one corrects the gage readings for temperature changes and for sensitivity to transverse strains (i.e., strains not along the axis for which the strain is to be measured).

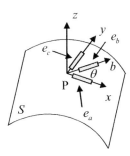

Figure 5.7 A strain gage rosette placed on a free surface.

142 Stress–Strain Relations

We will not discuss those issues here, but details can be found in a number of advanced strength of materials texts.

5.5 PROBLEMS

P5.1 Although aluminum alloys are often treated as if they are isotropic ($E = 69$ GPa, $v = 0.33$ are often used as typical values) pure aluminum is in fact a cubic material with $C_{11} = 103$ Gpa, $C_{12} = 55$ GPa, $C_{44} = 27.6$ GPa. If the state of strain at a point in a pure aluminum material with respect to a set of (x, y, z) axes in μstrain is

$$[e] = \begin{bmatrix} 300 & 50 & 0 \\ 50 & 200 & 0 \\ 0 & 0 & 100 \end{bmatrix}$$

determine:
(a) the principal strains and principal strain directions;
(b) the stress components with respect to the (x, y, z) axes;
(c) the principal stresses and principal directions;
(d) repeat part (b) assuming that the aluminum is isotropic (i.e., use the E, v values given above). Determine the principal stresses and principal stress directions. What are the differences between the principal values in the two cases?

P5.2 Along the orthotropic axes of a piece of birch-wood material, the stress–strain relations are

$$\begin{Bmatrix} e_{xx} \\ e_{yy} \\ e_{zz} \\ \gamma_{yz} \\ \gamma_{xz} \\ \gamma_{xy} \end{Bmatrix} = \begin{bmatrix} 72.5 & -36.25 & -36.25 & 0 & 0 & 0 \\ -36.25 & 942.5 & -652.5 & 0 & 0 & 0 \\ -36.25 & -652.5 & -1450 & 0 & 0 & 0 \\ 0 & 0 & 0 & 4350 & 0 & 0 \\ 0 & 0 & 0 & 0 & 1087.6 & 0 \\ 0 & 0 & 0 & 0 & 0 & 1015 \end{bmatrix} \begin{Bmatrix} \sigma_{xx} \\ \sigma_{yy} \\ \sigma_{zz} \\ \sigma_{yz} \\ \sigma_{xz} \\ \sigma_{xy} \end{Bmatrix}$$

where the strains are in μstrain and the stresses are in MPa. The x-axis here is along the wood grain, the y-axis is radial to the tree, and the z-axis is tangent to the growth rings in the tree. If the stresses (in MPa) at a point in the wood with respect to the (x, y, z) axes are

$$[\sigma] = \begin{bmatrix} 7 & 1.4 & 0 \\ 1.4 & 2.1 & 0 \\ 0 & 0 & -2.8 \end{bmatrix}$$

determine the following:
(a) the principal stresses and principal stress directions;
(b) the state of strain with respect to the (x, y, z) axes;
(c) the principal strains and principal strain directions.

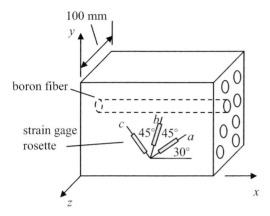

Figure P5.1 Measurement of strains on the surface of a composite plate.

P5.3 A 100 mm-thick composite plate is fabricated from boron-fiber reinforced epoxy, as shown in Fig. P5.1. A 0-45-90 strain gage rosette is placed on the stress-free x–y surface of this plate, as shown in Fig. P5.1. The nonzero elastic constants (all in GPa) for this material as measured in the (x, y, z) coordinates are given by

$$C_{11} = 209$$
$$C_{22} = C_{33} = 29.4$$
$$C_{12} = C_{13} = 24.6$$
$$C_{23} = 13.7$$
$$C_{44} = 5.9$$
$$C_{55} = C_{66} = 8.14$$

The strains (in μstrain) measured by the strain gages are

$$e_a = 385$$
$$e_b = 70$$
$$e_c = -285$$

(a) Based on these strain measurements, what are the e_{xx}, e_{yy}, and e_{xy} strains?
(b) What is the complete 3-D state of strain at the location of the rosette gage in (x, y, z) coordinates?
(c) What is the complete 3-D state of stress at the location of the rosette gage in the (x, y, z) coordinates?
(d) What is the change of thickness of the plate, assuming the strains are uniform throughout the plate?
(e) What are the principal strains and principal strain directions at the location of the rosette gage, and the corresponding principal stresses and principal stress directions?

P5.4 A rubber cylinder with Young's modulus, E_R, and Poisson's ratio, ν_R, is compressed inside a steel tube by an axial stress, σ_0, as shown in Fig. P5.2. The Young's modulus of

144   Stress–Strain Relations

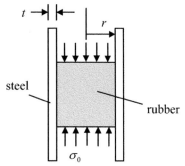

Figure P5.2 A rubber cylinder being compressed inside a thin steel tube.

the steel is E_S and its Poisson's ratio is v_S. Determine an expression for the pressure exerted between the rubber and the steel tube if:
(a) the elasticity of the steel is considered, and
(b) the steel is rigid ($E_S = \infty$).
(c) Take $E_R = 0.02$ GPa, $v_R = 0.5$, $E_S = 200$ GPa, $v_S = 0.3$, $t = 4$ mm, $r = 100$ mm. Calculate the pressure in terms of σ_0 for cases (a) and (b). How good is the assumption of rigidity? You can assume the tube is sufficiently thin so that the hoop stress in the tube can be considered to be uniform across the thickness.

P5.5 Consider the following 3-D state of strain (in μstrain) with respect to a set of (x, y, z) axes:

$$[e] = \begin{bmatrix} 200 & -300 & 0 \\ -300 & 100 & 200 \\ 0 & 200 & 100 \end{bmatrix}$$

(a) If this state of strain exists in a component made of stainless steel with $E = 200$ GPa, $v = 0.3$, determine the principal stresses and principal stress directions.
(b) Alternatively, consider stainless steel as a transversely anisotropic material (which it is) where the stress–strain relationship with respect to a set of (x, y, z) material axes is given by

$$\begin{Bmatrix} \sigma_{xx} \\ \sigma_{yy} \\ \sigma_{zz} \\ \sigma_{yz} \\ \sigma_{xz} \\ \sigma_{xy} \end{Bmatrix} = \begin{bmatrix} C_{11} & C_{12} & C_{13} & 0 & 0 & 0 \\ C_{12} & C_{11} & C_{13} & 0 & 0 & 0 \\ C_{13} & C_{13} & C_{33} & 0 & 0 & 0 \\ 0 & 0 & 0 & C_{44} & 0 & 0 \\ 0 & 0 & 0 & 0 & C_{44} & 0 \\ 0 & 0 & 0 & 0 & 0 & (C_{11}-C_{12})/2 \end{bmatrix} \begin{Bmatrix} e_{xx} \\ e_{yy} \\ e_{zz} \\ \gamma_{yz} \\ \gamma_{xz} \\ \gamma_{xy} \end{Bmatrix}$$

where

$C_{11} = 262.7$ GPa, $C_{33} = 216$ GPa, $C_{44} = 129$ GPa
$C_{12} = 98.2$ GPa, $C_{13} = 145$ GPa

For the same state of strain as in part (a), determine the principal stresses and principal stress directions.

P5.6 A 0-45-90 strain gage rosette is placed on the free surface of a linear elastic isotropic component whose Young's modulus and Poisson's ratio are $E = 150$ GPa, $v = 0.3$. The angle the gage makes with respect to the longitudinal (x-axis) of the component is $25°$ so the gage angles from the x-axis are $(\theta_a, \theta_b, \theta_c) = (25, 70, 115)°$, respectively. The measured strains are $(e_a, e_b, e_c) = (77.5, 51.8, 82.5)$ μstrain, respectively.

(a) Determine the state of strain at the gage location and from the stress–strain relations the corresponding state of stress. Using this state of stress, determine the principal stresses and principal stress directions.

(b) From the state of strain at the gage location, determine the principal strains and principal strain directions directly. Use the stress–strain relations to determine the corresponding principal stresses.

P5.7 The strains (in μstrain) measured on a free surface of a component with a rosette are $e_a = 400$, $e_b = 300$, $e_c = -50$ for $\theta_a = 30°$, $\theta_b = -30°$, $\theta_c = 90°$. Determine the magnitude of the maximum engineering shear strain at the gage. Let Poisson's ratio $v = 1/3$.

6 Governing Equations and Boundary Conditions

Previous chapters have examined a wide variety of topics including equilibrium, compatibility, strain–displacement relations, and stress–strain relations. When these elements are combined, we can form up different complete sets of governing differential and algebraic equations. In order to solve those sets of equations we must also specify the conditions that arise from having known loads or geometric constraints. These are called the boundary conditions for the problem. As discussed in Chapter 1, boundary conditions are one of the four pillars of stress analyses. In this chapter we will examine some of the choices we have for formulating complete sets of the governing equations and how those governing equations can be combined with appropriate boundary conditions to solve for the stresses and deformations. We will also discuss the principle of Saint-Venant, which gives us some flexibility in how we specify the boundary conditions.

In general, we can use either the displacements or the stresses as the fundamental unknowns. Solving for the stresses is often done with the use of auxiliary potential functions to help simplify the governing equations. We will see an example of that simplification for solving stress-based 2-D problems with the use of an Airy stress function.

The governing equations for deformable bodies are combinations of partial differential and algebraic equations so that obtaining solutions of complex problems is difficult and not formulated in a way that is well suited to numerical computation. However, we will see that problems can alternatively be expressed in terms of algebraic matrix–vector governing equations that can be easily solved with computers. Both displacement-based and stress-based approaches will be outlined that can solved with linear algebra. These approaches will be shown to be the counterparts of the governing differential/algebraic equations discussed originally. A classical deformable body problem, Navier's table problem, will be used as an example of these purely algebraic methods.

6.1 Governing Equations in Three Dimensions

In order to have a complete set of governing equations we must (1) identify the unknown variables, and (2) produce a set of differential/algebraic equations that are equal in number to the number of unknowns present. In this section we will list some of the common combinations for three-dimensional (3-D) problems.

The most general way to develop a complete set of equations is by taking all the unknowns present (the stresses, the strains, and the displacements) as the unknowns. In that case we can form up the following set of governing equations.

3-D Governing Set 1

	No. of Equations	No. of Unknowns
Equilibrium $\frac{\partial \sigma_{xx}}{\partial x} + \frac{\partial \sigma_{yx}}{\partial y} + \cdots + f_x = 0 \cdots$	3	6 stresses
Stress–strain $\sigma_{xx} = \frac{E}{(1+\nu)(1-2\nu)}[(1-\nu)e_{xx} + \cdots]$	6	6 strains
Strain–displacement $e_{xx} = \frac{\partial u_x}{\partial x} \cdots$	6	3 displacements
TOTAL	15	15

The compatibility equations do not appear because if the strain–displacement relations are appropriately differentiated, the displacements can be eliminated, resulting in the compatibility equations. Thus, the compatibility equations are still implicitly present. Here there are a very large number of unknowns. However, we can place the strain–displacement relations into the stress–strain equations, which would express the stresses directly in terms of the displacement gradients, and then place those stress–displacement gradient relations into the equilibrium equations to arrive at three equations for the three displacements only. To see this reduction, consider a general linear anisotropic material, where the equations of this governing set can be written as

$$\sum_{i=1}^{3} \frac{\partial \sigma_{ij}}{\partial x_i} + f_j = 0 \quad (j = 1, 2, 3)$$

$$\sigma_{ij} = \sum_{k=1}^{3} \sum_{l=1}^{3} C_{ijkl} e_{kl} \quad (i, j = 1, 2, 3) \quad (6.1)$$

$$e_{kl} = \frac{1}{2}\left(\frac{\partial u_k}{\partial x_l} + \frac{\partial u_l}{\partial x_k}\right) \quad (k, l = 1, 2, 3)$$

Placing the strain–displacement relations into the stress–strain relations and using the symmetry of the elastic constants $(C_{ijkl} = C_{ijlk})$ gives

$$\sigma_{ij} = \sum_{k=1}^{3} \sum_{l=1}^{3} C_{ijkl} \frac{\partial u_k}{\partial x_l} \quad (6.2)$$

Placing Eq. (6.2) into the equilibrium equations then gives three equations for the three unknown displacements forming an alternative set of equations:

$$\sum_{i=1}^{3} \sum_{k=1}^{3} \sum_{l=1}^{2} C_{ijkl} \frac{\partial^2 u_k}{\partial x_i \partial x_l} + f_j = 0 \quad (j = 1, 2, 3) \quad (6.3)$$

When the material is a linear isotropic material, these equations become *Navier's equations* for the displacements:

$$\mu \sum_{k=1}^{3} \frac{\partial^2 u_j}{\partial x_k \partial x_k} + (\lambda + \mu) \sum_{k=1}^{3} \frac{\partial^2 u_k}{\partial x_k \partial x_j} + f_j = 0 \quad (j = 1, 2, 3) \tag{6.4}$$

where the Lamé constants (λ, μ) are $\mu = G$, $\lambda = Ev/(1+v)(1-2v)$ in terms of the shear modulus, Young's modulus, and Poisson's ratio constants (G, E, v), respectively, where we recall $G = E/2(1+v)$. These displacement equations, therefore, provide an alternate complete set of governing equations. For the general anisotropic case of Eq. (6.3) we have the following.

3-D Governing Set 2

	No. of Equations	No. of Unknowns
Equilibrium	3	3 displacements
$\sum_{i=1}^{3}\sum_{k=1}^{3}\sum_{l=1}^{2} C_{ijkl} \frac{\partial^2 u_k}{\partial x_i \partial x_l} + f_j = 0 \quad (j=1,2,3)$		
TOTAL	3	3

A displacement formulation of the governing equations is very attractive since there are very few unknowns (the three displacements), and these unknowns can be obtained directly from the three equilibrium equations of Eq. (6.4). A displacement formulation is also a commonly used approach for solving complex problems numerically. Historically, the finite differences method was one of the first numerical methods to be used. In this case, differential operators such as those appearing in Eq. (6.4) are approximated by finite differences of displacements at discrete points, resulting in a matrix system of algebraic difference equations that are then solved under appropriate boundary conditions. An alternate numerical method is to use a set of fundamental solutions to Navier's equations to recast the differential equations of Eq. (6.4) into a set of integral equations involving the displacements on the boundary of the deformable body and then to approximate those integral equations as a system of linear equations which are solved numerically. Displacements and stresses throughout the body can then be obtained from the known boundary displacements. This approach is called the boundary integral equation method or the boundary element method. A third approach uses work–energy methods (see Chapter 8) as an alternative way to satisfy the equilibrium equations. In this case the equilibrium equations (written in terms of the displacements) are replaced by a set of equivalent matrix–vector equations $[K]\{u\} = \{P\}$, where $\{u\}$ is a vector of displacements at discrete locations (called nodes) in a body and $\{P\}$ is a vector of known discrete loads. Once these nodal displacements are known, they can be used to obtain the stresses in the body. The matrix $[K]$ is known as the *stiffness matrix*. This method, which is called the stiffness-based finite element method, is by far the most commonly used numerical method for solving for the displacements and stresses in deformable bodies.

In Chapter 9 we will outline the finite element method and the boundary element method in more detail. We will, however, not give a comprehensive treatment of those displacement-based methods. There are many books available that are excellent sources ([1], [2]). We will not describe the finite difference method in this book as this approach is now used infrequently but you can find more details in some advanced strength of materials or elasticity texts (see [3], for example).

Another alternative for generating a complete set of governing equations is to use the stress–strain relations in the compatibility equations to write those equations in terms of the stresses. The details are rather lengthy so they are given in detail in Appendix B for the case of an isotropic linear elastic material. The compatibility equations in this form are called the *Beltrami–Michell compatibility equations* given by

$$\nabla^2 \sigma_{il} + \frac{1}{1+v}\sum_{k=1}^{3}\frac{\partial^2 \sigma_{kk}}{\partial x_i \partial x_l} = -\frac{v}{1-v}\delta_{il}\sum_{j=1}^{3}\frac{\partial f_j}{\partial x_j} - \left(\frac{\partial f_i}{\partial x_l} + \frac{\partial f_l}{\partial x_i}\right) \quad (i,l=1,2,3) \tag{6.5}$$

where v is Poisson's ratio, f_i are the components of the body force, and δ_{il} is the Kronecker delta symbol (see Eq. (2.34)). Because the stresses are symmetric, Eq. (6.5) represents six equations, but again only three of them are independent. Thus, we can form a complete set of governing equations as follows.

3-D Governing Set 3

		No. of Equations	No. of Unknowns
Equilibrium	$\frac{\partial \sigma_{xx}}{\partial x} + \frac{\partial \sigma_{yx}}{\partial y} + \cdots + f_x = 0 \cdots$	3	6 stresses
Compatibility	$\frac{\partial^2 \sigma_{xy}}{\partial x^2} + \frac{\partial^2 \sigma_{xy}}{\partial y^2} + \cdots = 0 \cdots$	3 (independent)	–
TOTAL		6	6

In this formulation, one can introduce a set of auxiliary functions (also called stress potential functions), which automatically satisfy the equilibrium equations, thus reducing the number of equations and unknowns. The *Maxwell stress function* and *Morera stress function* are examples that are often discussed in texts on elasticity theory. In this book we will only see examples of such functions in simpler contexts, including the use of the *Airy stress function* for planar problems (which will be discussed later in this chapter and in Chapter 7) and the *Prandtl stress function* for torsion problems (see Chapter 11). Later in this chapter we will see that we can also develop a discrete stress-based formulation that combines the equilibrium and compatibility equations similar to the 3-D Governing Set 3. As discussed briefly in Section 6.6, this discrete stress formulation can also be developed as a force-based finite element method, a method that will be discussed in more detail in Chapter 9.

6.2 Governing Equations in Two Dimensions

6.2.1 Plane Stress

In two-dimensional (2-D) problems, there are very similar choices that can be made for the sets of governing equations. For example, for plane-stress problems we can take the stresses, strains, and displacements as the unknowns for the following set.

2-D Governing Set 1 (Plane Stress)

	No. of Equations	No. of Unknowns
Equilibrium $\dfrac{\partial \sigma_{xx}}{\partial x} + \dfrac{\partial \sigma_{yx}}{\partial y} + f_x = 0$ $\dfrac{\partial \sigma_{xy}}{\partial x} + \dfrac{\partial \sigma_{yy}}{\partial y} + f_y = 0$	2	3 stresses
Stress–strain $\sigma_{xx} = \dfrac{E}{1-v^2}(e_{xx} + v e_{yy})$ $\sigma_{yy} = \dfrac{E}{1-v^2}(e_{yy} + v e_{xx})$ $\sigma_{xy} = G\gamma_{xy}$	3	3 strains
Strain–displacement $e_{xx} = \dfrac{\partial u_x}{\partial x}$ $e_{yy} = \dfrac{\partial u_y}{\partial y}$ $\gamma_{xy} = \dfrac{\partial u_x}{\partial y} + \dfrac{\partial u_y}{\partial x}$	3	2 displacements
TOTAL	8	8

The plane-stress stress–strain relations can be written compactly as

$$\sigma_{mn} = \frac{Ev}{1-v^2} \sum_{k=1}^{2} e_{kk} \delta_{mn} + 2G e_{mn} \quad (m, n = 1, 2) \tag{6.6}$$

The in-plane strain–displacement relations and the equilibrium equations are likewise

$$e_{mn} = \frac{1}{2}\left(\frac{\partial u_m}{\partial x_n} + \frac{\partial u_n}{\partial x_m} \right) \quad (m, n = 1, 2) \tag{6.7}$$

and

$$\sum_{m=1}^{2} \frac{\partial \sigma_{mn}}{\partial x_m} + f_n = 0 \quad (n = 1, 2) \tag{6.8}$$

Combining the strain–displacement and stress–strain relations gives

$$\sigma_{mn} = \frac{Ev}{1-v^2} \sum_{k=1}^{2} \frac{\partial u_k}{\partial x_k} \delta_{mn} + G\left(\frac{\partial u_m}{\partial x_n} + \frac{\partial u_n}{\partial x_m}\right) \quad (m,n=1,2) \tag{6.9}$$

and placing this equation into the equilibrium equations, we find, after some algebra,

$$G \sum_{k=1}^{2} \frac{\partial^2 u_n}{\partial x_k \partial x_k} + \frac{E}{2(1-v)} \sum_{k=1}^{2} \frac{\partial^2 u_k}{\partial x_n \partial x_k} + f_n = 0 \quad (n=1,2) \tag{6.10}$$

which are Navier's equations for plane stress. These equations form a second set of governing equations.

2-D Governing Set 2 (Plane Stress)

	No. of Equations	No. of Unknowns
Equilibrium	2	2 displacements
$G \sum_{k=1}^{2} \frac{\partial^2 u_n}{\partial x_k \partial x_k} + \frac{E}{2(1-v)} \sum_{k=1}^{2} \frac{\partial^2 u_k}{\partial x_n \partial x_k} + f_n = 0$		
TOTAL	2	2

Finally, consider the plane-stress case where we use the stresses as the basic unknowns. In this case there is only one compatibility equation

$$2 \frac{\partial^2 e_{xy}}{\partial x \partial y} - \frac{\partial^2 e_{xx}}{\partial y^2} - \frac{\partial^2 e_{yy}}{\partial x^2} = 0 \tag{6.11}$$

and the two equilibrium equations (using the symmetry of the shear stress)

$$\begin{aligned} \frac{\partial \sigma_{xx}}{\partial x} + \frac{\partial \sigma_{xy}}{\partial y} + f_x &= 0 \\ \frac{\partial \sigma_{xy}}{\partial x} + \frac{\partial \sigma_{yy}}{\partial y} + f_y &= 0 \end{aligned} \tag{6.12}$$

The stress–strain relations are (for the strains written in terms of the stresses)

$$\begin{aligned} e_{xx} &= \frac{1}{E}(\sigma_{xx} - v\sigma_{yy}) \\ e_{yy} &= \frac{1}{E}(\sigma_{yy} - v\sigma_{xx}) \\ e_{xy} &= \frac{1}{2G}\sigma_{xy} = \frac{(1+v)}{E}\sigma_{xy} \end{aligned} \tag{6.13}$$

so that placing these stress–strain relations into the compatibility equation gives

$$2(1+v)\frac{\partial^2 \sigma_{xy}}{\partial x \partial y} - \frac{\partial^2 \sigma_{xx}}{\partial y^2} + v\frac{\partial^2 \sigma_{yy}}{\partial y^2} - \frac{\partial^2 \sigma_{yy}}{\partial x^2} + v\frac{\partial^2 \sigma_{xx}}{\partial x^2} = 0 \tag{6.14}$$

But if we take $(1+v)$ times the x-derivative of the first equilibrium equation and $(1+v)$ times the y-derivative of the second equilibrium equation and add the two resulting equations together, we find

$$2(1+v)\frac{\partial^2 \sigma_{xy}}{\partial x \partial y} = -(1-v)\left(\frac{\partial^2 \sigma_{xx}}{\partial x^2} + \frac{\partial^2 \sigma_{yy}}{\partial y^2}\right) - (1+v)\frac{\partial f_x}{\partial x} - (1+v)\frac{\partial f_y}{\partial y} \tag{6.15}$$

If we place Eq. (6.15) into Eq. (6.14) we obtain, finally, the plane-stress equivalent of the Beltrami–Michell equations:

$$\frac{\partial^2 (\sigma_{xx} + \sigma_{yy})}{\partial x^2} + \frac{\partial^2 (\sigma_{xx} + \sigma_{yy})}{\partial y^2} = -(1+v)\left(\frac{\partial f_x}{\partial x} + \frac{\partial f_y}{\partial y}\right) \tag{6.16}$$

which is often written in the more compact form

$$\nabla^2 (\sigma_{xx} + \sigma_{yy}) = -(1+v)\nabla \cdot \mathbf{f} \tag{6.17}$$

where $\nabla^2 = \partial^2/\partial x^2 + \partial^2/\partial y^2$ is the Laplacian operator and $\nabla = \mathbf{e}_x \partial/\partial x + \mathbf{e}_y \partial/\partial y$ is the gradient operator, with $\mathbf{f} = f_x \mathbf{e}_x + f_y \mathbf{e}_y$ being the vector body force. Thus, with the compatibility equation written in terms of the stresses, we have the governing set of equations.

2-D Governing Set 3 (Plane Stress)

	No. of Equations	No. of Unknowns
Equilibrium	2	3 stresses
$\frac{\partial \sigma_{xx}}{\partial x} + \frac{\partial \sigma_{yx}}{\partial y} + f_x = 0$		
$\frac{\partial \sigma_{xy}}{\partial x} + \frac{\partial \sigma_{yy}}{\partial y} + f_y = 0$		
Compatibility	1	—
$\nabla^2 (\sigma_{xx} + \sigma_{yy}) = -(1+v)\nabla \cdot \mathbf{f}$		
TOTAL	3	3

The use of a stress potential function in this case is particularly effective to simplify these equations and reduce their number. Consider, for example, if we assume that the body forces can be written in terms of a potential function, V, i.e.,

$$f_x = -\frac{\partial V}{\partial x}, \quad f_y = -\frac{\partial V}{\partial y} \tag{6.18}$$

Common forces such as constant forces or the force of gravity, for example, can be represented in this manner. Also, let the stresses be given in terms of V and a potential function, ϕ, called the *Airy stress function*, where

$$\sigma_{xx} = \frac{\partial^2 \phi}{\partial y^2} + V$$
$$\sigma_{yy} = \frac{\partial^2 \phi}{\partial x^2} + V \qquad (6.19)$$
$$\sigma_{xy} = -\frac{\partial^2 \phi}{\partial x \partial y}$$

Then the equilibrium equations are satisfied identically and the compatibility equation becomes

$$\nabla^4 \phi = -(1 - \nu)\nabla^2 V \qquad (6.20)$$

where $\nabla^4 = \partial^4/\partial x^4 + 2\partial^4/\partial x^2 \partial y^2 + \partial^4/\partial y^4$ is called the *biharmonic operator*, which we could also write as $\nabla^4 = (\partial^2/\partial x^2 + \partial^2/\partial y^2)(\partial^2/\partial x^2 + \partial^2/\partial y^2) = \nabla^2 \nabla^2$, i.e., it is the Laplacian operator applied twice. So far, we have been using Cartesian coordinates, but this is not essential. For example, in polar coordinates (r, θ) the equilibrium equations are (see Eq. (3.73))

$$\frac{\partial \sigma_{rr}}{\partial r} + \frac{\sigma_{rr} - \sigma_{\theta\theta}}{r} + \frac{1}{r}\frac{\partial \sigma_{\theta r}}{\partial \theta} + f_r = 0$$
$$\frac{\partial \sigma_{r\theta}}{\partial r} + \frac{2\sigma_{r\theta}}{r} + \frac{1}{r}\frac{\partial \sigma_{\theta\theta}}{\partial \theta} + f_\theta = 0 \qquad (6.21)$$

If we express the body force in terms of a potential, $V(r, \theta)$, we have

$$f_r = -\frac{\partial V}{\partial r}, \quad f_\theta = -\frac{1}{r}\frac{\partial V}{\partial \theta} \qquad (6.22)$$

so that if we write the stresses in terms of the Airy stress function, $\phi(r, \theta)$, as

$$\sigma_{rr} = \frac{1}{r}\frac{\partial \phi}{\partial r} + \frac{1}{r^2}\frac{\partial^2 \phi}{\partial \theta^2} + V$$
$$\sigma_{\theta\theta} = \frac{\partial^2 \phi}{\partial r^2} + V \qquad (6.23)$$
$$\sigma_{r\theta} = -\frac{\partial}{\partial r}\left(\frac{1}{r}\frac{\partial \phi}{\partial \theta}\right)$$

the equilibrium equations will again be satisfied identically. Now consider the compatibility equation. We can write that equation in a completely coordinate invariant form as

$$\nabla^2 I_1 = (1 + \nu)\nabla^2 V \qquad (6.24)$$

where $I_1 = \sigma_{xx} + \sigma_{yy} = \sigma_{rr} + \sigma_{\theta\theta}$ is the first stress invariant for plane stress. But

$$\sigma_{rr} + \sigma_{\theta\theta} = \frac{\partial^2 \phi}{\partial r^2} + \frac{1}{r}\frac{\partial \phi}{\partial r} + \frac{1}{r^2}\frac{\partial^2 \phi}{\partial \theta^2} + 2V \quad (6.25)$$
$$= \nabla^2 \phi + 2V$$

so that from Eq. (6.24) we have again

$$\nabla^2 \phi = -(1-\nu)\nabla^2 V \quad (6.26)$$

and the biharmonic operator is once more the Laplacian operator in polar coordinates applied twice:

$$\nabla^2 = \left(\frac{\partial^2}{\partial r^2} + \frac{1}{r}\frac{\partial}{\partial r} + \frac{1}{r^2}\frac{\partial^2}{\partial \theta^2}\right)\left(\frac{\partial^2}{\partial r^2} + \frac{1}{r}\frac{\partial}{\partial r} + \frac{1}{r^2}\frac{\partial^2}{\partial \theta^2}\right) \quad (6.27)$$
$$= \nabla^2\nabla^2$$

As we will see in Chapter 7, we can obtain the solution to a number of important canonical problems with the use of the Airy stress function. Thus, this is an important choice of governing equations.

2-D Governing Set 4 (Plane Stress)

	No. of Equations	No. of Unknowns
Compatibility $\nabla^4\phi = -(1-\nu)\nabla^2 V$	1	1
TOTAL	1	1

6.2.2 Plane Strain

Recall that the case of plane strain is where the displacements are given as $u_x = u_x(x,y)$, $u_y = u_y(x,y)$, $u_z = 0$, which produces a state of strain where $e_{xx} = \partial u_x/\partial x = e_{xx}(x,y)$, $e_{yy} = \partial u_y/\partial x = e_{yy}(x,y)$, $\gamma_{xy} = (\partial u_x/\partial y + \partial u_y/\partial x) = \gamma_{xy}(x,y)$, $\gamma_{xz} = \gamma_{yz} = \gamma_{zz} = 0$. From the 3-D stress–strain relations (see Eq. (5.5)) it follows that we also have $\sigma_{xx} = \sigma_{xx}(x,y)$, $\sigma_{yy} = \sigma_{yy}(x,y)$, $\sigma_{zz} = \sigma_{zz}(x,y)$, $\sigma_{xy} = \sigma_{xy}(x,y)$ so that the equilibrium equations reduce to

$$\frac{\partial \sigma_{xx}}{\partial x} + \frac{\partial \sigma_{yx}}{\partial y} + f_x = 0$$
$$\frac{\partial \sigma_{xy}}{\partial x} + \frac{\partial \sigma_{yy}}{\partial y} + f_y = 0 \quad (6.28)$$

which are identical to the plane-stress case. In this case the stress–strain relations are

$$\sigma_{xx} = \frac{E}{(1+v)(1-2v)}\left[(1-v)e_{xx} + ve_{yy}\right]$$

$$\sigma_{yy} = \frac{E}{(1+v)(1-2v)}\left[(1-v)e_{yy} + ve_{xx}\right] \qquad (6.29)$$

$$\sigma_{xy} = G\gamma_{xy}$$

so that in terms of the stresses, strains, and displacements the governing equations are as follows.

2-D Governing Set 1 (Plane Strain)

	No. of Equations	No. of Unknowns
Equilibrium $\frac{\partial \sigma_{xx}}{\partial x} + \frac{\partial \sigma_{yx}}{\partial y} + f_x = 0$ $\frac{\partial \sigma_{xy}}{\partial x} + \frac{\partial \sigma_{yy}}{\partial y} + f_y = 0$	2	3 stresses
Stress–strain $\sigma_{xx} = \frac{E}{(1+v)(1-2v)}\left[(1-v)e_{xx} + ve_{yy}\right]$ $\sigma_{yy} = \frac{E}{(1+v)(1-2v)}\left[(1-v)e_{yy} + ve_{xx}\right]$ $\sigma_{xy} = G\gamma_{xy}$	3	3 strains
Strain–displacement $e_{xx} = \frac{\partial u_x}{\partial x}$ $e_{yy} = \frac{\partial u_y}{\partial y}$ $\gamma_{xy} = \frac{\partial u_x}{\partial y} + \frac{\partial u_y}{\partial x}$	3	2 displacements
TOTAL	8	8

If we compare this plane-strain set of equations to the plane-stress case, we see that they are identical in form, but the elastic constants that appear are different. This means that we can transform a plane-stress to a plane-strain problem, and vice versa, simply by redefining the elastic constants. For example, if we have a plane-stress solution valid for (E, v, G) and set

$$G' = G$$
$$v = \frac{v'}{1-v'} \rightarrow v' = \frac{v}{1+v} \qquad (6.30)$$
$$\frac{E}{(1-v^2)} = \frac{E'(1-v')}{(1+v')(1-2v')} \rightarrow E' = \frac{E(1+2v)}{(1+v)^2}$$

we will obtain a plane-strain solution valid for (E', v', G'). Similarly, if we have a plane-strain solution valid for (E, v, G) and set

$$G' = G$$
$$v' = \frac{v}{1-v}$$
$$\frac{E(1-v)}{(1+v)(1-2v)} = \frac{E'}{\left[1-(v')^2\right]} \rightarrow E' = \frac{E}{(1-v^2)}$$ (6.31)

we will have a corresponding plane-stress solution valid for (E', v', G'). Table 6.1 summarizes these relationships. [Note: the roles of (E', v', G') and (E, v, G) have been reversed in the table in order that the solution being generated always has the values (E, v, G).] For example, consider Navier's equations for plane stress (Eq. (6.10)), which we will rewrite as

$$G' \sum_{k=1}^{2} \frac{\partial^2 u_n}{\partial x_k \partial x_k} + \frac{E'}{2(1-v')} \sum_{k=1}^{2} \frac{\partial^2 u_k}{\partial x_n \partial x_k} + f_n = 0 \quad (n=1,2)$$ (6.32)

Then using Table 6.1 we have

$$G' = G$$
$$\frac{E'}{2(1-v')} = \frac{E/(1-v^2)}{2(1-v/(1-v))} = \frac{E}{2(1+v)(1-2v)}$$ (6.33)

so that Navier's equations for plane strain become

$$G \sum_{k=1}^{2} \frac{\partial^2 u_n}{\partial x_k \partial x_k} + \frac{E}{2(1+v)(1-2v)} \sum_{k=1}^{2} \frac{\partial^2 u_k}{\partial x_n \partial x_k} + f_n = 0 (n=1,2)$$ (6.34)

We could have also obtained this result directly from the 3-D Navier's equation (Eq. (6.4)) by noting that plane-strain assumption $u_3 = 0$ eliminates all the terms involving u_3 and its derivatives, so that 3-D form remains the same but the summations only run from one to two.

Table 6.1 Transforming from a plane-stress solution to a plane-strain solution, and vice versa

Given a plane-stress solution for (E', v', G')	substituting $E' = \frac{E}{(1-v^2)}$ $v' = \frac{v}{(1-v)}$ $G' = G$	will give a plane-strain solution for (E, v, G)
Given a plane-strain solution for (E', v', G')	substituting $E' = \frac{E(1+2v)}{(1+v)^2}$ $v' = \frac{v}{(1+v)}$ $G' = G$	will give a plane-stress solution for (E, v, G)

In terms of the Lamé constants, $\lambda + \mu = E/2(1+v)(1-2v)$ and $\mu = G$ so we see we again arrive at Eq. (6.34) for plane strain and the governing set of equations are as follows.

2-D Governing Set 2 (Plane Strain)

	No. of Equations	No. of Unknowns
Equilibrium	2	2 displacements
$G \sum_{k=1}^{2} \dfrac{\partial^2 u_j}{\partial x_k \partial x_k} + \dfrac{E}{2(1+v)(1-2v)} \sum_{k=1}^{2} \dfrac{\partial^2 u_k}{\partial x_n \partial x_j} + f_j = 0$		
TOTAL	2	2

Now consider the governing set 3 for plane stress, which consists of the equilibrium and compatibility equations. In that case the only elastic constant term appearing for the plane-stress case is $(1+v')$ for a plane-stress Poisson ratio value v' so that, from Table 6.1,

$$(1+v') = 1 + \frac{v}{(1-v)} = \frac{1}{(1-v)} \quad (6.35)$$

so that the governing set for plane strain is as follows.

2-D Governing Set 3 (Plane Strain)

	No. of Equations	No. of Unknowns
Equilibrium	2	3 stresses
$\dfrac{\partial \sigma_{xx}}{\partial x} + \dfrac{\partial \sigma_{yx}}{\partial y} + f_x = 0$		
$\dfrac{\partial \sigma_{xy}}{\partial x} + \dfrac{\partial \sigma_{yy}}{\partial y} + f_y = 0$		
Compatibility	1	-
$\nabla^2(\sigma_{xx} + \sigma_{yy}) = -\dfrac{1}{(1-v)} \nabla \cdot \mathbf{f}$		
TOTAL	3	3

Finally, for the governing set using the Airy stress function, the only elastic constant term is $(1-v')$ for a plane-stress Poisson ratio value v' so

$$(1-v') = 1 - \frac{v}{(1-v)} = \frac{(1-2v)}{(1-v)} \quad (6.36)$$

and the plane-strain governing set is as follows.

2-D Governing Set 4 (Plane Strain)

	No. of Equations	No. of Unknowns
Compatibility $\nabla^4 \phi = -\dfrac{(1-2v)}{(1-v)} \nabla^2 V$	1	1
TOTAL	1	1

Note that when there are no body forces or when $\nabla^2 V = 0$ the compatibility equation for either plane stress or plane strain becomes simply $\nabla^2 \phi = 0$, which is called the biharmonic equation. Since the elastic constants in this case are absent, if the boundary conditions only involve the stresses we can solve for the stresses and the stress solutions will be independent of the elastic constants. This will not be true, however, for the strains and the displacements.

We can transform solutions for the in-plane stresses and displacements from plane stress to plane strain, and vice versa, but remember that the out-of-plane stresses and strains are inherently different in the two cases. In the plane-stress case, we have $\sigma_{zz} = 0$, but the corresponding normal strain is

$$e_{zz} = \frac{-v}{(1-v)}\left(e_{xx} + e_{yy}\right) \tag{6.37}$$

In contrast, for plane strain $e_{zz} = 0$, but the normal stress is

$$\sigma_{zz} = v\left(\sigma_{xx} + \sigma_{yy}\right) \tag{6.38}$$

Since these out-of-plane stresses or strains are obtained from the behavior of the in-plane quantities this difference does not affect how we solve either a plane-stress or plane-strain problem.

6.3 Boundary Conditions

The governing sets of equations are combinations of differential and algebraic equations in general that must be satisfied throughout the deformable body. In order to solve these sets, we must also specify conditions on the boundary of the body. Generally, these boundary conditions arise because the body is being supported in some fashion and because there are loads being applied to the surface of the body. The applied loads are usually characterized by specifying the stress vector acting on a portion, S_t, of the total surface, S, of the body, i.e., $\mathbf{T}^{(n)}(\mathbf{x}_s) = \mathbf{T}^*(\mathbf{x}_s)$, when the point $\mathbf{x}_s$ is on S_t, where $\mathbf{T}^*$ is a known quantity. Often the supports are considered to rigidly fix the body over the portion of the surface, S_u, where the body is being supported, i.e., the displacement $\mathbf{u}(\mathbf{x}_s) = 0$ when point $\mathbf{x}_s$ is on S_u. It is assumed that the total surface of the body, S, is $S = S_t + S_u$ so that these stress vector

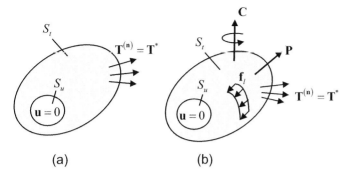

Figure 6.1 (a) A deformable body where applied loads are described by the stress vector acting over a portion S_t of the surface of the body and the displacement is zero over the remaining portion, S_u. (b) Cases of applied loads such as a concentrated force, **P**, a concentrated couple, **C**, and a force/unit length, $\mathbf{f}_l$, distributed along a line on the surface. All of these loads can be treated as special distributions of the stress vector, $\mathbf{T}^{(\mathbf{n})}$.

and displacement values characterize the behavior of the solution for every point on the surface (see Fig. 6.1a). By making the stress vector be very large over a small area of the surface S_t we can simulate a concentrated force, **P**, or a concentrated couple, **C**. Similarly, we can concentrate the stress vector along a line on the surface to simulate a line load (force/unit length), $\mathbf{f}_l$. These cases are illustrated in Fig. 6.1b together with a stress vector distribution, $\mathbf{T}^*$. Later, we will examine deformations of a body under some of these different types of loads, but for the present we can consider all of them as simply special cases of specifying a known stress vector distribution, $\mathbf{T}^*(\mathbf{x}_s)$, on S_t. The zero-displacement (rigid) boundary condition is a highly idealized support condition that may not be satisfied in practice, so one could relax such a boundary condition by providing some elasticity to the support through some "spring-like" conditions where the boundary displacement and stresses are related to each other. Likewise, one could relax the rigid specification by allowing some known displacement, $\mathbf{u}^*(\mathbf{x}_s)$, at the support, i.e., setting $\mathbf{u}(\mathbf{x}_s) = \mathbf{u}^*(\mathbf{x}_s)$ on S_u. Finally, we should note that, in some cases, the boundary conditions may involve combinations of the displacement and stresses at the boundary. However, to have a well-posed boundary value problem it is not permissible to specify any combination. For example, consider breaking the stress vector at the surface into its normal- and tangential-stress components by letting

$$\mathbf{T}^{(n)} = \sigma_{nn}\mathbf{n} + \sigma_{nt}\mathbf{t} + \sigma_{ns}\mathbf{s} \tag{6.39}$$

where σ_{nn} is the normal stress acting in the unit normal direction, **n**, and $(\sigma_{nt}, \sigma_{ns})$ are the shear stresses acting in the tangential directions $(\mathbf{t}, \mathbf{s})$, respectively, along the surface. Likewise, let the displacement, **u**, at the surface be represented by its normal and tangential components, (u_n, u_t, u_s) (see Fig. 6.2):

$$\mathbf{u} = u_n\mathbf{n} + u_t\mathbf{t} + u_s\mathbf{s} \tag{6.40}$$

We can specify combinations of these stress and displacement components on the boundary as long as we do not specify a pair of displacements and stresses in the same direction. For example, we could set $\sigma_{nt} = \sigma_{ns} = 0$ and $u_n = 0$ and this would be acceptable. Such a boundary condition is called a *rigid-smooth boundary condition*. Physically, this condition might occur, for example, where a relatively flexible body is in contact with a much more rigid

Figure 6.2 (a) Normal- and tangential-stress components of the stress vector acting on a surface. (b) The corresponding normal- and tangential-displacement components.

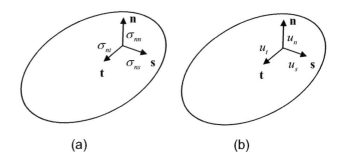

and smooth surface. In the rigid-smooth case, there is nothing to prevent slipping of the body in the tangential directions (u_t and u_s are unknown) and so no tangential shear stresses are developed at the surface, but the rigid surface does prevent displacement in the normal direction, which will produce an unknown normal stress σ_{nn} at the boundary. To see why specifying pairs of displacements and stresses in the same direction is not acceptable, consider, for example, setting $u_s = 0, \sigma_{ns} = 0$. If the displacement component in the s-direction is specified to be zero at a point on the surface, a corresponding unknown shear stress, σ_{ns}, must be developed at that point to counteract the displacement in the s-direction produced by other loads acting on the body. However, the shear stress required is known only after we solve the governing equations for all the displacements and stresses in the body and we cannot specify that shear stress to be zero (or any other value), since this would, in general, not be consistent with the value actually required by the solution. We have used physical reasoning to show why only particular combinations of stress and displacement components are permissible, but one can also use a mathematical argument and show that it is only under these permissible combinations that we can guarantee a unique solution (see Chapter 8). Similarly, specifying displacements (having either zero or nonzero values) on one part of the boundary and the stress vector on the remaining part or specifying spring-like boundary conditions will be boundary conditions for which the solution to the governing equations will be unique [4].

Although there are many possibilities for specifying boundary conditions, in this book we will typically only use special cases of the following set of boundary conditions described originally, namely

$$\begin{aligned} \mathbf{u}(\mathbf{x}_s) = \mathbf{u}^*(\mathbf{x}_s) = 0 & \quad \text{on } S_u \\ \mathbf{T}^{(\mathbf{n})}(\mathbf{x}_s) = \mathbf{T}^*(\mathbf{x}_s) & \quad \text{on } S_t \end{aligned} \tag{6.41}$$

when obtaining solutions and discussing solution procedures. Figure 6.3 shows an example of these types of boundary conditions for a cantilever beam where the components of the stress vector and displacement are given explicitly.

6.3.1 Saint-Venant's Principle

In many cases, boundary conditions cannot be known exactly or the modeled boundary conditions might be different from the actual values. These uncertainties might appear to

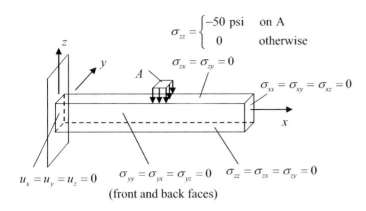

Figure 6.3 Boundary conditions for a cantilever beam.

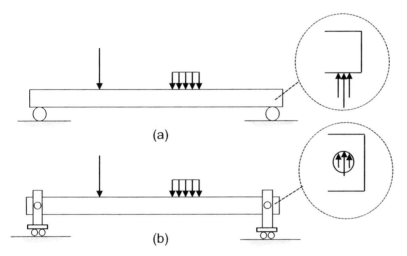

Figure 6.4 A simply-supported beam supported either by (a) smooth rollers, or (b) by pinned supports. In case (a) there are contact stresses over the base of the beam from the roller, while in case (b) there are contact stresses acting on the top of the hole from the inserted pin.

call in question many of the solutions we obtain. In elementary strength of materials courses, such issues are present but usually ignored. For example, consider the two simply supported beams shown in Figs. 6.4a, b. The applied loading is identical for both beams, but the beam in Fig. 6.4a sits on roller supports while the beam in Fig. 6.4b is supported by smooth pins. The stress boundary conditions are obviously quite different for these two cases (as seen in the insets shown in Figs. 6.4a, b) but those differences are ignored in a strength of materials solution for the bending and shear stresses since the support end forces and moments are the same for either case. Thus, in these problems it is implicitly assumed that the stresses in the beam are the same for both of these cases, at least as long as we are not too close to the ends where the detailed stresses reflect the boundary conditions acting on the local geometry at the supports. A "principle" originally due to Adhémar Jean Claude Barré de Saint-Venant (who will henceforth be called simply Saint-Venant for obvious reasons) can often be used to help justify such assumptions. *Saint-Venant's principle* states:

Statically equivalent systems of surface tractions (stress vectors) applied to a small portion of the surface enclosing an elastic body are elastically equivalent (i.e., they produce the same stress and strain distributions) except in the immediate neighborhood of the region where these surface tractions are applied.

Statically equivalent systems of tractions are two different stress vector distributions (which we will label I and II) that produce the same net force and moment with respect to any point, P, i.e.,

$$\left(\sum \mathbf{F}\right)_I = \left(\sum \mathbf{F}\right)_{II}$$
$$\left(\sum \mathbf{M}_P\right)_I = \left(\sum \mathbf{M}_P\right)_{II}$$
(6.42)

At the end supports in Figs. 6.4a, b the net vertical forces and moments are the same so that, according to Saint-Venant's principle, we could expect that the bending and shear stresses throughout most of the beam (i.e., not too close to the supports) will be unaffected.

Saint-Venant's principle can also be easily illustrated by considering the case of uniaxial loading of a bar, which is one of the first problems considered in strength of materials courses. If the applied axial force is P (acting through the centroid of the cross-section) and the cross-sectional area of the bar is A, the strength of materials stress solution is just $\sigma_{xx} = P/A$, a constant value throughout the bar. If, at the ends of the bar, the applied load is uniformly distributed across the area A, then the strength of materials stress solution is, in fact, an exact solution to the governing equations, satisfying both equilibrium, compatibility and the boundary conditions. This case is shown in Fig. 6.5a (note that the stresses and applied loads are compressive). If we make an imaginary cut through the bar of Fig. 6.5a at any section, regardless of how close the cut may be to the ends we obtain the same constant stress distribution, as shown by the dashed cut seen in Fig. 6.5a. However, now consider the case where the stress distribution at the ends is concentrated over a very small area at the centroid (Fig. 6.5b) but where the total force produced by this stress distribution is still P. Then for cross-sectional cuts near the ends, the stresses will not be uniform and will be larger

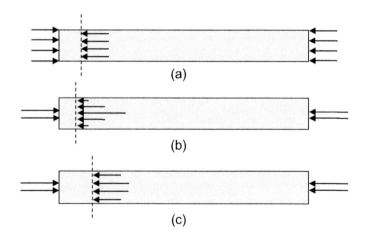

Figure 6.5 A bar under uniaxial loading where (a) the end boundary stresses are applied uniformly over the ends. (b, c) A case where the boundary stresses are concentrated over a small area near the centroid of the cross-section, showing the interior stresses at two different locations.

than P/A near where the stresses are applied, as shown in Fig. 6.5b. However, this non-uniformity of the stress distribution becomes smaller as we go further from the ends, and we approach a uniform stress distribution having magnitude P/A as shown in Fig.6.5c when we are only about a diameter away from the ends.

The "principle" of Saint-Venant is not a fundamental principle such as equilibrium, and it can be shown to be violated in some cases. Also, this principle, as previously stated, is not quantitative since it includes nonprecise terms such as "immediate neighborhood." Nevertheless, it can often be used successfully to relax how precisely we must specify the boundary conditions in many problems.

6.4 Governing Equations in Matrix–Vector Form

We separately developed the equations of equilibrium for the stresses and the compatibility equations for the strains, but there is a close connection between the differential operators appearing in those equations that we want to highlight here. Similarly, there is a relationship between the differential operators of the equilibrium equations and the operators appearing in the strain–displacement relations. In this section we will use those relationships to write the governing equations in equivalent matrix–vector forms. At first glance these may seem to be simply formal mathematical relationships but bear with me, as their importance will become evident later in Chapter 9 where we will see that these relationships between matrix differential operators can be transformed into equivalent algebraic matrix relationships – relationships that can be used to describe the solution of the governing equations and boundary conditions solely in linear matrix–vector algebra terms. Such linear algebra formulations are ideally suited to obtaining numerical computer solutions of complex deformable body problems and complex structures such as trusses, frames, plates, shells, etc. We will see a simple example of such a formulation in the next section.

Consider the first set of governing equations for 3-D problems that we considered (3-D Governing Set 1). The equations of equilibrium can be rewritten in a matrix–vector form as

$$\begin{bmatrix} \partial/\partial x & 0 & 0 & 0 & \partial/\partial z & \partial/\partial y \\ 0 & \partial/\partial y & 0 & \partial/\partial z & 0 & \partial/\partial x \\ 0 & 0 & \partial/\partial z & \partial/\partial y & \partial/\partial x & 0 \end{bmatrix} \begin{Bmatrix} \sigma_{xx} \\ \sigma_{yy} \\ \sigma_{zz} \\ \sigma_{yz} \\ \sigma_{xz} \\ \sigma_{xy} \end{Bmatrix} + \begin{Bmatrix} f_x \\ f_y \\ f_z \end{Bmatrix} = 0 \qquad (6.43)$$

or, more succinctly

$$[\mathcal{L}_E]\{\sigma\} + \{f\} = 0 \qquad (6.44)$$

where $[\mathcal{L}_E]$ is the 3×6 matrix of derivatives appearing in Eq. (6.43), $\{\sigma\}$ is the 6×1 row vector of stresses, and $\{f\}$ is the 3×1 body force vector. We have already placed the stress–strain relations in a matrix–vector form as

$$\begin{Bmatrix} \sigma_{xx} \\ \sigma_{yy} \\ \sigma_{zz} \\ \sigma_{yz} \\ \sigma_{xz} \\ \sigma_{xy} \end{Bmatrix} = \begin{bmatrix} C_{11} & C_{12} & C_{13} & C_{14} & C_{15} & C_{16} \\ C_{21} & C_{22} & C_{23} & C_{24} & C_{25} & C_{26} \\ C_{31} & C_{32} & C_{33} & C_{34} & C_{35} & C_{36} \\ C_{41} & C_{42} & C_{43} & C_{44} & C_{45} & C_{46} \\ C_{51} & C_{52} & C_{53} & C_{54} & C_{55} & C_{56} \\ C_{61} & C_{62} & C_{63} & C_{64} & C_{65} & C_{66} \end{bmatrix} \begin{Bmatrix} e_{xx} \\ e_{yy} \\ e_{zz} \\ \gamma_{yz} \\ \gamma_{xz} \\ \gamma_{xy} \end{Bmatrix} \quad (6.45)$$

Writing the strains in terms of the stresses

$$\begin{Bmatrix} e_{xx} \\ e_{yy} \\ e_{zz} \\ \gamma_{yz} \\ \gamma_{xz} \\ \gamma_{xy} \end{Bmatrix} = \begin{bmatrix} D_{11} & D_{12} & D_{13} & D_{14} & D_{15} & D_{16} \\ D_{21} & D_{22} & D_{23} & D_{24} & D_{25} & D_{26} \\ D_{31} & D_{32} & D_{33} & D_{34} & D_{35} & D_{36} \\ D_{41} & D_{42} & D_{43} & D_{44} & D_{45} & D_{46} \\ D_{51} & D_{52} & D_{53} & D_{54} & D_{55} & D_{56} \\ D_{61} & D_{62} & D_{63} & D_{64} & D_{65} & D_{66} \end{bmatrix} \begin{Bmatrix} \sigma_{xx} \\ \sigma_{yy} \\ \sigma_{zz} \\ \sigma_{yz} \\ \sigma_{xz} \\ \sigma_{xy} \end{Bmatrix} \quad (6.46)$$

In matrix–vector form

$$\{\sigma\} = [C]\{e\} \quad \text{or} \quad \{e\} = [D]\{\sigma\} \quad (6.47)$$

in terms of the 6 × 6 elastic constants and compliance matrices discussed in Chapter 5 and the 6 × 1 strain vector, $\{e\}$, written in terms of the normal strains and the engineering shear strains. The strain–displacement relationship in matrix–vector form becomes

$$\begin{Bmatrix} e_{xx} \\ e_{yy} \\ e_{zz} \\ \gamma_{yz} \\ \gamma_{xz} \\ \gamma_{xy} \end{Bmatrix} = \begin{bmatrix} \partial/\partial x & 0 & 0 \\ 0 & \partial/\partial y & 0 \\ 0 & 0 & \partial/\partial z \\ 0 & \partial/\partial z & \partial/\partial y \\ \partial/\partial z & 0 & \partial/\partial x \\ \partial/\partial y & \partial/\partial x & 0 \end{bmatrix} \begin{Bmatrix} u_x \\ u_y \\ u_z \end{Bmatrix} \quad (6.48)$$

However, if we compare Eq. (6.48) with the equilibrium equations, Eq. (6.43), we see that the transpose of the equilibrium matrix of derivatives is just the matrix appearing here, i.e., we have

$$\{e\} = [\mathcal{L}_E]^T \{u\} \quad (6.49)$$

Thus, the governing equations for the 3-D Set 1 become simply the following.

Matrix–Vector 3-D Governing Set 1

$$[\mathcal{L}_E]\{\sigma\} + \{f\} = 0$$
$$\{\sigma\} = [C]\{e\} \quad \text{or} \quad \{e\} = [D]\{\sigma\} \quad (6.50)$$
$$\{e\} = [\mathcal{L}_E]^T [\{u\}$$

Note that, in this form, it is very easy to express the 3-D governing equations in terms of the displacements since we can place the strain–displacement relationship into the stress–strain equation and then place that result into the equilibrium equations to arrive at the following.

Matrix–Vector 3-D Governing Set 2

$$[\mathcal{L}_E][C][\mathcal{L}_E]^T\{u\} + \{f\} = 0 \quad (6.51)$$

These are the equivalent to Navier's equations for a general anisotropic media.

Now, consider the compatibility equations, which we can also write in a matrix–vector form where

$$\begin{bmatrix} 0 & -\partial^2/\partial z^2 & -\partial^2/\partial y^2 & \partial^2/\partial y \partial z & 0 & 0 \\ -\partial^2/\partial z^2 & 0 & -\partial^2/\partial x^2 & 0 & \partial^2/\partial x \partial z & 0 \\ -\partial^2/\partial y^2 & -\partial^2/\partial x^2 & 0 & 0 & 0 & \partial^2/\partial x \partial y \\ 0 & 0 & 2\partial^2/\partial x \partial y & -\partial^2/\partial x \partial z & -\partial^2/\partial y \partial z & \partial^2/\partial z^2 \\ 2\partial^2/\partial y \partial z & 0 & 0 & \partial^2/\partial x^2 & -\partial^2/\partial x \partial y & -\partial^2/\partial x \partial z \\ 0 & 2\partial^2/\partial x \partial z & 0 & -\partial^2/\partial x \partial y & \partial^2/\partial y^2 & -\partial^2/\partial y \partial z \end{bmatrix} \begin{Bmatrix} e_{xx} \\ e_{yy} \\ e_{zz} \\ \gamma_{yz} \\ \gamma_{xz} \\ \gamma_{xy} \end{Bmatrix} = 0 \quad (6.52)$$

or, more succinctly

$$[\mathcal{L}_S]\{e\} = 0 \quad (6.53)$$

The 6×6 compatibility matrix is also related to the transpose of the 3×6 matrix of equilibrium derivatives. By simple matrix multiplication it is easy (but rather lengthy) to show that

$$[\mathcal{L}_S][\mathcal{L}_E]^T = 0 \quad (6.54)$$

where the "zero" is actually a 6×3 matrix of zeros in Eq. (6.54). Taking the transpose of Eq. (6.54) it is also true that

$$[\mathcal{L}_E][\mathcal{L}_S]^T = 0 \quad (6.55)$$

where the "zero" is now a 3×6 matrix of zeros.

In the 3-D Governing Set 1, the compatibility equations did not appear since it was stated that the strain–displacement relations implicitly satisfy compatibility. With our present results it is easy to see this since, if we multiply the strain in the strain–displacement relationship of Eq. (6.49) by the matrix compatibility operator, we have, using Eq. (6.54),

$$[\mathcal{L}_S]\{e\} = [\mathcal{L}_S][\mathcal{L}_E]^T\{u\} = 0 \quad (6.56)$$

We can form up a set of governing equations similar to 3-D Governing Set 3 involving only the stresses by using the equilibrium equations and placing the stress–strain relations into these compatibility equations to arrive at the following.

> **Matrix–Vector 3-D Governing Set 3**
>
> $$[\mathcal{L}_E]\{\sigma\} + \{f\} = 0$$
> $$[\mathcal{L}_S][D]\{\sigma\} = 0$$
> (6.57)

But note that these are not the Beltrami–Michell compatibility equations since those equations, as derived in Appendix B, use a set of six compatibility equations different from Eq. (6.52) and also use the equilibrium equations to write the Beltrami–Michell equation in its final form, as seen in Eq. (6.5). In writing the compatibility equations, there is always some flexibility since we can use different sets of six equations formed from the original 81 equations (see Eq. (4.50)). However, regardless of the set we use, there are still only three independent compatibility relations contained in the six equations. In 2-D problems, the Beltrami–Michell equations are typically the equations of choice since, when combined with the Airy stress potential, the biharmonic equation appears, which is a well-studied equation.

We can form similar matrix–vector forms for 2-D plane-stress or plane-strain problems, but we will not cover those details here.

6.5 Equivalent Algebraic Matrix–Vector Governing Equations for Structures

The matrix–vector differential and algebraic governing equations and boundary conditions associated with deformable bodies described above have equivalent purely algebraic counterparts for structural problems that have long been studied. By making that equivalence more explicit we can learn about two general ways to solve problems. A simple problem that has been used for many years to illustrate a deformable structure is called Navier's table problem, as shown in Fig. 6.6a. A table with four legs (labeled 1, 2, 3, 4 in Fig. 6.6a) is a statically indeterminate problem since, if we make a cut through the legs, we see that there are four unknown internal forces (F_1, F_2, F_3, F_4) and corresponding compressive stresses $(\sigma_1, \sigma_2, \sigma_3, \sigma_4)$ acting (where $F_i = \sigma_i A (i = 1, 2, 3, 4)$, and A is the common cross-sectional area of the legs) but only three equilibrium equations. Thus, Navier's table problem must be treated as a deformable body in order to be solved. We will assume we can neglect the weight of the table top, which is assumed to be rigid, so that the only applied force is the load P, which is applied eccentrically at distances (e_x, e_y) from the center C of the table, as shown in Fig. 6.6a. The equilibrium equations are

6.5 Equivalent Algebraic Matrix–Vector Governing Equations

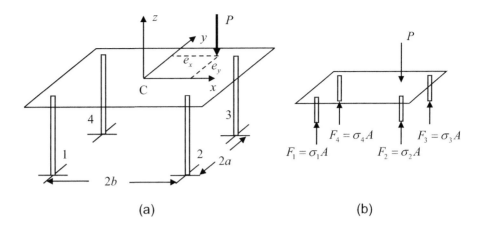

Figure 6.6 (a) An eccentrically loaded four-legged table (Navier's table problem). (b) The forces and corresponding stresses acting in the legs in equilibrium with the applied load, P.

$$\sum F_z = 0 \quad F_1 + F_2 + F_3 + F_4 - P = 0$$
$$\sum M_{Cx} = 0 \quad (F_3 + F_4)a - (F_1 + F_2)a - Pe_y = 0 \qquad (6.58)$$
$$\sum M_{Cy} = 0 \quad (F_1 + F_4)b - (F_2 + F_3)b + Pe_x = 0$$

which can be written in matrix–vector form in terms of the stresses as

$$\begin{bmatrix} 1 & 1 & 1 & 1 \\ -a & -a & a & a \\ b & -b & -b & b \end{bmatrix} \begin{Bmatrix} \sigma_1 \\ \sigma_2 \\ \sigma_3 \\ \sigma_4 \end{Bmatrix} = \begin{Bmatrix} P/A \\ Pe_y/A \\ -Pe_x/A \end{Bmatrix} \qquad (6.59)$$

or, more symbolically,

$$[\tilde{E}]\{\sigma\} = \{\sigma^*\} \qquad (6.60)$$

where $[\tilde{E}]$ is the 3×4 equilibrium matrix in Eq. (6.59) and

$$\{\sigma\} = \begin{Bmatrix} \sigma_1 \\ \sigma_2 \\ \sigma_3 \\ \sigma_4 \end{Bmatrix}, \quad \{\sigma^*\} = \begin{Bmatrix} P/A \\ Pe_y/A \\ -Pe_x/A \end{Bmatrix} \qquad (6.61)$$

where $\{\sigma\}$ is a stress vector and $\{\sigma^*\}$ can be considered to be a generalized applied "stress" vector (note, however, that some terms in $\{\sigma^*\}$ are not dimensionally stresses, but the equilibrium matrix consistent with these "stresses" is also not dimensionally homogeneous).

Now, let's consider the four strains in the legs which we call (e_1, e_2, e_3, e_4). Since the forces (F_1, F_2, F_3, F_4) were assumed to be compressive forces in the legs, the stresses also will be compressive, so that to be consistent we will also take the strains to be compressive strains

Figure 6.7 Deformations caused by the rotations of the table top, which is assumed to be rigid. The angles (θ_x, θ_y) are the rotations about the (x, y) axes, respectively. (a) Looking down the negative y-axis. (b) Looking down the positive x-axis.

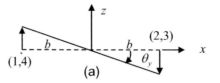
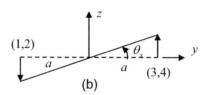

corresponding to shortening of the legs. Since we assumed that the table top was rigid, the top can only experience a translation in the plus or minus z-direction as well as rotations about the x- and y-axes (see Fig. 6.7). If we let w be the translation of the table in the negative z-direction and the angles (θ_x, θ_y) be the rotations about the (x, y) axes, respectively, then the total compressive strains in the legs are

$$\begin{aligned} e_1 &= (w - b\theta_y + a\theta_x)/L \\ e_2 &= (w + b\theta_y + a\theta_x)/L \\ e_3 &= (w + b\theta_y - a\theta_x)/L \\ e_4 &= (w - b\theta_y - a\theta_x)/L \end{aligned} \qquad (6.62)$$

which can be written in matrix–vector form as

$$\begin{Bmatrix} e_1 \\ e_2 \\ e_3 \\ e_4 \end{Bmatrix} = \begin{bmatrix} 1 & -a & b \\ 1 & -a & -b \\ 1 & a & -b \\ 1 & a & b \end{bmatrix} \begin{Bmatrix} w/L \\ \theta_x/L \\ \theta_y/L \end{Bmatrix} \qquad (6.63)$$

or, more symbolically,

$$\{e\} = [\tilde{E}]^T \{q\} \qquad (6.64)$$

where

$$\{e\} = \begin{Bmatrix} e_1 \\ e_2 \\ e_2 \\ e_4 \end{Bmatrix}, \quad \{q\} = \begin{Bmatrix} w/L \\ \theta_x/L \\ \theta_y/L \end{Bmatrix} \qquad (6.65)$$

Equation (6.64) is the relationship between the strain vector, $\{e\}$, and a generalized "displacement" vector, $\{q\}$. Again, we see the elements of this generalized displacement

6.5 Equivalent Algebraic Matrix–Vector Governing Equations

vector are not dimensionally homogeneous. The transpose of the equilibrium matrix appears in the generalized strain–displacement relationship of Eq. (6.64), which is consistent with what we saw earlier in the governing differential equations. Now consider the analogous stress–strain relations for this problem. The table legs are just members under uniaxial loads so, from Hooke's law, $\sigma_i = Ee_i$ ($i = 1, 2, 3, 4$), where E is Young's modulus. In matrix–vector form we have

$$\{\sigma\} = [C]\{e\} \tag{6.66}$$

where the generalized elastic constants matrix, $[C]$, is

$$[C] = \begin{bmatrix} E & 0 & 0 & 0 \\ 0 & E & 0 & 0 \\ 0 & 0 & E & 0 \\ 0 & 0 & 0 & E \end{bmatrix} \tag{6.67}$$

In summary, we have the equilibrium, strain–displacement, and stress–strain equations for the table to be

$$\begin{aligned} [\tilde{E}]\{\sigma\} &= \{\sigma^*\} \\ \{e\} &= [\tilde{E}]^T \{q\} \\ \{\sigma\} &= [C]\{e\} \end{aligned} \tag{6.68}$$

which are very similar in form to the 3-D Matrix–Vector Governing Set 1, except now the equilibrium matrix operator $[\mathcal{L}_E]$ is replaced by algebraic equilibrium matrix $[\tilde{E}]$. You might wonder about the boundary conditions. They have already been implicitly satisfied as part of the formulation of the governing equations in Eq. (6.68). There are displacement boundary $u_1 = u_2 = u_3 = u_4 = 0$ conditions at the floor (see Fig. 6.8) but these have already been satisfied by writing the strains in terms of the displacements of the table top only, instead of the difference between the table top displacements and the floor displacements. Similarly, the stress boundary conditions have also been accounted for by directly writing

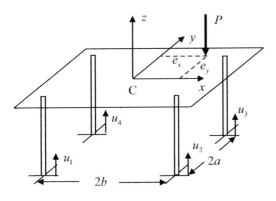

Figure 6.8 Navier's table problem, showing the displacements at the floor, all of which are assumed to be zero.

the unknown stresses in terms of the applied stresses $\{\sigma^*\}$. Thus, Eq. (6.68) contains a complete set of governing equations and boundary conditions for this problem.

6.5.1 Displacement Formulation

There are two ways to combine these governing equations to solve this problem. The first, a displacement formulation, is to place the stress–strain and strain–displacement relations into the equilibrium equations, yielding

$$[\tilde{E}][C][\tilde{E}]^T \{q\} = \{\sigma^*\} \qquad (6.69)$$

which is the purely algebraic equivalent of Navier's equation for the table problem. We can easily solve this problem numerically by placing explicit values into Eq. (6.69), but the problem is simple enough that we can also obtain the solution symbolically. MATLAB® can help us to do the necessary symbolic manipulations.

First, we define as symbolic the variables P, A, E, ex, ey, a, and b, which are the applied force, the cross-sectional area, Young's modulus, the eccentricities of the load (e_x, e_y), and the dimensions (a, b). We then form up the equilibrium matrix, Em = $[\tilde{E}]$, a known vector of stresses, pv = $\{\sigma^*\}$, and the elastic constants matrix Cm = $[C]$:

```
syms P A E ex ey a b
Em = [ 1 1 1 1; -a -a a a; b -b -b b];
pv = [ P/A ; P*ey/A ; -P*ex/A];
Cm = [ E 0 0 0;0 E 0 0;0 0 E 0;0 0 0 E];
```

Next, we form the matrix of coefficients in the displacement formulation, $M = [\tilde{E}][\tilde{C}][\tilde{E}]^T$, solve the system of linear equations, and simplify the final expression:

```
M = Em*Cm*Em.' ;
q = inv(M)*pv ;    % we can also write this as q = M\pv;
q = simplify(q)
```

Note the use of the MATLAB expression .' for the pure matrix transpose. The generalized displacement vector is

```
q =    P/(4*A*E)
       (P*ey)/(4*A*E*a^2)
      -(P*ex)/(4*A*E*b^2)
```

which we can multiply by L and rewrite as

$$w = \frac{PL}{4AE}$$

$$\theta_x = \frac{PLe_y}{4AEa^2} \qquad (6.70)$$

$$\theta_y = -\frac{PLe_x}{4AEb^2}$$

6.5 Equivalent Algebraic Matrix–Vector Governing Equations

If $e_x = e_y = 0$ then $w = PL/4AE$ and $\theta_x = \theta_y = 0$, so the table top does not tilt, as expected. To obtain the vector of compressive stresses in the legs sv = $\{\sigma\}$ from these displacements, we must write the stress–displacement relations:

$$\{\sigma\} = [C][\tilde{E}]^T \{q\} \tag{6.71}$$

which in MATLAB® is

```
sv = Cm*Em.'*q
sv =    P/(4*A)  -  (P*ey)/(4*A*a)  -  (P*ex)/(4*A*b)
        P/(4*A)  -  (P*ey)/(4*A*a)  +  (P*ex)/(4*A*b)
        P/(4*A)  +  (P*ey)/(4*A*a)  +  (P*ex)/(4*A*b)
        P/(4*A)  +  (P*ey)/(4*A*a)  -  (P*ex)/(4*A*b)
```

This result can be rewritten as

$$\begin{aligned}
\sigma_1 &= \frac{P}{4A}(1 - r_y - r_x) \\
\sigma_2 &= \frac{P}{4A}(1 - r_y + r_x) \\
\sigma_3 &= \frac{P}{4A}(1 + r_y + r_x) \\
\sigma_1 &= \frac{P}{4A}(1 + r_y - r_x)
\end{aligned} \tag{6.72}$$

where $r_x = e_x/b$, $r_y = e_y/a$. Equivalently, multiplying by A we obtain the forces in the legs. If we have $e_x = e_y = 0$ so the load is applied at the center of the table, then we have simply $\sigma_1 = \sigma_2 = \sigma_3 = \sigma_4 = P/4A$.

We can summarize the displacement formulation as a five-step process.

(1) Form up the equations of equilibrium

$$[\tilde{E}]\{\sigma\} = \{\sigma^*\}$$

(2) Write the strain–displacement relationship

$$\{e\} = [\tilde{E}]^T \{q\}$$

(3) Use the stress–strain relations to write the stress–displacement relationship

$$\{\sigma\} = [C]\{e\} \rightarrow \{\sigma\} = [C][\tilde{E}]^T\{q\}$$

(4) Place the stress–displacement relations into equilibrium and solve for the displacements

$$[\tilde{E}][C][\tilde{E}]^T\{q\} = \{\sigma^*\} \rightarrow \{q\}$$

(5) Calculate the stresses from the stress–displacement relations

$$\{\sigma\} = [C][\tilde{E}]^T\{q\}$$

6.5.2 Stress (Force) Formulation

A displacement formulation is a commonly used approach to solve structural problems. However, we can also formulate the table problem in terms of only the stresses (or, equivalently, the forces) by combining the equilibrium equations with the compatibility equations analogous to the approach used in the Matrix–Vector Governing Set 3. We have already seen that the compatibility equation can be obtained from the strain–displacement relations by eliminating the displacements to obtain relationships between the strains themselves. Since we have algebraic relations here, that elimination is easy to perform. From Eq. (6.63) we see by subtracting the first two relations and then adding the difference of the last two relations, we find

$$e_1 - e_2 + e_3 - e_4 = 0 \tag{6.73}$$

which we can write in matrix–vector form as

$$\begin{bmatrix} 1 & -1 & 1 & -1 \end{bmatrix} \begin{Bmatrix} e_1 \\ e_2 \\ e_3 \\ e_4 \end{Bmatrix} = 0 \tag{6.74}$$

or, more succinctly,

$$[\tilde{S}]\{e\} = 0 \tag{6.75}$$

where here $[\tilde{S}]$ is a 1×4 compatibility matrix. We can verify that the compatibility and the equilibrium matrices here satisfy the same relationship we found for their matrix operators, namely $[\tilde{S}][\tilde{E}]^T = [\tilde{E}][\tilde{S}]^T = 0$, by writing the latter of these two relationships, which yields

$$\begin{bmatrix} 1 & 1 & 1 & 1 \\ -a & -a & a & a \\ b & -b & -b & b \end{bmatrix} \begin{Bmatrix} 1 \\ -1 \\ 1 \\ -1 \end{Bmatrix} = \begin{Bmatrix} 0 \\ 0 \\ 0 \end{Bmatrix} \tag{6.76}$$

To combine the compatibility equations with the equilibrium equations we must write the compatibility equations in terms of the stresses, using the stress–strain relations written as $\{e\} = [D]\{\sigma\}$ in terms of the compliance matrix, $[D]$, where here

$$[D] = \begin{bmatrix} 1/E & 0 & 0 & 0 \\ 0 & 1/E & 0 & 0 \\ 0 & 0 & 1/E & 0 \\ 0 & 0 & 0 & 1/E \end{bmatrix} \tag{6.77}$$

But in this case the compatibility equations just become

$$[\tilde{S}][D]\{\sigma\} = \begin{bmatrix} 1/E & -1/E & 1/E & -1/E \end{bmatrix}\{\sigma\} = 0 \tag{6.78}$$

and the common factor of $1/E$ could be eliminated if we so desire. If we combine this compatibility equation with the equilibrium equations, we have

$$\begin{bmatrix} 1 & 1 & 1 & 1 \\ -a & -a & a & a \\ b & -b & -b & b \\ 1/E & -1/E & 1/E & -1/E \end{bmatrix} \begin{Bmatrix} \sigma_1 \\ \sigma_2 \\ \sigma_3 \\ \sigma_4 \end{Bmatrix} = \begin{Bmatrix} P/A \\ Pe_y/A \\ -Pe_x/A \\ 0 \end{Bmatrix} \qquad (6.79)$$

Using the same symbolic variables described previously, we can solve these equations in MATLAB® as:

```
Hm = [ 1 1 1 1; -a -a a a; b -b -b b; 1/E -1/E 1/E -1/E];
% equilibrium matrix
sv2 = [ P/A; P*ey/A; -P*ex/A; 0];          % right hand side
% of equilibrium equations
st2 = inv(Hm)*sv2;            % solve for stress vector st2.
% Can also use st2 = Hm\sv2;
simplify(st2)

ans =   -(P*(a*ex - a*b + b*ey))/(4*A*a*b)
         (P*(a*b + a*ex - b*ey))/(4*A*a*b)
         (P*(a*b + a*ex + b*ey))/(4*A*a*b)
         (P*(a*b - a*ex + b*ey))/(4*A*a*b)
```

which gives the same stresses again organized in a slightly different form. Note that Young's modulus is missing from these results, as we would expect, since it could be eliminated before solving. Since the stresses are known we can calculate the strains from the stress–strain relations and then use the strain–displacement relation to back-calculate the generalized displacements, if necessary. The stress formulation is nice in that, once the compatibility equations are known, we can solve for the stresses directly, which are often the quantities of most interest. We could summarize the stress formulation as a six-step process.

(1) Form up the equilibrium equations

$$[\tilde{E}]\{\sigma\} = \{\sigma^*\}$$

(2) Write the strain–displacement relationship

$$\{e\} = [\tilde{E}]^T \{q\}$$

(3) Eliminate the generalized displacements to obtain the compatibility equations

$$\{e\} = [\tilde{E}]^T \{q\} \rightarrow [\tilde{S}]\{e\} = 0$$

(4) Use the stress–strain relations to write the compatibility equations in terms of the stresses

$$\{e\} = [D]\{\sigma\} \to [\tilde{S}][D]\{\sigma\} = 0$$

(5) Append these compatibility equations to the equilibrium equations and solve for the stresses

$$\begin{bmatrix} \tilde{E} \\ - \\ [\tilde{S}][D] \end{bmatrix} \{\sigma\} = \begin{Bmatrix} \sigma^* \\ - \\ 0 \end{Bmatrix} \to \{\sigma\}$$

(6) Back-calculate the displacements, if needed, from the stress–displacement relations. This process will be described in more detail in Chapter 9, but here we can simply describe the process symbolically as:

$$\{\sigma\} = [C][\tilde{E}]^T \{q\} \to \{q\}$$

6.6 Structural Analysis – A Brief Preview

In our discussion of Navier's table problem, we used the stresses and strains to describe the problem so that we could make a close connection to the governing equations covered in the first sections of this chapter. Note that we did have to use generalized displacements, $\{q\}$, since the problem was best described in terms of both displacements and rotations. Similarly, engineers in structural problems often prefer to use generalized forces, $\{F\}$ (internal forces or moments or torques), instead of stresses. Generalized strains (deformations), $\{\Delta\}$, are also often used instead of the ordinary strains. For example, in Navier's table problem we could have replaced the strains in the strain-generalized displacement relations by the shortening of the legs $\Delta_i = e_i L (1 = 1, 2, 3, 4)$ instead of the strains themselves. The basic building blocks for a structural analysis, however, remain the same: the equilibrium equations for the generalized forces, the deformation (generalized strain)–generalized displacement relations and the generalized force–deformation relations, i.e.,

$$[\tilde{E}]\{F\} = \{P^*\}$$
$$\{\Delta\} = [\tilde{E}]^T \{q\} \tag{6.80}$$
$$\{\Delta\} = [G]\{F\} \quad \text{or} \quad \{F\} = [G]^{-1}\{\Delta\}$$

where $\{P^*\}$ are known generalized forces applied to the structure and $[G]$ is a flexibility matrix for the structure which plays the same role for the generalized forces and generalized strains that the compliance matrix does for the stresses and strains. These equations can be used to form and solve for the generalized displacements directly as done in the table problem

$$[\tilde{E}][G]^{-1}[\tilde{E}]^T \{q\} = \{P^*\} \tag{6.81}$$

or, for the generalized forces, if the compatibility equations

$$[\tilde{S}]\{\Delta\} = 0 \tag{6.82}$$

are obtained by eliminating the generalized displacements in the deformation-generalized displacements and then combining these compatibility equations, written in terms of the generalized forces, with the equilibrium equations to solve for the generalized forces:

$$\begin{bmatrix} \tilde{E} \\ - \\ [\tilde{S}][G] \end{bmatrix} \{F\} = \begin{Bmatrix} P^* \\ - \\ 0 \end{Bmatrix} \tag{6.83}$$

Historically, however, the displacement approach has not been implemented with Eq. (6.81) but with a closely related formulation based on the "stiffness" method of finite elements, as discussed briefly in Section 6.1. In terms of our present discussion, in a stiffness finite element approach the generalized forces would be expressed in terms of generalized displacements as

$$\{F\} = [K]\{q\} \tag{6.84}$$

where $[K]$ is a stiffness matrix for the structure. Typically, work–energy methods (see Chapter 8 and Chapter 9 for more details) are used to relate the generalized displacements to known generalized applied forces:

$$[K]\{q\} = \{P^*\} \tag{6.85}$$

In this approach the stiffness matrix is developed directly and is not formed from a combination of equilibrium and flexibility matrices although we see we could consider the matrix appearing in Eq. (6.81) as a pseudo stiffness matrix, K_{pseudo}, where

$$[K_{pseudo}] = [\tilde{E}][G]^{-1}[\tilde{E}]^T \tag{6.86}$$

In some problems the stiffness matrix and the pseudo stiffness matrix are identical but in general they are different. The stiffness-based finite element method, as mentioned earlier, has become the de facto method of choice for solving deformable body problems. In this method a deformable structure is broken into small elements (which is where the term "finite element" comes from) and the stiffness matrices are defined for those elements. The stiffness matrix for the entire structure can then be assembled from those elemental matrices, and Eq. (6.85) solved for the unknown generalized displacements once the known loads are expressed as known generalized forces, $\{P^*\}$. Generalized forces or stresses can then be calculated from the displacements. In contrast, when problems are solved with Eq. (6.83), where the generalized forces are the fundamental unknowns, this is called a force-based method. In this force-based method we can also assemble the equilibrium equations for an entire structure from the equilibrium equations for small elements, so that it also can be treated as a force-based finite element method.

There are, of course, many details involved in implementing stiffness-based or force-based finite elements that we have not discussed. We will get into the weeds with some of those

6.7 PROBLEMS

P6.1 The following state of stress

$$[\sigma] = \begin{bmatrix} Ay + Bz & Cz & -Cy \\ Cz & 0 & 0 \\ -Cy & 0 & 0 \end{bmatrix}$$

is supposed to exist in a cylindrical bar of length L and radius r, as shown in Fig. P6.1, where the bar is loaded only at its ends. Here, A, B, and C are constants and the cylinder is made from an isotropic elastic material. Show that this can be a legitimate state of stress by

(a) showing that it satisfies the equations of equilibrium everywhere in the cylinder,
(b) showing that it satisfies stress-free boundary conditions on the surface of the cylinder,
(c) showing that it produces strains that are compatible.
(d) What other conditions would have to be met before we could say that these stresses were an *exact* solution to a particular problem?
(e) If the cylinder is relatively long ($L \gg r$) use the idea of Saint-Venant and assume it is only important that these stresses generate the appropriate end forces and moments. What must those end forces and moments be in terms of A, B, C, and r?

P6.2 Consider three elastic rods that are all attached to a rigid floor and connected to a rigid bar, which is initially horizontal before being eccentrically loaded with a

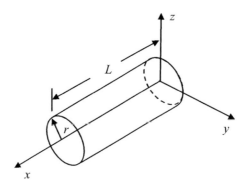

Figure P6.1 A cylindrical geometry.

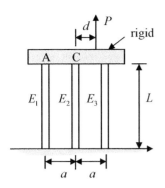

Fig. P6.2 Three elastic rods connected to an eccentrically loaded rigid bar.

force, P (see Fig. P6.2). Neglect the weight of all the members. The cross-sectional area of each rod is A, and they all have the same length, L, but different Young's moduli (E_1, E_2, E_3), as shown. This is a statically indeterminate problem that we can solve for the tensile forces (F_1, F_2, F_3) in the rods by combining equilibrium, compatibility, and the force–deformation (generalized stress–strain) relations.

(a) Summing forces vertically and taking moments about the center point C on the rigid bar, obtain the equilibrium equations for this problem in the form

$$[\tilde{E}]\{F\} = \{P^*\}$$

(b) If we let $(\Delta_1, \Delta_2, \Delta_3)$ be the generalized strains (elongations) of the rods, the generalized strain–displacement relations can be written as

$$\{\Delta\} = [\tilde{E}]^T \{q\}$$

where $\{q\}$ contains two unknown generalized displacements (q_1, q_2). From these strain–displacement relations, determine the physical meaning of these generalized displacements.

(c) By eliminating the displacements from the strain–displacement relations obtain the compatibility equation $[\tilde{S}]\{\Delta\} = 0$. Verify that $[\tilde{S}][\tilde{E}]^T = 0$.

(d) Relate the generalized strains to the forces through the relations $\{\Delta\} = [G]\{F\}$, where here the flexibility matrix $[G]$ contains the flexibilities of the rods $g_i = L/AE_i (i = 1, 2, 3)$ along its diagonal. Placing these relations into the compatibility equation, and then appending the resulting equation to the equilibrium equations, solve for the forces (F_1, F_2, F_3) in the bars in terms of (g_i, P, a, d).

(e) Repeat parts (a)–(d) if our equilibrium equations are instead obtained by summing forces vertically and taking moments about point A on the rigid bar. Verify that you obtain the same solution for the forces and show how the point about which the moment is taken is related to the displacements present in the strain–displacement relations.

P6.3 Consider the torsion of a circular rod that composed of two sections, AB and BC, that are welded together at B and are fixed between two rigid supports at A and C (see Fig. P6.3a). The shear moduli, polar area moments, and lengths of sections AB and BC are $(\mu_i, J_i, L_i)(i = 1, 2)$, respectively, as shown in Fig. P6.3a. [Note: we have used μ instead of G for the shear modulus since G has been used to denote the flexibility matrix.] A torque, T, acts at the junction B between AB and BC and there are two unknown end reaction torques (T_A, T_C) as shown in Fig. P6.3b. The twist at the applied torque is θ_B. This is a statically indeterminate problem that we can solve for the internal torques (M_1, M_2) acting in sections AB and BC as shown in Fig. P6.3c by combining equilibrium, compatibility and the torque–twist relations.

Figure P6.3 (a) Two bars welded together at B and fixed at sections A and C. (b) The two bar assembly showing the applied torque T, the end reaction torques (T_A, T_C) and the angle of twist, θ_B, at the applied torque. (c) The internal moments (M_1, M_2) in the bars (assumed to act in the directions shown).

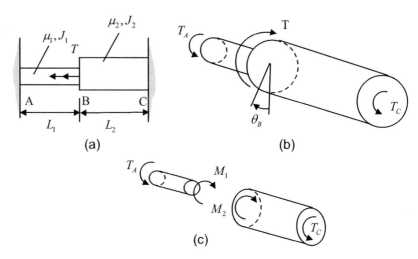

(a) Write down the equilibrium equation relating (M_1, M_2) to the applied torque, T in the form

$$[\tilde{E}]\{M\} = \{T^*\}$$

(b) If we let (Δ_1, Δ_2) be the generalized strains (which in this case are taken as the twists) of sections AB and BC the generalized strain–displacement relation can be written as

$$\{\Delta\} = [\tilde{E}]^T \{q\}$$

From this strain–displacement relation determine the physical meaning of the generalized displacement $\{q\}$.

(c) By eliminating the generalized displacement from the strain–displacement relation obtain the compatibility equation in the form $[\tilde{S}]\{\Delta\} = 0$. Verify that $[\tilde{S}][\tilde{E}]^T = 0$.

(d) Relate the generalized strains to the internal torques through the relations $\{\Delta\} = [G]\{M\}$, where the flexibility matrix $[G]$ contains the torsional flexibilities of the sections AB and BC $g_i = L_i/\mu_i J_i (i = 1, 2)$ along its diagonal. Placing these

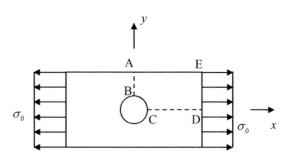

Figure P6.4 A thin plate with a central hole under uniaxial tension.

relations into the compatibility equation, and then appending the resulting equation to the equilibrium equations, solve for the internal torques (M_1, M_2) in terms of (g_i, T).

P6.4 Consider a thin plate with a central hole and loaded by a uniform normal stress on its ends, as shown in Fig. P6.4. If we analyze only the first quarter of this plate, region ABCDE, what are the boundary conditions we must satisfy on the boundaries of this region? Assume plane-stress conditions.

P6.5 Show that on a stress-free surface, the boundary conditions can be expressed in terms of the Airy stress function as the conditions, $\phi = 0$ and $\partial \phi / \partial n = 0$, where n is in a direction normal to the surface.

References

1. J. N. Reddy, *Introduction to the Finite Element Method*, 4th edn (New York, NY: McGraw-Hill, 2018)
2. M. Bonnet, *Boundary Integral Equation Methods for Solids and Fluids* (New York, NY: John Wiley, 1999)
3. A. C. Ugural and S. K. Fenster, *Advanced Mechanics of Materials and Applied Elasticity*, 5th edn (Boston, MA: Prentice Hall, 2012)
4. C. E. Pearson, *Theoretical Elasticity* (Cambridge, MA: Harvard University Press, 1959)

7 Analytical Solutions

Most problems involving complex-shaped deformable bodies require a numerical solution of the governing equations and boundary conditions. However, there are a number of important simple problems that can be solved analytically and that reveal the nature of the stresses and deformations in geometries of practical significance. In this chapter we will examine a number of such problems, using some of the various sets of governing equations discussed in the previous chapter.

7.1 Displacement-Based Solutions

Reducing the governing equations to Navier's equations is one of the important choices outlined in the previous chapter. In this section we will examine cylindrical geometries under radial pressure loads in configurations where explicit solutions to Navier's equations can be obtained.

7.1.1 Axisymmetric Solutions in Cylindrical Geometries

One case where we can solve the governing equations analytically is the case of axisymmetric solutions in cylindrical coordinates where there is no dependence of the solution on the angle θ and where we can assume that the stresses developed satisfy

$$\sigma_{rr} = \sigma_{rr}(r), \quad \sigma_{\theta\theta} = \sigma_{\theta\theta}(r) \\ \sigma_{r\theta} = \sigma_{rz} = \sigma_{z\theta} = \sigma_{zz} = 0 \tag{7.1}$$

In this case the equilibrium equations which in cylindrical coordinates with no body forces are

$$\frac{\partial \sigma_{rr}}{\partial r} + \frac{\sigma_{rr} - \sigma_{\theta\theta}}{r} + \frac{1}{r}\frac{\partial \sigma_{r\theta}}{\partial \theta} + \frac{\partial \sigma_{rz}}{\partial z} = 0 \\ \frac{\partial \sigma_{r\theta}}{\partial r} + \frac{2\sigma_{r\theta}}{r} + \frac{1}{r}\frac{\partial \sigma_{\theta\theta}}{\partial \theta} + \frac{\partial \sigma_{\theta z}}{\partial z} = 0 \\ \frac{\partial \sigma_{zr}}{\partial r} + \frac{\sigma_{zr}}{r} + \frac{1}{r}\frac{\partial \sigma_{z\theta}}{\partial \theta} + \frac{\partial \sigma_{zz}}{\partial z} = 0 \tag{7.2}$$

reduce to only one equation

$$\frac{\partial \sigma_{rr}}{\partial r} + \frac{\sigma_{rr} - \sigma_{\theta\theta}}{r} = 0 \tag{7.3}$$

The stress state given by Eq. (7.1) is one of plane stress so that the stress–strain relations are the plane-stress equations

$$\sigma_{rr} = \frac{E}{1-v^2}(\varepsilon_{rr} + v\varepsilon_{\theta\theta})$$

$$\sigma_{\theta\theta} = \frac{E}{1-v^2}(\varepsilon_{\theta\theta} + v\varepsilon_{rr}) \quad (7.4)$$

$$\varepsilon_{zz} = \frac{-v}{E}(\sigma_{rr} + \sigma_{\theta\theta})$$

To guarantee that we automatically satisfy compatibility, we will work directly with the displacements as the fundamental unknowns for this problem. Because of the assumed radial symmetry, we expect that $u_\theta = 0$, i.e., the θ-displacement is zero. Since the stresses are uniform in the z-direction, we also expect that the radial displacement will be uniform so that $u_r = u_r(r)$. The strains in cylindrical coordinates are

$$\varepsilon_{rr} = \frac{\partial u_r}{\partial r}, \quad \varepsilon_{\theta\theta} = \frac{1}{r}\frac{\partial u_\theta}{\partial \theta} + \frac{u_r}{r}, \quad \varepsilon_{zz} = \frac{\partial u_z}{\partial z}$$

$$\varepsilon_{r\theta} = \frac{1}{2}\left(\frac{1}{r}\frac{\partial u_z}{\partial \theta} + \frac{\partial u_\theta}{\partial r} - \frac{u_\theta}{r}\right)$$

$$\varepsilon_{rz} = \frac{1}{2}\left(\frac{\partial u_z}{\partial r} + \frac{\partial u_r}{\partial z}\right) \quad (7.5)$$

$$\varepsilon_{\theta z} = \frac{1}{2}\left(\frac{1}{r}\frac{\partial u_z}{\partial \theta} + \frac{\partial u_\theta}{\partial z}\right)$$

which, under our assumptions, reduce to

$$\varepsilon_{rr} = \frac{du_r}{dr}, \quad \varepsilon_{\theta\theta} = \frac{u_r}{r}, \quad \varepsilon_{zz} = \frac{\partial u_z}{\partial z}$$

$$\varepsilon_{r\theta} = \varepsilon_{\theta z} = \varepsilon_{rz} = 0 \quad (7.6)$$

Placing these strain–displacement relations into the stress–strain relations of Eq. (7.4) gives

$$\sigma_{rr} = \frac{E}{1-v^2}\left(\frac{du_r}{dr} + v\frac{u_r}{r}\right)$$

$$\sigma_{\theta\theta} = \frac{E}{1-v^2}\left(\frac{u_r}{r} + v\frac{du_r}{dr}\right) \quad (7.7)$$

$$\varepsilon_{zz} = \frac{\partial u_z}{\partial z} = \frac{-v}{E}(\sigma_{rr} + \sigma_{\theta\theta})$$

so that when the stresses of Eq. (7.7) are placed into the equilibrium equation, Eq. (7.3), we find

$$\frac{d^2 u_r}{dr^2} + \frac{1}{r}\frac{du_r}{dr} - \frac{u_r}{r^2} = 0 \quad (7.8)$$

This is Navier's equation for this radially symmetric problem. You may wonder why we did not obtain this equation directly from the plane-stress Navier's equations outlined in the previous chapter. The reason is that, in this problem, some of our assumptions are based on stresses and some are based on displacements so that it is more straightforward to obtain the stresses as a function of the reduced displacements, as done in Eq. (7.7), and then use those equations in the equilibrium equations to deduce Navier's equations. It is easy to solve Eq. (7.8) for the displacement since we can write it equivalently in the form

$$\frac{d}{dr}\left[\frac{1}{r}\frac{d}{dr}(ru_r)\right] = 0 \qquad (7.9)$$

Integrating once on r gives

$$\frac{d}{dr}(ru_r) = C'_1 r \qquad (7.10)$$

where C'_1 is a constant of integration. Integrating Eq. (7.10) once more then yields

$$u_r = \frac{C'_1}{2}r + \frac{C_2}{r} = C_1 r + \frac{C_2}{r} \qquad (7.11)$$

where $C_1 = C'_1/2$, C_2 are both constants of integration. If Eq. (7.11) is placed back into the stress–displacement relations, Eq. (7.7), the stresses and axial strain are

$$\sigma_{rr} = \frac{E}{1-v}C_1 - \frac{E}{1+v}\frac{C_2}{r^2}$$

$$\sigma_{\theta\theta} = \frac{E}{1-v}C_1 + \frac{E}{1+v}\frac{C_2}{r^2} \qquad (7.12)$$

$$\varepsilon_{zz} = \frac{-2v}{1-v}C_1 = \text{constant}$$

If we define two new constants (A, B) as

$$A = \frac{E}{1-v}C_1, \quad B = \frac{E}{1+v}C_2 \qquad (7.13)$$

then we can rewrite the stresses and axial strain simply as

$$\sigma_{rr} = A - \frac{B}{r^2}$$

$$\sigma_{\theta\theta} = A + \frac{B}{r^2} \qquad (7.14)$$

$$\varepsilon_{zz} = \frac{\partial u_z}{\partial z} = \frac{-2vA}{E} = \text{constant}$$

Equation (7.11) and Eq. (7.14) yield the displacements and stresses for our axisymmetric problem. They satisfy exactly the governing equations of equilibrium, compatibility, and the stress–strain relations. If we can also exactly satisfy the boundary conditions, then we will have an exact "elasticity" solution.

7.1.2 Thick-Wall Pressure Vessel

Consider the problem shown in Fig. 7.1, where a cylindrical vessel is loaded by uniform pressures (p_i, p_e) on the inside and outside of the cylinder, respectively. We will also assume the ends of the cylinder are stress free. A real pressure vessel will have its ends closed in some fashion, but we will deal with that case shortly. For the case shown in Fig. 7.1, the boundary conditions are: (1) for $0 \leq \theta \leq 2\pi, 0 \leq z \leq L$ on the inner and outer surfaces we have

$$\begin{aligned} \sigma_{rr}(r,\theta,z)|_{r=r_i} &= -p_i \\ \sigma_{r\theta}(r,\theta,z)|_{r=r_i} &= \sigma_{rz}(r,\theta,z)|_{r=r_i} = 0 \\ \sigma_{rr}(r,\theta,z)|_{r=r_e} &= -p_e \\ \sigma_{r\theta}(r,\theta,z)|_{r=r_e} &= \sigma_{rz}(r,\theta,z)|_{r=r_e} = 0 \end{aligned} \qquad (7.15)$$

and (2) for $0 \leq \theta \leq 2\pi, r_i \leq r \leq r_e$ on the cylinder ends

$$\begin{aligned} \sigma_{zz}(r,\theta,z)|_{z=0} &= \sigma_{zr}(r,\theta,z)|_{z=0} = \sigma_{z\theta}(r,\theta,z)|_{z=0} = 0 \\ \sigma_{zz}(r,\theta,z)|_{z=L} &= \sigma_{zr}(r,\theta,z)|_{z=L} = \sigma_{z\theta}(r,\theta,z)|_{z=L} = 0 \end{aligned} \qquad (7.16)$$

But from our original assumptions on the stresses, Eq. (7.1), all of these boundary conditions are satisfied identically except the two conditions

$$\begin{aligned} \sigma_{rr}(r)|_{r=r_i} &= -p_i \\ \sigma_{rr}(r)|_{r=r_e} &= -p_e \end{aligned} \qquad (7.17)$$

which, from Eq. (7.14), gives

$$\begin{aligned} A - \frac{B}{r_i^2} &= -p_i \\ A - \frac{B}{r_e^2} &= -p_e \end{aligned} \qquad (7.18)$$

Solving for (A, B) then we have

$$A = \frac{r_i^2 p_i - r_e^2 p_e}{r_e^2 - r_i^2}, \quad B = \frac{r_i^2 r_e^2 (p_i - p_e)}{r_e^2 - r_i^2} \qquad (7.19)$$

These constants then yield complete solutions for the stresses and displacement of the cylinder:

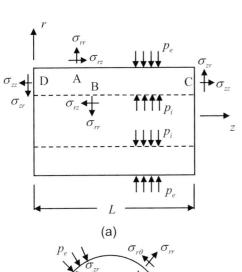

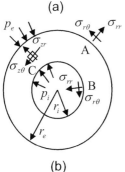

Figure 7.1 A thick-wall cylindrical pressure vessel subjected to internal and external pressures along its length (only parts of the pressure distributions are shown) showing the stresses acting (a) in a longitudinal plane, and (b) in a cross-sectional plane. Generic points where these stresses act are shown as points (A, B, C, D).

Analytical Solutions

$$\sigma_{rr} = \frac{p_i r_i^2}{r_e^2 - r_i^2}\left(1 - \frac{r_e^2}{r^2}\right) - \frac{p_e r_e^2}{r_e^2 - r_i^2}\left(1 - \frac{r_i^2}{r^2}\right)$$

$$\sigma_{\theta\theta} = \frac{p_i r_i^2}{r_e^2 - r_i^2}\left(1 + \frac{r_e^2}{r^2}\right) - \frac{p_e r_e^2}{r_e^2 - r_i^2}\left(1 - \frac{r_i^2}{r^2}\right)$$

$$\varepsilon_{zz} = \frac{du_z}{dz} = \frac{-2\nu}{E}\left(\frac{r_i^2 p_i - r_e^2 p_e}{r_e^2 - r_i^2}\right) \rightarrow u_z = \frac{-2\nu}{E}\left(\frac{r_i^2 p_i - r_e^2 p_e}{r_e^2 - r_i^2}\right)z + C$$

$$u_r = \frac{(1-\nu)}{E}\left(\frac{r_i^2 p_i - r_e^2 p_e}{r_e^2 - r_i^2}\right)r + \frac{(1+\nu)}{E}\frac{r_e^2 r_i^2 (p_i - p_e)}{r_e^2 - r_i^2}\frac{1}{r}$$

(7.20)

where we have integrated the strain, e_{zz}, once on z to obtain the z-displacement in terms of a constant of integration, C.

The solution of Eq. (7.20) is an exact solution to our pressurized cylinder problem since it satisfies all the governing equations and boundary conditions. (Note that the homework problems P3.1 and P3.2 in Chapter 3 considered only satisfaction of the equilibrium equations and boundary conditions.) One can show that for a pressurized cylinder subject to an internal pressure only, the highest stresses occur on the inner wall where we have

$$\sigma_{rr}|_{r=r_i} = \frac{p_i r_i^2}{r_e^2 - r_i^2}\left(1 - \frac{r_e^2}{r_i^2}\right) = -p_i$$

$$\sigma_{\theta\theta}|_{r=r_i} = \frac{p_i r_i^2}{r_e^2 - r_i^2}\left(1 + \frac{r_e^2}{r_i^2}\right)$$

(7.21)

For a thin-wall cylinder these stresses become

$$\sigma_{rr}|_{r=r_i} = -p_i$$

$$\sigma_{\theta\theta}|_{r=r_i} = \frac{p_i r_i^2}{r_e^2 - r_i^2}\left(1 + \frac{r_e^2}{r_i^2}\right) = \frac{p_i}{t}\left(\frac{r_e^2 + r_i^2}{r_e + r_i}\right) = \frac{p_i}{t}\left(\frac{r_e + r_i}{2} + \frac{t^2}{2(r_e + r_i)}\right)$$

$$\approx \frac{p_i}{t}\left(\frac{r_e + r_i}{2}\right) = \frac{p_i r_m}{t}$$

(7.22)

in terms of the thickness $t = r_e - r_i$ and the mean radius $r_m = (r_e + r_i)/2$. This agrees with the result for the thin-wall cylindrical pressure vessel hoop stress usually obtained in strength of materials courses (where the small radial stress component is normally ignored).

Now, let's come back to the question of the boundary conditions at the ends, since the "open ends" conditions that our exact solution satisfies are not those usually found in practice. Normally the ends of a pressure vessel are sealed, as shown in Fig. 7.2, by placing end caps on the cylinder. Figure 7.2a shows where a flat plate is welded to the cylinder, while Fig. 7.2b shows spherical end caps. In either case the end caps have a different radial stiffness than the cylinder so that the tendency of the cylinder to expand is not identical to the end caps. This will cause shear stresses and bending stresses to develop in the region near where they are connected. However, the pressure acting on the end caps also will cause the cylinder to be stretched in the axial z-direction throughout its length under an axial stress and corresponding axial load. Using the principle of Saint-Venant, we might expect that if we are not too close to the end connection (dashed lines in Fig. 7.2), the cylinder will experience a uniform axial stress, σ_{zz}, across its thickness. Thus, consider the case of a flat end cap, as shown in Fig 7.3. Summing forces in the axial direction gives

$$\sigma_{zz}\pi(r_e^2 - r_i^2) = p_i\pi r_i^2 - p_e\pi r_i^2 \tag{7.23}$$

which gives

$$\sigma_{zz} = \frac{p_i r_i^2 - p_e r_e^2}{r_e^2 - r_i^2} \tag{7.24}$$

This can also be shown to be the axial stress for other types of end caps. From generalized Hooke's law

$$\frac{u_r}{r} = e_{\theta\theta} = \frac{\sigma_{\theta\theta}}{E} - \frac{v(\sigma_{rr} - \sigma_{zz})}{E} \tag{7.25}$$

But the solution for the open-ends case satisfies plane-stress conditions where

$$e_{\theta\theta} = \frac{\sigma_{\theta\theta}}{E} - \frac{v\sigma_{rr}}{E} \tag{7.26}$$

so we see that adding end caps produces an additional axial stress term, which in turn affects the radial displacement in the cylinder by adding the term

$$(u_r)_{\sigma_{zz}} = \frac{-v\sigma_{zz}r}{E} \tag{7.27}$$

with σ_{zz} given by Eq. (7.24). The stresses $\sigma_{rr}, \sigma_{\theta\theta}$, are unaffected. The axial strain, e_{zz}, is also changed so the total axial strain is now, by superposition with the previous strains,

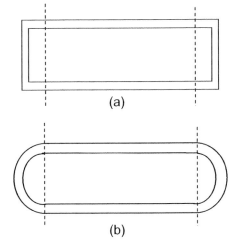

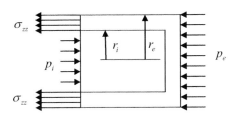

Figure 7.2 A cylindrical pressure vessel closed at its ends by (a) a flat plate, and (b) a half-sphere end cap.

Figure 7.3 Stress and pressures acting in the axial direction of a closed cylinder.

$$e_{zz} = (e_{zz})_{\sigma_{zz}} + (e_{zz})_{\sigma_{rr},\sigma_{\theta\theta}}$$

$$= \frac{\sigma_{zz}}{E} - \frac{2v}{E}\left(\frac{p_i r_i^2 - p_e r_e^2}{r_e^2 - r_i^2}\right) \qquad (7.28)$$

$$= \left(\frac{1-2v}{E}\right)\left(\frac{p_i r_i^2 - p_e r_e^2}{r_e^2 - r_i^2}\right)$$

In summary, the stresses and displacements for the closed pressure vessel are

$$\sigma_{rr} = \frac{p_i r_i^2}{r_e^2 - r_i^2}\left(1 - \frac{r_e^2}{r^2}\right) - \frac{p_e r_e^2}{r_e^2 - r_i^2}\left(1 - \frac{r_i^2}{r^2}\right)$$

$$\sigma_{\theta\theta} = \frac{p_i r_i^2}{r_e^2 - r_i^2}\left(1 + \frac{r_e^2}{r^2}\right) - \frac{p_e r_e^2}{r_e^2 - r_i^2}\left(1 + \frac{r_i^2}{r^2}\right)$$

$$\sigma_{zz} = \frac{p_i r_i^2 - p_e r_e^2}{r_e^2 - r_i^2} \qquad (7.29)$$

$$u_r = \frac{(1-2v)}{E}\left(\frac{r_i^2 p_i - r_e^2 p_e}{r_e^2 - r_i^2}\right)r + \frac{(1+v)r_e^2 r_i^2 (p_i - p_e)}{E} \frac{1}{r_e^2 - r_i^2} \frac{1}{r}$$

$$u_z = \left(\frac{1-2v}{E}\right)\left(\frac{r_i^2 p_i - r_e^2 p_e}{r_e^2 - r_i^2}\right)z + C$$

Note that these are no longer exact solutions for the stresses in the cylinder since we have used the Saint-Venant principle to arrive at these results. Stresses and deformations in the end caps are obviously much more difficult to describe.

We should make a few remarks about the assumptions we made originally in this problem which led to the solution for the open cylinder being a plane-stress solution. Plane stress is normally thought to apply to a very thin body, not a long cylinder under internal loads. However, we showed that our solution was in fact an exact solution to the governing equations and boundary conditions regardless of the length of the cylinder. Physically, in the open cylinder case the cylinder is free to expand uniformly in the z-direction regardless of how long the cylinder is and the ends are stress-free so a σ_{zz} is not needed anywhere in the cylinder. Thus, there is no contradiction with having a state of plane stress for the long, open cylinder.

Example 7.1 Thick-Wall Pressure Vessel

A closed cylindrical vessel has an inner radius $r_i = 10$ mm and an outer radius $r_e = 20$ mm and is subjected to an inside pressure p_i only. The properties of the vessel are $E = 210$ GPa, $v = 0.29$. A strain gage placed on the outside of the vessel measures a circumferential strain $e_{\theta\theta} = 400$ μ when the inside pressure is applied. What is the internal pressure p_i?

The stresses at outside of the vessel are

$$\sigma_{rr} = 0$$

$$\sigma_{\theta\theta} = \frac{2p_i r_i^2}{r_e^2 - r_i^2} = \frac{(2)(100)}{400 - 100} p_i = 0.667 p_i$$

$$\sigma_{zz} = \frac{p_i r_i^2}{r_e^2 - r_i^2} = \frac{100}{400 - 100} p_i = 0.333 p_i$$

The measured hoop strain is

$$e_{\theta\theta} = \frac{1}{E}(\sigma_{\theta\theta} - \nu\sigma_{rr} - \nu\sigma_{zz})$$

$$\rightarrow 400 \times 10^{-6} = p_i \frac{1}{210 \times 10^3}[0.667 - (0.29)(0.333)]$$

so that solving for p_i we find $p_i = 147$ MPa.

7.1.3 Shrink-Fits

Another problem where we can use the axisymmetric solutions of a cylinder with free ends subjected to inner or outer pressures is where two cylinders are pressed together by a "shrink-fit" process where, say, the outer cylinder is heated and allowed to expand, and then fitted over an inner cylinder (having a slightly larger outer radius than that of the inner radius of the outer cylinder at room temperature) and allowed to cool, creating a final state where there is a large residual pressure between the two cylinders. This final state of loading is shown in Fig. 7.4, where the outer cylinder is initially assumed to have an outer radius, r_e, and an inner radius, r_i, and the inner cylinder initially has an outer radius, $r_i + \Delta$, and an inner radius, r_u, all as measured at room temperature. The quantity Δ is called the

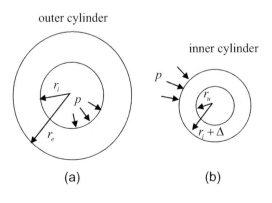

Figure 7.4 A shrink-fit problem where an outer cylinder, shown in (a), has a slightly smaller inner radius, r_i, than the outer radius of the inner cylinder, $r_i + \Delta$, shown in (b), and where Δ is the interference and where p is the interference pressure when they are pressed together in the shrink-fit process.

interference. The mutual pressure between the two cylinders, p, is called the interference pressure. For either cylinder the stresses and the radial displacements are given by the forms derived previously, namely the stresses from Eq. (7.14) and the radial displacement from Eq. (7.11), using the relationship between the constants in Eq. (7.13):

$$\sigma_{rr} = A - \frac{B}{r^2}$$
$$\sigma_{\theta\theta} = A + \frac{B}{r^2} \tag{7.30}$$
$$u_r = \frac{1}{E}\left[(1+\nu)\frac{B}{r} + (1-\nu)Ar\right]$$

We will let the outer cylinder have a Young's modulus and Poisson's ratio of (E_1, ν_1), respectively, and the inner cylinder will have the corresponding properties (E_2, ν_2). Then the stresses in the outer cylinder will be given as

$$\sigma_{rr} = A - \frac{B}{r^2}$$
$$\sigma_{\theta\theta} = A + \frac{B}{r^2} \tag{7.31}$$

and, for the inner cylinder,

$$\sigma_{rr} = A' - \frac{B'}{r^2}$$
$$\sigma_{\theta\theta} = A' + \frac{B'}{r^2} \tag{7.32}$$

Solutions for both the inner and outer cylinders subjected to the interference pressure, p, can be obtained from our general solution to the thick-wall cylinder problem with free ends (see Eq. (7.20)), giving

$$A = \frac{pr_i^2}{r_e^2 - r_i^2}$$
$$B = \frac{pr_i^2 r_e^2}{r_e^2 - r_i^2}$$
$$A' = \frac{-pr_i^2}{r_i^2 - r_u^2} \tag{7.33}$$
$$B' = \frac{-pr_i^2 r_u^2}{r_i^2 - r_u^2}$$

Thus, the stresses are known once we find the interference pressure, p. These stresses do not involve the elastic constants but the displacements do. If we let $u_r^{(o)}$ be the displacement of the outer cylinder and $u_r^{(i)}$ be the displacement of the inner cylinder, then

$$u_r^{(o)} = \frac{1}{E_1}\left[(1+v_1)\frac{B}{r} + (1-v_1)Ar\right]$$
$$u_r^{(i)} = \frac{1}{E_2}\left[(1+v_2)\frac{B'}{r} + (1-v_2)A'r\right]$$
(7.34)

The final state of the interface between the two cylinders after the shrink-fit process is shown in Fig. 7.5. The outer cylinder experiences a stretching corresponding to $u_r^{(o)}$ and the inner cylinder experiences a shortening, $-u_r^{(i)}$. The interface between the two cylinders after the shrink-fit lies somewhere between r_i and $r_i + \Delta$, but the interference is usually quite small so we can assume from Fig. 7.5 that

$$u_r^{(o)} - u_r^{(i)} = \Delta \quad at\ r \cong r_i \quad (7.35)$$

If we place Eq. (7.34) into Eq. (7.35), we can obtain the relationship between the pressure, p, and the interference, Δ. The result is algebraically rather complex, so consider the important special case where $E_1 = E_2 = E$, $v_1 = v_2 = v$, i.e., the materials are the same for both cylinders. In that case we find from Eq. (7.35)

$$\frac{p}{E}\left[\frac{(1+v)r_e^2 r_i + (1-v)r_i^3}{r_e^2 - r_i^2}\right] - \frac{p}{E}\left[\frac{-(1+v)r_u^2 r_i - (1-v)r_i^3}{r_i^2 - r_u^2}\right] = \Delta \quad (7.36)$$

Solving for Δ in terms of p:

$$\Delta = \frac{p}{E}\left[\frac{2r_i^3(r_e^2 - r_u^2)}{(r_e^2 - r_i^2)(r_i^2 - r_u^2)}\right] \quad (7.37)$$

or, solving for p,

$$p = E\Delta\left[\frac{(r_e^2 - r_i^2)(r_i^2 - r_u^2)}{2r_i^3(r_e^2 - r_u^2)}\right] \quad (7.38)$$

In these shrink-fit solutions, we again used the plane-stress solutions found for open cylinders. One might question the plane-stress conditions in this case, however, because in the shrink-fit process there may be axial frictional forces developed that restrain the axial deformations of the cylinders. If, in fact, we assume that there is no axial deformation allowed during the shrink-fit, then we have a plane-strain situation instead. From the results of Chapter 6, we can obtain the plane-strain result corresponding to Eq. (7.38) by making the replacement $E \to E/(1-v^2)$.

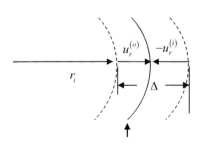

final position of the interface

Figure 7.5 The displacements experienced at the interface between the outer cylinder and inner cylinder in a shrink-fit.

Example 7.2 Shrink-Fit

Consider the shrink-fit between a cylindrical jacket and a solid cylinder ($r_u = 0$). (a) What is the relationship between the interference pressure, p, and the interference, Δ? (b) If the jacket and the solid cylinder both have the same length L before the shrink fit, what is the expression for their difference in lengths afterwards? Assume that there is no axial stress developed between the two bodies during the shrink-fit.

(a) From Eq. (7.38) we have, with $r_u = 0$

$$p = E\Delta \left[\frac{(r_e^2 - r_i^2)}{2 r_i r_e^2} \right]$$

(b) The stresses in the solid cylinder are (see Eq. (7.32) and Eq. (7.33)) $\sigma_{rr}^s = \sigma_{\theta\theta}^s = -p$, $\sigma_{zz} = 0$ and the axial strain is

$$e_{zz}^s = \frac{1}{E} (\sigma_{zz} - v\sigma_{rr} - v\sigma_{\theta\theta}) = \frac{2vp}{E}$$

For the jacket we have $e_{zz}^j = \frac{1}{E}\left(0 - v\sigma_{rr}^j - v\sigma_{\theta\theta}^j\right) = \frac{-v\left(\sigma_{rr}^j + \sigma_{\theta\theta}^j\right)}{E}$ but from Eq. (7.32) and Eq. (7.33):

$$\sigma_{rr}^j + \sigma_{\theta\theta}^j = 2A = p \frac{2 r_i^2}{r_e^2 - r_i^2}$$

so that we obtain $e_{zz}^j = \frac{-2v p r_i^2}{E(r_e^2 - r_i^2)}$. Thus, for their differences in length we have

$$\Delta L = L\left(e_{zz}^s - e_{zz}^j\right) = \frac{2v p r_e^2 L}{E(r_e^2 - r_i^2)}$$

7.2 Airy Stress Function Solutions in Cartesian Coordinates

One of the other choices of governing equations described in Chapter 6 treated the stresses as the fundamental unknowns, which required satisfying the equilibrium and compatibility equations. As discussed in that chapter, the Airy stress function is often used to simplify these equations, resulting in a single equation for plane-stress and plane-strain problems. In this section we will discuss obtaining solutions in 2-D problems using an Airy stress function approach.

7.2.1 Solutions of the Biharmonic Equation in Cartesian Coordinates

Plane-stress or plane-strain problems where there are no body forces both satisfy the biharmonic equation for the Airy stress function. In Cartesian coordinates we have

$$\nabla^4 \phi = \frac{\partial^4 \phi}{\partial x^4} + 2\frac{\partial^4 \phi}{\partial x^2 \partial y^2} + \frac{\partial^4 \phi}{\partial y^4} = 0 \tag{7.39}$$

One can obtain some useful solutions by representing the Airy stress function as the sum of polynomials where

$$\phi = \sum_{m=1}^{M} \sum_{n=1}^{N} A_{mn} x^m y^n \tag{7.40}$$

and A_{mn} are constants. Polynomials up to third order (cubic polynomials) satisfy the biharmonic equation without any restriction on the constants. Fourth-order and higher polynomials require the coefficients to be related.

There are many interesting problems that can be solved with such polynomials, but in this book our coverage will be limited to a single example. In using these functions, we have three objectives. First, we want to show a nontrivial example of a problem solved with such polynomials. Second, we want to highlight the importance of the boundary conditions. Third, and finally, we want to compare a solution with a corresponding solution from strength of materials. The simply-supported beam problem discussed in the following section will meet those three objectives.

7.2.2 A Simply Supported Beam

Consider a rectangular beam of height, $2h$, length, $2L$, and width, b, that has a uniform distributed pressure, p_0, acting on it top surface and is simply supported at its two ends, as shown in Fig. 7.6. This is a geometry suited to the use of Cartesian coordinates and we expect from the strength of materials solution for this problem that the stresses can be described with relatively few polynomial terms. The polynomial we will use for this problem is a fifth-order polynomial in terms of six coefficients:

$$\phi = ax^2 + bx^2 y + cy^3 + dx^4 y + ex^2 y^3 + fy^5 \tag{7.41}$$

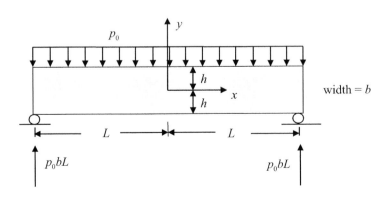

Figure 7.6 A simply supported beam of height $2h$, length $2L$, and width b subjected to a uniform pressure, p_0, acting on its top face. The reaction forces at the ends are also shown.

which has terms that are only even in x and odd in y. The stresses generated by these polynomials are

$$\sigma_{xx}(x,y) = 6cy + 6ex^2y + 20fy^3$$
$$\sigma_{yy}(x,y) = 2a + 2by + 12dx^2y + 2ey^3 \quad (7.42)$$
$$\sigma_{xy}(x,y) = -(2bx + 4dx^3 + 6exy^2)$$

where the stress functions $(\sigma_{xx}, \sigma_{yy})$ are even in x and odd in y while the shear stress, σ_{xy}, is odd in x and even in y. This behavior is what we see in a strength of materials solution for this problem so it might be expected that the functions in Eq. (7.41) are a reasonable choice. There are more detailed ways in which we can try to choose such functions (see [1], for example) but we will not go into them here. Since we have used a fifth-order polynomial, there will be relations that have to be satisfied among these six coefficients. If we place the polynomial of Eq. (7.41) into the biharmonic equation, we find $(24d + 24e + 120f)y = 0$, which must be satisfied for all y, so we have

$$24d + 24e + 120f = 0 \quad (7.43)$$

Thus, there are actually five free constants in our solution which we must use to satisfy the boundary conditions. On the top and bottom of the beam we have the conditions

$$\sigma_{yy}(x,h) = -p_0$$
$$\sigma_{xy}(x,h) = 0$$
$$\sigma_{yy}(x,-h) = 0 \quad (7.44)$$
$$\sigma_{xy}(x,-h) = 0$$

The boundary conditions on the ends are, however, more problematic because, as we discussed in Chapter 6, different simple support systems produce different stresses. Thus, we will use the idea contained in the principle of Saint-Venant that static equivalency is the essential condition that must be satisfied and simply require the stresses to produce the correct support forces and moments. Thus, at the ends we will specify so-called *Saint-Venant boundary conditions*, namely at $x = -L$ the shear stress must produce the end reaction force (see Fig. 7.6), and zero axial force and moment

$$\int_{-h}^{+h} \sigma_{xy} b \, dy = -p_0 bL$$

$$\int_{-h}^{+h} \sigma_{xx} b \, dy = 0 \quad (7.45)$$

$$\int_{-h}^{+h} \sigma_{xx} y b \, dy = 0$$

where the minus sign is present since σ_{xy} is taken positive acting in the negative y-direction on the face of the beam at $x = -L$. Similarly, at $x = L$

$$\int_{-h}^{+h} \sigma_{xy} b\, dy = P_y = p_0 bL$$

$$\int_{-h}^{+h} \sigma_{xx} b\, dy = 0 \tag{7.46}$$

$$\int_{-h}^{+h} \sigma_{xx} y b\, dy = 0$$

However, since we are using an Airy stress function that satisfies equilibrium throughout the beam and the boundary conditions are satisfied on the top and bottom of the beam exactly, if we satisfy these Saint-Venant boundary conditions on just one end of the beam they will be automatically satisfied at the other end. Thus, a complete set of boundary conditions for this problem is given by Eq. (7.44) and Eq. (7.46). Consider applying the boundary conditions of Eq. (7.44). We find from the expressions for the stresses, Eq. (7.42) and the first equation in Eq. (7.44)

$$\sigma_{yy}(x,h) = -p_0$$
$$\rightarrow 2a + 2bh + 12dx^2 h + 2eh^3 = -p_0 \tag{7.47}$$

which must be satisfied for all x. This can only be the case if we have

$$d = 0$$
$$2a + 2bh + 2eh^3 = -p_0 \tag{7.48}$$

From the other three boundary conditions in Eq. (7.44) we find

$$\begin{aligned}
\sigma_{xy}(x,h) = 0 &\rightarrow b = -3eh^2 \\
\sigma_{yy}(x,-h) = 0 &\rightarrow 2a - 2bh - 2eh^3 = 0 \\
\sigma_{xy}(x,-h) = 0 &\rightarrow b = -3eh^2
\end{aligned} \tag{7.49}$$

where two of these boundary conditions lead to identical equations. We can solve the five independent equations contained in Eq. (7.43), Eq. (7.48), and Eq. (7.49) for the five constants (a,b,d,e,f) as

$$a = -\frac{p_0}{4}, \quad b = -\frac{3p_0}{8h}, \quad d = 0, \quad e = \frac{p_0}{8h^3}, \quad f = -\frac{p_0}{40h^3} \tag{7.50}$$

The only remaining constant to find is the constant c. The boundary conditions left to satisfy are the Saint-Venant boundary conditions, where we will use those conditions at $x = L$. First, we have

$$\int_{-h}^{+h} \sigma_{xy}(L,y) b\, dy = p_0 bL \quad \rightarrow \quad -(4bh + 4eh^3) bL = p_0 bL \tag{7.51}$$

which is already satisfied by the constants in Eq. (7.50). Next, we have

$$\int_{-h}^{+h} \sigma_{xx}(L,y)bdy = 0 \qquad (7.52)$$

which is automatically satisfied since the stress σ_{xx} is odd in y. Thus, we are left with the last boundary condition

$$\int_{-h}^{+h} \sigma_{xx}(L,y)ybdy = 0 \quad \rightarrow \quad 4ch^3 + 2eL^2h^3 + 8fh^5 = 0 \qquad (7.53)$$

which gives the last coefficient as

$$c = \frac{p_0}{8h}\left(\frac{2}{5} - \frac{L^2}{h^2}\right) \qquad (7.54)$$

You can verify that these constants also satisfy the Saint-Venant boundary conditions at $x = -L$, as we stated they should.

The stresses for the beam are now

$$\sigma_{xx}(x,y) = \frac{p_0}{h}\left(\frac{6}{20} - \frac{3L^2}{4h^2}\right)y + \frac{3p_0}{4h^3}x^2y - \frac{p_0}{2h^3}y^3$$

$$\sigma_{yy}(x,y) = -\frac{p_0}{2} - \frac{3p_0}{4h}y + \frac{p_0}{4h^3}y^3 \qquad (7.55a)$$

$$\sigma_{xy}(x,y) = \frac{3p_0}{4h}x - \frac{3p_0}{4h^3}xy^2$$

which we will rewrite and group as

$$\sigma_{xx}(x,y) = \frac{3p_0}{4h^3}(x^2 - L^2)y + p_0\left(\frac{3y}{10h} - \frac{y^3}{2h^3}\right)$$

$$\sigma_{yy}(x,y) = -p_0\left(\frac{1}{2} + \frac{3y}{4h} - \frac{y^3}{4h^3}\right) \qquad (7.55b)$$

$$\sigma_{xy}(x,y) = \frac{3p_0}{4h}x\left(1 - \frac{y^2}{h^2}\right)$$

This is in a form that we can now directly compare to a strength of materials solution where the bending stress and shear stress are given as

$$\sigma_{xx}^{strength} = -\frac{M(x')y}{I}$$

$$\sigma_{xy}^{strength} = -\frac{V(x')Q(y)}{Ib} \qquad (7.56)$$

Here M is the bending moment and V is the shear force acting in the beam, both acting as shown in Fig. 7.7, and the distance $x' = x + L$ is measured from the left-hand end of the beam. For our rectangular beam we have $I = 2bh^3/3$ and $Q(y) = bh^2(1 - y^2/h^2)/2$. From equilibrium of the element shown in Fig. 7.7, we have $V(x') = -p_0 bx$, $M(x') = -p_0 b(x^2 - L^2)$, as expressed in terms of the x-coordinate, so that Eq. (7.55b) can be written finally as

$$\sigma_{xx}(x,y) = \frac{M(x')y}{I} + p_0\left(\frac{3y}{10h} - \frac{y^3}{2h^3}\right)$$

$$\sigma_{yy}(x,y) = -p_0\left(\frac{1}{2} + \frac{3y}{4h} - \frac{y^3}{4h^3}\right) \quad (7.57)$$

$$\sigma_{xy}(x,y) = -\frac{V(x')Q(y)}{Ib}$$

Our solution has two additional stresses not contained in the strength of materials solution. They are

$$\sigma_{xx}^{add} = p_0\left(\frac{3y}{10h} - \frac{y^3}{2h^3}\right)$$

$$\sigma_{yy}^{add} = -p_0\left(\frac{1}{2} + \frac{3y}{4h} - \frac{y^3}{4h^3}\right) \quad (7.58)$$

We have seen the additional normal stress, σ_{yy}^{add}, before in Chapter 3, where we showed that the equilibrium equations required such a stress to be present and we derived an expression for this stress in Eq. (3.27) that is identical to the current expression if we make the following changes from the solution in Chapter 3: $z \to y$, $t \to b$, $h \to 2h$, $q(x) \to p_0 b$. Our current expression for σ_{yy}^{add}, which is based on satisfying both equilibrium and compatibility, is plotted in Fig. 7.8. The other additional stress, σ_{xx}^{add}, is entirely new. This term is plotted in Fig. 7.9. This stress is called a self-equilibrated stress since it does not produce any net axial force or moment. The axial force is obviously zero since the stress is an odd function of y, so its integral over the cross-section of the beam is automatically zero (also see Fig. 7.8). The moment is also zero, which is easy to verify by direct integration. If Saint-Venant's principle was satisfied, we would expect this self-equilibrated stress would produce zero stress outside a small region at the ends, but this is clearly not the case as this stress is present throughout the beam. Thus, this is an example where Saint-Venant's principle is violated. However, the maximum strength of materials bending stress, which occurs at top/bottom faces at the center of the beam,

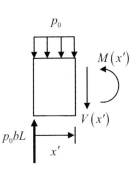

Figure 7.7 Element of the simply supported beam in Fig. 7.6, showing the internal shear force and bending moment acting in the beam. The coordinate x' is measured from the left end of the beam.

Figure 7.8 The distribution of the normalized normal stress, σ_{yy}, across the height of the simply supported beam of Fig. 7.6.

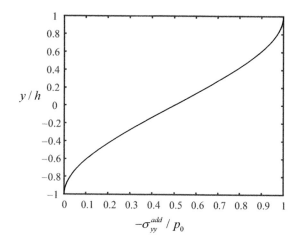

Figure 7.9 The distribution of the normalized normal stress, σ_{xx}, across the height of the simply supported beam of Fig. 7.6, showing the regions of positive and negative values.

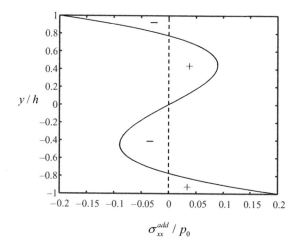

is still the dominant stress in this solution since the maximum strength of materials bending stress and maximum additional stresses in the beam are

$$\max\left(\sigma_{xx}^{strength}\right) = \frac{3p_0 L^2}{4h^2}, \quad \max\left(\sigma_{xx}^{add}\right) = \frac{p_0}{5}, \quad \max\left(\sigma_{yy}^{add}\right) = p_0 \qquad (7.59)$$

so that we have the ratios

$$\frac{\max\left(\sigma_{xx}^{add}\right)}{\max\left(\sigma_{xx}^{strength}\right)} = \frac{4}{15}\frac{h^2}{L^2}, \quad \frac{\max\left(\sigma_{yy}^{add}\right)}{\max\left(\sigma_{xx}^{strength}\right)} = \frac{4}{3}\frac{h^2}{L^2} \qquad (7.60)$$

For long, slender beams where $h/L \ll 1$, both of these additional stresses can be neglected.

7.3 Airy Stress Function Solutions in Polar Coordinates

There are a number of important problems that are best described in polar coordinates whose solutions can be obtained with the use of the Airy stress functions. In this section we will examine two cases of this type – the stress concentration around a hole in a plate, and the stresses in a curved beam. The Airy stress function solution for the curved beam will be compared to an equivalent strength of materials solution.

7.3.1 Michell's Solutions

For zero body forces the Airy stress function satisfies the biharmonic equation. If one seeks solutions to the biharmonic equation in polar coordinates of the form $\phi = f(r)g(\theta)$, then there are a number of such solutions, called separation of variables solutions, that one can obtain. We will not go through the details of the separation of variables solution process. Instead, we will simply write down a number of the important solutions of this type due to Michell [1], given as

$$\begin{aligned}\phi = &\, a_0 \ln r + b_0 r^2 + c_0 r^2 \ln r + d_0 r^2 \theta + e_0 \theta \\ &+ \left(a_1 r + b_1 r^3 + \frac{c_1}{r} + d_1 r \ln r\right) \cos\theta + j_1 r\theta \cos\theta \\ &+ \left(e_1 r + f_1 r^3 + \frac{g_1}{r} + h_1 r \ln r\right) \sin\theta + k_1 r\theta \sin\theta \\ &+ \sum_{n=2}^{\infty} \left(a_n r^n + b_n r^{n+2} + c_n r^{-n} + d_n r^{-n+2}\right) \cos n\theta \\ &+ \sum_{n=2}^{\infty} \left(e_n r^n + f_n r^{n+2} + g_n r^{-n} + h_n r^{-n+2}\right) \sin n\theta \end{aligned} \quad (7.61)$$

Most of these solutions depend on both r and θ, but we see there is a subset of axisymmetric solutions where θ is absent given by

$$\phi = a_0 \ln r + b_0 r^2 + c_0 r^2 \ln r \quad (7.62)$$

These solutions produce the stresses

$$\begin{aligned}\sigma_{rr} &= \frac{a_0}{r^2} + 2b_0 + c_0(2\ln r + 1) \\ \sigma_{\theta\theta} &= -\frac{a_0}{r^2} + 2b_0 + c_0(2\ln r + 3) \\ \sigma_{r\theta} &= 0 \end{aligned} \quad (7.63)$$

Note that the terms involving the coefficients (a_0, b_0) are of the same form that we used in Eq. (7.14) to solve the pressurized vessel and shrink-fit problems, so those problems could also be solved with the use of these Airy stress functions. Since these stresses satisfy the compatibility equation, we can relate the stresses to the strains and then integrate those strains to find the displacements. The details are rather lengthy so we will just give the final results. For plane stress, the displacements (u_r, u_θ) are

$$u_r = \frac{1}{E}\left[-\frac{(1+v)a_0}{r} + 2(1-v)b_0 r - (1+v)c_0 r\right.$$

$$\left. + 2(1-v)c_0 r \ln r\right] + C_1 \sin\theta + C_2 \cos\theta$$

$$u_\theta = \frac{4c_0 r\theta}{E} + C_1 \sin\theta - C_2 \cos\theta + C_3 r \tag{7.64}$$

while for the case of plane strain (which we can obtain from the plane-stress solution as described in Chapter 6):

$$u_r = \frac{1}{E}\left[-\frac{(1+v)a_0}{r} + 2(1-2v)(1+v)b_0 r - (1+v)c_0 r\right.$$

$$\left. + 2(1-2v)(1+v)c_0 r \ln r\right] + C_1 \sin\theta + C_2 \cos\theta$$

$$u_\theta = \frac{4(1-v^2)c_0 r\theta}{E} + C_1 \sin\theta - C_2 \cos\theta + C_3 r \tag{7.65}$$

where the constants (C_1, C_2) are constants of integration that describe a rigid body translation, and the constant C_3 is a constant of integration that describes a rigid body rotation. These rigid body terms produce no stresses or strains but may be needed to satisfy the boundary conditions of a problem. In the pressure vessel and shrink-fit problems, we had assumed $u_r = u_r(r), u_\theta = 0$ so that such rigid body terms were absent. If our body includes the origin, the stresses and displacements are unbounded at $r = 0$ unless we set $a_0 = c_0 = 0$. If the body does include a hole centered at $r = 0$ (i.e., it is multiply connected) then solutions for the displacements involving both a_0 and c_0 are acceptable since the point $r = 0$ is not a part of the body but the solutions are not single-valued since $u_\theta(r, 0) \neq u_\theta(r, 2\pi)$. This should not be surprising since we saw in our discussions of compatibility in Chapter 4 that for multiply connected bodies, the compatibility equations by themselves are not sufficient to ensure we have single-valued displacements. If we provide an additional criterion here such as

$$\int_0^{2\pi} du_\theta = \int_0^{2\pi} \frac{\partial u_\theta}{\partial \theta} d\theta = 0 \tag{7.66}$$

then $c_0 = 0$ and the displacements are single-valued.

7.3.2 Stress Concentration at a Circular Hole in a Large Plate

One problem that can be described with Michell's solutions is the case of an infinitely large thin plate containing a hole of radius a (Fig. 7.10) and under a uniform stress $\sigma_{xx} = \sigma_0$ at infinity. This is a plane-stress problem where the only nonzero stresses are $(\sigma_{rr}, \sigma_{\theta\theta}, \sigma_{r\theta})$.

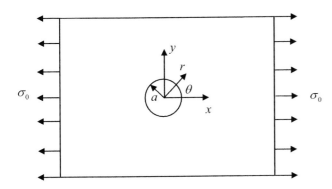

Figure 7.10 An infinitely large plate containing a hole of radius a and subjected to a uniform axial stress, σ_0, acting in the x-direction.

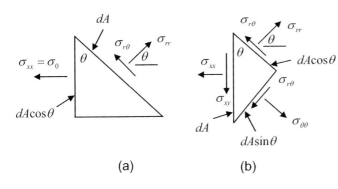

Figure 7.11 (a) An element in the plate at a large distance from the hole where the stresses σ_{yy} and σ_{xy} are zero. (b) An element at an arbitrary point in the plate, where there is a full state of stress $(\sigma_{rr}, \sigma_{\theta\theta}, \sigma_{r\theta})$.

To solve this problem, we need to decide which parts of the Michell solutions will be needed. Consider a small element at infinity (see Fig. 7.11a) which has the stresses $\sigma_{xx} = \sigma_0$, σ_{rr}, and $\sigma_{r\theta}$ acting on its faces ($\sigma_{yy} = 0$ at infinity so it is not shown). From the force equilibrium of this element in the r- and θ-directions it follows that

$$\sigma_{rr} = \sigma_0 \cos^2\theta = \frac{\sigma_0}{2}(1 + \cos 2\theta)$$

$$\sigma_{r\theta} = -\sigma_0 \sin\theta \cos\theta = -\frac{\sigma_0}{2}\sin 2\theta$$
(7.67)

where we have also used double angle formulas of trigonometry. But from the stress invariant $I_1 = \sigma_{xx} + \sigma_{yy} = \sigma_0 = \sigma_{rr} + \sigma_{\theta\theta}$ we also find

$$\sigma_{\theta\theta} = \frac{\sigma_0}{2}(1 - \cos 2\theta)$$
(7.68)

We want the stresses at infinity to have these forms in polar coordinates and because of the symmetry of the circular hole, we expect that there will also be axisymmetric stresses generated from the presence of the hole, so we will try an Airy stress function of the form

$$\phi = a_0 \ln r + b_0 r^2 + c_0 r^2 \ln r \\ + \left(a_2 r^2 + b_2 r^4 + a_2' r^{-2} + b_2'\right)\cos 2\theta$$
(7.69)

but we must set $c_0 = 0$ to have single-valued displacements. The stresses generated are

$$\sigma_{rr} = \frac{a_0}{r^2} + 2b_0 - \left(2a_2 + 6a'_2 r^{-4} + 4b'_2 r^{-2}\right)\cos 2\theta$$

$$\sigma_{r\theta} = 2\left(a_2 + 3b_2 r^2 - 3a'_2 r^{-4} - b'_2 r^{-2}\right)\sin 2\theta \tag{7.70}$$

$$\sigma_{\theta\theta} = \frac{-a_0}{r^2} + 2b_0 + \left(2a_2 + 12b_2 r^2 + 6a'_2 r^{-4}\right)\cos 2\theta$$

We must set $b_2 = 0$ to have stresses that are only functions of θ as $r \to \infty$, as demanded by Eq. (7.67) and Eq. (7.68). We then can match the stresses in Eq. (7.70) as $r \to \infty$ with the remaining stresses in Eq. (7.67) and Eq. (7.68) to obtain

$$a_2 = -\frac{\sigma_0}{4}, \quad b_0 = \frac{\sigma_0}{4} \tag{7.71a}$$

From the boundary conditions $\sigma_{rr} = \sigma_{r\theta} = 0$ at $r = a$ we find

$$a_0/a^2 + \sigma_0/2 - \left(-\sigma_0/2 + 6a'_2/a^4 + 4b'_2/a^2\right)\cos 2\theta = 0$$

$$\left(-\sigma_0/2 - 6a'_2/a^4 - 2b'_2/a^2\right)\sin 2\theta = 0$$

which, equating the constant term and terms involving θ separately equal to zero, gives the remaining constants as

$$a_0 = -\sigma_0 a^2/2$$
$$b'_2 = \sigma_0 a^2/2 \tag{7.71b}$$
$$a'_2 = -\sigma_0 a^4/4$$

and the stresses in the plate are, therefore, explicitly

$$\sigma_{\theta\theta}(r,\theta) = \frac{\sigma_0}{2}\left(1 + \frac{a^2}{r^2}\right) - \frac{\sigma_0}{2}\left(1 + \frac{3a^4}{r^4}\right)\cos 2\theta$$

$$\sigma_{r\theta}(r,\theta) = \frac{\sigma_0}{2}\left(1 - \frac{3a^4}{r^4} + \frac{2a^2}{r^2}\right)\sin 2\theta \tag{7.72}$$

$$\sigma_{rr}(r,\theta) = \frac{\sigma_0}{2}\left(1 - \frac{a^2}{r^2}\right) + \frac{\sigma_0}{2}\left(1 + \frac{3a^4}{r^4} - \frac{4a^2}{r^2}\right)\cos 2\theta$$

It is useful to consider the stress σ_{xx}/σ_0 as this stress goes to a constant value of one far from the hole so that one can see the effect that the hole has in modifying this uniform stress at infinity. This stress can be calculated from the stresses in Eq. (7.72) by using the element shown in Fig. 7.11b and examining the force equilibrium in the x-direction. We find

$$\sigma_{xx} = \sigma_{rr}\cos^2\theta - 2\sigma_{r\theta}\sin\theta\cos\theta + \sigma_{\theta\theta}\sin^2\theta \tag{7.73}$$

A 2-D contour plot of this stress is shown in Fig. 7.12a. On the top and bottom points of the hole ($x = 0, y \pm a$) a maximum stress of $\sigma_{xx} = 3\sigma_0$ is present (Fig. 7.12b). Along the x-axis at the hole surface, the stress $\sigma_{xx} = 0$ by virtue of the boundary conditions and there are small

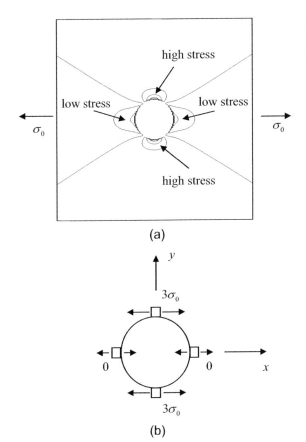

Figure 7.12 (a) A contour plot of σ_{xx}/σ_0 in the plate. (b) The largest and smallest values of σ_{xx}/σ_0 and where they occur on the periphery of the hole.

stresses near those points. The value $\sigma_{xx} = 3\sigma_0$ is the largest stress anywhere in the plate so the stress concentration factor of the hole (ratio of the maximum stress to the stress at infinity) is 3.0. Similarly, notches, fillets, and noncircular holes produce stress concentrations so the stress may be increased significantly over its nominal value away from the geometry changes. Analytical and numerical evaluations of stress concentration factors for many standard geometries have been calculated and tabulated in handbooks. Note that, in this case, the stress concentration factor is independent of the size of the hole. However, if an edge of a more realistic finite-sized plate is within about four radii from the hole, then the influence of the edge may start to affect the stress concentration for an infinite plate calculated here. Another plot of interest is that of the hoop stress, $\sigma_{\theta\theta}$, along the periphery of the hole at $r = a$, since large tensile values of this stress can initiate cracking at the hole. From Eq. (7.72) we have (see Fig. 7.13):

$$\sigma_{\theta\theta}(a, \theta) = \sigma_0(1 - 2\cos 2\theta) \tag{7.74}$$

The extreme values of $\sigma_{\theta\theta}/\sigma_0$ and their location along the edge of the hole are shown in Fig. 7.13. At the x-axis, the hoop stress has its largest compressive values $(-\sigma_0)$ while at the

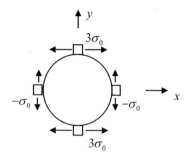

Figure 7.13 The largest and smallest values of the hoop stress and their locations.

y-axis, the hoop stress has its largest tensile values ($3\sigma_0$), which is the most significant stress since cracks are initiated by large tensile, not compressive stresses. The $3\sigma_0$ values also appear in Fig. 7.12b since $\sigma_{\theta\theta} = \sigma_{xx}$ at these locations.

7.3.3 Pure Bending of a Curved Beam

The axisymmetric Airy stress functions and resulting stresses of Eq. (7.62) and Eq. (7.63) can also be used to solve the problem of pure bending of a curved beam of rectangular cross-section, as shown in Fig. 7.14. It is convenient to define a new set of constants $c_3 = a_0$, $2b_0 + c_0 = c_1 - c_2 \ln a$, and $c_2 = 2c_0$ and rewrite the stresses in Eq. (7.63) as

$$\sigma_{rr} = c_1 + c_2 \ln \frac{r}{a} + \frac{c_3}{r^2}$$

$$\sigma_{\theta\theta} = c_1 + c_2 \left(\ln \frac{r}{a} + 1 \right) - \frac{c_3}{r^2} \quad (7.75)$$

$$\sigma_{r\theta} = 0$$

Since the shear stress $\sigma_{r\theta} = 0$, the boundary conditions at both $r = a$ and $r = b$ are $\sigma_{rr} = 0$, which gives the two equations

$$c_3 = -a^2 c_1$$

$$c_1 \left(\frac{a^2}{b^2} - 1 \right) = c_2 \ln \frac{b}{a} \quad (7.76)$$

On the ends of the beam, we use the Saint-Venant boundary conditions that the stresses must produce the appropriate forces and moments, giving

$$\int_a^b \sigma_{r\theta} t \, dr = 0$$

$$\int_a^b \sigma_{\theta\theta} t \, dr = 0 \quad (7.77)$$

$$\int_a^b \sigma_{\theta\theta} r t \, dr = M$$

where t is the thickness of the beam. The first equation in Eq. (7.77) is satisfied identically. The second equation gives

7.3 Airy Stress Function Solutions in Polar Coordinates

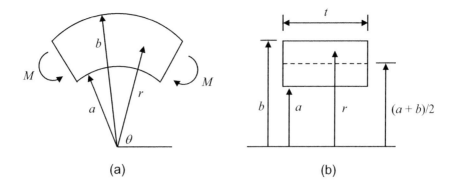

Figure 7.14 (a) Pure bending of a rectangular cross-section curved beam. (b) The beam cross-section.

$$(c_1 + c_2)(b - a) + c_3\left(\frac{1}{b} - \frac{1}{a}\right) + c_2 a\left[\frac{r}{a}\ln\frac{r}{a} - \frac{r}{a}\right]_a^b = 0$$

$$\rightarrow (c_1 + c_2)(b - a) + c_3\left(\frac{1}{b} - \frac{1}{a}\right) + c_2 a\left(\frac{b}{a}\ln\frac{b}{a} - \frac{b}{a} + 1\right) = 0$$

(7.78)

But placing Eq. (7.76) into Eq. (7.78), we find

$$c_1(b - a) - c_1\frac{a}{b}(a - b) - \frac{c_1}{b}(b - a)(b + a) = 0$$

$$\rightarrow 1 + \frac{a}{b} - \frac{1}{b}(a + b) = 0$$

(7.79)

so that the second equation in Eq. (7.77) is also satisfied. From Eq. (7.76) we can write both c_1 and c_3 in terms of c_2 as

$$c_1 = \frac{b^2 \ln(b/a)}{a^2 - b^2} c_2$$

$$c_3 = \frac{a^2 b^2 \ln(b/a)}{b^2 - a^2} c_2$$

(7.80)

and we can find c_2 from the third equation in Eq. (7.77). The three integral terms in that equation are

$$tc_1 \int_a^b r\, dr = -\frac{tc_2 b^2}{2} \ln(b/a)$$

$$tc_2 \int_a^b r(\ln(r/a) + 1)dr = \frac{tc_2 b^2}{2}\ln(b/a) + \frac{tc_2(b^2 - a^2)}{4}$$

(7.81)

$$-tc_3 \int_a^b dr/r = \frac{tc_2 a^2 b^2 [\ln(b/a)]^2}{a^2 - b^2}$$

so the third equation in Eq. (7.77) becomes

$$tc_2 \left[\frac{(b^2 - a^2)}{4} - \frac{a^2 b^2 \left[\ln (b/a)^2 \right]}{(b^2 - a^2)} \right] = M \qquad (7.82)$$

which can be solved for c_2 in the form

$$c_2 = \frac{4M \left(1 - \frac{a^2}{b^2} \right)}{tb^2 N} \qquad (7.83)$$

where

$$N = \left(1 - \frac{a^2}{b^2} \right)^2 - 4 \frac{a^2}{b^2} \left(\ln \left(\frac{b}{a} \right) \right)^2 \qquad (7.84)$$

Using the known constants in Eq. (7.75) gives finally

$$\sigma_{rr} = \frac{4M}{tb^2 N} \left[\left(1 - \frac{a^2}{b^2} \right) \ln \left(\frac{r}{a} \right) - \left(1 - \frac{a^2}{r^2} \right) \ln \left(\frac{b}{a} \right) \right]$$

$$\sigma_{\theta\theta} = \frac{4M}{tb^2 N} \left[\left(1 - \frac{a^2}{b^2} \right) \left\{ 1 + \ln \left(\frac{r}{a} \right) \right\} - \left(1 + \frac{a^2}{r^2} \right) \ln \left(\frac{b}{a} \right) \right] \qquad (7.85)$$

Figure 7.15 plots both of these stresses across the beam cross-section for the case $b/a = 2$. Unlike with the straight beam, we see that the bending stress, $\sigma_{\theta\theta}$, has a nonlinear distribution across the cross-section. The radial stress, σ_{rr}, is small and can typically be neglected, as is the case for the straight beam. If we simply ignore the curvature and treat the beam as straight, we can use the standard flexure expression and write the bending stress, σ_b, as (see Fig. 7.14b)

$$\sigma_b = \frac{12Ma}{b^3 t} \left[\frac{r/a - (1 + b/a)/2}{(1 - a/b)^3} \right] \qquad (7.86)$$

Figure 7.15 Normal stresses (normalized) in a curved beam under pure bending for $b/a = 2$.

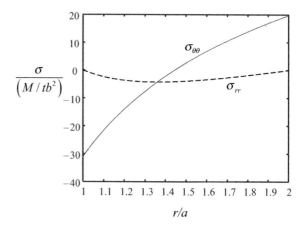

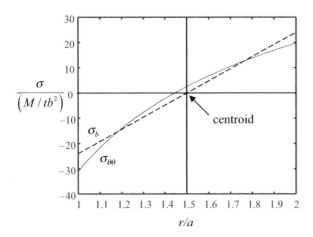

Figure 7.16 Comparison of the normalized bending stress for a curved beam (solid line) and a straight beam (dashed line) for the case $b/a = 2$.

A comparison of σ_b with the curved beam $\sigma_{\theta\theta}$ is shown in Fig. 7.16. It can be seen that although the stress distributions are similar, the neutral axis of the curved beam is no longer at the centroid of the cross-section, which is another difference between the straight and curved beam cases.

7.3.4 Comparison with a Curved Beam Strength of Materials Solution

One can use assumptions on the deformation of a curved beam similar to those used for a straight beam to obtain a strength of materials solution. In this section we will obtain the strength of materials solution and compare it to the solution found in the previous section. We will again examine pure bending of a curved beam and assume the cross-section is symmetric but, unlike the Airy stress function solution, the beam need not be rectangular. We will also assume plane cross-sections remain plane after deformation and that the only significant stress in the beam is $\sigma_{\theta\theta}$. The geometry and geometrical parameters to be used are shown in Fig. 7.17a. The radius to the centroid is R and the radius to the neutral axis is R_n. The segment of the beam shown subtends an angle $\Delta\theta$. We will again let r be the radius to a general fiber located at P in the beam. The hoop strain in the beam for the fiber located at P, assuming that plane cross-sections remain plane so that they rotate through a small angle, $\Delta\phi$ (see Fig. 7.17b), is

$$e_{\theta\theta} = \frac{\Delta l}{l} = \frac{(R_n - r)\Delta\phi}{r\Delta\theta} = \omega\left(\frac{R_n}{r} - 1\right) \qquad (7.87)$$

where

$$\omega = \frac{\Delta\phi}{\Delta\theta} \qquad (7.88)$$

From Hooke's law we have

$$\sigma_{\theta\theta} = E e_{\theta\theta} = E\omega\left(\frac{R_n}{r} - 1\right) \qquad (7.89)$$

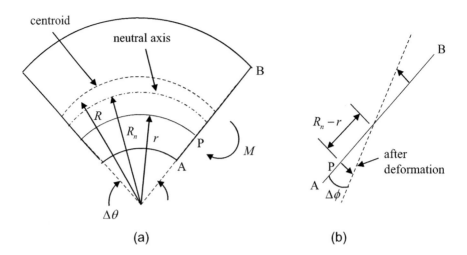

Figure 7.17 (a) An element of a curved beam in pure bending. (b) The deformation, assuming plane sections remain plane.

Requiring that this stress does not produce an axial force, N, we have

$$N = \int_A \sigma_{\theta\theta} dA = E\omega \left[R_n \int_A \frac{dA}{r} - \int_A dA \right] \tag{7.90}$$

$$= E\omega(R_n A_m - A) = 0$$

where A_m, which has the dimensions of a length, is defined as

$$A_m = \int_A \frac{dA}{r} \tag{7.91}$$

It follows from Eq. (7.90) that the radius of the neutral axis is

$$R_n = \frac{A}{A_m} \tag{7.92}$$

which is, in general, not at the centroid. For the bending moment we have

$$M = -\int_A \sigma_{\theta\theta}(R - r)dA = E\omega \int_A \left(\frac{R_n}{r} - 1 \right)(R - r)dA$$

$$= -E\omega \left[R_n R \int_A \frac{dA}{r} - R \int_A dA - R_n \int_A dA + \int_A r dA \right] \tag{7.93}$$

$$\to M = E\omega R_n(A - RA_m)$$

7.3 Airy Stress Function Solutions in Polar Coordinates

where the final result in Eq. (7.93) is obtained since the second and fourth integrals in Eq. (7.93) cancel from the definition of the location of the centroid. Solving Eq. (7.93) for $E\omega$ and placing that result back into Eq. (7.89), the flexure stress is given by

$$\sigma_{\theta\theta} = \frac{M(A - rA_m)}{A(A - RA_m)r} \tag{7.94}$$

To compare this result with the Airy stress function solution, we have for the rectangular beam

$$A_m = t \ln\left(\frac{b}{a}\right), \quad A = t(b-a), \quad R = (a+b)/2 \tag{7.95}$$

so that the strength of materials solution becomes, after some rearrangement,

$$\sigma_{\theta\theta} = \frac{M}{tb^2} \frac{2\left(\frac{b^2}{a^2}\right)\left[\left(\frac{b}{a} - 1\right) - \frac{r}{a}\ln\left(\frac{b}{a}\right)\right]}{\left(\frac{b}{a} - 1\right)\left[2\left(\frac{b}{a} - 1\right) - \left(\frac{b}{a} + 1\right)\ln\left(\frac{b}{a}\right)\right]\left(\frac{r}{a}\right)} \tag{7.96}$$

We can compare this result and the corresponding straight beam solution with the Airy stress solution obtained previously. It is convenient to use the slenderness ratio R/h to make that comparison, where h is the thickness of the beam in the r-direction and R is the distance to the centroid (Fig. 7.18a). Table 7.1 shows the ratios of the maximum bending stresses in the beam calculated (a) with the curved beam strength of materials and Airy stress function solutions and (b) with the straight beam strength of materials solution and the Airy stress function solution. It can be seen for large slenderness ratios ($R/h > 5$ approximately) one can neglect curvature effects and use the ordinary straight beam flexural stress result. For smaller

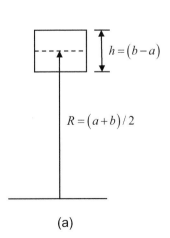
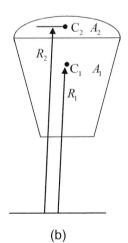

Figure 7.18 (a) Definition of the terms in the slenderness ratio, R/h, for a rectangular beam. (b) The areas and radii to the centroids (C_1, C_2) of a composite area.

Table 7.1 Comparison of the ratios of the maximum bending stresses for (a) the curved beam strength of materials solution divided by the Airy stress function solution, and (b) for the straight beam strength of materials solution divided by the Airy stress function solution

R/h	(a) Curved strength/Airy stress solutions	(b) Straight strength/Airy stress solutions
0.65	1.0455	0.4390
0.75	1.0124	0.5262
1.0	0.9970	0.6545
1.5	0.9961	0.7737
2.0	0.9973	0.8313
3.0	0.9986	0.8881
5.0	0.9994	0.9331

slenderness ratios we see that there is little difference between the curved beam strength of materials solution and the Airy stress function solution as long as the R/h value is not too small. Thus, generally we can use the straight beam strength formula for $R/h > 5$ and the curved beam strength formula for $R/h > 0.5$.

The curved beam strength of materials solution can be used for more complex cross-sections composed of composite areas, A_i (see Fig. 7.18b), where

$$A = \sum A_i$$
$$A_m = \sum A_{mi} \qquad (7.97)$$
$$R = \frac{\sum R_i A_i}{\sum A_i}$$

It should be noted that normal forces and shear forces are also present in many curved beam problems, and curved beam strength of materials solutions can account for those forces as well (see [2], for example). However, we will not give those details here.

7.3.5 Concentrated Force on a Wedge

Another case that we will examine with a Michell solution is the case of a concentrated force P acting on a wedge of thickness t and opening angle 2α (Fig. 7.19a). The boundary conditions for $r > 0$, $\theta = \pm \alpha$ are $\sigma_{\theta\theta} = \sigma_{r\theta} = 0$. If we try an Airy stress function $\phi = Cr\theta \sin\theta$, where C is a constant, the stresses are given as

$$\sigma_{rr} = \frac{2C}{r} \sin\theta$$
$$\sigma_{\theta\theta} = \sigma_{r\theta} = 0 \qquad (7.98)$$

so the boundary conditions on the faces of the wedge are satisfied. To find the constant C, consider the radial stress acting on a circular cut-out at the wedge tip (Fig. 7.19b).

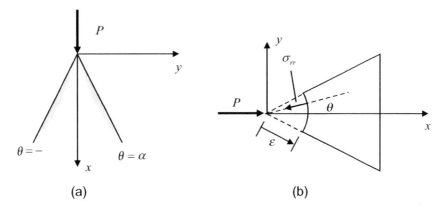

Figure 7.19 (a) Concentrated force on a wedge. (b) Geometry near the wedge tip.

This radial stress will not produce any force in the y-direction or moment about the wedge tip, but it should produce the force P in the x-direction, so we must have, at $r = \varepsilon$,

$$F_x = \int_{-\alpha}^{+\alpha} (-\sigma_{rr}\cos\theta)t\varepsilon d\theta = P \tag{7.99}$$

which gives

$$-2Ct\int_{-\alpha}^{+\alpha}\cos^2\theta d\theta = -2Ct\left(\frac{\theta}{2} + \frac{\sin 2\theta}{4}\right)\bigg|_{-\alpha}^{+\alpha} = P$$
$$\to C = \frac{-P/t}{2\alpha + \sin 2\alpha} \tag{7.100}$$

The radial stress is then given explicitly as

$$\sigma_{rr} = \frac{-2P\cos\theta}{t(2\alpha + \sin 2\alpha)r} \tag{7.101}$$

This stress is singular near $r = 0$ because the applied force is idealized as a concentrated force.

7.3.6 Concentrated Force on a Planar Surface

An important special case of the wedge problem is when $\alpha = \pi/2$ so that the concentrated force acts on a plane surface (Fig. 7.20a) and where

$$\sigma_{rr} = \frac{-2P\cos\theta}{\pi t}\frac{}{r} \tag{7.102}$$

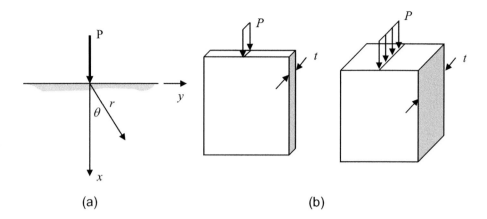

Figure 7.20 (a) Concentrated force on a planar surface. (b) Plane-stress and plane-strain problems.

Since this solution is independent of the elastic constants it can be used both for a plane-stress problem where the load is distributed over the thickness of a large thin plate or a plane-strain problem where the load is distributed on the surface of a long body (see Fig. 7.20b). Of course, the displacements in the plane-stress and plane-strain problems are different.

The equation for a circle of diameter d can be written as $r = d \cos \theta$ so stress distribution of Eq. (7.102) has the property that the radial stress is a constant on such circles as shown in Fig. 7.21. If we take a set of axes (x', y') in the (r, θ) directions, we see that the state of stress is just a uniaxial stress $\sigma_{x'x'} = \sigma_{rr}$. Thus, the magnitude of the maximum in-plane shear stress is just $|\sigma_{rr}|/2$. If we plot the contours of this maximum shear stress in MATLAB® we see these circles clearly in Fig. 7.22, where the circles become more and more dense near the load because the stresses are singular there.

We can use the stress transformation relations to obtain the stresses along the x- and y-directions as

$$\sigma_{xx} = \sigma_{rr} \cos^2 \theta = -\frac{2P}{\pi t} \frac{\cos^3 \theta}{r}$$

$$\sigma_{yy} = \sigma_{rr} \sin^2 \theta = -\frac{2P}{\pi t} \frac{\sin^2 \theta \cos \theta}{r} \quad (7.103)$$

$$\sigma_{xy} = \sigma_{rr} \sin \theta \cos \theta = -\frac{2P}{\pi t} \frac{\sin \theta \cos^2 \theta}{r}$$

or, since

$$\cos \theta = x/r$$
$$\sin \theta = y/r \quad (7.104)$$
$$r^2 = x^2 + y^2$$

we have equivalently

$$\sigma_{xx} = -\frac{2P}{\pi t}\frac{x^3}{(x^2+y^2)^2}$$

$$\sigma_{yy} = -\frac{2P}{\pi t}\frac{y^2 x}{(x^2+y^2)^2} \qquad (7.105)$$

$$\sigma_{xy} = -\frac{2P}{\pi t}\frac{yx^2}{(x^2+y^2)^2}$$

This concentrated load solution is very useful as a building block for obtaining the stresses due to an arbitrary vertical force distribution acting on the plane surface. To see this, let us displace the load so it acts at a distance y_1 along the y-axis and let the force P be a force $dP = p(y_1)t dy_1$ due to a pressure $p(y_1)$ acting on a small area $t dy_1$ (Fig. 7.23). Then if we superimpose such solutions due to an arbitrary pressure acting on the entire surface we have from Eq. (7.105) with the replacement $y \to y - y_1$

$$\sigma_{xx}(x,y) = -\frac{2}{\pi}\int \frac{x^3 p(y_1) dy_1}{\left(x^2+(y-y_1)^2\right)^2}$$

$$\sigma_{yy}(x,y) = -\frac{2}{\pi}\int \frac{x(y-y_1)^2 p(y_1) dy_1}{\left(x^2+(y-y_1)^2\right)^2} \qquad (7.106)$$

$$\sigma_{xy}(x,y) = -\frac{2}{\pi}\int \frac{x^2(y-y_1) p(y_1) dy_1}{\left(x^2+(y-y_1)^2\right)^2}$$

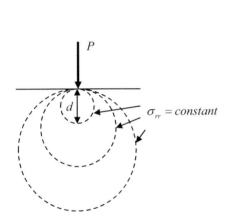

Figure 7.21 Circles of constant radial stress.

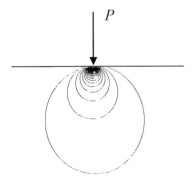

Figure 7.22 A contour plot of the maximum shear stress due to the concentrated load.

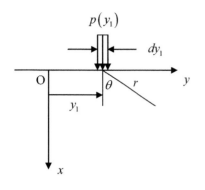

Figure 7.23 Geometry for a concentrated load when the load is due to a pressure distribution $p(y_1)$ acting over and small area on the surface.

7.3.7 Elastic Bodies in Contact

One problem where Eq. (7.106) finds application is the problem of two bodies in contact, such as the two cylinders shown in Fig. 7.24a. When these cylinders are pressed together with a force P a high pressure between the bodies is developed over a very small area. This pressure can cause the materials to yield and fail so it is important to know the maximum pressure exerted over the contact area and the size of the contact region. As shown in Fig. 7.24a the two cylinders can have different shear moduli and Poisson's ratios given by (G_1, ν_1) and (G_2, ν_2) as well as different radii (R_1, R_2). The length of both cylinders is assumed to be t. The equal and opposite pressure distributions are shown in Fig. 7.24b as well as the half-length, a, of the contact region. The solution to this problem can be found by using a set of assumptions called *Hertz contact theory* [3]. We will not go over the details of such a solution but will present some of the major results. Even though the surfaces in contact are curved, the contact area is typically very small so that the surfaces can be considered to be locally plane (see Fig. 7.24b), thus allowing us to use Eq. (7.106) to calculate the stresses. The pressure $p(y_1)$ is found to be the distribution

$$p(y_1) = \frac{p_0}{a}\sqrt{a^2 - y_1^2} \qquad (7.107)$$

where p_0 is the maximum pressure. This maximum pressure and the contact half-distance, a, can be found explicitly as

$$p_0 = \left\{\frac{2P}{\pi t}\left[\frac{(R_1 + R_2)}{(k_1 + k_2)R_1 R_2}\right]\right\}^{1/2}$$

$$a = \left[\frac{2P}{\pi t}\frac{(k_1 + k_2)R_1 R_2}{(R_1 + R_2)}\right]^{1/2} \qquad (7.108)$$

where

$$k_1 = \frac{(1 - \nu_1)}{G_1}$$

$$k_2 = \frac{(1 - \nu_2)}{G_2} \qquad (7.109)$$

The stresses in a cylinder are thus given from Eq. (7.106) as

7.3 Airy Stress Function Solutions in Polar Coordinates

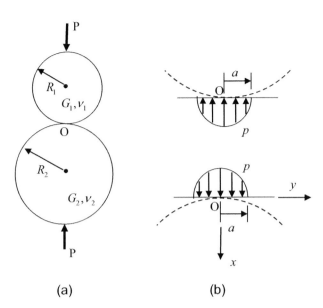

Figure 7.24 (a) Two elastic cylinders pressed together. (b) The surfaces in contact where a pressure p is exerted over the contact area. Locally the two surfaces are considered to be planar.

$$\sigma_{xx}(x,y) = -\frac{2p_0}{\pi a} \int_{-a}^{+a} \frac{x^3\sqrt{a^2-y_1^2}\,dy_1}{\left(x^2+(y-y_1)^2\right)^2}$$

$$\sigma_{yy}(x,y) = -\frac{2p_0}{\pi a} \int_{-a}^{+a} \frac{x(y-y_1)^2\sqrt{a^2-y_1^2}\,dy_1}{\left(x^2+(y-y_1)^2\right)^2} \quad (7.110)$$

$$\sigma_{xy}(x,y) = -\frac{2p_0}{\pi a} \int_{-a}^{+a} \frac{x^2(y-y_1)\sqrt{a^2-y_1^2}\,dy_1}{\left(x^2+(y-y_1)^2\right)^2}$$

These expressions can be integrated numerically to find the normalized maximum in-plane shearing stress, τ_{max}/p_0, where

$$\tau_{max} = \sqrt{(\sigma_{xx}-\sigma_{yy})^2/4 + \sigma_{xy}^2} \quad (7.111)$$

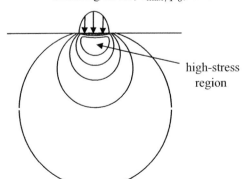

The results are shown in the 2-D contour plot of Fig. 7.25. Away from the surface, the plot looks much like that of a point source. However, here a

Figure 7.25 Contour plot of the normalized maximum shearing stress, τ_{max}/p_0, due to the contact pressure distribution.

local maximum normalized shear stress in the body occurs at a distance approximately $0.78a$ below the surface (within the region indicated below the distributed load) and has a value of about $0.3p_0$. In practice this means that failure will also likely occur first below the surface and result in a chunk of the material being spalled off, a type of failure that is commonly seen in such contact problems.

Example 7.1 Steel Cylinder Pressed on a Steel Plate

To illustrate the type of numerical values that can be found in contact problems, consider the case of a steel cylinder of radius $R_1 = 0.25$ in. being pressed by a force/unit length of $P/t = 50$ lb/in. while contact with a steel plate (Fig. 7.26). Let $G = 12 \times 10^6$ psi, $v = 0.3$ for both the cylinder and the plate. Note that in this case for the plate we have $R_2 = \infty$. The half-length contact distance is therefore

$$a = \left[\frac{4P(1-v)}{\pi t}\frac{}{G}R_1\right]^{1/2}$$

$$= \left[\frac{(4)(50 \text{ lb/in.})(0.7)}{\pi(12 \times 10 \text{ lb/in.}^2)}(0.25 \text{ in.})\right]^{1/2} \quad (7.112)$$

$$= 9.64 \times 10^{-4} \text{ in.}$$

while the maximum pressure is

$$p_0 = \left[\frac{P}{\pi t R_1}\frac{G}{(1-v)}\right]^{1/2} = \left[\frac{(50 \text{ lb/in.})}{\pi(0.25)}\frac{(12 \times 10^6 \text{ lb/in.}^2)}{(0.7)}\right]^{1/2} \quad (7.113)$$

$$= 33\,036 \text{ psi}$$

showing that the contact area is indeed quite small and that it is easy to develop very high stresses in such problems.

7.4 Stress Singularities

Stress singularities appear in concentrated load problems because of the assumption that the load is idealized as acting at a point. In the 2-D problems just discussed, we saw that the stresses had a $1/r$ singular behavior near $r = 0$. Stress singularities can also appear if we assume the geometry has sharp changes. Thus, let us consider the two-dimensional problem of a sharp corner (notch) in a plate that is being loaded in some

Figure 7.26 A steel cylinder being pressed in contact with a steel plate.

7.4 Stress Singularities

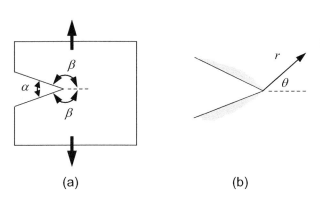

Figure 7.27 (a) Notch in a loaded plate. (b) The polar coordinates (r, θ) at the notch tip.

fashion, as shown in Fig. 7.27a. We want to examine the stresses in the neighborhood of the notch tip in terms of the polar coordinates (r, θ) as seen in Fig. 7.27b. We could use an Airy stress function approach (see [4], for example), but instead we will describe the problem in terms of displacements, as done by Karp and Karal ([5], [6]). The details are rather lengthy so we will just outline the steps here. Near the notch tip the displacements are assumed to be given as

$$u_r = r^m f(\theta)$$
$$u_\theta = r^m g(\theta) \tag{7.114}$$

where f and g are arbitrary functions. The corresponding stresses will then behave like r^{m-1} near the notch tip so that we will seek solutions where $0 < m < 1$, in which case the stresses will go to infinity at the notch tip but will not be as singular as in a concentrated load problem. Placing Eq. (7.114) into the equations of equilibrium written in terms of the displacements (Navier's equations in polar coordinates) we find solutions of the form

$$u_r = r^m \{A_1 \cos[(1+m)\theta] + A_2 \sin[(1+m)\theta] + A_3 \cos[(1-m)\theta] + A_4 \sin[(1-m)\theta]\}$$
$$u_\theta = r^m \{A_2 \cos[(1+m)\theta] - A_1 \sin[(1+m)\theta] + \gamma A_4 \cos[(1-m)\theta] - \gamma A_3 \sin[(1-m)\theta]\} \tag{7.115}$$

where

$$\gamma = \frac{3(1-\nu)}{3-5\nu} \tag{7.116}$$

in terms of Poisson's ratio ν. These displacement expressions are then used to calculate the stresses, which are placed into the boundary conditions:

$$\sigma_{rr} = \sigma_{\theta\theta} = 0 \quad \theta = +\beta$$
$$\sigma_{rr} = \sigma_{\theta\theta} = 0 \quad \theta = -\beta \tag{7.117}$$

These are a set of homogeneous equations that have a nontrivial solution only if the determinant of their coefficients is zero, a condition that gives the two conditions

$$\frac{\sin(2m\beta)}{2m\beta} = \mp \frac{\sin(2\beta)}{2\beta} \tag{7.118}$$

which are equations for the power m. However, an examination of the roots of these two equations shows that the equation with the minus sign in Eq. (7.118) gives the most singular stresses for $0 < m < 1$ so its roots will dominate the stress behavior for r small. Thus, we will examine the roots of

$$\frac{\sin(2m\beta)}{2m\beta} = -\frac{\sin(2\beta)}{2\beta} \tag{7.119}$$

which we will write in terms of the opening angle, α, of the notch, where $\alpha = 2\pi - 2\beta$, giving

$$\sin(m[2\pi - \alpha]) = m \sin(\alpha) \tag{7.120}$$

A MATLAB® function stress_singularity was written to find the roots of Eq. (7.120) and plot the results, which are given in Fig. 7.28. It can be seen that a sharp crack ($\alpha = 0$) has a

Figure 7.28 The behavior of the stress singularity as a function of the opening angle of a notch.

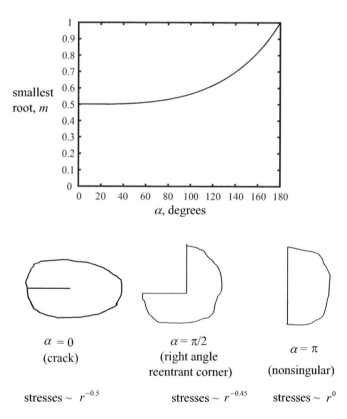

singular behavior of $r^{-1/2}$ for the stresses and other cases also have a singular behavior as long as $\alpha < \pi$, i.e., they represent reentrant corners.

It may appear that such singular behavior is purely a mathematical consequence of the highly idealized geometry, but such behavior also has an important physical significance for the behavior of cracks. In fact, the field of fracture mechanics, which describes cracks and their relationship to failure, has arisen from analysis of the stresses associated with such singularities. We will describe some basic elements of fracture mechanics in Chapter 13.

7.5 PROBLEMS

P7.1 Figure P7.1 shows a hydraulic conduit composed of a steel liner in a concrete pipe. The material properties of the steel and concrete are:

$$E_{steel} = 30 \times 10^6 \text{ psi}, E_{concrete} = 4 \times 10^6 \text{ psi}$$

$$\nu_{steel} = 0.3, \nu_{concrete} = 0.2$$

The inner radius of the concrete pipe is a and its outer radius is b. The thickness of the steel liner is t, as shown. The pipe carries an internal pressure, p.

(a) If $a = 3$ in., $b = 8$ in., $t = 0.2$ in., $p = 100$ psi, determine and plot the radial and hoop stress in the steel liner and the concrete pipe, using the thick-wall pressure vessel expressions.

(b) How much do your answers to part (a) change if you use the strength of materials expressions for a thin-wall pressure vessel for the steel liner, where we neglect the radial stress and assume the hoop stress is uniform across the liner?

P7.2 A large plate made of aluminum ($E_{Al} = 72$ GPa, $\nu_{Al} = 0.33$) contains a circular hole of radius, a (Fig. P7.2). A steel plug ($E_s = 194$ GPa, $\nu_s = 0.30$) of radius $a + \Delta$ is shrunk-fit into this plate.

(a) Determine the stresses $\sigma_{rr}, \sigma_{\theta\theta}$ at the contact surface in terms of $(a, \Delta, E_{Al}, \nu_{Al}, E_s, \nu_s)$. [Hint: you can use the thick-wall pressure vessel expressions for either the inner radius $= 0$ or the outer radius $= \infty$ to consider the plug or plate, respectively.]

(b) Determine the maximum allowable interference, Δ, if the maximum allowable shear stress in the aluminum plate is 50 MPa. Take $a = 40$ mm.

P7.3 A cantilever rectangular beam of length, L, height, $2H$, and width, b, carries a uniform shear stress, τ_0, on its top surface, as shown in Fig. P7.3.

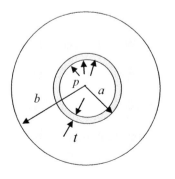

Figure P7.1 A steel liner of a concrete conduit where a uniform pressure, p, acts within the liner.

Figure P7.2 A large aluminum plate and an oversized steel plug.

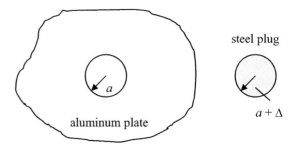

Figure P7.3 A cantilever beam where a uniform shear stress, τ_0, acts on its top surface.

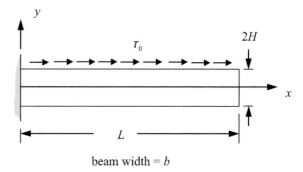

(a) Determine the internal axial force, shear force, and bending moment in this beam and the axial stress and shear stress as determined from a strength of materials approach.

(b) Using the Airy stress function given by

$$\phi = axy + bxy^2 + cxy^3 + dy^2 + ey^3$$

determine the constants (a, b, c, d, e) so that the appropriate boundary conditions are satisfied. How do your results differ from the strength of materials solution?

P7.4 Consider a triangular retaining wall subject to a hydrostatic pressure, p, from the surrounding soil that varies linearly with depth, as shown in Fig. P7.4. Neglecting the weight of the wall, consider an Airy stress function

$$\phi = \frac{a}{6}x^3 + \frac{b}{2}x^2y + \frac{c}{2}xy^2 + \frac{d}{6}y^3$$

where (a, b, c, d) are constants.

(a) What are the exact boundary conditions on ϕ that must be satisfied on the vertical wall?

(b) What are the exact boundary conditions on ϕ that must be satisfied on the inclined wall? [Hint: consider the x- and y-components of the traction vector.]

(c) What are the Saint-Venant boundary conditions on the horizontal base?

(d) Determine the constants (a, b, c, d) that satisfy this problem and the corresponding stresses.

(e) Sketch out the normal and shear stresses at the base ($y = L$). Without solving explicitly for the strength of materials solution for this problem, comment on whether the stresses you have sketched appear to be consistent with a strength of materials solution.

P7.5 The rectangular concrete wall shown in Fig. P7.5 is subjected to a linearly varying pressure, $p = ky$ on both of its faces. The weight of the concrete can be taken into account by including the body force due to gravity, which can be written in terms of potential function, V, given by $V = -\gamma y$, where γ is the weight/unit volume of the concrete. Using an Airy stress function

$$\phi = ax^2 + by^2 + cy^3$$

(a) write down the exact boundary conditions that must be satisfied at the top and on the sides of the wall.
(b) Using the boundary conditions from part (a), determine the constants (a, b, c) in terms of γ and k. What are the stresses at the base (at $y = L$)?
(c) Because of symmetry, we expect that $u_x = 0$ at $x = 0$. Also, at $y = L$ we expect $u_y = 0$. Using these displacement conditions, determine the displacement components in the wall in terms of (γ, k, E, ν, L), where E and ν are the Young's modulus and Poisson's ratio, respectively, for the concrete. Assume the wall is in a state of plane strain.

P7.6 Consider the Airy stress function $\phi = Ar^2(1 + \cos(2\theta))$. What is the state of stress in Cartesian coordinates generated by this function? How does your answer change if
$\phi = Ar^2(1 - \cos(2\theta))$?

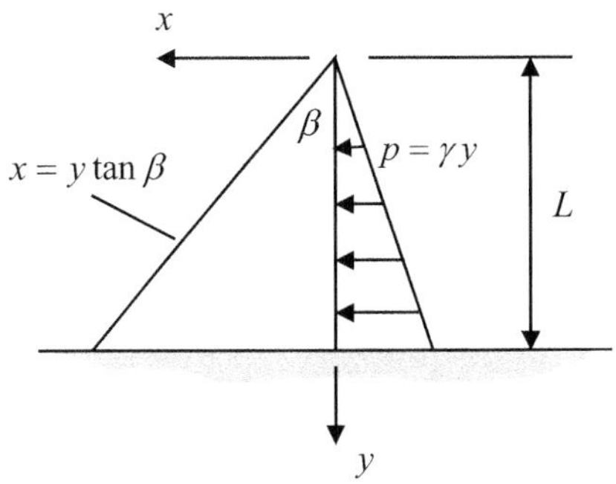

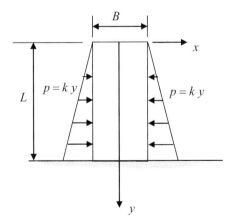

Figure P7.4 A retaining wall subjected to a linearly varying soil pressure.

Figure P7.5 A concrete wall subjected to its own weight and linearly varying pressures on its sides.

P7.7 Consider a rigid disk of radius a that is bonded to a large elastic plate, as shown in Fig. P7.6a, where the disk is subjected to a torque, T. The thickness of the plate is t.
  (a) Using a free body diagram of a cut section of the plate of radius r, as shown in Fig. P7.6b, and assuming the shear stress, τ, is uniform on the cut section, determine the shear stress as a function of (T, t, r).
  (b) From your results in part (a), what Michell type of Airy stress function (see Eq. (7.61)) would be a solution for this problem?

P7.8 Consider a circular hole in a large plate, where the remote stress field is the biaxial field shown in Fig. P7.7a.
  (a) On a diagram similar to that seen in Fig. 7.13, show the extreme values of the normalized hoop stress on the hole.
  (b) Repeat part (a) for the biaxial stress field shown in Fig. P7.7b.
  (c) Repeat part (a) for the biaxial stress field shown in Fig. P7.7c. What more can you say about this case?

P7.9 Consider a circular hole in a large plate where the remote stress field is the pure shear stress shown in Fig. P7.8a.
  (a) On a diagram similar to that seen in Fig. 7.13, show the extreme values of the normalized hoop stress and where they occur on the hole.
  (b) Repeat part (a) when the remote stress field is the more general state of stress shown in Fig. P7.8b.

Figure P7.6 (a) Rigid disk of radius a embedded in a large plate and subjected to a torque, T. (b) A free body diagram of a circular section of the plate of radius r.

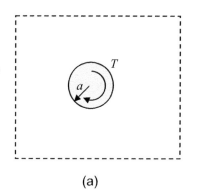
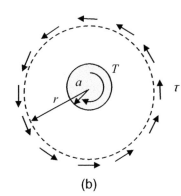

Figure P7.7 A hole in a plate under the remote stress fields shown in (a), (b), and (c).

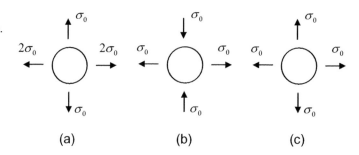

P7.10 The strength of materials flexure stress expression of Eq. (7.94) was for a curved beam under a pure moment. In practice, curved beams will also have axial forces and shear forces present. Consider, for example the curved member under a pair of forces, P, in Fig. P7.9. The largest moment in the member will occur at section A–A, but at that section there will also be an axial force, N, acting at the centroid of the cross-section. It is easy to show that in this case the flexure stress in the curved beam is given by the sum of extensional and bending contributions where

$$\sigma_{\theta\theta} = \frac{N}{A} + \frac{M(A - rA_m)}{A(A - RA_m)r}$$

If the curved member of Fig. P7.9 has a square 50×50 mm cross-section, find the maximum tensile and compressive stresses acting on section A–A if $P = 5.9$ kN. Plot the total stress across the cross-section.

P7.11 Consider the beam in Fig. P7.10a where the applied load and the reaction forces are all applied through rollers. In strength of materials it is normally assumed that these forces are concentrated forces so that the shear force diagram experiences "jumps" at the supports and the applied load as shown by the solid line in Fig. P7.10b. In reality, these loads are spread over contact areas and the force changes are, therefore, spread out over finite lengths.

(a) If the applied force $P = 4$ kN, the width of the beam (and rollers) is 100 mm, the diameter of the each of the two support rollers is 50 mm, and the diameter of the

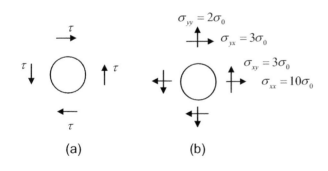

Figure P7.8 (a) Hole in a plate under a remote stress that is a pure shear stress. (b) The case where the remote stress is a more general 2-D state of stress.

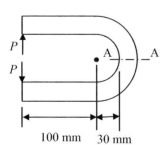

Figure P7.9 A pair of equal and opposite forces applied to a curved member.

roller under the load is 25 mm, estimate the finite lengths over which the shear force diagram changes at the applied load and supports. Assume both rollers and the beam are made out of steel ($E = 210$ GPa, $v = 0.29$).

(b) In Fig. P7.10b, the dashed lines indicate the shear forces varying at the loads. Plot the actual distribution of the shear force within the contact area at the left support. [Hint: first find the load, P, in terms of the length a and the maximum pressure, p_0, in the stress distribution.]

P7.12 Consider one of the Michell solutions for the Airy stress function, $\phi = Cr\theta \cos\theta$, where C is a constant.

(a) Show that this function is the solution of a wedge-shaped plate of thickness, t, with a concentrated force P at the tip of the wedge, as shown in Fig. P7.11, and find explicitly the radial stress in the wedge.

(b) Determine the normal stress, σ_{xx}, and the shear stress, σ_{xy}, in polar coordinates and then obtain approximate values for these stresses, written in terms of the Cartesian coordinates (x, y), when the wedge half-angle, α, is very small. Compare these stresses with the strength of materials expressions obtained by considering this wedge as a beam whose height varies linearly in x. Where there are discrepancies, comment on which expression you would trust more, and why.

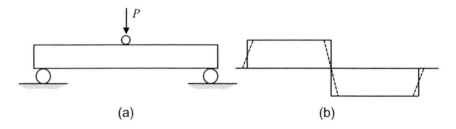

(a)   (b)

Figure P7.10 (a) Simply supported beam loaded and supported through rollers. (b) The idealized shear force diagram for concentrated loads (solid line) and for loads distributed over the contact area between the beam and the rollers.

Figure P7.11 A wedge-shaped plate loaded by a vertical force.

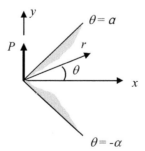

References

1. J. R. Barber, *Elasticity*, 3rd edn (Switzerland: Springer, 2010)
2. A. P. Boresi and R. J. Schmidt, *Advanced Mechanics of Materials*, 6th edn (New York, NY: John Wiley, 2002)
3. A. C. Ugural and S. K. Fenster, *Advanced Mechanics of Materials and Applied Elasticity*, 6th edn (Boston, MA: Pearson, 2020)
4. M. H. Sadd, *Elasticity – Theory, Applications, and Numerics* (Burlington, MA: Elsevier, 2005)
5. S. N. Karp and F. C. Karal, "Elastic field behavior in the neighborhood of a crack of arbitrary angle," *Comm. Pure and Appl. Math*, **15**, (1962), 413–421
6. F. C. Karal and S. N. Karp, "Stress behavior in the neighborhood of sharp corners," *Geophysics*, **29**, (1964), 360–369

8 Work–Energy Concepts

Work–energy concepts are important for two reasons. First, they provide an alternative way to guarantee equilibrium and compatibility, which are two key elements in all stress analyses. Second, energy methods have become the basis of formulating numerical methods, so they are at the heart of the field of computational mechanics, which will be discussed in Chapter 9. In this chapter we will discuss two types of internal energy in deformable bodies – strain energy and complementary strain energy. Although we will show that these internal energies are equal for linear elastic bodies, we will see that they play distinct roles in terms of work–energy relations.

8.1 Work Concepts

You are likely familiar with the concept of work involving forces where the differential work, dW, done by a force, $\mathbf{F}$, taken through a small displacement, $d\mathbf{u}$, is just $dW = \mathbf{F} \cdot d\mathbf{u}$. The work concept for a deformable body, which is discussed in the following section, follows this same form except the forces can be distributed over areas in the form of stresses, or over volumes in terms of body forces. These distributions can be expressed in various ways, which make the work concept for a deformable body appear rather complex. This complexity, however, will also lead to very useful and powerful tools, as we will see, for analyzing the deformations and stresses present.

8.1.1 Work for a Deformable Body

Consider a deformable body occupying a volume V whose surface is S. If this body experiences a small displacement, $d\mathbf{u}(\mathbf{x})$, the work, dW, done by the stress vector, $\mathbf{T}^{(\mathbf{n})}(\mathbf{x}_s)$, acting on the body surface and a body force, $\mathbf{f}(\mathbf{x})$, is given by (see Fig. 8.1):

$$dW = \int_S \mathbf{T}^{(\mathbf{n})}(\mathbf{x}_s) \cdot d\mathbf{u}(\mathbf{x}_s) dS + \int_V \mathbf{f}(\mathbf{x}) \cdot d\mathbf{u}(\mathbf{x}) dV \tag{8.1}$$

As in Chapter 6, this stress vector may represent surface loads distributed over the surface as well as line loads and concentrated forces and/or moments as special cases (see Fig. 6.1). Concentrated forces and moments in particular play an important role in certain work–energy theorems, and these will be discussed later. If we express the stress vector in terms of the Cartesian stress components and correspondingly the displacement vector and body force in terms of their Cartesian components, i.e.,

8.1 Work Concepts

$$\mathbf{T}^{(\mathbf{n})} = \sum_{i=1}^{3}\sum_{j=1}^{3} \sigma_{ij} n_i \mathbf{e}_j$$

$$\mathbf{f} = \sum_{j=1}^{3} f_j \mathbf{e}_j \qquad (8.2)$$

$$d\mathbf{u} = \sum_{j=1}^{3} du_j \mathbf{e}_j$$

then we have

$$\mathbf{T}^{(\mathbf{n})} \cdot d\mathbf{u} = \sum_{i=1}^{3}\sum_{j=1}^{3} \sigma_{ij} n_i du_j$$

$$\mathbf{f} \cdot d\mathbf{u} = \sum_{j=1}^{3} f_j du_j \qquad (8.3)$$

which, when placed into Eq. (8.1) gives

$$dW = \sum_{i=1}^{3}\sum_{j=1}^{3} \int_S (\sigma_{ij} du_j) n_i dS + \sum_{j=1}^{3} \int_V f_j du_j dS \qquad (8.4)$$

However, the surface integral can be transformed into an integral over the volume, V, since by Gauss's theorem (for any well-behaved g, where g can be a scalar vector, or matrix)

$$\int_S g n_i dS = \int_V \frac{\partial g}{\partial x_i} dV \qquad (8.5)$$

so that we can rewrite Eq. (8.4) as

$$\begin{aligned}
dW &= \sum_{j=1}^{3} \int_V \left(\sum_{i=1}^{3} \frac{\partial}{\partial x_i} (\sigma_{ij} du_j) + f_j du_j \right) dV \\
&= \sum_{j=1}^{3} \int_V \left(\sum_{i=1}^{3} \frac{\partial \sigma_{ij}}{\partial x_i} + f_j \right) du_j dV + \sum_{i=1}^{3}\sum_{j=1}^{3} \int_V \sigma_{ij} \frac{\partial}{\partial x_i} (du_j) dV
\end{aligned} \qquad (8.6)$$

The first volume integral in Eq. (8.6) vanishes because the equilibrium equations must be satisfied, and the integrand of the second integral can be written as

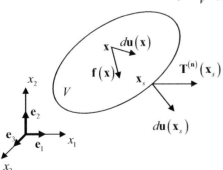

Figure 8.1 A deformable body acted upon by a body force and stress vector and small displacements generated by those forces.

$$\sum_{i=1}^{3}\sum_{j=1}^{3}\sigma_{ij}\frac{\partial}{\partial x_i}(du_j) = \sum_{i=1}^{3}\sum_{j=1}^{3}\sigma_{ij}d\left(\frac{\partial u_j}{\partial x_i}\right)$$

$$= \sum_{i=1}^{3}\sum_{j=1}^{3}\sigma_{ij}(de_{ij} + d\omega_{ij}) \quad (8.7)$$

$$= \sum_{i=1}^{3}\sum_{j=1}^{3}\sigma_{ij}de_{ij}$$

where the sum of the terms involving the change of the local rotation, $d\omega_{ij}$, vanishes because the stresses are symmetric while the rotations are antisymmetric. Thus, Eq. (8.6) becomes

$$dW = \int_V \left(\sum_{i=1}^{3}\sum_{j=1}^{3}\sigma_{ij}de_{ij}\right)dV \quad (8.8)$$

8.2 Work–Strain Energy

The work done by all the external forces acting on an elastic body is equal to the change of the internal potential energy, U, of the body, an energy of stored deformation, also called the *strain energy*. We can write this work–energy principle in differential form as

$$dW = dU = d\int_V u_0 dV = \int_V (du_0)dV \quad (8.9)$$

where u_0 is the strain energy per unit volume, i.e., the *strain energy density*. Comparing Eq. (8.9) and Eq. (8.8), we see that the change in this strain energy density is given as

$$du_0 = \sum_{i=1}^{3}\sum_{j=1}^{3}\sigma_{ij}de_{ij} \quad (8.10)$$

If we write the stresses in terms of the strains, then we can express the strain energy density as a function of the strains only, which we write symbolically as $u_0 = u_0(e)$, and from which it follows that

$$du_0(e) = \sum_{i=1}^{3}\sum_{j=1}^{3}\frac{\partial u_0(e)}{\partial e_{ij}}de_{ij} \quad (8.11)$$

so that comparing Eq. (8.11) and Eq. (8.10) we find

$$\sigma_{ij} = \frac{\partial u_0(e)}{\partial e_{ij}} \quad (i,j = 1,2,3) \quad (8.12)$$

Thus, just as a conservative force can be written in terms of the spatial gradient of a potential energy, the stresses in an elastic body that has a strain energy density function can be written as the strain gradient of that strain energy density. It is sometimes also convenient to write the stresses as components of a stress vector, σ_I $(I = 1, 2, \ldots, 6)$, and the strains as components of a strain vector, e_I $(I = 1, 2, \ldots, 6)$. In that case the increase in the strain energy density can be written as

$$du_0 = \sum_{I=1}^{6} \sigma_I de_I \tag{8.13}$$

and the stress vector as

$$\sigma_I = \frac{\partial u_0(e)}{\partial e_I} \quad (I = 1, 2, \ldots, 6) \tag{8.14}$$

8.2.1 Linear Elastic Material

Equations (8.12) and (8.14) show that if we can obtain the strain energy density for a material, then we can use that strain energy density to obtain the stress–strain relations. In general, those stress–strain relations will be nonlinear. However, if we assume that the strains in the material are small, it makes sense to expand the strain energy density in a power series. Keeping only quadratic terms at most we have

$$u_0(e) = A + \sum_{I=1}^{6} B_I e_I + \frac{1}{2} \sum_{I=1}^{6} \sum_{J=1}^{6} C_{IJ} e_I e_J \tag{8.15}$$

If, when the strains are all zero the strain energy density is zero and the stresses are also zero, then $A = B_I = 0$ and the strain energy density is expressible entirely in terms of the 36 elastic constants, C_{IJ}. However, from the second derivatives we find

$$C_{IJ} = \frac{\partial^2 u_0(e)}{\partial e_I \partial e_J} = \frac{\partial^2 u_0(e)}{\partial e_J \partial e_I} = C_{JI} \tag{8.16}$$

Thus, the matrix of elastic constants is symmetric and there are only 21 constants for the most general linear elastic material where

$$\sigma_I = \sum_{J=1}^{6} C_{IJ} e_J \quad (I = 1, 2, \ldots, 6) \tag{8.17}$$

which, we recall, can also be written in terms the C_{ijkl} constants as

$$\sigma_{ij} = \sum_{k=1}^{3} \sum_{l=1}^{3} C_{ijkl} e_{kl} \quad (i, j = 1, 2, 3) \tag{8.18}$$

Note that from Eq. (8.15) the strain energy density for a linear elastic material can be written as

$$u_0 = \frac{1}{2}\sum_{I=1}^{6}\sum_{J=1}^{6} C_{IJ}e_I e_J \qquad (8.19a)$$

or, using Eq. (8.17)

$$u_0 = \frac{1}{2}\sum_{I=1}^{6} \sigma_I e_I \qquad (8.19b)$$

If we write the strain in Eq. (8.19b) in terms of the stress through the compliance matrix, D_{IJ}, then we also have

$$u_0 = \frac{1}{2}\sum_{I=1}^{6}\sum_{J=1}^{6} D_{IJ}\sigma_I \sigma_J \qquad (8.19c)$$

In the first form in Eq. (8.19a) we see that we have written the strain energy density in terms of the strains only, i.e., $u_0 = u_0(e)$, while in the second form in Eq. (8.19b) the strain energy density is being written as a function of both stresses and strains, i.e., $u_0 = u_0(\sigma, e)$. In the third form of Eq. (8.19c) only stresses are involved so $u_0 = u_0(\sigma)$. We can also write these forms in terms of C_{ijkl} elastic constants and D_{ijkl} elastic compliances as

$$u_0 = \frac{1}{2}\sum_{i=1}^{3}\sum_{j=1}^{3}\sum_{k=1}^{3}\sum_{l=1}^{3} C_{ijkl} e_{ij} e_{kl} \qquad (8.20a)$$

$$u_0 = \frac{1}{2}\sum_{i=1}^{3}\sum_{j=1}^{3} \sigma_{ij} e_{ij} \qquad (8.20b)$$

$$u_0 = \frac{1}{2}\sum_{i=1}^{3}\sum_{j=1}^{3}\sum_{k=1}^{3}\sum_{l=1}^{3} D_{ijkl} \sigma_{ij} \sigma_{kl} \qquad (8.20c)$$

It is useful to graphically illustrate these strain energy concepts for the simple case of a state of uniaxial stress. In this case, Fig. 8.2a shows the stress–strain curve for a general elastic material where the stress σ_{xx} is plotted versus the strain e_{xx}. The differential of the strain energy density is shown in that figure as well as the total strain energy density, which is just the area under the stress–strain curve. Figure 8.2b shows the special case of a linear elastic material where, in the uniaxial case, $\sigma_{xx} = E e_{xx}$ and E is Young's modulus. In that case, as shown, the total strain energy density is the shaded triangular area under the stress–strain curve, which can be written in the equivalent forms $u_0 = E e_{xx}^2/2 = \sigma_{xx}^2/2E = \sigma_{xx} e_{xx}/2$ depending on whether the strain energy density is written in terms of the strain, the stress, or in terms of both stress and strain. The factor of one half in these expressions arises from integrating the differential of the strain energy density from zero to the final stress or strain state when the stress–strain law is linear. For example, we have $u_0 = \int du_0 = \int \sigma'_{xx} de'_{xx} = E \int e'_{xx} de'_{xx} = E e_{xx}^2/2$, where e_{xx} is the final strain. In this case $u_0 = u_0(e)$ but for $u_0 = u_0(\sigma)$ a similar integration yields $u_0 = \sigma_{xx}^2/2E$.

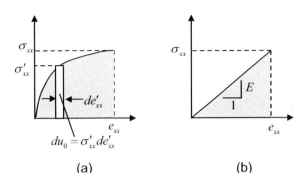

Figure 8.2 (a) A one-dimensional (1-D) stress–strain curve, showing the graphical interpretation of the differential of the strain energy density, du_0. (b) The special case of a linear stress–strain law. In both cases the total strain energy density, u_0, is the shaded area under the stress–strain curve from zero to a final state of stress and strain of (σ_{xx}, e_{xx}).

The mixed form of $u_0(\sigma, e) = \sigma_{xx} e_{xx}/2$ also arises directly from the area formula for the shaded triangular area in Fig. 8.2b.

So far, we have considered the work–strain energy principle in terms of differential changes of the strain energy and differential work (see Eq. (8.1)). For a linearly elastic material we can write a similar work–energy relationship in terms of total strain energy and total work terms. To see this, use Eq. (8.20b) and write the total strain energy as

$$U = \frac{1}{2} \int_V \sum_{i=1}^{3} \sum_{j=1}^{3} \sigma_{ij} e_{ij} dV \tag{8.21}$$

Using the symmetry of the stress matrix, Gauss's theorem, and the equilibrium equations we can rewrite the strain energy as

$$U = \frac{1}{2} \int_V \sum_{i=1}^{3} \sum_{j=1}^{3} \sigma_{ij} \frac{\partial u_j}{\partial x_i} dV$$

$$= \frac{1}{2} \int_V \sum_{i=1}^{3} \sum_{j=1}^{3} \frac{\partial}{\partial x_i} (\sigma_{ij} u_j) dV - \frac{1}{2} \int_V \sum_{i=1}^{3} \sum_{j=1}^{3} \frac{\partial \sigma_{ij}}{\partial x_i} u_j dV$$

$$= \frac{1}{2} \int_S \sum_{i=1}^{3} \sum_{j=1}^{3} \sigma_{ij} n_i u_j dS + \frac{1}{2} \int_V \sum_{j=1}^{3} f_j u_j dV \tag{8.22}$$

$$= \frac{1}{2} \int_S \sum_{j=1}^{3} T_j^{(n)} u_j dS + \frac{1}{2} \int_V \sum_{j=1}^{3} f_j u_j dV$$

$$= \frac{1}{2} \int_S \mathbf{T}^{(n)} \cdot \mathbf{u} \, dS + \frac{1}{2} \int_V \mathbf{f} \cdot \mathbf{u} \, dV$$

which says that the total strain energy stored in the material is equal to the total work done by the stress vector and the body force acting on the body. This result is called *Clapeyron's theorem*. The factor of one half is present again because in loading the elastic body from its undeformed state to one where the displacement vector is **u**, the stresses are proportional to the strains in the loading process.

8.2.2 Isotropic Case

As seen in Chapter 5, if a linear elastic material is also isotropic then the material coefficients can be expressed in terms of only two elastic constants such as Young's modulus, E, and Poisson's ratio, v, where

$$C_{11} = C_{22} = C_{33} = \frac{E(1-v)}{(1+v)(1-2v)}$$

$$C_{12} = C_{21} = C_{13} = C_{31} = C_{23} = C_{32} = \frac{Ev}{(1+v)(1-2v)} \quad (8.23)$$

$$C_{44} = C_{55} = C_{66} = \frac{1}{2}(C_{11} - C_{12}) = \frac{E}{2(1+v)} \equiv G$$

all others $= 0$

Placing these values into Eq. (8.19a)) gives

$$u_0(e) = \frac{1}{2}\frac{E}{(1+v)(1-2v)}\left[(1-v)\left(e_{xx}^2 + e_{yy}^2 + e_{zz}^2\right) + 2v\left(e_{xx}e_{yy} + e_{xx}e_{zz} + e_{yy}e_{zz}\right)\right]$$
$$+ \frac{1}{2}G\left(\gamma_{xy}^2 + \gamma_{xz}^2 + \gamma_{yz}^2\right) \quad (8.24)$$

- or, in terms of principal strains,

$$u_0(e) = \frac{1}{2}\frac{E}{(1+v)(1-2v)}\left[(1-v)\left(e_{p1}^2 + e_{p2}^2 + e_{p3}^2\right) + 2v\left(e_{p1}e_{p2} + e_{p1}e_{p3} + e_{p2}e_{p3}\right)\right] \quad (8.25)$$

which can also be written in terms of the strain invariants as

$$u_0(e) = \frac{1}{2}\frac{E(1-v)}{(1+v)(1-2v)}J_1^2 - 2GJ_2 \quad (8.26)$$

where (see Eq. (4.36))

$$J_1 = e_{p1} + e_{p2} + e_{p3}$$
$$J_2 = e_{p1}e_{p2} + e_{p1}e_{p3} + e_{p2}e_{p3} \quad (8.27)$$

Another way to express the strain energy density is in terms of the strain matrix components, e_{ij}. First rewrite Eq. (8.24) as

$$u_0(e) = \frac{G}{(1-2v)}\left[(1-v)\left(e_{xx}^2 + e_{yy}^2 + e_{zz}^2\right) + 2v\left(e_{xx}e_{yy} + e_{xx}e_{zz} + e_{yy}e_{zz}\right)\right.$$
$$\left. + 2(1-2v)\left(e_{xy}^2 + e_{xz}^2 + e_{yz}^2\right)\right] \quad (8.28a)$$

and rearrange the terms as

$$u_0(e) = \frac{G}{(1-2v)}\left[(1-2v)\left(e_{xx}^2 + e_{yy}^2 + e_{zz}^2 + 2e_{xy}^2 + 2e_{xz}^2 + 2e_{yz}^2\right)\right.$$
$$\left. + v\left(e_{xx}^2 + e_{yy}^2 + e_{zz}^2 + 2e_{xx}e_{yy} + 2e_{xx}e_{zz} + 2e_{yy}e_{zz}\right)\right] \quad (8.28b)$$

which then can be expressed finally in the compact form

$$u_0(e) = G\left[\sum_{i=1}^{3}\sum_{j=1}^{3} e_{ij}e_{ij} + \frac{v}{(1-2v)}\left(\sum_{i=1}^{3} e_{ii}\right)^2\right] \quad (8.29)$$

If we write the strains in terms of the stresses, then we can also express the strain energy density solely in terms of the stresses, i.e., $u_0 = u_0(\sigma)$. The end result is

$$u_0(\sigma) = \frac{1}{2E}\left[\sigma_{xx}^2 + \sigma_{yy}^2 + \sigma_{zz}^2 - 2v\left(\sigma_{xx}\sigma_{yy} + \sigma_{xx}\sigma_{zz} + \sigma_{yy}\sigma_{zz}\right)\right]$$
$$+ \frac{1}{2G}\left(\sigma_{xy}^2 + \sigma_{xz}^2 + \sigma_{yz}^2\right) \quad (8.30)$$

In terms of the principal stresses we have

$$u_0(\sigma) = \frac{1}{2E}\left[\sigma_{p1}^2 + \sigma_{p2}^2 + \sigma_{p3}^2 - 2v\left(\sigma_{p1}\sigma_{p2} + \sigma_{p1}\sigma_{p3} + \sigma_{p2}\sigma_{p3}\right)\right] \quad (8.31)$$

or, for the stress invariants I_1, I_2,

$$u_0(\sigma) = \frac{1}{2E}\left[I_1^2 - 2(1+v)I_2\right] = \frac{1}{2E}I_1^2 - \frac{1}{2G}I_2 \quad (8.32)$$

where (see Eq. (2.61))

$$I_1 = \sigma_{p1} + \sigma_{p2} + \sigma_{p3}$$
$$I_2 = \sigma_{p1}\sigma_{p2} + \sigma_{p1}\sigma_{p3} + \sigma_{p2}\sigma_{p3} \quad (8.33)$$

We can also express the strain energy in terms of the stress matrix components, σ_{ij}. Starting from Eq. (8.30), we can rewrite that equation as

$$u_0(\sigma) = \frac{1}{2E}\left[\sigma_{xx}^2 + \sigma_{yy}^2 + \sigma_{zz}^2 + 2(1+v)\left(\sigma_{xy}^2 + \sigma_{xz}^2 + \sigma_{yz}^2\right)\right.$$
$$\left. - 2v\left(\sigma_{xx}\sigma_{yy} + \sigma_{xx}\sigma_{zz} + \sigma_{yy}\sigma_{zz}\right)\right] \quad (8.34)$$

and then rearrange it to obtain

$$u_0(\sigma) = \frac{1}{2E}\left[(1+v)\left(\sigma_{xx}^2 + \sigma_{yy}^2 + \sigma_{zz}^2 + 2\sigma_{xy}^2 + 2\sigma_{xz}^2 + 2\sigma_{yz}^2\right)\right.$$
$$\left. - v\left(\sigma_{xx}^2 + \sigma_{yy}^2 + \sigma_{zz}^2 + 2\sigma_{xx}\sigma_{yy} + 2\sigma_{xx}\sigma_{zz} + 2\sigma_{yy}\sigma_{zz}\right)\right] \quad (8.35)$$

which then can be written finally in matrix form as

$$u_0(\sigma) = \frac{1}{2E}\left[(1+v)\sum_{i=1}^{3}\sum_{j=1}^{3}\sigma_{ij}\sigma_{ij} - v\left(\sum_{i=1}^{3}\sigma_{ii}\right)^2\right] \quad (8.36)$$

8.2.3 Distortional Strain Energy

Strain energy is a very useful quantity when discussing failure concepts since intuitively we expect a material will fail if we require it to store too much energy. (See Chapter 13 for a discussion of failure.) For failure with respect to slip (yielding), it has been observed experimentally that such failure cannot be produced by a hydrostatic state of stress, i.e., where we apply a pure pressure to the material. Since the strain energy in general contains a hydrostatic energy component, we first need to remove that hydrostatic component of the strain energy before we can use that energy in a failure theory for slip. To accomplish this removal, note that, for a state of stress consisting of a pure pressure, p, we have $\sigma_{p1} = \sigma_{p2} = \sigma_{p3} = -p$ which produces from Eq. (8.31) a strain energy density

$$u_0 = u_p = \frac{3(1-2v)}{2E}p^2 \quad (8.37)$$

For a general state of stress, if we replace the pressure, p, by the average of the three principal stresses, i.e.,

$$p = -\frac{\sigma_{p1} + \sigma_{p2} + \sigma_{p3}}{3} \quad (8.38)$$

then we can define the hydrostatic component as

$$u_p = \frac{(1-2v)}{6E}\left(\sigma_{p1} + \sigma_{p2} + \sigma_{p3}\right)^2 \quad (8.39)$$

and we can also define the *distortional strain energy density*, u_d, as

$$u_d = u_0 - u_p \quad (8.40)$$

Using Eq. (8.31) and Eq. (8.39), after some algebra we find

$$u_d = \frac{(1+v)}{6E}\left[\left(\sigma_{p1} - \sigma_{p2}\right)^2 + \left(\sigma_{p1} - \sigma_{p3}\right)^2 + \left(\sigma_{p2} - \sigma_{p3}\right)^2\right] \quad (8.41)$$

If you recall, in Chapter 2 we discussed the stresses acting on a particular plane called the octahedral plane. There we found an explicit expression for the magnitude of the total shear stress acting on the octahedral plane, $|(\tau_s)_{oct}|$, Eq. (2.82), which shows that the distortional strain energy density seen in Eq. (8.41) is just

$$u_d = \frac{3(1+v)}{2E}|(\tau_s)_{oct}|^2 \quad (8.42)$$

8.3 Complementary Strain Energy

In applying work–energy principles, we will see that a quantity called the *complementary strain energy density*, u_0^c, also appears, where this density is defined as

$$u_0^c = \sum_{I=1}^{6} \sigma_I e_I - u_0 \tag{8.43a}$$

or, equivalently,

$$u_0^c = \sum_{i=1}^{3} \sum_{j=1}^{3} \sigma_{ij} e_{ij} - u_0 \tag{8.43b}$$

If we consider the one-dimensional (1-D) state of stress again then, as Fig. 8.3a shows, the complementary strain energy density is the shaded area shown above the stress–strain curve relating σ_{xx} to e_{xx}. A small change of this complementary strain energy density is therefore given by the chain rule of differentiation as

$$\begin{aligned} du_0^c &= \sum_{I=1}^{6} e_I d\sigma_I + \sum_{I=1}^{6} \sigma_I de_I - du_0 \\ &= \sum_{I=1}^{6} e_I d\sigma_I \end{aligned} \tag{8.44a}$$

where we have used Eq. (8.13). This change is also shown graphically in Fig. 8.3a for the 1-D state of stress case. We can also write this relationship in terms of the stress and strain matrices as

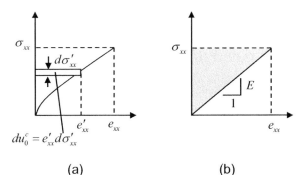

Figure 8.3 (a) A 1-D stress–strain curve, showing the graphical interpretation of the differential of the complementary strain energy density, du_0^c. (b) The special case of a linear stress–strain law. In both cases the total complementary strain energy density, u_0^c, is the shaded area above the stress–strain curve from zero to a final state of stress and strain of (σ_{xx}, e_{xx}).

$$du_0^c = \sum_{i=1}^{3}\sum_{j=1}^{3} e_{ij} d\sigma_{ij} \tag{8.44b}$$

If we now express the strains as a function of the stresses so that the complementary strain energy density is a function of the stresses only, we have

$$du_0^c(\sigma) = \sum_{I=1}^{6} \frac{\partial u_0^c}{\partial \sigma_I} d\sigma_I \tag{8.45}$$

Comparing Eq. (8.45) with Eq. (8.44a) we find

$$e_I = \frac{\partial u_0^c(\sigma)}{\partial \sigma_I} \quad (I = 1, 2, \ldots, 6) \tag{8.46a}$$

or, equivalently, we can obtain

$$e_{ij} = \frac{\partial u_0^c(\sigma)}{\partial \sigma_{ij}} \tag{8.46b}$$

Comparing Eq. (8.46a) with Eq. (8.14) or comparing Eq. (8.46 b) with Eq. (8.12), it follows that u_0 and u_0^c play complementary roles where the stresses and strains are concerned – hence the name complementary energy density for u_0^c.

8.3.1 Linear Elastic Material

If we have a linear elastic material, then it follows from Eq. (8.19b) and Eq. (8.43a) that $u_0^c = u_0$, where in the "mixed" form for these densities we have

$$u_0(\sigma, e) = u_0^c(\sigma, e) = \frac{1}{2}\sum_{I=1}^{6} \sigma_I e_I \tag{8.47}$$

All the other expressions obtained earlier for the strain energy density are also valid for the complementary strain energy density as well. In the 1-D case (see Fig. 8.3b) it is easy to see that these densities are identical since the area below the linear stress–strain "curve" is equal to the area above that "curve."

8.4 Strain Energy for Strength of Materials Problems

In the previous sections we have discussed the concept of strain energy for the general case of a deformable body. In this section, we will obtain explicit total strain energy expressions for the simple problems of axial extension/compression, bending, and torsion treated in an introductory strength of materials course. In all these cases the material is assumed to have linear elastic behavior.

8.4.1 Axial Loads

In the case of the uniaxial loading of a bar (see Fig. 8.4a), the stress, strain, and stress–strain relations are

$$\sigma_{xx} = \frac{F_x(x)}{A}$$
$$e_{xx} = \frac{du_x}{dx} \qquad (8.48)$$
$$\sigma_{xx} = Ee_{xx}$$

where F_x is the internal axial load, u_x is the axial displacement, and A and E are the cross-sectional area and Young's modulus, respectively, of the bar. As discussed in Chapter 3, the equilibrium equation for the bar is

$$\frac{dF_x}{dx} = -q_x \qquad (8.49)$$

where $q_x(x)$ is a distributed load/unit length acting on the bar. In terms of the displacement, the equilibrium equation becomes

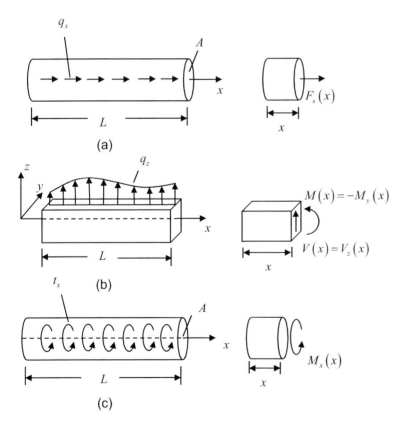

Figure 8.4 (a) An axial-load geometry showing a distributed force/unit length, $q_x(x)$, acting on the bar and the internal force, F_x. (b) Bending of a beam showing the distributed vertical force/unit length, $q_z(x)$ acting along the beam and the internal bending moment, $M = -M_y$, and shear force, $V = V_z$, acting in the beam. (c) Torsion of a circular shaft showing a distributed torque/unit length, $t_x(x)$, acting along the shaft and the internal twisting moment (torque), M_x.

$$\frac{d}{dx}\left(EA\frac{du_x}{dx}\right) = -q_x \tag{8.50}$$

Since there is only one stress the total strain energy of the bar, U, is

$$\begin{aligned} U &= \frac{1}{2}\int_0^L \frac{\sigma_{xx}^2}{E} A dx \\ &= \frac{1}{2}\int_0^L \frac{(F_x)^2}{AE} dx \end{aligned} \tag{8.51}$$

where L is the length of the bar. Similarly, in terms of the strain or displacement

$$\begin{aligned} U &= \frac{1}{2}\int_0^L E e_{xx}^2 A dx \\ &= \frac{1}{2}\int_0^L EA\left(\frac{du_x}{dx}\right)^2 dx \end{aligned} \tag{8.52}$$

which also follows directly from Eq. (8.51) and the force–displacement relation $F_x = EA du_x/dx$ for the bar.

8.4.2 Bending

In engineering beam theory, the normal stress, normal strain, and normal stress–strain relation for a beam are (see Fig. 8.4b):

$$\begin{aligned} \sigma_{xx} &= -\frac{M(x)z}{I} \\ e_{xx} &= -\frac{M(x)z}{EI} = -\frac{d^2w}{dx^2}z \\ \sigma_{xx} &= E e_{xx} \end{aligned} \tag{8.53}$$

where $M = -M_y$ is the internal bending moment about the *negative* y-axis, $I = I_{yy}$ is the area moment of the cross-section about the y-axis, and w is the displacement of the neutral axis of the beam in the z-direction. In Eq. (8.53) we have also used the moment–curvature relationship

$$M(x) = EI\kappa(x) = EI\frac{d^2w}{dx^2} \tag{8.54}$$

where the curvature of the beam, κ, is just the second derivative of the displacement, w, when the slope of the beam is small, as assumed in engineering beam theory. Force and moment equilibrium equations for the beam are

$$\frac{dM}{dx} = -V$$
$$\frac{dV}{dx} = -q_z \tag{8.55}$$

where $V = V_z$ is the internal shear force in the beam (see Fig. 8.4b) and $q_z(x)$ is a distributed load/unit length acting in the z-direction. The equilibrium equation for the moment can be written directly in terms of the distributed load by eliminating the shear force, giving

$$\frac{d^2 M}{dx^2} = q_z \tag{8.56}$$

or, using the moment–curvature relation,

$$\frac{d^2}{dx^2}\left(EI\frac{d^2 w}{dx^2}\right) = q_z \tag{8.57}$$

The total strain energy due to bending for a bar of length, L, is

$$U = \frac{1}{2}\int_0^L \int_A \frac{\sigma_{xx}^2}{E} dA\, dx$$

$$= \frac{1}{2}\int_0^L \left\{\int_A z^2 dA\right\} \frac{M^2}{EI^2} dx \tag{8.58}$$

$$= \frac{1}{2}\int_0^L \frac{M^2}{EI} dx$$

or, using the moment–curvature relation in Eq. (8.58), the strain energy in terms of the displacement of the neutral axis is

$$U = \frac{1}{2}\int_0^L EI\left(\frac{d^2 w}{dx^2}\right)^2 dx \tag{8.59}$$

There is also internal strain energy associated with the shear stress in the beam. In engineering beam theory, this shear stress is $\tau = \sigma_{xz}$ given by

$$\tau = \frac{V(x)Q(z)}{I\, t(z)} \tag{8.60}$$

so that

$$U = \frac{1}{2}\int_0^L \int_A \frac{\tau^2}{G} dA\, dx$$

$$= \frac{1}{2}\int_0^L \left\{ \frac{A}{I^2}\int_A \frac{Q^2}{t^2} dA \right\} \frac{V^2}{GA} dx \tag{8.61}$$

$$= \frac{1}{2}\int_0^L k \frac{V^2}{GA} dx$$

where the dimensionless constant, k, is

$$k = \frac{A}{I^2}\int_A \frac{Q^2}{t^2} dA \tag{8.62}$$

This constant is a function of the geometry of the cross-section only. For example, for a rectangular cross-section $k = 1.2$. In bending problems involving long and slender beams, the strain energy associated with the bending (flexural) stress is considerably larger than this strain energy due to the shear stress. Thus, in those cases the strain energy of the shear stresses can be neglected.

8.4.3 Torsion

The shear stress, shear strain, and stress–strain relations for the torsion of a circular shaft are given by

$$\tau = \frac{M_x(x)r}{J}$$

$$\gamma = \frac{d\phi(x)}{dx} r \tag{8.63}$$

$$\tau = G\gamma$$

where τ is the total shear stress in the shaft (see Eq. (1.38)) and M_x is the internal moment about the x-axis (torque) (see Fig. 8.4c). The total shear strain, γ, given in Eq. (8.63) follows from the torque–twist relation given by (see Eq. (1.34))

$$M_x(x) = GJ \frac{d\phi(x)}{dx} \tag{8.64}$$

and Hooke's law, where $\phi(x)$ is the twist of the cross-section about the x-axis of the shaft, and J and G are the polar area moment of the cross-section and the shear modulus. The equation of equilibrium for the shaft is (see Eq. (3.49))

$$\frac{dM_x}{dx} = -t_x \tag{8.65}$$

where $t_x(x)$ is the distributed torque/unit length acting along the shaft. In terms of the twist, the equilibrium equation becomes

$$\frac{d}{dx}\left(GJ\frac{d\phi}{dx}\right) = -t_x \tag{8.66}$$

Using these results, the strain energy of the shaft in terms of the torque is

$$\begin{aligned} U &= \frac{1}{2}\int_0^L \int_A \frac{\tau^2}{G} dA\, dx \\ &= \frac{1}{2}\int_0^L \left\{\int_A r^2 dA\right\} \frac{M_x^2}{GJ^2} dx \\ &= \frac{1}{2}\int_0^L \frac{M_x^2}{GJ} dx \end{aligned} \tag{8.67}$$

or, in terms of the twist, using the torque–twist relation in Eq. (8.67)

$$U = \frac{1}{2}\int_0^L GJ\left(\frac{d\phi}{dx}\right)^2 dx \tag{8.68}$$

8.5 Principle of Virtual Work and Minimum Potential Energy

Satisfying the governing differential equilibrium equations is only one way in which we can ensure equilibrium is satisfied in a deformable body. In this section we will see that an alternative approach is through work–energy principles and the introduction of the concept of virtual work and potential energy.

8.5.1 General Deformable Body

Assume the equilibrium equations for a deformable body are satisfied. If we multiply those equations by a set of weighting functions, δu_j ($j = 1, 2, 3$), which represent small changes of "virtual" displacements (i.e., not the actual displacements), u_j ($j = 1, 2, 3$), of the body and integrate over the volume, V, of the body we obtain

$$\int_V \sum_{i=1}^3 \left(\sum_{j=1}^3 \frac{\partial \sigma_{ij}}{\partial x_i} + f_j\right) \delta u_j dV = 0 \tag{8.69}$$

which we can rewrite as

$$\int_V \sum_{i=1}^{3}\sum_{j=1}^{3} \frac{\partial}{\partial x_i}(\sigma_{ij}\delta u_j)dV - \int_V \sum_{i=1}^{3}\sum_{j=1}^{3} \sigma_{ij}\frac{\partial \delta u_j}{\partial x_i}dV + \int_V \sum_{i=1}^{3} f_j \delta u_j dV = 0 \qquad (8.70)$$

We can use Gauss's theorem on the first term in Eq. (8.70) and interchange the real differential changes with the virtual changes, i.e., $\partial(\delta u_i)/\partial x_j = \delta(\partial u_i/\partial x_j)$, to obtain

$$\int_S \sum_{i=1}^{3}\sum_{j=1}^{3} \sigma_{ij} n_i \delta u_j dS - \int_V \sum_{i=1}^{3}\sum_{j=1}^{3} \sigma_{ij} \delta\left(\frac{\partial u_j}{\partial x_i}\right) dV + \int_V \sum_{j=1}^{3} f_j \delta u_j dV = 0 \qquad (8.71a)$$

which we can also write as

$$\int_V \sum_{i=1}^{3}\sum_{j=1}^{3} \sigma_{ij} \delta e_{ij} dV = \int_S \sum_{j=1}^{3} T_j^{(n)} \delta u_j dS + \int_V \sum_{j=1}^{3} f_j \delta u_j dV \qquad (8.71b)$$

where we have introduced the stress vector and used the symmetry of the stress matrix to write the left-hand side of Eq. (8.71b) in terms of virtual strain changes. However, recall from Eq. (8.12) when we write the strain energy density in terms of the strains only, we have

$$\sigma_{ij}(e) = \frac{\partial u_0(e)}{\partial e_{ij}} \quad (i,j = 1, 2, 3) \qquad (8.72)$$

and so

$$\int_V \sum_{i=1}^{3}\sum_{j=1}^{3} \sigma_{ij}(e) \delta e_{ij} dV = \int_V \sum_{i=1}^{3}\sum_{j=1}^{3} \frac{\partial u_0(e)}{\partial e_{ij}} \delta e_{ij} dV$$
$$= \int_V \delta u_0(e) dV = \delta U(e) \qquad (8.73)$$

where $U(e)$ is the total strain energy when written as a function of the strains only and $\delta U(e)$ is the virtual change of this strain energy due to the virtual displacements δu_j (and correspondingly virtual strains). This leads to *the principle of virtual work*, which is

$$\delta U(e) = \int_S \sum_{j=1}^{3} T_j^{(n)} \delta u_j dV + \int_V \sum_{j=1}^{3} f_j \delta u_j dV \qquad (8.74)$$
$$= \delta W_v$$

where the right-hand side of Eq. (8.74) is the virtual work, δW_v, done by the stress vector and body force during the virtual displacements, δu_j (compare to Eq. (8.1) where we have actual displacements and work terms). Thus, we have shown that, if equilibrium is satisfied throughout the body, then the principle of virtual work is satisfied. This virtual work principle can also be written compactly as

8.5 Virtual Work and Minimum Potential Energy

$$\delta U(e) = \int_S \mathbf{T}^{(\mathbf{n})} \cdot \delta \mathbf{u} \ dV + \int_V \mathbf{f} \cdot \delta \mathbf{u} \ dV \tag{8.75}$$

We can simply reverse the steps just outlined and go from Eq. (8.75) to Eq. (8.69). If we assume the principle virtual work is satisfied for all virtual displacements then the $\delta \mathbf{u}$ appearing in Eq. (8.69) is arbitrary, which in turn requires that the integrand of Eq. (8.69) itself must be zero, i.e., the equilibrium equations must be satisfied at every point in the body. In summary, we can say that satisfying equilibrium for all points in a deformable body and satisfying the principle of virtual work for all virtual displacements are equivalent, i.e.,

equilibrium $\Rightarrow$ principle of virtual work

principle of virtual work $\Rightarrow$ equilibrium

As we will see, using this equivalence can give us an alternate way to satisfy the governing equations of equilibrium either exactly, or approximately.

One can make an even stronger connection between the principal of virtual work and the solution to the problem where the displacements are specified as known values $u_j = u_j^*$ on a portion of the surface, S_u, and the stresses are specified as known values $\sigma_{ji} n_j = T_i^*$ on the remaining portion, S_t, where the total surface $S = S_u + S_t$. Namely, one can show the following.

If a displacement field, u_j, is found such that (1) it satisfies the prescribed boundary displacement values $u_j = u_j^*$ on S_u and where $\delta u_j = 0$ on S_u, and (2) if the principle of virtual work in the form

$$\delta U(e) = \int_{S_t} \mathbf{T}^* \cdot \delta \mathbf{u} \ dV + \int_V \mathbf{f}^* \cdot \delta \mathbf{u} \ dV \tag{8.76}$$

is satisfied for all δu_j which vanish on S_u then u_j satisfies both equilibrium and the stress boundary conditions $\sum_{j=1}^{3} \sigma_{ji} n_j = T_i^*$ on S_t, i.e., it is the solution to the governing equations and boundary conditions.

In Eq. (8.76), $(\mathbf{T}^*, \mathbf{f}^*)$ are the prescribed stress-vector values and body force values in the problem. We will not prove the result stated above, but the proof follows the same steps needed when using Eq. (8.75) to show that equilibrium is satisfied, or you can find the details in [2]. Virtual displacements that satisfy $\delta u_j = 0$ on S_u are said to satisfy the "essential" boundary conditions. In contrast, the boundary conditions $\sigma_{ji} n_j = T_i^*$ on S_t are said to be the "natural" boundary conditions. These terms come from considering these problems with the techniques of variational calculus [3].

In using the principle of virtual work, Eq. (8.75), it is often useful to also consider virtual displacements where $\delta u_j = 0$ on S_u, since otherwise Eq. (8.75) will include an unknown "reaction" stress vector, $\mathbf{R}^{(n)}$, on S_u corresponding to the stresses that must act to enforce the displacement boundary conditions there. By choosing virtual displacements that vanish on S_u, this unknown stress vector can be eliminated when using Eq. (8.75). However, these reaction stresses can be found once the problem is solved for the stresses in the body.

Now, let us introduce the potential energy of the deformable body, Π, defined as

$$\Pi = U(e) - \int_S \sum_{j=1}^{3} T_j^{(n)} u_j dS - \int_V \sum_{j=1}^{3} f_j u_j dV \qquad (8.77)$$

where $T_j^{(n)}$ and f_j are the actual stress vector and body force while u_j is an arbitrary displacement field and the strains in $U(e)$ are obtained from that displacement. For a linear elastic material, we will show the following.

The minimum value of the potential energy, $\Pi(u_j)$, is attained for the true displacement field, u_j.

This is called the *theorem of minimum potential energy*. To prove this theorem let u_j be the true displacement field and $u_j + \delta u_j$ be any other displacement. Also, let $(e_I, \delta e_I)$ be the vectors of strains associated with $u_j, \delta u_j$, respectively. Then, using the integral of the strain energy density for a linear elastic material given in Eq. (8.19a) to represent the total strain energy, we have

$$\Pi(u_j + \delta u_j) - \Pi(u_j) = \int_V \sum_{I=1}^{6} \sum_{J=1}^{6} \sigma_I \delta e_I dV - \int_S \sum_{j=1}^{3} T_j^{(n)} \delta u_j dS - \int_V \sum_{j=1}^{3} f_j \delta u_j dV$$
$$+ \frac{1}{2} \int_V \sum_{I=1}^{6} \sum_{J=1}^{6} C_{IJ} \delta e_I \delta e_J dV \qquad (8.78)$$

where σ_I is the vector of stresses due to the true displacement, u_j. But from the principal of virtual work the sum of the first three terms on the right-hand side of Eq. (8.78) vanish. Thus, we find

$$\Pi(u_j + \delta u_j) - \Pi(u_j) = \frac{1}{2} \int_V \sum_{I=1}^{6} \sum_{J=1}^{6} C_{IJ} \delta e_I \delta e_J dV \qquad (8.79)$$

This difference is positive and Π is indeed a minimum if the matrix of elastic constants, C_{IJ}, are positive definite so that the right-hand side of Eq. (8.79) is the corresponding positive definite strain energy due to the displacement δu_j. From the expression we obtained earlier for the total strain energy of an isotropic linear elastic material written as a function of the strains (see Eq. (8.29)) we see that the total strain energy is indeed a positive definite quantity in that case, i.e., it is a positive quantity and zero only if all the strains and/or stresses are identically zero. If the material is not linearly elastic then the theorem of minimum potential energy is still true as long as the matrix of second derivatives of the strain energy, $\partial^2 U / \partial e_I \partial e_J$, is positive definite but we will not show that generalization here.

The positive definiteness of the strain energy also allows us to show that solution to the typical boundary value problem for linear elastic deformable bodies (consisting of the governing equations and boundary conditions) is unique. From Eq. (8.22) we found that as long as equilibrium is satisfied for a symmetrical stress matrix

$$U(e) = \frac{1}{2} \int_S \mathbf{T}^{(n)} \cdot \mathbf{u} \, dS + \frac{1}{2} \int_V \mathbf{f} \cdot \mathbf{u} \, dV \tag{8.80}$$

for the displacement, $\mathbf{u}$. Now, assume there are two solutions $(\mathbf{u}^{(1)}, \mathbf{u}^{(2)})$ to the same problem and the corresponding stresses are $\left(\sigma_{ij}^{(1)}, \sigma_{ij}^{(2)}\right)$. Then for a linear elastic material the displacement $\mathbf{u} = \mathbf{u}^{(1)} - \mathbf{u}^{(2)}$ and stress $\sigma_{ij} = \sigma_{ij}^{(1)} - \sigma_{ij}^{(2)}$ are also solutions to the governing equations with zero body forces. Using those solutions in Eq. (8.80) gives

$$U(e) = \frac{1}{2} \int_S \mathbf{T}^{(n)} \cdot \mathbf{u} \, dS \tag{8.81}$$

If the stress vector is specified over the entire surface S then, since both solutions must satisfy the same stress boundary conditions, we have $\mathbf{T}^{(n)} = 0$ so the total strain energy must be zero. Similarly, if the stress vector was specified over part of the surface, S_t, and the displacement specified over the remainder of the surface, S_u, we would have $\mathbf{T}^{(n)} = 0$ on S_t and $\mathbf{u} = 0$ on S_u, so that the strain energy would be zero again. Obviously, if the displacement is specified over the entire surface then the strain energy is also zero. In the more general case, which we discussed in Chapter 6, as long as one of the components of either the stress vector or the displacement in the product $\mathbf{T}^{(n)} \cdot \mathbf{u}$ are specified on all or part of the surface, the surface integral in Eq. (8.80) will be zero so the strain energy will be zero. For all these typical boundary conditions $U(e) = 0$ so if the total strain energy is positive definite the strains must satisfy $e_{ij} = e_{ij}^{(1)} - e_{ij}^{(2)} = 0$ and the strains in the two solutions will be identical. The stress–strain relations then also require that the stresses in the two solutions be identical. If there are no displacement boundary conditions, the vanishing of the strains means that the displacements in the two solutions can differ at most by a rigid body displacement. If there are displacement boundary conditions, then typically these rigid body displacements must vanish and so the displacements in the two solutions are also identical.

Example 8.1 Virtual Work and a Truss

As a simple example of how the virtual work principle can be used as a replacement for satisfying equilibrium directly, consider the two-bar truss shown in Fig. 8.5a subjected to the loads (P_x, P_y). The two bars, labeled (1) and (2), are assumed to have the same cross-sectional areas, A, the same lengths, L, and the same Young's modulus, E, and are pinned at

Figure 8.5 (a) A two-bar truss. (b) The free body diagram of the forces at pin C.

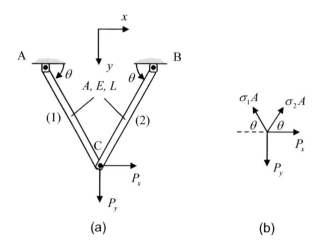

points A, B, and C. Since each of the bars is under uniaxial loads, the stresses (σ_1, σ_2) in bars (1) and (2), respectively, are constants, as are the corresponding axial strains, (e_1, e_2). This is a statically determinant problem that we can solve for the stresses by simply examining force equilibrium at pin C, as shown in Fig. 8.5b. From equilibrium in the x- and y-directions we find

$$\overset{+}{\rightarrow} \sum F_x = 0$$

$$-\sigma_1 A \cos\theta + \sigma_2 A \cos\theta + P_x = 0$$

$$\overset{+}{\downarrow} \sum F_y = 0 \tag{8.82}$$

$$-\sigma_1 A \sin\theta - \sigma_2 A \sin\theta + P_y = 0$$

which are two equations for the two unknown stresses (σ_1, σ_2). We can write these equilibrium equations in matrix form as

$$\begin{bmatrix} A\cos\theta & -A\cos\theta \\ A\sin\theta & A\sin\theta \end{bmatrix} \begin{Bmatrix} \sigma_1 \\ \sigma_2 \end{Bmatrix} = \begin{Bmatrix} P_x \\ P_y \end{Bmatrix} \tag{8.83}$$

Now, let us consider the use of virtual work. We can view the three pins A, B, and C as the "boundary" of the structure, where pins A and B satisfy zero displacements and pin C has the loads applied, so that pins A and B act as the surface S_u of a continuous body where the displacements are specified and pin C acts as the surface S_t where the stress vector is applied. The two bars themselves act as the volume V for the continuous body. The displacements of the two bars can be described by the pin displacements as shown in Fig. 8.6a, and the forces acting on the structure can be described by the applied loads (P_x, P_y) and the reaction forces (A_x, A_y) and (B_x, B_y). Since the strains are constants for

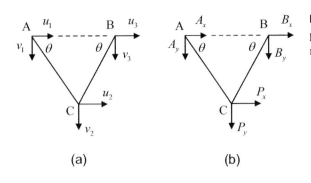

Figure 8.6 (a) Displacements of the truss defined at pins A, B, and C. (b) The applied forces and reaction forces.

the two bars, they are just equal to the bar elongations divided by their lengths. Thus, the strains can be obtained directly from the pin displacements in Fig. 8.6a as

$$e_1 = \frac{(u_2 - u_1)\cos\theta + (v_2 - v_1)\sin\theta}{L}$$
$$e_2 = \frac{(u_3 - u_2)\cos\theta + (v_2 - v_3)\sin\theta}{L}$$
(8.84)

The real displacements, of course, satisfy the displacement boundary conditions $u_1 = v_1 = 0$ and $u_3 = v_3 = 0$ but the virtual displacements do not have to satisfy those conditions. If we choose virtual displacements that do not satisfy the displacement boundary conditions, then the reaction forces at A and B will do virtual work as well as the applied loads. If our focus is on obtaining the unknown stresses (σ_1, σ_2), these reaction forces are not needed so we can eliminate them in the principle of virtual work by choosing virtual displacements $(\delta u_2, \delta v_2)$ with $\delta u_1 = \delta v_1 = 0$ and $\delta u_3 = \delta v_3 = 0$. The corresponding virtual strains will then be

$$\delta e_1 = \frac{\delta u_2 \cos\theta + \delta v_2 \sin\theta}{L}$$
$$\delta e_2 = \frac{-\delta u_2 \cos\theta + \delta v_2 \sin\theta}{L}$$
(8.85)

We can place these strains into the principle of virtual work but it is useful to examine the separate contributions in Eq. (8.85) from δu_2 and δv_2. Consider the virtual displacement δv_2 as shown in Fig. 8.7a. This will produce an elongation $\delta v_2 \sin\theta$ in bar (2) as shown in Fig. 8.7b so the strain in that bar will be $\delta e_2 = \delta v_2 \sin\theta / L$ and by symmetry this will also be the strain in bar (1). From the principle of virtual work, we have

$$\delta U = \delta W_v$$

$$\int_{bar(1)} \sigma_1 \delta e_1 dV + \int_{bar(2)} \sigma_2 \delta e_2 dV = P_y \delta v_2$$

$$\sigma_1 \left(\frac{\delta v_2 \sin\theta}{L}\right) AL + \sigma_2 \left(\frac{\delta v_2 \sin\theta}{L}\right) AL = P_y \delta v_2$$

$$(\sigma_1 A \sin\theta + \sigma_2 A \sin\theta - P_y) \delta v_2 = 0$$
(8.86)

Figure 8.7 (a) Virtual displacement of the truss that satisfies the displacement boundary conditions at A and B. (b) The deformation of bar (2), showing the elongation and corresponding strain. By symmetry bar (1) has the same strain.

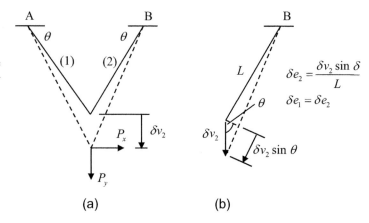

which, since the virtual displacement δv_2 is arbitrary, just gives the second equilibrium equation in Eq. (8.83). In entirely the same fashion for the virtual displacement δu_2 as shown in Fig. 8.8a we have (see Figs. 8.8b, c)

$$\delta U = \delta W_v$$

$$\int_{bar(1)} \sigma_1 \delta e_1 dV + \int_{bar(2)} \sigma_2 \delta e_2 dV = P_x \delta u_2$$

$$\sigma_1 \left(\frac{\delta u_2 \cos \theta}{L} \right) AL + \sigma_2 \left(\frac{-\delta u_2 \cos \theta}{L} \right) AL = P_x \delta u_2 \qquad (8.87)$$

$$(\sigma_1 A \cos \theta - \sigma_2 A \cos \theta - P_x) \delta u_2 = 0$$

which, since δu_2 is arbitrary, gives the first equilibrium equation in Eq. (8.83).

Now consider what happens when the virtual displacements do not satisfy the displacement boundary conditions. In that case the virtual strains and displacements are related through the strain–displacement relations of Eq. (8.84), and the principle of virtual work, following similar steps as before, gives

$$\sigma_1 \left(\frac{(\delta u_2 - \delta u_1) \cos \theta + (\delta v_2 - \delta v_1) \sin \theta}{L} \right) AL$$

$$+ \sigma_2 \left(\frac{(\delta u_3 - \delta u_2) \cos \theta + (\delta v_2 - \delta v_3) \sin \theta}{L} \right) AL \qquad (8.88)$$

$$= A_x \delta u_1 + A_y \delta v_1 + P_x \delta u_2 + P_y \delta v_2 + B_x \delta u_3 + B_y \delta v_3$$

Collecting terms that multiply the same virtual displacements and using the fact that the virtual displacements are arbitrary so the coefficients of those virtual displacements must vanish, we arrive at a set of equations which can be placed in matrix–vector form for the unknown reactions and stresses in the bars as

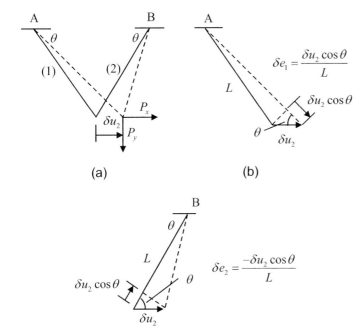

Figure 8.8 (a) Virtual displacement that satisfies the displacement boundary conditions at A and B. (b) The deformation of bar (1) showing the elongation and corresponding strain. (c) The deformation of bar (2) showing the shortening and corresponding strain.

$$\begin{bmatrix} -1 & 0 & -A\cos\theta & 0 & 0 & 0 \\ 0 & -1 & -A\sin\theta & 0 & 0 & 0 \\ 0 & 0 & A\cos\theta & -A\cos\theta & 0 & 0 \\ 0 & 0 & A\sin\theta & A\sin\theta & 0 & 0 \\ 0 & 0 & 0 & A\cos\theta & -1 & 0 \\ 0 & 0 & 0 & -A\sin\theta & 0 & -1 \end{bmatrix} \begin{Bmatrix} A_x \\ A_y \\ \sigma_1 \\ \sigma_2 \\ B_x \\ B_y \end{Bmatrix} = \begin{Bmatrix} 0 \\ 0 \\ P_x \\ P_y \\ 0 \\ 0 \end{Bmatrix} \qquad (8.89)$$

The first equation in Eq. (8.89) is just the equilibrium equation in the x-direction at pin A while the second equation is the equilibrium equation at pin A in the y-direction. Similarly, the last two equations in Eq. (8.89) are the equilibrium equations at pin B in the x- and y-directions, respectively. The middle two equations in Eq. (8.89) are the same equilibrium equations derived previously at pin C in terms of the applied loads. We have left these middle equations in terms of the stresses to make the connection more explicit to the principle of virtual work as originally derived, but in practice it is better to write these equations in terms of the internal forces, which will also make the equilibrium matrix in Eq. (8.89) dimensionally homogeneous. Equation (8.89) represents six equations in six unknowns, which can be solved so that the problem remains statically determinant even when we include the unknown reaction forces. But if we solve for the stresses from the equilibrium equations at pin C, we can always go back and find the reaction forces at pins A and B from those known stresses so we see that choosing virtual displacements that satisfy the displacement boundary conditions simplifies the problem greatly. In problem P8.1, you are asked to use the symbolic algebra capabilities of MATLAB® to solve the system of equations in Eq. (8.89). You should find that the solution can be written compactly as:

$$A_x = -\frac{1}{2}(P_x + P_y \cot\theta), \quad B_x = \frac{1}{2}(P_y \cot\theta - P_x)$$

$$A_y = -\frac{1}{2}(P_y + P_x \tan\theta), \quad B_y = \frac{1}{2}(P_x \tan\theta - P_y) \tag{8.90}$$

$$\sigma_1 = \frac{1}{2A}\left(\frac{P_x}{\cos\theta} + \frac{P_y}{\sin\theta}\right), \quad \sigma_2 = \frac{1}{2A}\left(\frac{P_y}{\sin\theta} - \frac{P_x}{\cos\theta}\right)$$

We could have also written the strain energy entirely in terms of the displacements through the use of the stress–strain and strain–displacement relations. Since the displacements satisfy the displacement boundary conditions, the stresses then are

$$\sigma_1 = E\frac{u_2\cos\theta + v_2\sin\theta}{L}$$

$$\sigma_2 = E\frac{-u_2\cos\theta + v_2\sin\theta}{L} \tag{8.91}$$

so that the principle of virtual work, when it is written entirely in terms of displacements, gives the equilibrium equations as (compare with the stress equations of Eq. (8.83)):

$$\begin{bmatrix} \frac{2EA}{L}\cos^2\theta & 0 \\ 0 & \frac{2EA}{L}\sin^2\theta \end{bmatrix} \begin{Bmatrix} u_2 \\ v_2 \end{Bmatrix} = \begin{Bmatrix} P_x \\ P_y \end{Bmatrix} \tag{8.92}$$

where the matrix of coefficients is called a *stiffness matrix* for the truss. Because this matrix is diagonal, the displacements are obtained directly and then the stresses can be obtained from Eq. (8.91). This truss problem is statically determinant so there is little difference if we solve for the stresses as the fundamental unknowns or the displacements. But in statically indeterminate problems, the equilibrium equations for the stresses are insufficient by themselves to solve for the stresses and we must combine the equilibrium and compatibility equations to reach a solution, as seen in Navier's table problem in Chapter 6. However, if we use the principle of virtual work written directly in terms of the displacements, even for statically indeterminate problems we can generate a stiffness matrix for the structure that defines a set of equations for the displacements that can be solved for those displacements and then the stresses back-calculated. We will examine both these approaches in more detail in Chapter 9.

8.6 Principle of Complementary Virtual Work and Minimum Complementary Potential Energy

The complementary strain energy also plays an important role for deformable bodies. Just as the principle of virtual work can act as a substitute to the equilibrium equations, the principle of complementary virtual work can replace the strain–displacement relations and hence also acts a substitute for the compatibility equations.

8.6.1 General Deformable Body

Assume that the strain–displacement relations are satisfied. If we multiply those relations by a symmetrical weighting matrix, $\delta\sigma_{ij}(i,j = 1,2,3)$, which represents small virtual changes of the stresses σ_{ij}, and integrates over the volume V of a deformable body we have

$$\int_V \sum_{i=1}^{3}\sum_{j=1}^{3} \left(e_{ij} - \frac{1}{2}\left(\frac{\partial u_i}{\partial x_j} + \frac{\partial u_j}{\partial x_i}\right)\right) \delta\sigma_{ij} dV = 0 \tag{8.93}$$

which, by virtue of the symmetry of the virtual stresses ($\delta\sigma_{ij} = \delta\sigma_{ji}$), we can write as

$$\int_V \sum_{i=1}^{3}\sum_{j=1}^{3} \left(e_{ij} - \frac{\partial u_i}{\partial x_j}\right) \delta\sigma_{ij} dV = 0 \tag{8.94}$$

and which can be rearranged as

$$\int_V \sum_{i=1}^{3}\sum_{j=1}^{3} e_{ij}\delta\sigma_{ij} dV - \int_V \sum_{i=1}^{3}\sum_{j=1}^{3} \frac{\partial}{\partial x_j}\left(u_i\delta\sigma_{ij}\right) dV + \int_V \sum_{i=1}^{3}\sum_{j=1}^{3} u_i \frac{\partial \delta\sigma_{ij}}{\partial x_j} dV = 0 \tag{8.95}$$

Using Gauss's theorem, the second terms in Eq. (8.95) can be placed on the boundary and that equation can be expressed as

$$\int_V \sum_{i=1}^{3}\sum_{j=1}^{3} e_{ij}\delta\sigma_{ij} dV - \int_S \sum_{i=1}^{3} \delta T_i^{(n)} u_i dS - \int_V \sum_{i=1}^{3} \delta f_i u_i dV = 0 \tag{8.96}$$

where

$$\delta T_i^{(\mathbf{n})} = \sum_{j=1}^{3} \delta\sigma_{ij} n_j$$

$$\delta f_i = -\sum_{j=1}^{3} \frac{\partial \delta\sigma_{ij}}{\partial x_j} \tag{8.97}$$

are the virtual stress vector and virtual body force generated by the virtual stresses. However, recall the relationship between the strain and the complementary strain energy, Eq. (8.46 b), from which it follows

$$\int_V \sum_{i=1}^{3}\sum_{j=1}^{3} e_{ij}(\sigma_{ij})\delta\sigma_{ij}dV = \int_V \sum_{i=1}^{3}\sum_{j=1}^{3} \frac{\partial u_0^c}{\partial \sigma_{ij}}\delta\sigma_{ij}dV$$
$$= \int_V \delta u_0^c(\sigma)dV = \delta U^c(\sigma) \qquad (8.98)$$

where U^c is the total complementary strain energy of the body. Thus, we obtain *the principle of complementary virtual work*:

$$\delta U^c = \int_S \sum_{i=1}^{3} \delta T_i^{(n)} u_i dS + \int_V \sum_{i=1}^{3} \delta f_i u_i dV \qquad (8.99)$$
$$= \delta W_v^c$$

where δW_v^c is the differential complementary virtual work. This result can also be written more compactly as

$$\delta U^c = \int_S \delta \mathbf{T}^{(n)} \cdot \mathbf{u}\, dS + \int_V \delta \mathbf{f} \cdot \mathbf{u}\, dV \qquad (8.100)$$

We have shown that if the strain–displacement relationship (and hence compatibility) is satisfied throughout the body, then the principle of complementary virtual work must be satisfied. We can also reverse the steps in going from the strain–displacement relation to the principle of complementary virtual work and show that if the principle of complementary virtual work is satisfied for all stresses σ_{ij} whose virtual changes, $\delta\sigma_{ij}$, satisfy Eq. (8.97) then the strain that is generated by those stresses will satisfy the strain–displacement relations and hence be compatible strains. Again, we can summarize these results by saying that satisfying compatibility and satisfying the principle of complementary virtual work are equivalent, i.e.,

> compatibility ⇒ principle of complementary virtual work
>
> principle of complementary virtual work ⇒ compatibility

As in the virtual work case, one can make a stronger statement about what the principle of complementary virtual work implies for the problem when the stress vector is specified on S_t and the displacement is specified on S_u. That statement is as follows.

8.6 Complementary Virtual Work & Minimum Complementary Potential Energy

Let a symmetric state of stress, σ_{ij}, be found such that

$$\sum_{i=1}^{3} \sigma_{ij} n_i = T_j^* \quad \text{on } S_t$$

$$\sum_{i=1}^{3} \frac{\partial \sigma_{ij}}{\partial x_i} + f_j^* = 0 \quad \text{in } V$$

(where T_j^*, f_j^* are specified stress vector and body force components) and such that for all symmetrical $\delta\sigma_{ij}$ satisfying

$$\sum_{i=1}^{3} \delta\sigma_{ij} n_j = 0 \quad \text{on } S_t$$

$$\sum_{i=1}^{3} \frac{\partial \delta\sigma_{ij}}{\partial x_i} = 0 \quad \text{in } V$$

the principle of complementary virtual work holds, which in this case is

$$\delta U^c(\sigma) = \int_{S_u} \delta \mathbf{T}^{(n)} \cdot \mathbf{u} \, dS$$

(where, since the stress vector is specified on S_t, the integral over that part of the surface vanishes and, since the body force is also specified, we have $\delta f_j = 0$). Then the stress σ_{ij} satisfies the displacement boundary conditions $u_i = u_i^*$ on S_u and the strain–displacement relations, i.e., it is the solution to the governing equations and boundary conditions.

If we define the complementary potential energy, Π^c, as

$$\Pi^c = U^c(\sigma) - \int_S \mathbf{T}^{(n)} \cdot \mathbf{u} \, dS - \int_V \mathbf{f} \cdot \mathbf{u} \, dV \tag{8.101}$$

(where $\mathbf{u}$ is the true displacement and $\mathbf{T}^{(n)}, \mathbf{f}$ are the stress vector and body force corresponding to any stress field, σ_{ij}) and follow similar steps shown for the minimum potential energy theorem we can also prove that, for a linear elastic material *the principle of minimum complementary potential energy* is true, as follows.

The minimum value of the complementary potential energy, $\Pi^c(\sigma_{ij})$, is attained for the true stress field, σ_{ij}.

Again, if the material is not linearly elastic but the matrix of second derivatives of the complementary strain energy, $\partial^2 U^c / \partial \sigma_I \partial \sigma_J$, is positive definite the theorem of minimum complementary energy is still true. We will not prove the two complementary energy theorems described above, but you can find details in [2] and [4].

Example 8.2 Complementary Strain Energy and a Truss

To illustrate the application of the principle of complementary strain energy consider a truss again, but where now the truss has three members so that it is statically indeterminate (Fig. 8.9a). There are still only two equilibrium equations at pin C given from the free body diagram of Fig. 8.9b by

$$\overset{+}{\rightarrow} \sum F_x = 0$$

$$-\sigma_1 A \cos\theta + \sigma_2 A \cos\theta + P_x = 0$$

$$\overset{+}{\downarrow} \sum F_y = 0 \qquad (8.102)$$

$$-\sigma_3 A - \sigma_1 A \sin\theta - \sigma_2 A \sin\theta + P_y = 0$$

which we also could get from the principle of virtual work. In matrix–vector form these equations in terms of stresses are

$$\begin{bmatrix} \cos\theta & -\cos\theta & 0 \\ \sin\theta & \sin\theta & 1 \end{bmatrix} \begin{Bmatrix} \sigma_1 \\ \sigma_2 \\ \sigma_3 \end{Bmatrix} = \begin{Bmatrix} P_x/A \\ P_y/A \end{Bmatrix} \qquad (8.103)$$

Choosing virtual stresses $\delta\sigma_j$ that also satisfy these equilibrium equations, we have

$$\begin{bmatrix} \cos\theta & -\cos\theta & 0 \\ \sin\theta & \sin\theta & 1 \end{bmatrix} \begin{Bmatrix} \delta\sigma_1 \\ \delta\sigma_2 \\ \delta\sigma_3 \end{Bmatrix} = \begin{Bmatrix} \delta P_x/A \\ \delta P_y/A \end{Bmatrix} \qquad (8.104)$$

Figure 8.9 (a) Statically indeterminate truss where all the members have the same cross-sectional area, A, and Young's modulus, E, and bars (1) and (2) have lengths L (not shown) while bar (3) has a length $L \sin\theta$. (b) The free body diagram of the forces at pin C.

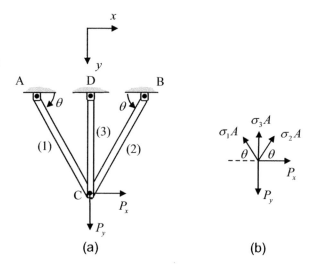

8.6 Complementary Virtual Work & Minimum Complementary Potential Energy

and the principle of complementary virtual work gives

$$\delta U^c = \delta W_v^c$$

$$\int_{bar(1)} e_1 \delta\sigma_1 dV + \int_{bar(2)} e_2 \delta\sigma_2 dV + \int_{bar(3)} e_3 \delta\sigma_3 dV = u_2 \delta P_x + v_2 \delta P_y \quad (8.105)$$

$$e_1 \delta\sigma_1 AL + e_2 \delta\sigma_2 AL + e_3 \delta\sigma_3 AL \sin\theta = u_2 \delta P_x + v_2 \delta P_y$$
$$= u_2(A\cos\theta\delta\sigma_1 - A\cos\theta\delta\sigma_2) + v_2(A\sin\theta\delta\sigma_1 + A\sin\theta\delta\sigma_2 + A\delta\sigma_3)$$

where we have used Eq. (8.104). Since the $\delta\sigma_j$ are arbitrary, we have

$$\begin{aligned} e_1 L &= \Delta_1 = u_2 \cos\theta + v_2 \sin\theta \\ e_2 L &= \Delta_2 = -u_2 \cos\theta + v_2 \sin\theta \\ e_3 L \sin\theta &= \Delta_3 = v_2 \end{aligned} \quad (8.106)$$

in terms of the strains, e_i, or, equivalently, the elongations, Δ_i, which we can also write in matrix–vector form as

$$\begin{Bmatrix} \Delta_1 \\ \Delta_2 \\ \Delta_3 \end{Bmatrix} = \begin{bmatrix} \cos\theta & \sin\theta \\ -\cos\theta & \sin\theta \\ 0 & 1 \end{bmatrix} \begin{Bmatrix} u_2 \\ v_2 \end{Bmatrix} \quad (8.107)$$

Thus, we have shown that the principle of complementary virtual work has given us the generalized strain–displacement relations of Eq. (8.107). We recognize the matrix in Eq. (8.107) as just the transpose of the equilibrium matrix in Eq. (8.104), following the same relationships seen in Chapter 6 for Navier's table problem.

Because there are three elongations (or strains) written in terms of two displacements, these elongations are related by a compatibility equation which is easy to obtain in this case since adding the first two equations in Eq. (8.106) and comparing the result to the third equation gives

$$\Delta_1 + \Delta_2 = 2\Delta_3 \sin\theta \quad (8.108)$$

or, in terms of the strains,

$$e_1 + e_2 = 2e_3 \sin^2\theta \quad (8.109)$$

which we can multiply by Young's modulus to obtain the compatibility equation in terms of the stresses as

$$\sigma_1 + \sigma_2 - 2\sigma_3 \sin^2\theta = 0 \quad (8.110)$$

Combining the equilibrium and compatibility equations (Eq. (8.103) and Eq. (8.110)), we then have three equations for the three stresses given as

$$\begin{bmatrix} \cos\theta & -\cos\theta & 0 \\ \sin\theta & \sin\theta & 1 \\ 1 & 1 & -2\sin^2\theta \end{bmatrix} \begin{Bmatrix} \sigma_1 \\ \sigma_1 \\ \sigma_3 \end{Bmatrix} = \begin{Bmatrix} P_x/A \\ P_y/A \\ 0 \end{Bmatrix} \quad (8.111)$$

which can be solved directly. This is a stress-based (or force-based) solution to the truss problem. To obtain a displacement-based solution, we will first write the equilibrium equations in terms of the internal forces $F_j = \sigma_j A$ as

$$\begin{bmatrix} \cos\theta & -\cos\theta & 0 \\ \sin\theta & \sin\theta & 1 \end{bmatrix} \begin{Bmatrix} F_1 \\ F_2 \\ F_3 \end{Bmatrix} = \begin{Bmatrix} P_x \\ P_y \end{Bmatrix} \quad (8.112)$$

which we will write symbolically as

$$[\tilde{E}]\{F\} = \{P\} \quad (8.113)$$

in terms of the equilibrium matrix $[\tilde{E}]$, the internal force vector $\{F\} = [F_1, F_2, F_3]^T$, and the applied load vector $\{P\} = [P_x, P_y]^T$. Then we can express the strain–displacement equations in terms of the forces and relate them to the displacements as

$$\{F\} = [C]\{\Delta\} = [C][\tilde{E}]^T\{u\} \quad (8.114)$$

where $\{u\} = [u_2, v_2]^T$ and

$$[C] = \begin{bmatrix} EA/L & 0 & 0 \\ 0 & EA/L & 0 \\ 0 & 0 & EA/L\sin\theta \end{bmatrix} \quad (8.115)$$

Placing Eq. (8.114) into the equilibrium equations, Eq. (8.113), we then find the displacement equations

$$[K]\{u\} = \{P\} \quad (8.116)$$

in terms of the pseudostiffness matrix $[K] = [\tilde{E}][C][\tilde{E}]^T$. We can symbolically solve either the combination of Eq. (8.116) and Eq. (8.114) or Eq. (8.111) with MATLAB® (see problem P8.2) for the stresses or forces in the truss. The result for the internal forces is

$$\begin{aligned} F_1 &= \frac{P_x}{2\cos\theta} + \frac{P_y \sin^2\theta}{2\sin^3\theta + 1} \\ F_2 &= \frac{P_y \sin^2\theta}{2\sin^3\theta + 1} - \frac{P_x}{2\cos\theta} \\ F_3 &= \frac{P_y}{2\sin^3\theta + 1} \end{aligned} \quad (8.117)$$

8.7 Work–Energy Principles and Discrete Forces and Moments

Consider an elastic body which carries a set of surface loads, some of which are distributed and some of which are discrete (concentrated forces and moments), and also body forces.

8.7 Work–Energy Principles and Discrete Forces and Moments 255

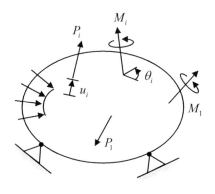

Figure 8.10 A deformable body acted upon by concentrated forces and moments. There may also be distributed forces or body forces (body forces not shown) acting on the body.

Let the displacement at the concentrated load, P_i ($i = 1, 2, \ldots, N$), in the direction of that force be given by u_i. Similarly, let the rotation at the concentrated moment, M_i ($i = 1, 2, \ldots, N$), in the direction determined by that moment from the right-hand rule be given by θ_i (see Fig. 8.10). We will assume that the strain energy of this body is expressed in terms of these displacements and rotations as $U = U(u_1, \ldots, u_N, \theta_1, \ldots, \theta_N)$. Note that the strain energy may also depend on other variables but, unlike the displacements and rotations at the concentrated loads and moments, these other variables will not be changed so we will not show them explicitly.

8.7.1 Virtual Work and Potential Energy

Now, imagine that we fix the concentrated forces and moments and make small changes to the displacements and rotations, $\delta u_i, \delta \theta_i$. These changes, in turn, will cause the strain energy to change by a small amount, δU. Then the principle of virtual work says that the work done by the forces and moments taken through these virtual displacements must be equal to the resulting virtual change in the strain energy, i.e.,

$$\delta U = \sum_{i=1}^{N} P \delta u_i + \sum_{i=1}^{M} M_i \delta \theta_i \tag{8.118}$$

Similarly, in terms of the principle of minimum potential energy, where the potential energy is defined here as

$$\Pi = U - \sum_{i=1}^{N} P_i u_i - \sum_{i=1}^{M} M_i \theta_i \tag{8.119}$$

since the potential energy must be stationary, we have

$$\delta \Pi = \delta U - \sum_{i=1}^{N} P_i \delta u_i - \sum_{i=1}^{M} M_i \delta \theta_i = 0 \tag{8.120}$$

which again gives the principle of virtual work.

Castigliano's First Theorem

Since the strain energy has been assumed to be written as a function of the displacements and rotations at the concentrated loads and moments, it follows that

$$\delta U = \sum_{i=1}^{N} \frac{\partial U}{\partial u_i} \delta u_i + \sum_{i=1}^{M} \frac{\partial U}{\partial \theta_i} \delta \theta_i \qquad (8.121)$$

so that placing Eq. (8.121) into either Eq. (8.118) or Eq. (8.120) gives

$$\sum_{i=1}^{N} \left(\frac{\partial U}{\partial u_i} - P_i\right) \delta u_i + \sum_{i=1}^{M} \left(\frac{\partial U}{\partial \theta_i} - M_i\right) \delta \theta_i = 0 \qquad (8.122)$$

which can only be satisfied for all possible virtual displacements and rotations if

$$\begin{aligned}\frac{\partial U}{\partial u_i} &= P_i \quad (i = 1, \ldots, N) \\ \frac{\partial U}{\partial \theta_i} &= M_i \quad (i = 1, \ldots, M)\end{aligned} \qquad (8.123)$$

which are a set of relations called *Castigliano's first theorem*.

Example 8.3 Castigliano's First Theorem

As a simple example of the use of Castigliano's first theorem, consider an axial-load problem where two bars are welded together between two rigid walls and are subjected to a concentrated force, P, at their connection (see Fig. 8.11). Let u_P be the displacement at the load P. The internal forces, stresses, and strains in the bars AB and BC are all constants so we assume the tensile strain in bar AB is $e_1 = u_P/L_1$ and the compressive strain in bar BC is $e_2 = -u_P/L_2$. Thus, from Eq. (8.52) the strain energy of the entire system is

$$U = \frac{1}{2}\frac{A_1 E_1}{L_1} u_P^2 + \frac{1}{2}\frac{A_2 E_2}{L_2} u_P^2 \qquad (8.124)$$

Castigliano's first theorem gives

$$P = \frac{\partial U}{\partial u_P} = \left(\frac{A_1 E_1}{L_1} + \frac{A_2 E_2}{L_2}\right) u_P \qquad (8.125)$$

which can be solved for the displacement as

$$u_P = \frac{P}{\left(\dfrac{A_1 E_1}{L_1} + \dfrac{A_2 E_2}{L_2}\right)} \qquad (8.126)$$

from which we can obtain the strains, stresses, and internal forces in both sections of the bar. It can also be verified by solving this problem directly that Eq. (8.126) is the exact result.

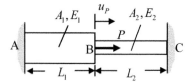

Figure 8.11 An assembly of two bars acted upon by a concentrated load P at B where the two bars are welded together. The ends A and C are rigidly fixed. The cross-sectional area, Young's modulus and length of bar AB and bar BC are (A_1, E_1, L_1) and (A_2, E_2, L_2), respectively. The displacement at B in the direction of the load P is u_P.

The Rayleigh–Ritz Method

In the previous example, we obtained an exact solution because the assumed deformation field that we used to calculate the strain energy coincided with the exact values. In more complicated problems, it may not be a simple matter to obtain exact deformation expressions. Nevertheless, if we can make a "reasonable" guess at the form of the deformations in terms of some simple functions and unknown coefficients, then the principle of minimum potential energy (or virtual work) gives us a method to determine those unknown coefficients in such a manner that equilibrium will be approximately satisfied. This method is called the *Rayleigh–Ritz method*. The procedure of the method is to assume that the displacements of the body can be represented parametrically in terms of a set of basis functions and n unknown parameters $(a_1, a_2, \ldots, a_n)$. For example, in a 1-D problem we might take the displacement in the form

$$u_x = a_0 + a_1 x + \cdots + a_{n-1} x^{n-1} \tag{8.127}$$

In the Rayleigh–Ritz method, it is assumed that the functions and constants are chosen so that any boundary conditions involving the displacements or rotations of the body (which are called the "essential" boundary conditions) are satisfied. Then the functions and coefficients are placed into the expression for the potential energy of the body so that this potential energy can be expressed in the form

$$\Pi = \Pi(a_0, \ldots, a_{n-1}) \tag{8.128}$$

The potential energy is made stationary by requiring that

$$\delta \Pi = \sum_{i=0}^{n-1} \frac{\partial \Pi}{\partial a_i} \delta a_i = 0 \tag{8.129}$$

which can only be satisfied for arbitrary changes of these coefficients if

$$\frac{\partial \Pi}{\partial a_i} = 0 \quad (i = 0, \ldots, a_{n-1}) \tag{8.130}$$

which are n equations to be solved for the n coefficients.

Example 8.4 The Rayleigh–Ritz Method

As an example of the Rayleigh–Ritz method, consider a cantilever beam loaded by an end moment, as shown in Fig. 8.12. The exact solution for the vertical deflection, $w(x)$, of the neutral axis of the beam is

$$w(x) = \frac{M_0 x^2}{2EI} \tag{8.131}$$

so that the end deflection is given by

$$w(L) = \frac{M_0 L^2}{2EI} \tag{8.132}$$

Figure 8.12 A cantilever beam acted upon by an end moment, M_0, and the displacement of the end at $x = L$, $w(L)$.

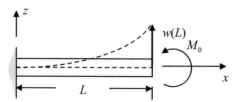

To obtain an approximate solution to this problem, let us assume

$$w(x) = a\left[1 - \cos\left(\frac{\pi x}{2L}\right)\right] \quad (8.133)$$

This function meets the requirements stated previously; i.e., it satisfies the essential conditions of zero displacement and slope (where, for small slopes, the slope is just the rotation of the neutral axis) at $x = 0$. The total potential energy is:

$$\Pi = U - M_0 \theta(L)$$

$$= \frac{EI}{2} \int_0^L \left(\frac{d^2 w}{dx^2}\right)^2 dx - M_0 \frac{dw(L)}{dx} \quad (8.134a)$$

Placing the approximate deflection expression into Eq. (8.134a) and carrying out the indicated differentiations and integrations, we find

$$\Pi = \frac{\pi^2 EI a^2}{64 L^3} - \frac{\pi M_0 a}{2L} \quad (8.134b)$$

Requiring $\partial \Pi / \partial a = 0$ yields $a = 16 M_0 L^2 / \pi^3 EI$ and, since $a = w(L)$ we obtain

$$w(L) = \frac{0.52 M_0 L^2}{EI} \quad (8.135)$$

which is very close to the exact solution of Eq. (8.132).

Although the Rayleigh–Ritz method is a very useful tool, for complicated 3-D problems it is impossible to make good guesses for what the deformations might be for the entire body, i.e., to choose *global* functions of approximations. However, suppose the body is broken into small elements over which *locally* the deformations can be reasonably assumed to have simple variations, and where these variations are written in terms of unknown parameters (called nodal variables). Then the principle of minimum potential energy (or virtual work) can be used to form up a set of linear equations for all these nodal variables (in the example just shown there was one nodal variable $a = w(L)$). Solving this linear system yields an approximate solution for the deformation of the entire body. This is the basic idea behind the finite element method, which will be discussed in more depth in Chapter 9.

8.7.2 Complementary Virtual Work and Complementary Potential Energy

Consider the body shown in Fig. 8.10 again, where the concentrated forces and moments were shown together with their corresponding displacements and rotations. Now, imagine these displacements and rotations are held fixed while we vary the applied concentrated forces and moments by small virtual amounts $(\delta P_i, \delta M_i)$, where these virtual changes satisfy the equations of equilibrium. For a *statically determinate* elastic system, we can write the complementary strain energy solely in terms of these applied loads as $U^c = U^c(P_1, \ldots, P_N, M_1, \ldots, M_M)$ so that virtual changes in the applied loads will cause the complementary strain energy to change by an amount δU^c. The principle of complementary virtual work states that

$$\delta U^c = \sum_{i=1}^{N} u_i \delta P_i + \sum_{i=1}^{M} \theta_i \delta M_i \tag{8.136}$$

For example, Fig. 8.13 shows a specific example of a statically determinate truss structure loaded in two dimensions where the uniaxial forces in all the truss members can be found from equilibrium in terms of P_1 so that the complementary strain energy can be written explicitly in terms of P_1, and the principle of complementary virtual work gives $\delta U^c = u_1 \delta P_1$. The reaction forces at supports A and B are also present but they do no complementary virtual work since the corresponding displacements in the directions of these forces are all zero. If we define a complementary energy, Π^c, for the system as

$$\Pi^c = U^c - \sum_{i=1}^{N} u_i P_i - \sum_{i=1}^{M} \theta_i M_i \tag{8.137}$$

then the principle of minimum complementary potential energy requires that the complementary energy be stationary, i.e.,

$$\delta \Pi^c = \delta U^c - \sum_{i=1}^{N} u_i \delta P_i - \sum_{i=1}^{M} \theta_i \delta M_i = 0 \tag{8.138}$$

Engesser's First Theorem and Castigliano's Second Theorem

Since, as stated earlier, the deformable body is assumed to be a statically determinate problem where the complementary strain energy can be written explicitly in terms of the applied forces and moments, it follows that the change in complementary strain energy can also be expressed as

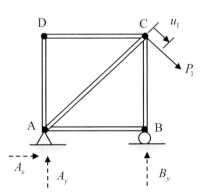

Figure 8.13 A statically determinate truss where the forces in the bars of the truss can all be written in terms of the applied force, P_1.

$$\delta U^c = \sum_{i=1}^{N} \frac{\partial U^c}{\partial P_i} \delta P_i + \sum_{i=1}^{M} \frac{\partial U^c}{\partial M_i} \delta M_i \qquad (8.139)$$

so that Eq. (8.136) or Eq. (8.138) give

$$\sum_{i=1}^{N} \left(\frac{\partial U^c}{\partial P_i} - u_i \right) \delta P_i + \sum_{i=1}^{M} \left(\frac{\partial U^c}{\partial M_i} - \theta_i \right) \delta M_i = 0 \qquad (8.140)$$

If the virtual changes in the applied loads are varied independently, the only way for Eq. (8.140) to be satisfied is if

$$\begin{aligned}\frac{\partial U^c}{\partial P_i} &= u_i \quad (i=1,\ldots,N) \\ \frac{\partial U^c}{\partial M_i} &= \theta_i \quad (i=1,\ldots,M)\end{aligned} \qquad (8.141)$$

a result that is called *Engesser's first theorem*. For a *linear* elastic material $U^c = U$, so that

$$\begin{aligned}\frac{\partial U}{\partial P_i} &= u_i \quad (i=1,\ldots,N) \\ \frac{\partial U}{\partial M_i} &= \theta_i \quad (i=1,\ldots,M)\end{aligned} \qquad (8.142)$$

which is usually called *Castigliano's second theorem*. Although we have assumed that the system under question is statically determinate when deriving these results, they are also true for statically indeterminate structures as we will show at the end of Section 8.7.3.

Example 8.5 Castigliano's Second Theorem

As a simple example of the use of Castigliano's second theorem, consider a cantilever beam subjected to an end force, P, as shown in Fig. 8.14. The bending moment in the beam is given by $M(x) = -Px$ so that the strain energy is

$$\begin{aligned}U &= \frac{1}{2} \int_0^L \frac{[M(x)]^2}{EI} dx \\ &= \frac{1}{6} \frac{P^2 L^3}{EI}\end{aligned} \qquad (8.143)$$

Castigliano's second theorem says that the deflection at the load P in the direction of P, u_P, is given by

$$u_P = \frac{\partial U}{\partial P} = \frac{1}{3} \frac{PL^3}{EI} \qquad (8.144)$$

which can be verified independently by integrating the moment–curvature relation.

As this example shows, we can obtain the deflection (or rotation) at the location of any applied force (or moment). However, we can use the concept of a *dummy load* to obtain the deflection (or rotation) at any point in the body. To see this, consider the cantilever beam of Fig. 8.14 again where now we want to obtain the deflection at the center of the beam. In order to use Castigliano's second theorem we place a dummy force, Q, at the center, as shown in Fig. 8.15. The bending moment in the beam is now

$$M(x) = \begin{cases} -Px & (0 < x < L/2) \\ -Px - Q(x - L/2) & (L/2 < x < L) \end{cases} \qquad (8.145)$$

so that the strain energy is

$$U = U(P, Q)$$
$$= \frac{1}{2EI} \left\{ \int_0^{L/2} P^2 x^2 dx + \int_{L/2}^L [Px + Q(x - L/2)]^2 dx \right\} \qquad (8.146)$$

From Castigliano's second theorem we have

$$u_Q = \frac{\partial U(P, Q)}{\partial Q} = \frac{1}{EI} \int_{L/2}^L [Px + Q(x - L/2)](x - L/2) dx \qquad (8.147)$$

This is the deflection at Q due to both P and Q but we want the deflection at Q due to P only, which we can obtain from Eq. (8.147) by simply setting $Q = 0$ to find

$$u_Q\big|_{Q=0} = \frac{\partial U(P, Q)}{\partial Q}\bigg|_{Q=0} = \frac{1}{EI} \int_{L/2}^L (Px)(x - L/2) dx$$
$$= \frac{5}{48} \frac{PL^3}{EI} \qquad (8.148)$$

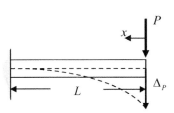

Figure 8.14 A cantilever beam acted upon by an end force, P, and the deflection at P in the direction of the force, Δ_P.

which is the desired answer. With the use of a dummy moment (couple) instead of a dummy force, the same procedure would have allowed us to obtain the local rotation at any point in the beam.

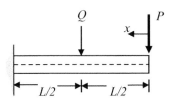

Figure 8.15 Use of a dummy load, Q, to obtain the deflection at the center of a cantilever beam.

There is an alternate way of viewing the dummy load method which leads to what is called the *unit-load method*. Consider, for example, a bending problem of the type we have been discussing where the beam is acted upon by a set of concentrated forces, P_i, and moments, M_i. Following the dummy load procedure, we can calculate

$$u_Q\big|_{Q=0} = \frac{\partial U(Q, P_i, M_i)}{\partial Q}\bigg|_{Q=0} \tag{8.149}$$

where the strain energy, U, is given by

$$U = \frac{1}{2EI}\int_0^L M^2 dx \tag{8.150}$$

so that

$$u_Q\big|_{Q=0} = \frac{1}{EI}\int_0^L M\big|_{Q=0} \frac{\partial M}{\partial Q}\bigg|_{Q=0} dx \tag{8.151}$$

But $M|_{Q=0} = M(P_i, M_i, x)$ is just the moment distribution without the dummy load present and, by superposition, we have the total moment distribution with the dummy load present given as

$$M(Q, P_i, M_i, x) = M(P_i, M_i, x) + M_1(Q, x) \tag{8.152}$$

where $M_1(Q, x)$ is the moment distribution due to only Q. Since we are dealing with a linear system

$$M_1(Q, x) = QM_1(Q=1, x) \tag{8.153}$$

where $M_1(Q=1, x)$ is the moment distribution due only to a unit force at Q. Using Eq. (8.152) and Eq. (8.153) it follows that

$$\frac{\partial M(Q, P_i, M_i, x)}{\partial Q}\bigg|_{Q=0} = M_1(Q=1, x) \tag{8.154}$$

so that Eq. (8.151) becomes

$$u_Q\big|_{Q=0} = \frac{1}{EI}\int_0^L M(P_i, M_i, x) M_1(Q=1, x) dx \tag{8.155}$$

which is an expression for desired displacement using the unit-load method. The advantage of using Eq. (8.155) over the original dummy load procedure is that in the unit-load method, we need calculate only (1) the original bending moment distribution (without Q) and (2) the bending moment distribution from the unit load itself, rather than having to compute a bending moment distribution when both the original loads and Q are present. As you can verify yourself, having the original loads and Q present leads to more complicated moment distribution expressions, but where much of the complexity ultimately disappears through the differentiation process and setting $Q = 0$ in the original dummy load method. Note that we can use a unit moment in exactly the same manner as done here with a unit force to calculate the rotation at any point in the beam instead.

Consider again the previous cantilever beam problem of Fig. 8.14 where we calculate the deflection at the center of the beam with the unit-load method. The required moment distributions are

$$M(P, x) = -Px \quad (0 < x < L)$$
$$M_1(Q = 1, x) = \begin{cases} 0 & (0 < x < L/2) \\ -(x - L/2) & (L/2 < x < L) \end{cases} \quad (8.156)$$

which you can verify leads to Eq. (8.148) for the center deflection.

8.7.3 Principle of Least Work

All of the previous uses of the principle of complementary virtual work have been for statically determinate problems where we could write the complementary strain energy in terms of the applied loads. For statically indeterminate problems, this is not possible, and we can only write the complementary strain energy in terms of the applied loads and a set of unknown forces and/or moments (called redundants) which arise from the overconstrained nature of statically indeterminate problems. For example, Fig. 8.16a shows a truss structure similar to the one shown in Fig. 8.13 but where there is an extra internal member BD. Similarly, Fig. 8.16b shows the same truss as in Fig. 8.13 but where there are now two pin supports instead of a pin and a roller. In both cases we cannot write the complementary strain energy in terms of the load P. However, we can use complementary energy and work concepts to solve for the redundants present in such problems. To illustrate the procedure,

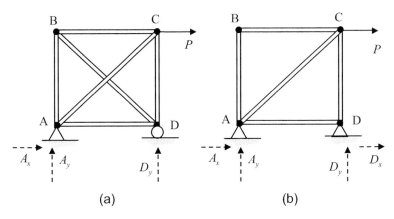

Figure 8.16 (a) A truss where the external reaction forces (A_x, A_y, D_y) can be obtained from equilibrium of the entire truss but where the truss is statically indeterminate because of too many internal truss members, leading to a redundant internal force, such as the force in member BD, for example. (b) A truss that is statically indeterminate because of being overly constrained at its supports, leading to a redundant support force such as the force D_y, for example.

we will use the specific example of Fig. 8.16b. There are four reaction force components (A_x, A_y, D_x, D_y) at A and D for the truss and only three equilibrium equations, so there is one excess unknown which we will take as a redundant force, D_y. We can then write the complementary strain energy as $U^c = U^c(P, D_y)$. Now, by the principle of complementary virtual work, if we vary the force D_y, we have

$$\delta U^c = \frac{\partial U^c}{\partial D_y} \delta D_y = \delta W_v^c = 0 \tag{8.157}$$

since there is no vertical displacement at the reaction force so that there is no corresponding complementary virtual work. Since this result must be true for any variation, δD_y, we have

$$\frac{\partial U^c(P, D_y)}{\partial D_y} = 0 \tag{8.158}$$

which is an equation that we can use to solve for D_y.

This same process works in more general cases where we have multiple known applied forces and moments (P_i, M_i) and similarly multiple redundant forces and moments (R_i, M_{Ri}) at locations where the generalized displacements or rotations are zero. In this case we can write

$$U^c = U^c(R_i, M_{Ri}, P_i, M_i) \tag{8.159}$$

If we vary the redundant forces and moments, the principle of complementary virtual work gives

$$\delta U^c = \sum_{i=1}^{n} \frac{\partial U^c}{\partial R_i} \delta R_i + \sum_{i=1}^{m} \frac{\partial U^c}{\partial M_{Ri}} \delta M_{Ri} = \delta W_v^c = 0 \tag{8.160}$$

where we have assumed that there are n redundant forces and m redundant moments. If we can vary these redundant forces and moments independently, then

$$\begin{aligned}\frac{\partial U^c}{\partial R_i} &= 0 \quad (i = 1, \ldots, n) \\ \frac{\partial U^c}{\partial M_{Ri}} &= 0 \quad (i = 1, \ldots, m)\end{aligned} \tag{8.161}$$

are $n + m$ independent equations that can be used to solve for the $n + m$ unknown redundants. Equation (8.161) is called the *theorem (or principle) of least work* (or *Engesser's second theorem*). Stated explicitly, this theorem says:

> Of all the possible values of the independent redundants (R_i, M_{Ri}) that satisfy equilibrium for a statically indeterminate elastic system, the correct values of the redundants are those that make the complementary strain energy stationary and hence satisfy Eq. (8.161).

The theorem is called the theorem of least work because $\delta W_v^c = 0$ (although, strictly speaking, this means only that the complementary virtual work has a stationary value). We have illustrated the principle of least work for the case where the redundants arise from an excess number of external constraints on a system but the principle is also valid when the redundancy is due to having too many internal forces, as is the case for the truss shown in Fig. 8.16a (see problem P8.3).

Example 8.6 The Principle of Least Work

As an example of the use of the principle of least work, consider the statically indeterminate beam shown in Fig. 8.17a. This problem is statically indeterminate to the first degree, i.e., there is one more unknown reaction forces than equations of equilibrium. We can take the unknown reaction force, R, at the roller support A as the single redundant. The bending moment in terms of the coordinate x shown in Fig. 8.17b is

$$M(x) = Rx - \frac{qx^2}{2} \tag{8.162}$$

so that the complementary strain energy is

$$U^c(R) = U(R) = \frac{1}{2EI} \int_0^L \left(Rx - \frac{qx^2}{2} \right)^2 dx \tag{8.163}$$

and from the principle of least work

$$\frac{\partial U^c}{\partial R} = \frac{1}{EI} \int_0^L \left(Rx - \frac{qx^2}{2} \right) x \, dx = 0 \tag{8.164}$$

which yields $R = 3qL/8$. This can be verified to be the correct answer by integrating the moment–curvature relationship twice to obtain the displacement, $w(x)$, of the neutral axis and then using the boundary conditions $w(0) = w(L) = 0$ and $dw(L)/dx = 0$ to solve for the constants of integration and the reaction force, R.

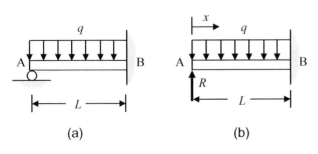

Figure 8.17 (a) Statically indeterminate beam subjected to a uniform distributed load, q. (b) The reaction force, R, at the roller support A, which can be taken as the redundant force for the structure.

Figure 8.18 (a) Statically indeterminate beam between two fixed supports. (b) The free body diagram of the beam showing the reaction forces and moments at the supports.

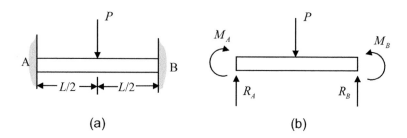

In choosing the redundants, *one must ensure that we can vary those redundants independently without violating equilibrium*, which is the condition we used to arrive at Eq. (8.161). For the beam between two fixed supports A and B, as shown in Fig. 8.18a, there are four unknown reaction forces and moments (Fig. 8.18b) and only two equations of equilibrium so that the problem is statically indeterminate to the second degree and we must choose two redundants. We cannot choose the reaction forces R_A and R_B as those redundants since force equilibrium requires that $R_A + R_B = P$ so that the virtual changes of these forces are related, i.e., $\delta R_A = -\delta R_B$. However, we could choose, for example, R_B, M_B as the redundants.

Earlier, we stated that Engesser's first theorem and Castigliano's second theorem are also true for statically indeterminate systems. The principle of least work shows why this is the case. In a statically indeterminate case, the complementary energy can be described only in terms of the known loads and moments and the redundants, i.e., $U^c = U^c(P_i, M_i, R_i, M_{Ri})$. The principle of complementary virtual work gives

$$\delta U^c = \sum_{i=1}^{N} \frac{\partial U^c}{\partial P_i} \delta P_i + \sum_{i=1}^{M} \frac{\partial U^c}{\partial M_i} \delta M_i + \left(\sum_{i=1}^{m} \frac{\partial U^c}{\partial R_i} \delta R_i + \sum_{i=1}^{n} \frac{\partial U^c}{\partial M_{Ri}} \delta M_{Ri} \right)$$

$$= \sum_{i=1}^{N} u_i \delta P_i + \sum_{i=1}^{M} \theta_i \delta M_i$$

(8.165)

where the displacements and rotations at the redundants are zero so they do no complementary virtual work. By the theorem of least work, the term in parentheses in Eq. (8.165) vanishes so we are again left with Eq. (8.140) which leads to Engesser's first theorem and Castigliano's second theorem. Although the redundants themselves are functions of the applied loads, i.e., $R_i = R_i(P_i, M_i)$, $M_{Ri} = M_{Ri}(P_i, M_i)$, it is not necessary to take derivatives of the applied loads and moments appearing in these redundants in computing partial derivatives of the complementary energy since by the chain rule we have, for example,

$$\frac{\partial U^c(P_i, M_i, R_i(P_i, M_i), M_{Ri}(P_i, M_i))}{\partial P_i} = \left. \frac{\partial U^c}{\partial P_i} \right|_{M_i, R_i, M_{Ri}=\text{constants}}$$

$$+ \sum_{k=1}^{m} \left(\left. \frac{\partial U^c}{\partial R_k} \right|_{P_i, M_i, M_{Ri}=\text{constants}} \right) \frac{\partial R_k}{\partial P_i}$$

$$+ \sum_{k=1}^{n} \left(\left. \frac{\partial U^c}{\partial M_{Rk}} \right|_{P_i, M_i, R_i=\text{constants}} \right) \frac{\partial M_{Rk}}{\partial P_i}$$

(8.166)

But by Eq. (8.161) all the partials with respect to the redundants must vanish, leaving

$$\frac{\partial U^c(P_i, M_i, R_i, M_{Ri})}{\partial P_i} = \left.\frac{\partial U^c}{\partial P_i}\right|_{M_i, R_i, M_{Ri} = \text{constants}} \quad (8.167)$$

and, similarly,

$$\frac{\partial U^c(P_i, M_i, R_i, M_{Ri})}{\partial M_i} = \left.\frac{\partial U^c}{\partial M_i}\right|_{P_i, R_i, M_{Ri} = \text{constants}} \quad (8.168)$$

Thus, in applying Engesser's first theorem or Castigliano's second theorem we can treat the redundants as constants in those theorems.

8.8 Reciprocity

A linear elastic body possesses the important property of *reciprocity* that can be used to advantage in many cases to find solutions to problems. Consider a linear elastic body that occupies a volume V whose surface is S, and let x be an arbitrary point in V and let x_s be an arbitrary point on S. Also, let $\sigma_{ij}^{(1)}, e_{ij}^{(1)}, u_j^{(1)}$ be the stresses, strains, and displacements in V for this body due to $T_j^{(1)}(\mathbf{x}_s), u_j^{(1)}(\mathbf{x}_s), f_j^{(1)}(\mathbf{x})$, which are the surface stress components, surface displacement components, and body force components (we can call this case (1)). Similarly, let $\sigma_{ij}^{(2)}, e_{ij}^{(2)}, u_j^{(2)}$ be the stresses, strains, and displacements due to $T_j^{(2)}(\mathbf{x}_s), u_j^{(2)}(\mathbf{x}_s), f_j^{(2)}(\mathbf{x})$ for the same body (we can call this case (2)). Then

$$\sum_{i=1}^{3}\sum_{j=1}^{3}\int_V \sigma_{ij}^{(1)} e_{ij}^{(2)} dV = \sum_{i=1}^{3}\sum_{j=1}^{3}\int_V \sigma_{ij}^{(2)} e_{ij}^{(1)} dV \quad (8.169)$$

which follows directly from placing the linear stress–strain relations

$$\sigma_{ij}^{(1)} = \sum_{i=1}^{3}\sum_{j=1}^{3} C_{ijkl} e_{kl}^{(1)}$$

$$\sigma_{ij}^{(2)} = \sum_{i=1}^{3}\sum_{j=1}^{3} C_{ijkl} e_{kl}^{(2)} \quad (8.170)$$

into Eq. (8.169) and using the symmetry of the elastic constants $C_{ijkl} = C_{klij}$.

Because of the symmetry of the stresses, for either case (1) or case (2) we have

$$\sum_{i=1}^{3}\sum_{j=1}^{3} \sigma_{ij} e_{ij} = \frac{1}{2}\sum_{i=1}^{3}\sum_{j=1}^{3} \sigma_{ij}\left(\frac{\partial u_i}{\partial x_j} + \frac{\partial u_j}{\partial x_i}\right) = \sum_{i=1}^{3}\sum_{j=1}^{3} \sigma_{ij}\frac{\partial u_j}{\partial x_i} \quad (8.171)$$

so that

$$\sum_{i=1}^{3}\sum_{j=1}^{3}\int_V \sigma_{ij}^{(1)} \frac{\partial u_j^{(2)}}{\partial x_i} dV = \sum_{i=1}^{3}\sum_{j=1}^{3}\int_V \sigma_{ij}^{(2)} \frac{\partial u_j^{(1)}}{\partial x_i} dV \qquad (8.172)$$

From the use of the chain rule of calculus and the equations of equilibrium, for either case we have

$$\sum_{i=1}^{3}\sum_{j=1}^{3} \sigma_{ij} \frac{\partial u_j}{\partial x_i} = \sum_{i=1}^{3}\sum_{j=1}^{3} \frac{\partial}{\partial x_i}(\sigma_{ij} u_j) + \sum_{j=1}^{3} f_j u_j \qquad (8.173)$$

so that Eq. (8.172) can be written as

$$\begin{aligned}\sum_{i=1}^{3}\sum_{j=1}^{3}\int_V \frac{\partial}{\partial x_i}\left(\sigma_{ij}^{(1)} u_j^{(2)}\right) dV + \sum_{j=1}^{3}\int_V f_j^{(1)} u_j^{(2)} dV \\ = \sum_{i=1}^{3}\sum_{j=1}^{3}\int_V \frac{\partial}{\partial x_i}\left(\sigma_{ij}^{(2)} u_j^{(1)}\right) dV + \sum_{j=1}^{3}\int_V f_j^{(2)} u_j^{(1)} dV\end{aligned} \qquad (8.174)$$

But, by the divergence theorem, the volume integrals in Eq. (8.174) can be transformed into integrals over the surface of the body, giving

$$\sum_{j=1}^{3}\int_S T_j^{(1)} u_j^{(2)} dS + \sum_{j=1}^{3}\int_V f_j^{(1)} u_j^{(2)} dV = \sum_{j=1}^{3}\int_S T_j^{(2)} u_j^{(1)} dS + \sum_{j=1}^{3}\int_V f_j^{(2)} u_j^{(1)} dV \qquad (8.175)$$

where $T_j^{(m)} = \sum_{i=1}^{3} \sigma_{ij}^{(m)} n_i$ ($m = 1,2$) are the surface stress vector components for the two cases. [Note: the notation in Eq. (8.175) is different than used previously where we expressed the stress vector components on a surface as $T_j^{(\mathbf{n})}$ (see Eq. (2.13)). Here we will drop the $(\mathbf{n})$ superscript, where $\mathbf{n}$ is the unit normal to the surface, so that $T_j \equiv T_j^{(\mathbf{n})}$. The superscript (m) in $T_j^{(m)}$ is now simply an indication of the case number.] Equation (8.175) is a statement of the *reciprocal theorem of Betti–Rayleigh*. Stated explicitly in words, this theorem says that the work done by the surface stresses and body forces of case (1) acting through the displacements of case (2) is equal to the work done by the surface stresses and body forces of case (2) acting through the displacements of case (1). For the special case when all the surface forces and moments (P_i, M_i) are concentrated and the body forces are absent, this theorem can be written as

$$\sum_{j=1}^{N} P_j^{(1)} u_j^{(2)} + \sum_{j=1}^{3} M_j^{(1)} \theta_j^{(2)} = \sum_{j=1}^{N} P_j^{(2)} u_j^{(1)} + \sum_{j=1}^{3} M_j^{(2)} \theta_j^{(1)} \qquad (8.176)$$

Example 8.7 Reciprocity

To illustrate a use of reciprocity, consider a cantilever beam acted upon by a distributed load, $q(x)$, as shown in Fig. 8.19a. Also, let the same beam be loaded by a concentrated load, P, at a distance, d, from the support, as shown in Fig. 8.19b. Then from the reciprocal theorem we have

$$-\int_0^L q(x) w^{(2)}(P, x, d) \, dx = P w^{(1)}(w, d) \tag{8.177}$$

where $w^{(2)}(P, x, d)$ is the z-displacement of the beam at x due to a load P acting at d and $w^{(1)}(q, d)$ is the displacement at d due to the distributed load $q(x)$. The minus sign exists on the left side of Eq. (8.177) because the distributed load acts in the negative z-direction. It is relatively easy to solve the concentrated load problem of Fig. 8.19b and obtain the z-displacement of the beam for that problem as

$$w^{(2)}(P, x, d) = \begin{cases} \dfrac{Px^2}{6EI}(3d - x) & (0 < x < d) \\ \dfrac{Pd^2}{6EI}(3x - d) & (d < x < L) \end{cases} \tag{8.178}$$

which gives

$$w^{(1)}(q, d) = \dfrac{-1}{6EI} \left\{ \int_0^d q(x) x^2 (3d - x) \, dx + \int_d^L q(x) d^2 (3x - d) \, dx \right\} \tag{8.179}$$

Equation (8.179) gives the z-displacement of the beam at any point d ($0 < d < l$) due to the arbitrary distributed load $q(x)$.

This example shows that the usefulness of the reciprocal theorem is based on its ability to relate the effects of two solutions on each other, a property that can be used to extract information about one of those solutions. The boundary element method, which is one of the

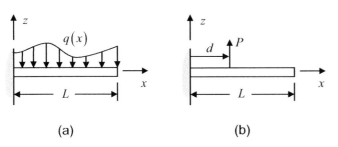

Figure 8.19 (a) Cantilever beam loaded by an arbitrary distributed load, $q(x)$. (b) The same beam loaded by a concentrated force, P, located at an arbitrary distance, d, from the fixed end.

important numerical methods described briefly in Chapter 6, is based on reciprocity and uses a fundamental solution (which is a solution to point sources acting in the body) to derive an integral equation for the solution of a stress analysis problem. The solution derived in Eq. (8.179) also is based on a point-source solution, Eq. (8.178), so it shares some of the same characteristics found in the boundary element method, which will be discussed in more detail in Chapter. 9.

8.9 PROBLEMS

P8.1 For the two-bar truss problem of Fig. 8.5, solve Eq. (8.89) for the bar forces and reactions using the symbolic algebra capabilities of MATLAB®. Verify that you obtain the solution given in Eq. (8.90).

P8.2 For the three-bar truss problem of Fig. 8.9, use the symbolic algebra capabilities of MATLAB® to solve for the bar forces using both a displacement-based approach and a force-based approach. Verify that you obtain the solution given in Eq. (8.117).

P8.3 Solve the statically indeterminate truss of Fig 8.16a for the redundant force, R, in bar BD using the principle of least work. Assume all bars have the same cross-sectional area, A, and Young's modulus, E.

P8.4 In statics courses one usually examines the equilibrium of rigid bodies where the assumption of rigidity means that there can be no elastic strain energy stored in the bodies. In this case the principle of virtual work gives simply $\delta W_v = 0$. As in the case of elastic bodies, the virtual work principle can be used in place of the equations of equilibrium. For example, consider the two rigid bars shown in Fig. P8.1. These bars, each of length L and weight W, are pinned to each other and supported by a pin at point A. At C, where the bar rides on a smooth surface, a couple, M, acts.

(a) Using the equations of equilibrium, determine the angle, θ, of the system in terms of M, W, and L.

(b) Using the principle of virtual work, determine the angle, θ, of the system in terms of M, W, and L.

P8.5 The "wishbone" frame shown in Fig. P8.2 is subjected to a horizontal force, P, at point B. The frame is pinned at point A and on rollers at point B. Determine expressions for the horizontal deflection and the rotation at B using Castigliano's second theorem and the unit load method, respectively. Assume that the bending stiffness, EI, is constant throughout the beam and consider the strain energy contributions for both axial extension and bending.

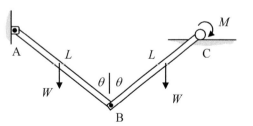

Figure P8.1 Two rigid bars in equilibrium under the action of their own weights and an applied couple, M.

P8.6 The bending stiffness, EI, is uniform throughout the frame shown in Fig P8.3. The frame is fixed (built-in) at point B and supported on rollers at point A. Using Castigliano's second theorem and the dummy load method, determine the horizontal displacement at point A in terms of the load, F, the stiffness, and the length, L. Consider the strain energy due to bending only.

P8.7 A symmetrical L-shaped bar of circular cross-section lies in a horizontal plane and is rigidly fixed at ends A and B to a vertical wall. The two legs of the bar are of equal length, L, and at right angles to each other. The stiffnesses EI and GJ are constant throughout. If a vertical force, F, acts on the bar as shown in Fig. P8.4 determine the support forces and moments at the wall supports and show them on a sketch. Assume the strain energy is due to bending and torsion only. For simplicity, also assume $EI = GJ$. [Hint: by symmetry, the vertical forces at A and B are equal.]

P8.8 A structure in the form of a three-sided rectangular frame of constant circular cross-section lies in a horizontal plane and is rigidly fixed at points A and B, as shown in Fig. P8.5. A force, P, is applied to the frame.

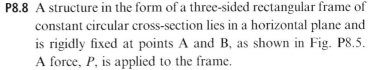

Figure P8.2 A frame pinned at point A and on rollers at point B, subjected to horizontal force, P.

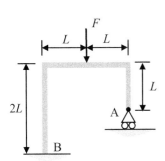

Figure P8.3 A frame that is clamped (built-in) at point B and on rollers at point A.

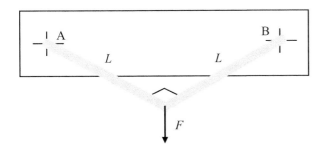

Figure P8.4 An L-shaped bar of circular cross-section that lies in a horizontal plane and is rigidly fixed at A and B and subjected to a vertical force, F.

(a) Using symmetry, determine explicitly the reaction forces and bending moments at points A and B. What does symmetry imply about the reaction torques at A and B?

(b) Using energy methods, determine the reaction torques at points A and B in terms of P, a, b, EI, and GJ. Consider only the strain energy due to bending and torsion.

P8.9 The frame shown in Fig. P8.6 is of constant cross-section and is subjected to a uniform distributed load of intensity q_0 lb/unit length. Using energy methods, determine the horizontal displacement of the frame at roller D in terms of q_0, H, L, and EI. Neglect the strain energy due to shear.

P8.10 A sign of weight W is hung from a bent bar of circular cross-section, which is built into a wall at point C (see Fig. P8.7) and has a constant EI and GJ along its entire length, where, for simplicity, take $GJ = 2EI/3$. Determine – in terms of W, L, and EI – the vertical sag of the bar at A (i.e., the relative vertical displacement at A relative to that at B). Consider only the strain energy due to bending and torsion.

P8.11 A sign of weight W is hung from a frame which has a constant EI along its entire length (see Fig. P8.8). Determine at the end point P:

(a) the downward deflection, Δ, in terms of W, L, and EI
(b) the clockwise rotation, θ, in terms of W, L, and EI.

Use the unit-load method and consider only the strain energy due to bending.

P8.12 A pair of forces of magnitude P each act on a rectangular frame which has a constant EI, as shown in Fig. P8.9. Determine the total horizontal elongation of the frame between the points at which the loads act. Consider only the strain energy due to bending.

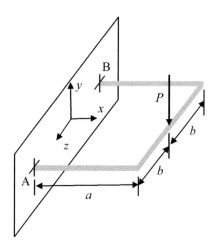

Figure P8.5 A rectangular frame that lies in the x–z plane and subjected to a force, P, in the negative y-direction.

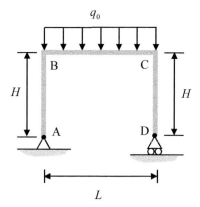

Figure P8.6 A frame subjected to a distributed load.

P8.13 The rectangular beam shown in Fig. P8.10 carries a uniform distributed force/unit length, q_0, acting at the bottom of the beam over its entire length. Its Young's modulus is E. Determine the horizontal and vertical deflections of the beam at point O in terms of E, q_0, L, b, and h. Point O is at the centroid of the cross-section at the free end. Consider the strain energy due to the bending moment and the axial force in the beam only.

P8.14 The L-shaped frame shown in Fig. P8.11 has a uniform EI value and is pinned at both ends so that it represents a statically indeterminate problem. Determine the horizontal component of the reaction force at point C in terms of the uniform distributed load intensity, q_0, and the lengths d, and l. Consider the strain energy due to the internal bending moment only.

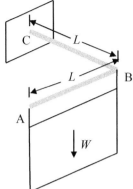

Figure P8.7 A sign of weight W hung from a bent bar at points A and B.

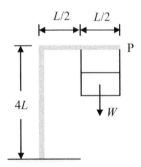

Figure P8.8 A sign of weight W hung from a frame.

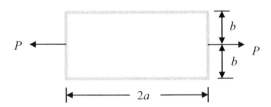

Figure P8.9 A rectangular frame acted upon by a pair of horizontal forces.

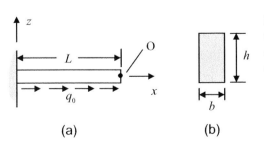

Figure P8.10 (a) Rectangular cantilever beam where a uniform axial distributed force/unit length acts on the bottom of the beam over its entire length. (b) The dimensions of the cross-section.

P8.15 A circular thin-wall tube is bent to form the symmetrical frame shown in Fig. P8.12 and is loaded by a pair of forces, each of magnitude P, acting at the centers of the top two horizontal sections. Let $a = b = c$ and $E = 3G$, where E is Young's modulus and G is the shear modulus. Using energy methods, determine:
(a) the internal bending moment in the frame at the applied loads in terms of P and a;
(b) the relative displacement of the frame along the line AB between the two loads in terms of P, a, E, and I, where I is the area moment of the thin circular tube;
(c) the rotation of the frame about the x-axis at B in terms of P, a, E, and I.

Neglect the strain energy in the frame due to the internal shear force and axial load.

P8.16 An end force, P, is applied to a thin cantilever beam which is bent into a circular shape, as shown in Fig. P8.13. Determine the radial displacement of the beam as a function of the angle, ϕ, the force, P, and the EI of the cross-section, where E is Young's modulus and I is the area moment. Use ordinary engineering beam theory as developed

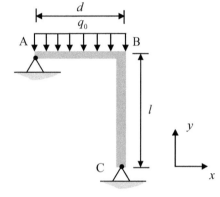

Figure P8.11 An L-shaped frame pinned at both ends and loaded by a uniform distributed force/unit length, q_0.

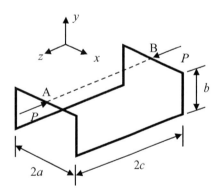

Figure P8.12 A bent frame acted upon by a pair of forces.

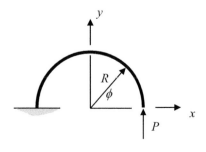

Figure P8.13 A thin beam bent into a circular shape and loaded by a force, P.

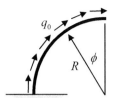

Figure P8.14 A thin beam bent into a circular shape. A uniform load/unit length, q_0, acts along the beam.

for a straight beam and consider the strain energy due to the internal bending moment only.

P8.17 A thin cantilever beam is bent into a circular shape as shown in Fig. P8.14. A uniform distributed load/ unit length, q_0, acts along its entire length.
  (a) Determine the horizontal and vertical deflections of the beam at its free end, $\phi = 0$, in terms of q_0, R, and the EI of the cross-section, where E is Young's modulus and I is the area moment. Use ordinary engineering beam theory as developed for a straight beam and consider the strain energy due to the internal bending moment only.
  (b) Repeat part (a) but compute the horizontal and vertical deflections of the beam at its free end, $\phi = 0$, in terms of q_0, R, and the EA of the cross-section, where E is Young's modulus and A is the cross-sectional area. Consider the strain energy due to the internal axial force only.

References

1. R. G Budynas, *Advanced Strength and Applied Stress Analysis*, 2nd edn (Boston, MA: McGraw-Hill, 1999)
2. C. E. Pearson, *Theoretical Elasticity* (Cambridge, MA: Harvard Univ. Press, 1959)
3. W. B. Bickford, *Advanced Mechanics of Materials* (Menlo Park, CA: Addison Wesley, 1998)
4. H. Reismann and P. S Pawlick, *Elasticity – Theory and Applications* (New York, NY: John Wiley, 1980)

9 Computational Mechanics of Deformable Bodies

Analytical solutions of deformable body problems are possible in cases involving simple geometries and loads. The complex problems commonly found in practice, however, can only be solved with approximate numerical methods. The method of finite elements has been highly developed over the past 60 years to become the numerical method of choice for analyzing deformable bodies as well as many other engineering problems. More precisely, the stiffness-based finite element method is the engineering tool used almost universally for stress and deformation analyses. In this chapter we will describe the basic elements of both stiffness-based and force-based finite elements. [Note: what we are calling here force-based finite elements is also called the Integrated Force Method to distinguish it from earlier force methods [1], [2].] This introduction is intended to pave the way for more in-depth studies of finite elements in higher-level courses and to show the importance of the energy methods discussed in the previous chapter. Force-based finite elements are not discussed in current advanced strength of materials texts but they have been presented in an elementary strength of materials setting [3]. The lack of discussion of force-based methods is partially due to the almost exclusive use of the stiffness approach in powerful commercial software codes but it is also rooted in the fact that force-based methods rely on appropriately combining equilibrium and compatibility while the stiffness method relies solely on equilibrium. Thus, it has taken longer to appreciate how that combination can be used effectively as a general computational tool. Force-based methods, however, are more closely related to the governing equations described in Chapter 6 than are stiffness-based methods, so there is much to be gained by understanding how force-based methods work.

We will first examine both stiffness-based and force-based finite elements for axial-load and bending problems (the discussion of torsion problems is omitted since they are identical to axial-load problems in their structure). These problems will be described using a matrix–vector approach that will then be generalized to general deformable body problems. Simple examples will be used that allow us to work through the application of these finite element methods in detail and to compare results with analytical solutions.

Another method that has proven to be an important alternative to the stiffness-based finite element approach is the boundary element method. This method, like force-based finite elements, has been largely overshadowed by stiffness-based finite elements, but it is an important method and a computational tool that has a useful niche in solving specialized problems such as cracked bodies. We will not provide a detailed description of boundary elements, but will outline the basic foundations of the method and contrast it to finite elements.

9.1 Numerical Solutions – Axial Loads

Many of the axial-load problems described in elementary strength of materials courses are simple enough that they can be solved analytically. Here, however, we want to lay the foundations for solving more complex problems numerically so we will use axial load problems as part of that foundation. Consider the problem shown in Fig. 9.1a where a bar of varying cross-sectional area, $A(x)$, is subjected to concentrated axial forces (such as P_j^*) and distributed forces, $q_x(x)$. The forces (R_1, R_2) could be either reaction forces at the ends where displacements are fixed or they could be applied end forces. We will also let the Young's modulus, $E(x)$, of the bar be a continuously varying function. We will examine both stiffness-based and forced-based finite element methods for solving this problem numerically.

9.1.1 Principles of Virtual Work and Complementary Virtual Work

In the finite element method, the bar is divided into M elements as shown in Fig. 9.1b. If we consider the (j)th element of the bar, as shown in Fig. 9.2a, this element will have end forces (P_j, P_{j+1}) acting in the x-direction, a distributed force/unit length, $q_x(\xi)$, acting in the element, as well as an internal axial force, $F_x(\xi) = \sigma_{xx}(\xi)A(\xi)$, where $\sigma_{xx}(\xi)$ is the axial stress and $A(\xi)$ is the cross-sectional area of the bar. The coordinate ξ is measured from the start of the element as shown in Fig. 9.1b. There are also end displacements of the element, (U_j, U_{j+1}) which are called the nodal displacements. Note there are $M + 1$ nodal displacements for the entire bar if there are M elements. The length of the element is l. We will

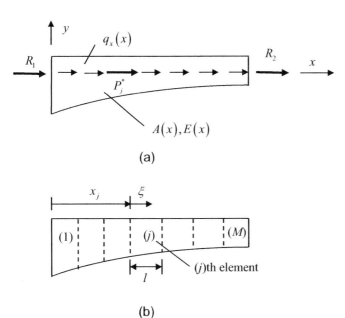

Figure 9.1 (a) A general axial-load problem where there are concentrated forces and a body force distribution acting on the bar. The bar can be of varying cross-sectional area and have a varying Young's modulus. (b) The bar divided into M finite elements.

Figure 9.2 (a) The (j)th element of the bar, showing the distributed force/unit length, $q_x^{(j)}(\xi)$, and the displacements (U_j, U_{j+1}) and forces (P_j, P_{j+1}) acting at the end nodes. (b) The same element showing the internal axial force, $F_x(\xi)$ (assumed to be tensile), in the element and its values at the end nodes.

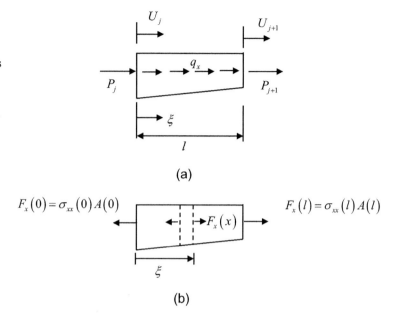

assume that these elements are chosen so that any concentrated force such as the P_j^* shown in Fig. 9.1a lies at the start or end of an element, i.e., at a node.

The principle of virtual work applied to the (j)th element says that the virtual change of the strain energy of the element, δU, is equal to the virtual work, δW_v, of the external forces acting on the element, i.e.,

$$\delta U = \delta W_v \tag{9.1}$$

From the general results developed in Chapter 8 for a deformable body (see Eq. (8.71b), for example) specialized for this uniaxial-load problem we have

$$\int_0^l \sigma_{xx}(\xi)\delta e_{xx}(\xi)A(\xi)d\xi = \int_0^l f_x(\xi)\delta u_x(\xi)A(\xi)d\xi + \sum_{m=j}^{j+1} P_m \delta U_m \tag{9.2}$$

where f_x is the body force (force/unit volume) and P_m ($m = j, j+1$) are the forces acting at nodes j and $j+1$ (Fig. 9.2a). The quantities $(\delta e_{xx}, \delta u_x, \delta U_m)$ are virtual changes of the strain, e_{xx}, and x-displacement, u_x, in the element and virtual changes of the nodal displacements, U_m ($m = j, j+1$), at the ends of the element, respectively. The internal force acting in the element is just $F_x(\xi) = \sigma_{xx}(\xi)A(\xi)$ and we can define the distributed force/unit length, $q_x(\xi)$, in terms of the body force term in the element as $q_x(\xi) = f_x(\xi)A(\xi)$ so that the principle of virtual work for the element becomes

$$\int_0^l F_x(\xi)\delta e_{xx}(\xi)d\xi = \int_0^l q_x(\xi)\delta u_x(\xi)d\xi + \sum_{m=j}^{j+1} P_m \delta U_m \tag{9.3}$$

[A note on the notation: various properties will vary from element to element so at times it is important to distinguish the element associated with that property. We will do this by using a superscript with a number or symbol written in parentheses, e.g. (j), to indicate the element. Numbers or symbols written in subscripts without parentheses, e.g., $j, j+1$, will indicate nodal values. Some quantities such as the area written as $A^{(j)}(\xi)$ are only functions of the element number, while some other quantities, such as the nodal displacements, U_j, have only nodal values and are not associated with a particular element. An end force such as P_{j+1}, however, might be the force at the right end of element (j), as seen in Fig. 9.2a, or it might be the end force at the start of element $(j+1)$. Thus, the P_{j+1} seen in Fig. 9.2a can be written as $P_{j+1}^{(j)}$ to indicate it is associated with the element (j) and acts at node $j+1$ and to distinguish it from $P_{j+1}^{(j+1)}$, which is an end force for element $(j+1)$ at the same node $j+1$. This notation is very explicit but it is not used except where needed for clarity. Also note that, for convenience, we have taken the start and end nodes of element (j) to be labeled as j and $j+1$, but such sequential numbering is not essential. We should also note that we have denoted the strain energy by U and the nodal displacements by U_m. This should not cause confusion since the strain energy is always a scalar, while the displacements are always vectors written with either an explicit subscript or written as a vector in matrix–vector notation such as $\{U\}$.]

Equation (9.3) also follows directly from the equation of equilibrium for the internal force in the bar following similar steps used in Chapter 8, starting from the local equilibrium equations for the stresses. Recall from Eq. (8.49) we have in the bar

$$\frac{dF_x}{d\xi} + q_x = 0 \tag{9.4}$$

If we multiply this equilibrium equation by a virtual displacement $\delta u_x(\xi)$ and integrate over the length of the element we have

$$\int_0^l \left(\frac{dF_x}{d\xi} + q_x\right)\delta u_x d\xi = 0 \tag{9.5}$$

But we have

$$\frac{dF_x}{d\xi}\delta u_x = \frac{d}{d\xi}(F_x \delta u_x) - F_x \frac{d}{d\xi}(\delta u_x)$$

$$= \frac{d}{d\xi}(F_x \delta u_x) - F_x \delta\left(\frac{du_x}{d\xi}\right) \tag{9.6}$$

$$= \frac{d}{d\xi}(F_x \delta u_x) - F_x \delta e_{xx}$$

so using Eq. (9.6) in Eq. (9.5) and carrying out the integration of the perfect differential we find

$$\int_0^l F_x(\xi)\delta e_{xx}(\xi)d\xi = \int_0^l q_x(\xi)\delta u_x(\xi)d\xi + F_x(l)\delta u_x(l) - F_x(0)\delta u_x(0) \tag{9.7}$$

We also have $\delta u_x(l) = \delta U_{j+1}$, $\delta u_x(0) = \delta U_j$ and $F_x(l) = P_{j+1}$, $F_x(0) = -P_j$ (see Figs. 9.2a, b) so that the principle of virtual work becomes

$$\int_0^l F_x(\xi)\delta e_{xx}(\xi)d\xi = \int_0^l q_x(\xi)\delta u_x(\xi)d\xi + P_{j+1}\delta U_{j+1} + P_j\delta U_j \qquad (9.8)$$

which is identical with Eq. (9.3).

Now let us turn our attention to the principle of complementary virtual work, which states that the virtual change of the complementary strain energy, δU^c, for an element is equal to the complementary virtual work done by the external forces, δW_v^c, acting on that element, i.e.,

$$\delta U^c = \delta W_v^c \qquad (9.9)$$

In applying this principle, we normally consider virtual changes of the stresses that satisfy the homogeneous equations of equilibrium, i.e., the virtual changes $\delta \sigma_{ij}$ satisfy Eq. (8.97) with $\delta f_i = 0$. In that case the principle applied to the element of Fig. 9.2a gives (see Eq. (8.96)):

$$\int_0^l e_{xx}(\xi)\delta\sigma_{xx}(\xi)A(\xi)d\xi = \sum_{m=j}^{j+1} U_m\delta P_m \qquad (9.10)$$

or, in terms of the internal force, the principle of complementary virtual work is:

$$\int_0^l e_{xx}(\xi)\delta F_x(\xi)d\xi = \sum_{m=j}^{j+1} U_m\delta P_m \qquad (9.11)$$

We can also develop Eq. (9.11) directly from the 1-D strain–displacement relationship for the element following similar steps carried out in Chapter 8 for a general state of strain. We have

$$e_{xx}(\xi) = \frac{du_x(\xi)}{d\xi} \qquad (9.12)$$

Multiplying this relationship by a virtual change of the internal force, $\delta F_x(\xi)$, and integrating over the element gives

$$\int_0^l \left(e_{xx} - \frac{du_x}{d\xi}\right)\delta F_x d\xi = 0 \qquad (9.13)$$

and we have

$$\begin{aligned}\frac{du_x}{d\xi}\delta F_x &= \frac{d}{d\xi}(u_x\delta F_x) - u_x\frac{d}{d\xi}(\delta F_x) \\ &= \frac{d}{d\xi}(u_x\delta F_x)\end{aligned} \qquad (9.14)$$

where we have used the fact that the virtual change of the internal force satisfies the homogeneous equilibrium equation

$$\frac{d}{d\xi}(\delta F_x) = 0 \tag{9.15}$$

(see Eq. (9.4) with $q_x = 0$). Using Eq. (9.14) in Eq. (9.13) and carrying out the integration of the perfect differential gives

$$\int_0^l e_{xx}(\xi)\delta F_x(\xi)d\xi = u_x(l)\delta F_x(l) - u_x(0)\delta F_x(0)$$
$$= U_{j+1}\delta P_{j+1} + U_j\delta P_j \tag{9.16}$$

which is identical to Eq. (9.11).

Having the principles of virtual work and complementary virtual work, we now will examine the use of these principles to develop stiffness-based and force-based finite element methods for solving axial-load problems numerically. We will start with the stiffness-based method, which is, as mentioned previously, the one in predominant use today.

9.2 Stiffness-Based Finite Elements for Axial-Load Problems

We can express the principle of virtual work, Eq. (9.3), solely in terms of the displacement of the element (and derivatives of that displacement) as

$$\int_0^l A(\xi)E(\xi)e_{xx}(\xi)\delta e_{xx}(\xi)d\xi = \int_0^l q_x(\xi)\delta u_x(\xi)d\xi + \sum_{m=j}^{j+1} P_m \delta U_m \tag{9.17}$$

In the stiffness-based finite element method, we write the displacement in the element in terms of known functions, $H_m(\xi)$, and a set of unknown nodal displacements, U_m. By differentiating the displacement, we can also write the strain in the element in terms of the known functions $B_m(\xi) = dH_m(\xi)/d\xi$ and the nodal displacements, i.e.,

$$u_x(\xi) = \sum_m H_m(\xi) U_m$$
$$e_{xx}(\xi) = \sum_m B_m(\xi) U_m \tag{9.18}$$

The functions $H_m(\xi)$ are called the *shape functions* that define the approximation of the displacement we wish to use for the element. If the element is very small or the loading is very simple, a linear variation of displacements (and, hence, a constant strain in the element) might be sufficient, a choice that we will make to keep the algebra simple in our discussions. Thus, in the element we will let

$$u_x(\xi) = U_j + \frac{(U_{j+1} - U_j)\xi}{l}$$
$$= U_j\left(1 - \frac{\xi}{l}\right) + U_{j+1}\left(\frac{\xi}{l}\right) \quad (0 < \xi < l) \tag{9.19}$$

where the shape functions $H_j(\xi), H_{j+1}(\xi)$ are

$$H_j(\xi) = 1 - \frac{\xi}{l}, \quad H_{j+1} = \frac{\xi}{l} \tag{9.20}$$

or, in matrix–vector forms,

$$u_x(\xi) = [H]\{U\} = \{U\}^T[H]^T \tag{9.21}$$

where

$$\{U\} = \begin{Bmatrix} U_j \\ U_{j+1} \end{Bmatrix}, \quad [H] = [1 - \xi/l \;\; \xi/l] \tag{9.22}$$

For this linear variation of displacement, the strain in the element can be written similarly in terms of strain shape functions $B_j(\xi), B_{j+1}(\xi)$ as

$$e_{xx} = \frac{\partial u_x}{\partial \xi} = \frac{U_{j+1} - U_j}{l}$$
$$= U_j\left(-\frac{1}{l}\right) + U_{j+1}\left(\frac{1}{l}\right) \tag{9.23}$$
$$= \sum_{m=j}^{j+1} B_m(\xi)\, U_m$$

In this case, the strain shape functions are just constants given by

$$B_j(\xi) = -\frac{1}{l}, \quad B_{j+1}(\xi) = \frac{1}{l} \tag{9.24}$$

and in matrix–vector forms we have

$$e_{xx} = [B]\{U\} = \{U\}^T[B]^T \tag{9.25}$$

where

$$[B] = [-1/l \;\; 1/l] \tag{9.26}$$

We are going to use these approximate expressions for the displacements in the principle of virtual work so that we also need corresponding forms for virtual displacements and virtual strains for this element. The matrix–vector forms are

$$\delta u_x = [H]\{\delta U\} = \{\delta U\}^T[H]^T$$
$$\delta e_{xx} = [B]\{\delta U\} = \{\delta U\}^T[B]^T \tag{9.27}$$

where

$$\{\delta U\} = \left\{ \begin{array}{c} \delta U_j \\ \delta U_{j+1} \end{array} \right\} \quad (9.28)$$

Using the matrix–vector forms involving either $\{U\}$ or $\{\delta U\}^T$ in Eq. (9.21), Eq. (9.25), and Eq. (9.27) then allows us to write the virtual work expression of Eq. (9.17) as

$$\{\delta U\}^T \int_0^l E(\xi)A(\xi)[B]^T[B]d\xi \,\{U\}$$

$$= \{\delta U\}^T \int_0^l [H]^T q_x d\xi + \{\delta U\}^T \{P\} \quad (9.29)$$

where

$$\{P\} = \left\{ \begin{array}{c} P_j \\ P_{j+1} \end{array} \right\} \quad (9.30)$$

Since Eq. (9.29) must be satisfied for all choices of $\{\delta U\}^T$ we must have

$$\int_0^l E(\xi)A(\xi)[B]^T[B]d\xi \,\{U\} = \int_0^l q_x(\xi)[H]^T d\xi + \{P\} \quad (9.31)$$

which is just a set of linear equations of the form

$$[K]\{U\} = \{Q\} + \{P\} \quad (9.32a)$$

or, equivalently, using the more explicit notation where a superscript (j) denotes the (j)th element

$$\left[K^{(j)}\right]\left\{U^{(j)}\right\} = \left\{Q^{(j)}\right\} + \left\{P^{(j)}\right\} \quad (9.32b)$$

where $\left[K^{(j)}\right]$ is called the *stiffness matrix* for the element and $\left\{Q^{(j)}\right\}$ is a set of nodal forces produced by the distributed force/length acting in the element. For our specific choice of linear displacement and hence constant strain we have

$$\left[K^{(j)}\right] = \int_0^{l^{(j)}} E^{(j)}(\xi) A^{(j)}(\xi) \begin{bmatrix} -1/l^{(j)} \\ 1/l^{(j)} \end{bmatrix} \begin{bmatrix} -1/l^{(j)} & 1/l^{(j)} \end{bmatrix} d\xi$$

$$= \frac{1}{\left(l^{(j)}\right)^2} \int_0^{l^{(j)}} E^{(j)}(\xi) A^{(j)}(\xi) d\xi \begin{bmatrix} 1 & -1 \\ -1 & 1 \end{bmatrix} \quad (9.33a)$$

If the area and Young's modulus are constants, and all elements have the same length l, then the stiffness matrix for every element is simply

$$[K^{(j)}] = \frac{EA}{l}\begin{bmatrix} 1 & -1 \\ -1 & 1 \end{bmatrix} = \begin{bmatrix} EA/l & -EA/l \\ -EA/l & EA/l \end{bmatrix} \quad (9.33b)$$

For the $\{Q^{(j)}\}$ vector associated with element (j) we have

$$\{Q^{(j)}\} = \left\{\begin{array}{c} Q_j^{(j)} \\ Q_{j+1}^{(j)} \end{array}\right\} = \left\{\begin{array}{c} \int_0^{l^{(j)}} q_x^{(j)}(\xi)\left(1 - \xi/l^{(j)}\right)d\xi \\ \int_0^{l^{(j)}} q_x^{(j)}(\xi)\left(\xi/l^{(j)}\right)d\xi \end{array}\right\} \quad (9.34a)$$

However, normally the distributed load is not given as a function of ξ but instead is given as $q_x^{(j)} = q_x^{(j)}(x)$, i.e., it is a function of position as measured from a fixed origin, whereas we recall ξ is measured from the beginning of the element (Fig. 9.1b). Assuming we have $q_x^{(j)} = q_x^{(j)}(x)$, then we must write $\{Q^{(j)}\}$ instead as

$$\{Q^{(j)}\} = \left\{\begin{array}{c} Q_j^{(j)} \\ Q_{j+1}^{(j)} \end{array}\right\} = \left\{\begin{array}{c} \int_0^{l^{(j)}} q_x^{(j)}(x_j + \xi)\left(1 - \xi/l^{(j)}\right)d\xi \\ \int_0^{l^{(j)}} q_x^{(j)}(x_j + \xi)\left(\xi/l^{(j)}\right)d\xi \end{array}\right\} \quad (9.34b)$$

where x_j is the location of the start of the element in the x-coordinate system (Fig. 9.1b). For the special case of a constant distributed force/unit length q_0 acting in an element of length l, it does not matter if we use Eq. (9.34a) or Eq. (9.34b). From either equation we find

$$\{Q^{(j)}\} = \left\{\begin{array}{c} Q_j^{(j)} \\ Q_{j+1}^{(j)} \end{array}\right\} = \left\{\begin{array}{c} q_0 l/2 \\ q_0 l/2 \end{array}\right\} \quad (9.35)$$

Equation (9.32b) is a stiffness-based expression for equilibrium of an element based on the use of the principle of virtual work. To obtain the equilibrium equations for the entire bar, we must assemble the elements, i.e., combine these equations appropriately. First, let us examine the nodal force terms on the right-hand side of Eq. (9.32b). Figure 9.3a shows the nodal forces generated by the distributed load acting on two adjacent elements individually, and Fig. 9.3b shows the combined nodal forces when these two elements are combined (using our more explicit notation where the element involved is indicated). For example, if $q_x = q_0 =$ constant over both elements and the elements have the same length (Fig. 9.4a), then adding the nodal contributions found from Eq. (9.35) we have nodal forces

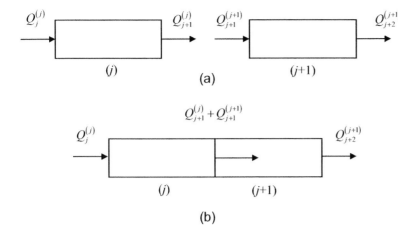

Figure 9.3 (a) The equivalent nodal forces on two adjacent elements that are produced by a distributed load. (b) The equivalent forces when these elements are combined.

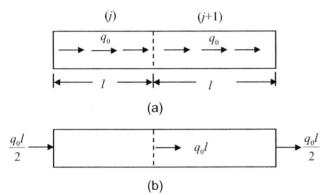

Figure 9.4 (a) A bar with a uniform distributed force/unit length, q_0, acting over two adjacent elements. (b) The equivalent nodal forces.

contributions shown in Fig. 9.4b when the two elements are combined. Now, consider the nodal forces contained in the end forces $\{P^{(j)}\}$ and $\{P^{(j+1)}\}$. Figure 9.5a again shows these forces for two adjacent elements (j) and (j+1). If an applied concentrated force, P^*_{j+1} is present at node $j+1$ then Fig. 9.5b shows a free body diagram of a small section of the bar at that node. From the equilibrium of that small section, we have

$$P^{(j)}_{j+1} + P^{(j+1)}_{j+1} = P^*_{j+1} \tag{9.36}$$

so that the nodal forces simply combine to the value of that concentrated force. This also shows that if no concentrated force is present, the sum of the end forces must vanish. This result also follows from the fact that these end forces arise from the internal axial forces present in the bar, as shown in Fig. 9.5c. At node $j+1$, these internal forces (assumed tensile) satisfy, from Fig. 9.5c,

$$F^{(j)}_x\left(l^{(j)}\right) - F^{(j+1)}_x(0) = P^*_{j+1} \tag{9.37}$$

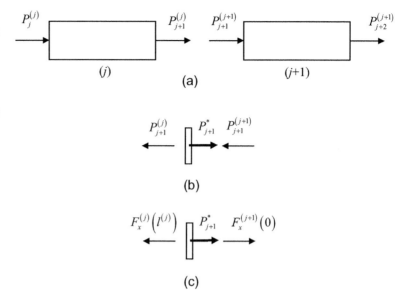

Figure 9.5 (a) The nodal forces produced by the internal forces in two adjacent elements. (b) The free body diagram of a small section of the bar when a concentrated force, P^*_{j+1}, is present at node $j+1$. (c) The same free body diagram expressed in terms of the internal forces in the bar.

so that whenever a concentrated force is not present, the internal force must be continuous. When a concentrated force is present at a node, then the internal force must exhibit a jump in value at that node as found from Eq. (9.37).

Now consider the combination of the elements of the stiffness matrix as well as the nodal forces. The entire bar will involve all the nodal displacements so, if we have M elements, we will end up with a set of $M + 1$ linear equations for the nodal displacements $(U_1, U_2, \ldots, U_{M+1})$ of the form

$$\begin{bmatrix} K_{1,1} & \cdots & K_{1,M+1} \\ \vdots & & \vdots \\ K_{1,j+1} & \ddots & K_{j+1,M+1} \\ \vdots & & \vdots \\ K_{M+1,1} & \cdots & K_{M+1,M+1} \end{bmatrix} \begin{Bmatrix} U_1 \\ \vdots \\ U_{j+1} \\ \vdots \\ U_{M+1} \end{Bmatrix} = \begin{Bmatrix} Q_1^{(1)} \\ \vdots \\ Q_{j+1}^{(j)} + Q_{j+1}^{(j+1)} \\ \vdots \\ Q_{M+1}^{(M)} \end{Bmatrix} + \begin{Bmatrix} P_1^{(1)} \\ \vdots \\ P_{j+1}^* \\ \vdots \\ P_{M+1}^{(M)} \end{Bmatrix} \quad (9.38)$$

All the body force nodal terms are known values as well as the applied forces, P_j^*, at the nodes $j = 2, \ldots, M$ (which may be zero or nonzero). The end forces are $P_1^{(1)} = R_1, P_{M+1}^{(M)} = R_2$ (see Fig. 9.1). These may be unknown reaction forces if the end nodes are restrained. If there is an applied force at an end, then the end value will just be that force. We will discuss later how to deal with any unknown reaction forces when solving the linear system of Eq. (9.38). To show how the stiffness matrix for the entire bar is assembled, it is convenient to consider a specific case such as a uniform bar (A, E, and l are all constants) with three elements (and four nodes). Then for each element, the equilibrium equations in Eq. (9.32b) contribute terms to the entire set of equations given

element 1
$$\begin{bmatrix} EA/l & -EA/l & 0 & 0 \\ -EA/l & EA/l & 0 & 0 \\ 0 & 0 & 0 & 0 \\ 0 & 0 & 0 & 0 \end{bmatrix} \begin{Bmatrix} U_1 \\ U_2 \\ U_3 \\ U_4 \end{Bmatrix} = \begin{Bmatrix} Q_1^{(1)} \\ Q_2^{(1)} \\ 0 \\ 0 \end{Bmatrix} + \begin{Bmatrix} P_1^{(1)} \\ P_2^{(1)} \\ 0 \\ 0 \end{Bmatrix}$$

Figure 9.6 The equilibrium equations for a bar composed of three elements.

element 2
$$\begin{bmatrix} 0 & 0 & 0 & 0 \\ 0 & EA/l & -EA/l & 0 \\ 0 & -EA/l & EA/l & 0 \\ 0 & 0 & 0 & 0 \end{bmatrix} \begin{Bmatrix} U_1 \\ U_2 \\ U_3 \\ U_4 \end{Bmatrix} = \begin{Bmatrix} 0 \\ Q_2^{(2)} \\ Q_3^{(2)} \\ 0 \end{Bmatrix} + \begin{Bmatrix} 0 \\ P_2^{(2)} \\ P_3^{(2)} \\ 0 \end{Bmatrix}$$

element 3
$$\begin{bmatrix} 0 & 0 & 0 & 0 \\ 0 & 0 & 0 & 0 \\ 0 & 0 & EA/l & -EA/l \\ 0 & 0 & -EA/l & EA/l \end{bmatrix} \begin{Bmatrix} U_1 \\ U_2 \\ U_3 \\ U_4 \end{Bmatrix} = \begin{Bmatrix} 0 \\ 0 \\ Q_3^{(3)} \\ Q_4^{(3)} \end{Bmatrix} + \begin{Bmatrix} 0 \\ 0 \\ P_3^{(3)} \\ P_4^{(3)} \end{Bmatrix}$$

in Eq. (9.38) that are shown in Fig. 9.6. When we add these three elemental contributions, we arrive at

$$\begin{bmatrix} EA/l & -EA/l & 0 & 0 \\ -EA/l & 2EA/l & -EA/l & 0 \\ 0 & -EA/l & 2EA/l & -EA/l \\ 0 & 0 & -EA/l & EA/l \end{bmatrix} \begin{Bmatrix} U_1 \\ U_2 \\ U_3 \\ U_4 \end{Bmatrix} = \begin{Bmatrix} Q_1^{(1)} \\ Q_2^{(1)} + Q_2^{(2)} \\ Q_3^{(2)} + Q_3^{(3)} \\ Q_4^{(3)} \end{Bmatrix} + \begin{Bmatrix} R_1 \\ P_2^* \\ P_3^* \\ R_2 \end{Bmatrix} \quad (9.39)$$

which shows the diagonal matrix stiffness terms of the individual elements adding when they overlap in the system stiffness matrix of Eq. (9.38). Although we have used three elements to illustrate this process, it is the same regardless of the number of elements present. For seven elements, each of length l, for example, we would find the system stiffness matrix

$$\frac{EA}{l} \begin{bmatrix} 1 & -1 & 0 & 0 & 0 & 0 & 0 & 0 \\ -1 & 2 & -1 & 0 & 0 & 0 & 0 & 0 \\ 0 & -1 & 2 & -1 & 0 & 0 & 0 & 0 \\ 0 & 0 & -1 & 2 & -1 & 0 & 0 & 0 \\ 0 & 0 & 0 & -1 & 2 & -1 & 0 & 0 \\ 0 & 0 & 0 & 0 & -1 & 2 & -1 & 0 \\ 0 & 0 & 0 & 0 & 0 & -1 & 2 & -1 \\ 0 & 0 & 0 & 0 & 0 & 0 & -1 & 1 \end{bmatrix}$$

which shows that for a large number of elements, the system stiffness matrix is sparsely populated with nonzero values only on or near the matrix diagonal. When solving large systems of linear equations defined by such a matrix, we can take advantage of this behavior to make the storage of the matrix and the solution process more efficient.

Example 9.1 Statically Determinate Bar – Stiffness Method

We have shown how to generate a system of linear equations using finite elements. To see how to solve such a system, consider the problem shown in Fig. 9.7a where a uniform bar (A and E are constants) carries both a uniform distributed force q_0 over the middle third of its length and an end force $P = q_0 L$. The bar is rigidly fixed to the wall at $x = 0$. This is a statically determinate problem that we can easily solve analytically. The solution is

$$\frac{AE}{q_0 l^2} u_x(x) = \begin{cases} 2\left(\frac{x}{l}\right) & \left(0 \le \frac{x}{l} \le 1\right) \\ 3\left(\frac{x}{l}\right) - \frac{1}{2}\left(\frac{x}{l}\right)^2 - \frac{1}{2} & \left(1 \le \frac{x}{l} \le 2\right) \\ \left(\frac{x}{l}\right) + \frac{3}{2} & \left(2 \le \frac{x}{l} \le 3\right) \end{cases} \quad (9.40)$$

To solve this problem with finite elements, we will use three elements (see Fig. 9.7b) so that we can use Eq. (9.39). In this case we have the system of equations

$$\frac{EA}{l} \begin{bmatrix} 1 & -1 & 0 & 0 \\ -1 & 2 & -1 & 0 \\ 0 & -1 & 2 & -1 \\ 0 & 0 & -1 & 1 \end{bmatrix} \begin{Bmatrix} U_1 \\ U_2 \\ U_3 \\ U_4 \end{Bmatrix} = \begin{Bmatrix} 0 \\ q_0 l/2 \\ q_0 l/2 \\ 0 \end{Bmatrix} + \begin{Bmatrix} R_1 \\ 0 \\ 0 \\ q_0 l \end{Bmatrix} \quad (9.41)$$

To correspond to the form of the exact solution let us define normalized nodal displacements $\tilde{U}_j = \frac{AE}{q_0 l^2} U_j$ so that Eq. (9.41) becomes simply

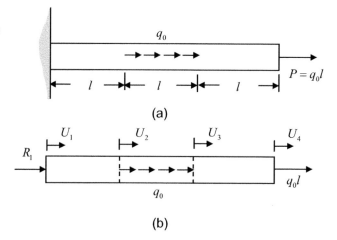

Figure 9.7 (a) A bar loaded by a uniform distributed force/unit length, q_0, over the middle third of its length and a concentrated end force $P = q_0 l$. (b) A finite element model of the bar using three elements showing the nodal displacements.

9.2 Stiffness-Based Finite Elements for Axial-Load Problems

$$\begin{bmatrix} 1 & -1 & 0 & 0 \\ -1 & 2 & -1 & 0 \\ 0 & -1 & 2 & -1 \\ 0 & 0 & -1 & 1 \end{bmatrix} \begin{Bmatrix} \tilde{U}_1 \\ \tilde{U}_2 \\ \tilde{U}_3 \\ \tilde{U}_4 \end{Bmatrix} = \begin{Bmatrix} R_1/q_0 l \\ 1/2 \\ 1/2 \\ 1 \end{Bmatrix} \qquad (9.42)$$

Because the problem is statically determinate, applying equilibrium to the entire bar gives $R_1/q_0 l = -2$, so it is tempting to place this value into Eq. (9.42) and try to solve for the displacements. However, this will fail since the determinant of the stiffness matrix is zero so the matrix does not have an inverse. For example, if we evaluate the determinant in MATLAB® we find:

```
Kg = [1 -1 0 0; -1 2 -1 0; 0 -1 2 -1; 0 0 -1 1]
Kg = 1    -1    0    0
    -1     2   -1    0
     0    -1    2   -1
     0     0   -1    1
det(Kg)
ans = 0
```

We cannot solve for the displacements from Eq. (9.42) because we have not used the boundary condition of zero displacement at $x = 0$. Without this condition, the bar is free to have an arbitrary rigid body displacement and we cannot hope to uniquely solve for the displacements. We can also see this in MATLAB® by letting the nodal displacements all have some arbitrary constant value and multiplying the system stiffness matrix by those constants:

```
u = [2; 2; 2; 2];
Kg*u
ans =  0
       0
       0
       0
```

These constant displacements produce zero forces so they could always be added to the actual solution without affecting the conditions of equilibrium. We can satisfy the boundary condition of zero displacement at the wall, i.e., $\tilde{U}_1 = 0$, by simply eliminating the first row and column of the stiffness matrix in Eq. (9.42). The remaining equations are then

$$\begin{bmatrix} 2 & -1 & 0 \\ -1 & 2 & -1 \\ 0 & -1 & 1 \end{bmatrix} \begin{Bmatrix} \tilde{U}_2 \\ \tilde{U}_3 \\ \tilde{U}_4 \end{Bmatrix} = \begin{Bmatrix} 1/2 \\ 1/2 \\ 1 \end{Bmatrix} \qquad (9.43)$$

which we can solve for the remaining displacements. In MATLAB®, for example, we can obtain the solution as follows:

```
K = Kg;                    % define a copy of the system stiffness
% matrix that we will modify
K(1,:) = [ ]               % eliminate the first row
K =   -1     2    -1     0
       0    -1     2    -1
       0     0    -1     1

K(:,1) = [ ]               % eliminate the first column
K =    2    -1     0
      -1     2    -1
       0    -1     1
P= [ 1/2  1/2  1]'         % define a force column vector { P}
P = 0.5000
    0.5000
    1.0000
u = K\P                    % solve the system of equations [ K]{ u} = { P}
u =    2.0000
       3.5000
       4.5000
```

so that the normalized displacements are $\tilde{U}_2 = 2$, $\tilde{U}_3 = 3.5$, $\tilde{U}_4 = 4.5$. If we multiply the original system stiffness matrix by these displacements together with the displacement $\tilde{U}_1 = 0$, we can obtain the normalized reaction force at $x = 0$ (as well as the original nodal forces):

```
ug = [ 0 ; u]              % make a column vector of all the nodal
% displacements
ug =       0
        2.0000
        3.5000
        4.5000
Kg*ug                      % multiply these displacements by the system
% stiffness matrix
ans =  -2.0000
        0.5000
        0.5000
        1.0000
```

which indeed gives the correct normalized reaction force. The exact displacement of the bar together with the finite element solution are plotted in Fig. 9.8. It can be seen that the linear displacement approximation for each element captures the solution quite well for this problem even with only three elements. This was possible since the exact solution of Eq. (9.40) has linear behavior in the first and third elements.

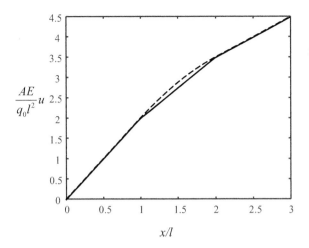

Figure 9.8 The displacement of the bar shown in Fig. 9.7a. Dashed line – exact solution, solid line – finite element solution.

To obtain the internal axial forces and stresses in the element in the general case we have

$$F_x(\xi) = E(\xi)A(\xi)e_{xx}(\xi) = E(\xi)A(\xi)\sum_{m=j}^{j+1} B_m(\xi)\, U_m$$

$$\sigma_{xx}(\xi) = E(\xi)e_{xx}(\xi) = E(\xi)\sum_{m=j}^{j+1} B_m(\xi)\, U_m$$
(9.44)

but where in our previous example all these terms are constants. We will leave the determination of the stresses to the reader.

In cases where the body forces are zero so that all the applied loads are due only to concentrated forces and the product EA is piecewise constant, the elements can be chosen such that the exact internal forces are constants in the elements and hence the strains will be constants (as assumed in our model with a linear shape functions for the displacement). Thus, a stiffness-based finite element model can reproduce an exact solution for problems such as the stepped bar shown in Fig. 9.9a. [Note: to be more precise, by "exact" we mean the solution will be identical to a solution that satisfies the strength of materials governing equations. However, remember those equations themselves are approximations to the underlying equations of elasticity.] As shown in Chapter 8 (Section 8.4), the torsion of circular shafts satisfy the same basic equations as axial loads if we make the replacements $F_x \rightarrow M_x$, $EA \rightarrow GJ$, $u \rightarrow \phi$, $q_x \rightarrow t_x$, where M_x is the internal axial torque, GJ is the shear modulus times the polar area moment of the cross-section, ϕ is the angle of twist, and t_x is the distributed torque/unit length. Thus, for $t_x = 0$ and when GJ has piecewise constant values for the shaft, using linear shape functions for the twist can also produce exact solutions. An example is shown in Fig. 9.9b. We will show in Section 9.6 that stiffness-based finite elements can also produce exact solutions for beams containing only concentrated forces (or moments) and where EI is piecewise constant such as the beam of Fig. 9.9c. All the examples shown in Fig. 9.9 are statically determinate problems, but this is not necessary and similar statically indeterminate structures also can have exact stiffness-based finite element solutions. Even more complex structures may have exact finite element solutions. A 2-D truss is a set of axial-load members (whose own weights are neglected) connected by pins and where there are only concentrated forces applied at those pins (see Fig. 9.10a). The internal force in each truss member is a constant so that each member can be treated as a single constant-strain element in a stiffness-based finite element solution. Similarly, a frame structure consisting of beam and truss parts and carrying only concentrated forces (see, for example, Fig. 9.10b) can be decomposed into finite elements and solved exactly. We have pointed out these examples because, although the real power of the stiffness-based finite element method is to find approximate numerical solutions to very complex 3-D geometries and loading situations, the method can also find exact solutions to many of the problems considered in strength of materials, a fact that may not be appreciated. There may also be some instances when exact solutions can be produced in cases with distributed loads.

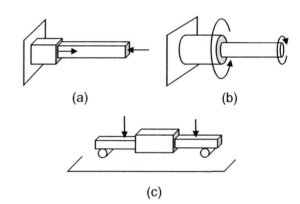

Figure 9.9 Cases where a stiffness-based finite element solution can be formulated to produce an "exact" solution. (a) An axial-load problem. (b) A torsion problem. (c) A beam problem.

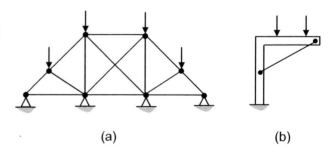

Figure 9.10 (a) A truss structure. (b) A frame structure. Both of these structures can be solved exactly by the stiffness-based finite element method.

Example 9.2 Statically Indeterminate Bar – Stiffness Method

Consider now the statically indeterminate problem of Fig. 9.11a, where a uniform bar is fixed at both ends and subjected to a uniform distributed force/unit length, q_0, over its entire length. We will use three constant-strain elements to solve this problem (see Fig. 9.11b) so that we can use Eq. (9.39), which in this case becomes

$$\frac{EA}{l}\begin{bmatrix} 1 & -1 & 0 & 0 \\ -1 & 2 & -1 & 0 \\ 0 & -1 & 2 & -1 \\ 0 & 0 & -1 & 1 \end{bmatrix}\begin{Bmatrix} U_1 \\ U_2 \\ U_3 \\ U_4 \end{Bmatrix} = \begin{Bmatrix} q_0 l/2 \\ q_0 l \\ q_0 l \\ q_0 l/2 \end{Bmatrix} + \begin{Bmatrix} R_1 \\ 0 \\ 0 \\ R_2 \end{Bmatrix} \quad (9.45)$$

which in terms of the normalized displacement $\tilde{U}_j = (EA/q_0 l^2) U_j$ becomes

$$\begin{bmatrix} 1 & -1 & 0 & 0 \\ -1 & 2 & -1 & 0 \\ 0 & -1 & 2 & -1 \\ 0 & 0 & -1 & 1 \end{bmatrix}\begin{Bmatrix} \tilde{U}_1 \\ \tilde{U}_2 \\ \tilde{U}_3 \\ \tilde{U}_4 \end{Bmatrix} = \begin{Bmatrix} 1/2 \\ 1 \\ 1 \\ 1/2 \end{Bmatrix} + \begin{Bmatrix} R_1/q_0 l \\ 0 \\ 0 \\ R_2/q_0 l \end{Bmatrix} = \begin{Bmatrix} R_1/q_0 l + 1/2 \\ 1 \\ 1 \\ R_2/q_0 l + 1/2 \end{Bmatrix} \quad (9.46)$$

To solve these equations, we must apply the displacement boundary conditions $\tilde{U}_1 = \tilde{U}_4 = 0$ by eliminating the first and fourth rows and columns of the stiffness matrix to arrive at

$$\begin{bmatrix} 2 & -1 \\ -1 & 2 \end{bmatrix}\begin{Bmatrix} \tilde{U}_2 \\ \tilde{U}_3 \end{Bmatrix} = \begin{Bmatrix} 1 \\ 1 \end{Bmatrix} \quad (9.47)$$

which has the solution $\tilde{U}_2 = \tilde{U}_3 = 1$ or, equivalently, $U_2 = U_3 = q_0 L^2/9EA$, where L is the total length of the bar. Multiplying the original stiffness matrix by these known nodal displacements and the boundary values $\tilde{U}_1 = \tilde{U}_4 = 0$ we then obtain $R_j/q_0 l + 1/2 = -1 \; (j = 1, 2)$. These steps in MATLAB® are:

```
Kg = [ 1 -1 0 0 ; -1 2 -1 0; 0 -1 2 -1; 0 0 -1 1];
% stiffness matrix
K = Kg;                    % copy of the stiffness matrix
% which we will modify

K(4,:) = [ ];  K(:,4) = [ ];  % delete largest column and row
% first so that when we apply the next boundary condition
K(1,:) = [ ];  K(:,1) = [ ];  % its row and column number is
% not changed
P= [ 1; 1];                % define a force column vector { P}
u =K\P         % solve the system of equations  [K]{u} = {P}
u = 1.0000
    1.0000
ug = [ 0; 1; 1; 0];        % define a column vector of all
% the nodal displacements
```

Figure 9.11 (a) A uniform distributed force/unit length, q_0, acting on a uniform bar that is fixed at both ends. (b) A finite element model for the bar using three elements and showing the nodal displacements U_j ($j = 1, 2, 3, 4$) and end reactions R_j ($j = 1, 2$).

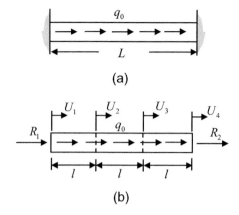

Figure 9.12 Displacement of the bar for the axial-load problem shown in Fig. 9.11a. Exact solution – dashed line, constant-strain finite element solution for three elements – solid line.

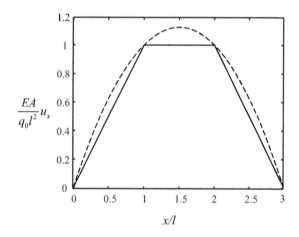

```
Kg*ug                          % multiply these displacements by
% the original stiffness matrix
ans =    -1
          1
          1
         -1
```

Solving for the reaction forces explicitly, we find $R_1 = R_2 = -3q_0 l/2 = -q_0 L/2$. These reaction forces obviously are in equilibrium with the total force, $q_0 L$, generated on the bar by the distributed load. We can compare these results with the exact solution for the displacement and stress, which are given by

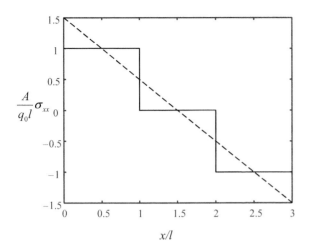

Figure 9.13 Stress in the bar for the axial-load problem shown in Fig. 9.11a. Exact solution – dashed line, constant-strain finite element solution for three elements – solid line.

$$\frac{EA}{q_0 l^2} u_x(x) = \frac{3}{2}\frac{x}{l} - \frac{1}{2}\left(\frac{x}{l}\right)^2$$

$$\frac{EA}{q_0 l^2} u_x(l) = \tilde{U}_2 = 1$$

$$\frac{EA}{q_0 l^2} u_x(2l) = \tilde{U}_3 = 1 \qquad (9.48)$$

$$\frac{A}{q_0 l}\sigma_{xx}(x) = \frac{3}{2} - \frac{x}{l}$$

showing that the nodal displacement values obtained are in fact exact. Figure 9.12 shows both the exact and finite element solutions for the displacements while Fig. 9.13 compares the stresses. Typically, the stiffness-based finite element approximation for the displacements is better than the approximation for stresses. Obviously, it will require many more constant strain elements than used here to obtain an adequate approximation for the stress.

9.3 Force-Based Finite Elements for Axial-Load Problems

The stiffness-based finite element method has been highly developed and is widely used. However, there is a stress-based or force-based finite element approach that is an important alternative method we will consider in this section. Again, we will use axial-load problems in order to discuss the details. This method will employ both the principle of virtual work and the principle of complementary virtual work. We will write the virtual work principle (see Eq. (9.8), for example) for a bar element as

$$\int_0^l F_x(\xi)\delta(\partial u_x(\xi)/\partial x)d\xi = \int_0^l q_x(\xi)\delta u_x(\xi)d\xi + P_{j+1}\delta U_{j+1} + P_j\delta U_j \quad (9.49)$$

and the principle of complementary virtual work as (see Eq. (9.16))

$$\int_0^l \frac{1}{E(\xi)A(\xi)} F_x(\xi)\delta F_x(\xi)d\xi = U_{j+1}\delta F_x(l) - U_j\delta F_x(0) \quad (9.50)$$
$$= U_{j+1}\delta P_{j+1} + U_j\delta P_j$$

Also, recall in deriving this form of the principle of complementary virtual work, the virtual variation of the internal force was required to satisfy the homogeneous equilibrium equation, Eq. (9.15), i.e.,

$$\frac{d}{d\xi}(\delta F_x) = 0 \quad (9.51)$$

so that any approximation we choose for this variation must also satisfy Eq. (9.51). The exact internal force itself satisfies

$$\frac{dF_x(\xi)}{d\xi} = -q_x(\xi) \quad (9.52)$$

Both virtual work principles have been written in terms of the internal force in the element and the displacement only. Although we will call this a force-based method since the internal forces will be the primary quantities we will solve for, Eq. (9.49) and Eq. (9.50) show that the method will also need to consider the displacements. In order to link the displacements and forces, strains and stress–strain relations will also be involved. We can satisfy Eq. (9.51) if we assume that the approximation used for the internal force itself also satisfies the homogeneous equilibrium equation. For this problem we need only assume that the internal force is a constant in each element, i.e., $F_x(\xi) = F_j$, since a constant force satisfies the homogeneous equilibrium equation identically. For the displacement we will assume the same linear variation used in the stiffness-based method so the displacement and strain will again be

$$u_x(\xi) = [H]\{U\}$$
$$e_{xx}(\xi) = [B]\{U\} \quad (9.53)$$

where the shape functions are again given by Eq. (9.19) and Eq. (9.24).

In explaining the force-based method, we will want to use forms that will correspond to the more general problems we will discuss later. In this axial-load case, the force, displacement, and strain in the bar are all scalars, whereas they will be vectors in more general problems. Similarly, while we have simply a scalar relationship between internal strain and force,

$e_{xx}(\xi) = F_x(\xi)/E(\xi)A(\xi)$, in the more general case we will have a matrix relating the vector strains and displacements. In addition, the scalar distributed load in the general case will become a vector. Thus, in all our following discussions we will use the following expressions:

$$\begin{aligned}
\{F_x(\xi)\} &\equiv F_x(\xi) = [N(\xi)]\{F\} \\
\{u_x(\xi)\} &\equiv u_x(\xi) = [H(\xi)]\{U\} \\
\{e_{xx}(\xi)\} &\equiv e_{xx}(\xi) = [B(\xi)]\{U\} \\
\{q_x(\xi)\} &\equiv q_x(\xi) \\
\{e_{xx}(\xi)\} &= [1/EA]\{F_x(\xi)\}
\end{aligned} \quad (9.54)$$

where $\{F\}$ is a vector of force coefficients while $\{U\}$ is a vector of nodal displacements. We will use these matrix–vector expressions with the understanding that, in the axial-load case with the choice of linear and constant shape functions for the displacement and force, we will evaluate these terms explicitly as

$$\begin{aligned}
[N(\xi)] &= 1 \\
\{F\} &= F_j \\
[H(\xi)] &= [1 - \xi/l \quad \xi/l] \\
[B(\xi)] &= [-1/l \quad 1/l] \\
[1/EA] &= 1/E(\xi)A(\xi)
\end{aligned} \quad (9.55)$$

The notation of Eq. (9.54) will then carry over to more general problems with other choices of shape functions and will also allow us to see more clearly the general structure of the terms involved in a problem. In more general cases, the requirement that the variation of the force expressions satisfy homogeneous equilibrium equations will be satisfied (as done here in the axial-load problems) by choosing a matrix of force shape functions, $[N(\xi)]$, that satisfy the homogeneous equations of those more general cases.

[Note: the vector of forces, $\{F\}$, that define the internal forces, unlike the displacements, $\{U\}$, are not necessarily values defined at nodes, so our previous convention that subscripts define nodal values is only true for the displacements. For the axial-load case, for example, there is only one internal force, F_j, for the (j)th element (see Fig. 9.14) and two nodal displacements (U_j, U_{j+1}). In other problems there can be more than one of these internal forces (or stresses) and more than two displacements at a node. The number, K, of F_k values for an element can be considered to be the number of assumed force degrees of freedom for that element, just as the number, M, of U_m values is the number of nodal displacement degrees

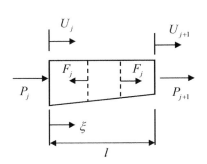

Figure 9.14 An element showing the internal force, F_j, acting in the element.

of freedom. For the axial-load case we are considering, $K = 1$, $M = 2$.]

In these generalized matrix–vector forms, then, the principles of virtual work and complementary virtual work for an element can be written as

$$\int_0^l \{\delta e_{xx}(\xi)\}^T \{F_x(\xi)\} d\xi = \int_0^l \{\delta u_x(\xi)\}^T \{q_x(\xi)\} d\xi + \{\delta U\}^T \{P\}$$

$$\int_0^l \{\delta F_x(\xi)\}^T [1/EA] \{F_x(\xi)\} d\xi = \{\delta P\}^T \{U\}$$

(9.56)

where explicitly in the axial-load case we have

$$\{U\} = \begin{Bmatrix} U_j \\ U_{j+1} \end{Bmatrix}, \quad \{P\} = \begin{Bmatrix} P_j \\ P_{j+1} \end{Bmatrix}$$

$$\{\delta U\} = \begin{Bmatrix} \delta U_j \\ \delta U_{j+1} \end{Bmatrix}, \quad \{\delta P\} = \begin{Bmatrix} \delta P_j \\ \delta P_{j+1} \end{Bmatrix}$$

(9.57)

Placing the matrix–vector expressions of Eq. (9.54) into Eq. (9.56) then gives

$$\{\delta U\}^T \int_0^l [B]^T [N] d\xi \{F\} = \{\delta U\}^T \int_0^l [H]^T \{q_x\} d\xi + \{\delta U\}^T \{P\}$$

(9.58)

and

$$\{\delta F\}^T \int_0^l [N]^T [1/EA][N] d\xi \{F\} = \{\delta P\}^T \{U\}$$

(9.59)

Consider first the virtual work principle, Eq. (9.58). Since the $\{\delta U\}^T$ in Eq. (9.58) is arbitrary, we must have the force vector $\{F\}$ satisfy the equilibrium equations

$$[\tilde{E}]\{F\} = \{Q\} + \{P\}$$

(9.60)

where the equilibrium matrix, $[\tilde{E}]$ for the element is defined as

$$[\tilde{E}] = \int_0^l [B]^T [N] d\xi$$

(9.61)

and the distributed force term, which is the same as in the stiffness-based method, is

$$\{Q\} = \int_0^l [H(\xi)]^T \{q_x(\xi)\} d\xi$$

(9.62)

If we use the specific values for our axial-load problem, the equilibrium matrix becomes

$$[\tilde{E}] = \begin{bmatrix} -1 \\ 1 \end{bmatrix} \quad (9.63)$$

and the equilibrium equation for the element, Eq. (9.60), is

$$\begin{bmatrix} -1 \\ 1 \end{bmatrix} F_j = \begin{Bmatrix} Q_j \\ Q_{j+1} \end{Bmatrix} + \begin{Bmatrix} P_j \\ P_{j+1} \end{Bmatrix} \quad (9.64)$$

where (recall Eq. (9.34b)) for q_x given in terms of a global x-coordinate

$$\{Q\} = \begin{Bmatrix} Q_j \\ Q_{j+1} \end{Bmatrix} = \begin{Bmatrix} \int_0^l q_x(x_j + \xi)(1 - \xi/l)d\xi \\ \int_0^l q_x(x_j + \xi)(\xi/l)d\xi \end{Bmatrix} \quad (9.65)$$

Now, consider the complementary virtual work principle, Eq. (9.59). The virtual variations of the internal force must satisfy the homogeneous equilibrium equations (no distributed force term) so from Eq. (9.60) we must have

$$[\tilde{E}]\{\delta F\} = \{\delta P\} \quad (9.66)$$

Placing the transpose of this relationship into Eq. (9.59) and using the fact that the resulting equation must be satisfied for all variations $\{\delta F\}^T$, we obtain

$$[G]\{F\} = [\tilde{E}]^T \{U\} \quad (9.67)$$

in terms of a flexibility matrix, $[G]$, given by

$$[G] = \int_0^l [N]^T [1/EA][N] d\xi \quad (9.68)$$

If we define a set of discrete generalized strains as $\{\Delta\} = [G]\{F\}$, then we see the virtual work and complementary virtual work principles yield

$$\begin{aligned} [\tilde{E}]\{F\} &= \{Q\} + \{P\} \\ \{\Delta\} &= [G]\{F\} = [\tilde{E}]^T \{U\} \end{aligned} \quad (9.69)$$

which are just the discrete equilibrium and strain–displacement relations for an element in the force-based finite element method. Explicitly, for the axial-load case and a uniform element, where E and A are constants, we have

$$\begin{bmatrix} -1 \\ 1 \end{bmatrix} F_j = \begin{Bmatrix} Q_j \\ Q_{j+1} \end{Bmatrix} + \begin{Bmatrix} P_j \\ P_{j+1} \end{Bmatrix}$$

$$\Delta_j = \left(\frac{l}{EA}\right) F_j = \begin{bmatrix} -1 & 1 \end{bmatrix} \begin{Bmatrix} U_j \\ U_{j+1} \end{Bmatrix} \tag{9.70}$$

where the generalized strain in the bar, Δ_j, is just the elongation of the element.

Having these results for a single element, we now want to assemble a set of elements and solve axial-load problems. For the equilibrium equations we find, using our previous example of three elements and again letting the superscript (j) denote the (j)th element:

$$\begin{bmatrix} -1 & 0 & 0 \\ 1 & -1 & 0 \\ 0 & 1 & -1 \\ 0 & 0 & 1 \end{bmatrix} \begin{Bmatrix} F_1 \\ F_2 \\ F_3 \end{Bmatrix} = \begin{Bmatrix} Q_1^{(1)} \\ Q_2^{(1)} + Q_1^{(2)} \\ Q_2^{(2)} + Q_1^{(3)} \\ Q_2^{(3)} \end{Bmatrix} + \begin{Bmatrix} P_1^{(1)} \\ P_2^{(1)} + P_1^{(2)} \\ P_2^{(2)} + P_1^{(3)} \\ P_2^{(3)} \end{Bmatrix} = \begin{Bmatrix} Q_1^{(1)} \\ Q_2^{(1)} + Q_1^{(2)} \\ Q_2^{(2)} + Q_1^{(3)} \\ Q_2^{(3)} \end{Bmatrix} + \begin{Bmatrix} R_1 \\ P_2^* \\ P_3^* \\ R_2 \end{Bmatrix} \tag{9.71}$$

where (R_1, R_2) are again the end forces acting on the bar and P_j^* are known concentrated forces acting at the inner nodes. In this case the assembly process is very simple since we are just adding columns, suitably displaced, to the complete equilibrium matrix. The generalized strain–displacement equations similarly become

$$\begin{Bmatrix} \Delta_1 \\ \Delta_2 \\ \Delta_3 \end{Bmatrix} = \begin{bmatrix} -1 & 1 & 0 & 0 \\ 0 & -1 & 1 & 0 \\ 0 & 0 & -1 & 1 \end{bmatrix} \begin{Bmatrix} U_1 \\ U_2 \\ U_3 \\ U_4 \end{Bmatrix} \tag{9.72}$$

where now we are adding rows to form the matrix of Eq. (9.72), which is just the transpose of the complete equilibrium matrix for the entire bar. In matrix–vector form we have the same form as Eq. (9.69), which was for a single element, namely

$$[\tilde{E}]\{F\} = \{P\}$$
$$\{\Delta\} = [\tilde{E}]^T \{U\} \tag{9.73}$$

where now $[\tilde{E}]$ is the complete equilibrium matrix for the body.

Example 9.3 Statically Determinate Bar – Force Method

As a very simple example of the use of the force-based method for a statically determinate problem, consider the bar shown in Fig. 9.15a. In this case we only need the equilibrium equations to solve for the forces. Breaking the bar up into three elements of length l each, we can use the assembled system of Eq. (9.71) for this case and obtain

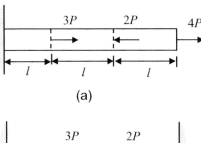

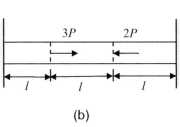

Figure 9.15 (a) A statically determinate uniform bar carrying a set of concentrated forces. (b) A statically indeterminate uniform bar carrying a set of concentrated forces.

$$\begin{bmatrix} -1 & 0 & 0 \\ 1 & -1 & 0 \\ 0 & 1 & -1 \\ 0 & 0 & 1 \end{bmatrix} \begin{Bmatrix} F_1 \\ F_2 \\ F_3 \end{Bmatrix} = \begin{Bmatrix} R_1 \\ 3P \\ -2P \\ 4P \end{Bmatrix} \quad (9.74)$$

where R_1 is the reaction at the fixed support. The internal forces in the bars themselves must be in equilibrium with the known applied forces so we can delete the first equation in Eq. (9.74) and solve the remaining system:

$$\begin{bmatrix} 1 & -1 & 0 \\ 0 & 1 & -1 \\ 0 & 0 & 1 \end{bmatrix} \begin{Bmatrix} F_1 \\ F_2 \\ F_3 \end{Bmatrix} = \begin{Bmatrix} 3P \\ -2P \\ 4P \end{Bmatrix} \quad (9.75)$$

We can symbolically solve this system easily in MATLAB®:

```
E = [-1 0 0;1 -1 0;0 1 -1; 0 0 1];    % define the full
% equilibrium matrix
Er = E;
Er(1,:) = [ ];   % eliminate the first row in E to obtain a
% reduced matrix

syms  P
Pr = [ 3*P;-2*P; 4*P];     % the symbolic force vector for the
% reduced system
F = inv(Er)*Pr             % solve for the internal forces (can
% also use F = Er\Pr)

F = 5*P
    2*P
    4*P
```

So we see $F_1 = 5P, F_2 = 2P, F_3 = 4P$. If we multiply the original equilibrium matrix by these internal forces, we find

```
E*F   % obtain reaction force and original loads
ans =    -5*P
          3*P
         -2*P
          4*P
```

Figure 9.16 Internal force distribution for the bar of Fig. 9.15a.

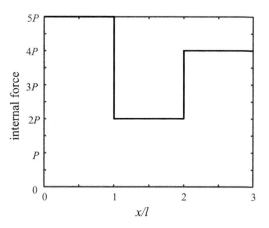

which shows that $R_1 = -5P$. We can also obtain this same value for the reaction force directly from equilibrium of the entire bar since this is a statically determinate problem. Figure 9.16 shows the internal force distribution in the bar. We also see that there are downward jumps in the internal force at the locations of the applied forces that are equal to the value of those forces in the x-direction. This is simply a consequence of applying equilibrium in a small region about those concentrated forces, as is often discussed in elementary strength of materials courses and which we have also used previously in the stiffness-based approach to relate the behavior of the internal forces to the applied concentrated forces (see Fig. 9.5c and Eq. (9.37)).

Statically determinate bar problems are, of course, very simple and can be solved directly by application of the equilibrium equations. Statically indeterminate bar problems are also simple but require more than equilibrium equations for their solution, as we will see in the next example.

Example 9.4 Statically Indeterminate Bar – Force Method and Force-Based Stiffness Method

To see the application of a force-based finite element approach to a statically indeterminate problem, consider the bar shown in Fig. 9.15b. Since we again have three elements, our equilibrium equations become

$$\begin{bmatrix} -1 & 0 & 0 \\ 1 & -1 & 0 \\ 0 & 1 & -1 \\ 0 & 0 & 1 \end{bmatrix} \begin{Bmatrix} F_1 \\ F_2 \\ F_3 \end{Bmatrix} = \begin{Bmatrix} R_1 \\ 3P \\ -2P \\ R_2 \end{Bmatrix} \qquad (9.76)$$

where there are now two unknown reaction forces (R_1, R_2). There are a total of four equations and five unknowns, and the problem is statically indeterminate. Even if we eliminate the two equations for the unknown end reactions, we have two equations in three unknowns:

$$\begin{bmatrix} 1 & -1 & 0 \\ 0 & 1 & -1 \end{bmatrix} \begin{Bmatrix} F_1 \\ F_2 \\ F_3 \end{Bmatrix} = \begin{Bmatrix} 3P \\ -2P \end{Bmatrix} \quad (9.77)$$

which we will write as $[\tilde{E}_f]\{F\} = \{P\}$ to indicate that we are working with a reduced "free" equilibrium matrix associated with free nodes, i.e., those without constraint forces. Thus, regardless of the unknown forces we choose, there is always one more unknown force than equilibrium equations, so we say the problem is indeterminate of degree one, requiring one more equation in order to solve. The additional equation we need we can get from the strain–displacement relation, $\{\Delta\} = [\tilde{E}]^T \{U\}$. which in this case is again (see Eq. (9.72)):

$$\begin{Bmatrix} \Delta_1 \\ \Delta_2 \\ \Delta_3 \end{Bmatrix} = \begin{bmatrix} -1 & 1 & 0 & 0 \\ 0 & -1 & 1 & 0 \\ 0 & 0 & -1 & 1 \end{bmatrix} \begin{Bmatrix} U_1 \\ U_2 \\ U_3 \\ U_4 \end{Bmatrix} \quad (9.78)$$

However, before we can use this relationship, we must apply the displacement boundary conditions $U_1 = U_4 = 0$, which we can do by simply eliminating the first and fourth rows of the matrix in Eq. (9.78), to obtain

$$\begin{Bmatrix} \Delta_1 \\ \Delta_2 \\ \Delta_3 \end{Bmatrix} = \begin{bmatrix} 1 & 0 \\ -1 & 1 \\ 0 & -1 \end{bmatrix} \begin{Bmatrix} U_2 \\ U_3 \end{Bmatrix} \quad (9.79)$$

which is still in the form $\{\Delta\} = [\tilde{E}_f]^T \{U_f\}$ but where now the unknown reaction forces have been eliminated from the equilibrium equations and the corresponding specified displacements have been eliminated from the strain–displacement relations so that only the free displacements are present. In our example we have the three equations

$$\begin{aligned} \Delta_1 &= U_2 \\ \Delta_2 &= U_3 - U_2 \\ \Delta_3 &= -U_3 \end{aligned} \quad (9.80)$$

These are nothing more than the relationships between the elongations of the elements and their end displacements when the displacement boundary conditions are satisfied, a result we could have written down by inspection. Since there are three elongations written in terms of only two displacements, there must be one compatibility equation that relates those elongations. We can use this compatibility equation to find the additional force equation(s) we need in a statically indeterminate problem in the following fashion. Because there are fewer free displacements than strains, we can always take combinations of the strain–displacement relations and eliminate the displacements to arrive at a set of compatibility equations, which we write as

$$[\tilde{S}]\{\Delta\} = 0 \quad (9.81)$$

where $[\tilde{S}]$ is a compatibility matrix relating the generalized strains $\{\Delta\}$. Applying this compatibility matrix to the strain–displacement relations $\{\Delta\} = [\tilde{E}_f]^T \{U_f\}$ we have

$$[\tilde{S}][\tilde{E}_f]^T \{U_f\} = 0 \tag{9.82}$$

but here $\{U_f\}$ are all free (unconstrained) displacements so that the only way we can satisfy Eq. (9.82) is if

$$[\tilde{S}][\tilde{E}_f]^T = [\tilde{E}_f][\tilde{S}]^T = 0 \tag{9.83}$$

where we have taken the transpose to obtain the second relation seen in Eq. (9.83). Later, we will see how Eq. (9.83) is part of the determination of the compatibility matrix in general. In the present case being considered, we can obtain the compatibility matrix by direct elimination. For example, simply by adding the three equations in Eq. (9.80) we find the compatibility equation

$$\Delta_1 + \Delta_2 + \Delta_3 = 0 \tag{9.84}$$

which is our compatibility equation $[\tilde{S}]\{\Delta\} = 0$, where here

$$[\tilde{S}] = [1 \quad 1 \quad 1] \tag{9.85}$$

It is easy to verify that

$$[\tilde{S}][\tilde{E}_f]^T = [1 \quad 1 \quad 1] \begin{bmatrix} 1 & 0 \\ -1 & 1 \\ 0 & -1 \end{bmatrix} = [0 \quad 0] \tag{9.86}$$

Physically, the compatibility equation, Eq. (9.84), says that the total elongation of the bar is zero, a result we could have written down by inspection.

To turn the compatibility equation into the additional equation(s) we need to solve the problem, we relate the unknown forces to the generalized strains through the flexibility matrices. Recall, for each element we had $\{\Delta^{(j)}\} = [G^{(j)}]\{F^{(j)}\}$ so we can combine these elements into a similar relationship for the entire system as

$$\{\Delta\} = [G]\{F\} \tag{9.87}$$

where, for example, for M elements we have

$$[G] = \begin{bmatrix} [G^{(1)}] & \cdots & 0 & 0 \\ \vdots & [G^{(2)}] & & 0 \\ 0 & & \ddots & 0 \\ 0 & 0 & \cdots & [G^{(M)}] \end{bmatrix} \tag{9.88}$$

which is simply the concatenation of the element flexibility matrices along the main diagonal of the $[G]$ matrix. In the case at hand, the element flexibility matrices are just identical scalars for each element, namely $\left[G^{(j)}\right] = l^{(j)}/E^{(j)}A^{(j)} = l/EA$, and there are three elements so we have

$$[G] = \frac{l}{EA}\begin{bmatrix} 1 & 0 & 0 \\ 0 & 1 & 0 \\ 0 & 0 & 1 \end{bmatrix} \quad (9.89)$$

Placing the strain–force relations of Eq. (9.87) into the compatibility equations, Eq. (9.81) we find

$$[\tilde{S}][G]\{F\} = 0 \quad (9.90)$$

which, in general, are the additional force equations required and which can be added to the equilibrium equations $[\tilde{E}_f]\{F\} = \{P\}$ to obtain a system of equations that can be solved for the forces, namely,

$$\begin{bmatrix} [\tilde{E}_f] \\ -- \\ [\tilde{S}][G] \end{bmatrix} \{F\} = \begin{Bmatrix} \{P\} \\ -- \\ \{0\} \end{Bmatrix} \quad (9.91)$$

In our case

$$[\tilde{S}][G]\{F\} = \frac{l}{EA}(F_1 + F_2 + F_3) = 0 \quad (9.92)$$

and we can eliminate the common flexibility term. The augmented equilibrium equations then become the solvable system:

$$\begin{bmatrix} 1 & -1 & 0 \\ 0 & 1 & -1 \\ 1 & 1 & 1 \end{bmatrix} \begin{Bmatrix} F_1 \\ F_2 \\ F_3 \end{Bmatrix} = \begin{Bmatrix} 3P \\ -2P \\ 0 \end{Bmatrix} \quad (9.93)$$

The solution is $F_1 = 4P/3$, $F_2 = -5P/3$, $F_3 = P/3$. We can go back to Eq. (9.80) and find the displacements as $U_2 = 4Pl/3AE$, $U_3 = -Pl/3AE$.

We can also combine the equilibrium and strain–displacement equations of the force-based method to solve directly for the displacements instead of the forces and produce a displacement-based finite element method. From Eq. (9.73) and Eq. (9.87), for the entire bar, for example, we have

$$[\tilde{E}]\{F\} = \{P\}$$
$$[G]\{F\} = [\tilde{E}]^T\{U\} \quad (9.94)$$

[Note: here the equilibrium matrix is the original equilibrium matrix seen in Eq. (9.76).] Solving for the force vector and placing it into the equilibrium equation then gives

$$[\tilde{E}][G]^{-1}[\tilde{E}]^T\{U\} = \{P\} \quad (9.95)$$

where $[\tilde{E}]$ is given in Eq. (9.76), $[G]^{-1} = (EA/l)[I]$, and where $[I]$ is the identity matrix so that $[\tilde{E}][G]^{-1}[\tilde{E}]^T = (EA/l)[\tilde{E}][\tilde{E}]^T$. Carrying out this matrix multiplication in MATLAB® we find (omitting the EA/l term):

```
E = [ -1 0 0; 1 -1 0; 0 1 -1; 0 0 1];
E* E'

ans = 1    -1    0    0
     -1     2   -1    0
      0    -1    2   -1
      0     0   -1    1
```

so we recognize that we have obtained the stiffness matrix for the system, i.e.,

$$[\tilde{E}][G]^{-1}[\tilde{E}]^T = [K] = \frac{EA}{l}\begin{bmatrix} 1 & -1 & 0 & 0 \\ -1 & 2 & -1 & 0 \\ 0 & -1 & 2 & -1 \\ 0 & 0 & -1 & 1 \end{bmatrix} \quad (9.96)$$

and our system of equations is explicitly

$$\frac{EA}{l}\begin{bmatrix} 1 & -1 & 0 & 0 \\ -1 & 2 & -1 & 0 \\ 0 & -1 & 2 & -1 \\ 0 & 0 & -1 & 1 \end{bmatrix}\begin{Bmatrix} U_1 \\ U_2 \\ U_3 \\ U_4 \end{Bmatrix} = \begin{Bmatrix} R_1 \\ 3P \\ -2P \\ R_2 \end{Bmatrix} \quad (9.97)$$

which is identical to the equations found in the stiffness-based finite element method. Applying the displacement boundary conditions and solving the remaining system, as done before, we find the same values previously obtained with the force-based method, namely $U_2 = 4Pl/3AE$ and $U_3 = -Pl/3AE$.

Let us finish this discussion of the force-based method by considering an example where distributed forces are present

---

Example 9.5 Bar with Distributed Loads – Force Method

Specifically, let us examine the problem of Fig. 9.11a, again using three elements. In this case we can use Eq. (9.71). From our discussion of the stiffness-based method, we have already evaluated the distributed force terms so we find

$$\begin{bmatrix} -1 & 0 & 0 \\ 1 & -1 & 0 \\ 0 & 1 & -1 \\ 0 & 0 & 1 \end{bmatrix}\begin{Bmatrix} F_1 \\ F_2 \\ F_3 \end{Bmatrix} = \begin{Bmatrix} q_0 l/2 \\ q_0 l \\ q_0 l \\ q_0 l/2 \end{Bmatrix} + \begin{Bmatrix} R_1 \\ 0 \\ 0 \\ R_2 \end{Bmatrix} \quad (9.98)$$

and we can eliminate the first and last equations to obtain

$$\begin{bmatrix} 1 & -1 & 0 \\ 0 & 1 & -1 \end{bmatrix}\begin{Bmatrix} F_1 \\ F_2 \\ F_3 \end{Bmatrix} = \begin{Bmatrix} q_0 l \\ q_0 l \end{Bmatrix} \quad (9.99)$$

The compatibility equation written in terms of the forces for three uniform elements was given in Eq. (9.92) so we can use that equation here to augment Eq. (9.99) and arrive at the system

$$\begin{bmatrix} 1 & -1 & 0 \\ 0 & 1 & -1 \\ 1 & 1 & 1 \end{bmatrix} \begin{Bmatrix} F_1 \\ F_2 \\ F_3 \end{Bmatrix} = \begin{Bmatrix} q_0 l \\ q_0 l \\ 0 \end{Bmatrix} \qquad (9.100)$$

which has the solution $F_1 = q_0 l$, $F_2 = 0$, $F_3 = -q_0 l$. It is easy to see that the nodal displacements are just $U_1 = U_4 = 0$, $U_2 = U_3 = q_0 l^2 / EA = q_0 L^2 / 9EA$. All of these results are identical to those obtained with the stiffness-based method. This should not be surprising because, in both cases, the forces and displacements assumed were the same (constant force and strain, linear variation of the displacements in each element).

9.3.1 A Summary and Discussion

The force-based finite element method can be summarized as follows: For statically indeterminate problems, using a combination of the principles of virtual work and complementary virtual work we can obtain discrete versions of the equilibrium and strain–displacement equations for the entire system written as

$$\begin{aligned} \left[\tilde{E}_f\right]\{F\} &= \{Q\} + \{P\} \\ \{\Delta\} &= \left[\tilde{E}_f\right]^T \{U_f\} \end{aligned} \qquad (9.101)$$

where $\left[\tilde{E}_f\right]$ is a "free" equilibrium matrix (i.e., one that does not involve unknown reaction forces) and where a set of generalized strains are related to the free (i.e., unknown) nodal displacements $\{U_f\}$ through the transpose of the same "free" equilibrium matrix. The internal forces are related to these generalized strains through a flexibility matrix, $[G]$:

$$\{\Delta\} = [G]\{F\} \qquad (9.102)$$

By eliminating the displacements from the strain–displacement equations (either by hand or by a more automatic procedure to be discussed shortly), we can find a compatibility matrix for the generalized strains where $[\tilde{S}]\{\Delta\} = 0$ and, placing Eq. (9.102) into these compatibility equations, we obtain

$$[\tilde{S}][G]\{F\} = 0 \qquad (9.103)$$

which can be appended to the equilibrium equations to arrive at a system that can be solved for the unknown forces:

$$\begin{bmatrix} [\tilde{E}_f] \\ - \\ [\tilde{S}][G] \end{bmatrix} \{F\} = \begin{Bmatrix} \{Q\} + \{P\} \\ - \\ \{0\} \end{Bmatrix} \qquad (9.104)$$

Equation (9.104) describes the final equations that must be solved in the force-based method. It requires the "free" equilibrium matrix, $[\tilde{E}_f]$, the compatibility matrix, $[\tilde{S}]$, the flexibility matrix, $[G]$, and the equivalent distributed force vector, $\{Q\}$, as well as the known concentrated forces, $\{P\}$. These equations are the discrete equivalents of the Matrix–Vector 3-D Governing Set 3 equations described in Chapter 6 (see Eq. (6.57)):

$$[\mathcal{L}_E]\{\sigma\} + \{f\} = 0$$
$$[\mathcal{L}_S][D]\{\sigma\} = 0 \tag{9.105}$$

which were obtained by combining the equilibrium equations for the stresses with the compatibility equations and the stress–strain relations:

$$[\mathcal{L}_E]\{\sigma\} + \{f\} = 0$$
$$[\mathcal{L}_S]\{e\} = 0 \tag{9.106}$$
$$\{e\} = [D]\{\sigma\}$$

while the final equations, Eq. (9.104), of the discrete force-based method came from combining

$$[\tilde{E}_f]\{F\} = \{Q\} + \{P\}$$
$$[\tilde{S}]\{\Delta\} = 0 \tag{9.107}$$
$$\{\Delta\} = [G]\{F\}$$

We see in the discrete approach that the matrices of differential operators have been replaced by equivalent matrices, i.e., $[\mathcal{L}_E] \to [\tilde{E}_f]$, $[\mathcal{L}_S] \to [\tilde{S}]$. We have placed a tilde over these matrices to emphasize that they represent discrete matrix replacements for operators.

Since we solve for the forces in a force-based approach it would be useful to have an explicit expression for calculating the nodal displacements. We can do this by noting that the forces are obtained by the solution of Eq. (9.104) which we will write as

$$[\tilde{E}_{system}]\{F\} = \begin{Bmatrix} \{Q\} + \{P\} \\ - \\ \{0\} \end{Bmatrix} \tag{9.108}$$

Now, recall from the generalized strain–displacement relations we had

$$\{\Delta\} = [\tilde{E}_f]^T \{U_f\} \tag{9.109}$$

where $[\tilde{E}_f]$ is the "free" equilibrium matrix. Here we have a nonsquare matrix relating the strains and displacements, but we can make the matrix square without changing Eq. (9.109) by adding additional columns to the matrix and placing corresponding additional zeros in the displacement vector. Specifically, we extend Eq. (9.109) as:

$$\{\Delta\} = \left[\tilde{E}_f^T \mid [G]^T [\tilde{S}]^T \right] \left\{ \begin{array}{c} \{U_f\} \\ - \\ \{0\} \end{array} \right\} \qquad (9.110)$$

$$= \left[\tilde{E}_{system} \right]^T \left\{ \begin{array}{c} \{U_f\} \\ - \\ \{0\} \end{array} \right\}$$

This relationship can be inverted to yield

$$\left\{ \begin{array}{c} \{U_f\} \\ - \\ \{0\} \end{array} \right\} = \left(\left[E_{system} \right]^T \right)^{-1} \{\Delta\} \qquad (9.111)$$

$$= \left(\left[E_{system} \right]^T \right)^{-1} [G]\{F\}$$

which gives us the free displacements (as well as some zeros) directly from a knowledge of the internal forces and other quantities we have already used in determining those forces.

Example 9.6 Statically Indeterminate Bar – Force Method and Calculation of Displacements

As an example of the use of Eq. (9.111), consider the bar between two fixed supports in Fig. 9.11a. Let us solve this problem symbolically in MATLAB® where we have let $q_0 = q$, $l = L$:

```
syms q L A E                        % define variables
Esys = [ 1 -1 0; 0 1 -1; 1 1 1]     % system matrix derived
% previously

Esys =   1      -1      0
         0       1     -1
         1       1      1

Q = [ q*L; q*L; 0]     % nodal forces due to distributed load
% vector { Q}
Q =    L*q
       L*q
        0
F = inv(Esys)*Q        % solve for internal force vector { F}  (can
% also use F  = Esys\Q)

F =    L*q
        0
      -L*q
```

```
G = (L/(E*A))*eye(3,3)    % flexibility matrix, [ G]
G = [ L/(A*E),         0,         0]
    [      0,   L/(A*E),         0]
    [      0,         0,   L/(A*E)]
U = inv(Esys.')*G*F      % solve for the two free displacements
% and a zero (can also use U = Esys.'\(G*F))
U =   (L^2*q)/(A*E)
      (L^2*q)/(A*E)
      0
```

which agrees with our previous results for the internal forces $F_1 = q_0 l, F_2 = 0, F_3 = -q_0 l$ and the two displacements $U_2 = U_3 = q_0 l^2 / EA$.

We have seen previously that we could combine the equilibrium equations with the generalized strain–displacement equations and the force-generalized strain equations to arrive at a set of equations to solve for the displacements. Let us examine that case further here. Recall, we can use

$$[\tilde{E}_f]\{F\} = \{Q\} + \{P\}$$
$$\{\Delta\} = [\tilde{E}_f]^T\{U_f\} \qquad (9.112)$$
$$\{\Delta\} = [G]\{F\}$$

to arrive at

$$[\tilde{E}_f][G]^{-1}[\tilde{E}_f]^T\{U_f\} = \{Q\} + \{P\} \qquad (9.113)$$

which is the discrete version of the Matrix–Vector 3-D Governing Set 2 of equations of Chapter 6 (see Eq. (6.51)) given by

$$[\mathcal{L}_E][C][\mathcal{L}_E]^T\{u\} + \{f\} = 0 \qquad (9.114)$$

which, as mentioned previously, are just Navier's equations for the displacements. Thus, Eq. (9.113) is just a discrete form equivalent to Navier's equations.

[Note that we can also write a set of equations similar to those in Eq. (9.112) in terms of the original equilibrium matrix, $[\tilde{E}]$, and all the displacements (free or otherwise), $\{U\}$. Then Eq. (9.113) will also involve all the equilibrium equations and displacements but there will be unknown reactions on the right-hand side of Eq. (9.113) that must be eliminated before we can arrive at a solution.]

We also obtained very similar looking expressions in the stiffness-based method, where we found

$$[K]\{U\} = \{Q\} + \{P\} \qquad (9.115)$$

9.3 Force-Based Finite Elements for Axial-Load Problems

and where $[K]$ is a stiffness matrix. However, Eq. (9.113) and Eq. (9.115) are not always identical. One can see this by looking at the expressions at the element level. An element stiffness, for example, can be written in a complete matrix–vector form as (see, for example, the stiffness matrix appearing in Eq. (9.31)):

$$\left[K^{(j)}\right] = \int_0^{l^{(j)}} \left[B^{(j)}\right]^T [EA] \left[B^{(j)}\right] d\xi \qquad (9.116)$$

where the matrix $[EA] = E^{(j)} A^{(j)}$, i.e., it is just a scalar for axial load problems. In contrast, at the element level the pseudostiffness matrix, $\left[K_{pseudo}^{(j)}\right] = \left[\tilde{E}^{(j)}\right] \left[G^{(j)}\right]^{-1} \left[\tilde{E}^{(j)}\right]^T$ is composed of the terms:

$$\left[\tilde{E}^{(j)}\right] = \int_0^{l^{(j)}} \left[B^{(j)}\right]^T \left[N^{(j)}\right] d\xi$$

$$\left[\tilde{E}^{(j)}\right]^T = \int_0^{l^{(j)}} \left[N^{(j)}\right]^T \left[B^{(j)}\right] d\xi \qquad (9.117)$$

$$\left[G^{(j)}\right] = \int_0^{l^{(j)}} \left[N^{(j)}\right]^T [1/EA] \left[N^{(j)}\right] d\xi$$

In the stiffness method, the stiffness matrix is formed from shape function matrices involving the strain only. Shape functions involving both the force and the strain are present in the pseudostiffness matrix, so there is no guarantee that these two stiffness matrices are the same. For a uniform axial-load element, however, we saw that when we use a linear strain shape function, the elemental stiffness for a uniform element was given by (see Eq. (9.33b)):

$$\left[K^{(j)}\right] = \frac{E^{(j)} A^{(j)}}{l^{(j)}} \begin{bmatrix} 1 & -1 \\ -1 & 1 \end{bmatrix} \qquad (9.118)$$

In a force-based method, if we also use a linear shape function for the strain in an element and a force shape function of unity in that same element, however, a uniform element has the terms in the pseudostiffness matrix given by

$$\left[\tilde{E}^{(j)}\right] = \begin{bmatrix} -1 \\ 1 \end{bmatrix}$$

$$\left[\tilde{E}^{(j)}\right]^T = \begin{bmatrix} -1 & 1 \end{bmatrix} \qquad (9.119)$$

$$\left[G^{(j)}\right] = \frac{l^{(j)}}{E^{(j)} A^{(j)}}$$

and we have

$$\left[K^{(j)}_{pseudo}\right] = \frac{E^{(j)}A^{(j)}}{l^{(j)}}\begin{bmatrix}-1\\1\end{bmatrix}\begin{bmatrix}-1 & 1\end{bmatrix} = \frac{E^{(j)}A^{(j)}}{l^{(j)}}\begin{bmatrix}1 & -1\\-1 & 1\end{bmatrix} \quad (9.120)$$

We see in this case the stiffness and pseudostiffness matrices for each element are identical. When we assemble these elements, the system stiffness and pseudostiffness matrices will also be identical, as we found in a previous example. In strength of materials torsion and bending problems, we will also find equality between the stiffness and pseudostiffness matrices but in other problems such as 2-D and 3-D elasticity problems, they will be different. As a consequence of these differences, the stiffness-based approach, which is the method commonly used today in finite element analyses, is not in general a discrete form of Navier's equations.

9.4 Generation of the Compatibility Equations

When solving statically indeterminate problems with a force-based finite element method, we saw a key step was the determination of the compatibility matrix, $[\tilde{S}]$. Since compatibility equations are relations between the strains that guarantee we can relate those strains to a set of displacements, the strain–displacement equations are the natural starting point for obtaining the compatibility matrix. In the discrete form we have the generalized strains, $\{\Delta\}$, and the free nodal displacements, $\{U_f\}$, which are related through the transpose of the "free" equilibrium matrix, $[\tilde{E}_f]$, i.e.,

$$\{\Delta\} = [\tilde{E}_f]^T\{U_f\} \quad (9.121)$$

The compatibility equations $[\tilde{S}]\{\Delta\} = 0$ define the relationship(s) we seek between the strains. One way to obtain these compatibility equations, which we have used previously in all our examples, is simply to eliminate the displacements by taking appropriate combinations of the equations in Eq. (9.121). For simple problems this elimination can be done by hand but we would like a more automated process. Recall, we also mentioned that applying the compatibility matrix to Eq. (9.121) gives

$$[\tilde{S}]\{\Delta\} = [\tilde{S}][\tilde{E}_f]^T\{U_f\} = 0 \quad (9.122)$$

which must be true for all displacements so that we find

$$[\tilde{S}][\tilde{E}_f]^T = [\tilde{E}_f][\tilde{S}]^T = 0 \quad (9.123)$$

[Note: in this section we will describe how to obtain compatibility equations for the "free" equilibrium matrix, $[\tilde{E}_f]$, but the same steps can be used to find compatibility equations for the original equilibrium matrix, $[\tilde{E}]$.]

9.4 Generation of the Compatibility Equations

We can also determine the compatibility matrix if we can find a matrix $[\tilde{S}]$ that satisfies Eq. (9.123). To see the steps needed, it is useful to choose a specific example, which we will solve with MATLAB®. Suppose we have an equilibrium matrix defined as

```
E = [ 1   1/2   0    0    0   1; 0   1/2   1   0   0 0; 0   0   0   1/2   1   0;
      0   0   -1   -1/2   0   0]

E =    1.0000    0.5000         0         0         0    1.0000
            0    0.5000    1.0000         0         0         0
            0         0         0    0.5000    1.0000         0
            0         0   -1.0000   -0.5000         0         0
```

We see that there are four rows and six columns, so this represents four equilibrium equations for six forces and is statically indeterminate to the second degree. Let us consider finding all the solutions to the homogeneous equilibrium equation

$$[\tilde{E}_f]\{F\} = \{0\} \tag{9.124}$$

The first step is to perform a row reduction on these equations and produce what is called a *row-reduced echelon form* of the equilibrium equations. This form can be obtained in MATLAB® with the function rref:

```
rref(E)

ans =   1    0    0    0   -1    1
        0    1    0    0    2    0
        0    0    1    0   -1    0
        0    0    0    1    2    0
```

In this form we see that leading nonzero coefficients in columns one through four (corresponding to the forces (F_1, F_2, F_3, F_4)) have a value of one. This means that the forces (F_5, F_6) are free forces in the sense that we can write all the forces in terms of them. To see this, let us write down the homogeneous equilibrium equations in this row-reduced echelon form:

$$\begin{aligned} F_1 - F_5 + F_6 &= 0 \\ F_2 + 2F_5 &= 0 \\ F_3 - F_5 &= 0 \\ F_4 + 2F_5 &= 0 \end{aligned} \tag{9.125}$$

which we can use to group and write all the forces as

$$\begin{Bmatrix} F_1 \\ F_2 \\ F_3 \\ F_4 \\ F_5 \\ F_6 \end{Bmatrix} = F_5 \begin{Bmatrix} 1 \\ -2 \\ 1 \\ -2 \\ 1 \\ 0 \end{Bmatrix} + F_6 \begin{Bmatrix} -1 \\ 0 \\ 0 \\ 0 \\ 0 \\ 1 \end{Bmatrix} \tag{9.126}$$

but since F_5 and F_6 are arbitrary, we could call them constants c_1 and c_2, respectively, and write

$$\begin{Bmatrix} F_1 \\ F_2 \\ F_3 \\ F_4 \\ F_5 \\ F_6 \end{Bmatrix} = c_1 \begin{Bmatrix} 1 \\ -2 \\ 1 \\ -2 \\ 1 \\ 0 \end{Bmatrix} + c_2 \begin{Bmatrix} -1 \\ 0 \\ 0 \\ 0 \\ 0 \\ 1 \end{Bmatrix} \qquad (9.127)$$

Equation (9.127) represents the general solution to the homogeneous equilibrium equations, showing that any combinations of the two column vectors appearing in Eq. (9.127) are solutions to those equations. Thus, we can consider these two column vectors as basis vectors for all the homogeneous solutions. If we form a matrix of these two column vectors and call it $[\tilde{S}]^T$, i.e.,

$$[\tilde{S}]^T = \begin{bmatrix} 1 & -1 \\ -2 & 0 \\ 1 & 0 \\ -2 & 0 \\ 1 & 0 \\ 0 & 1 \end{bmatrix} \qquad (9.128)$$

then Eq. (9.123) will be satisfied and we will have found the compatibility equations matrix. We can verify this statement in MATLAB®:

```
ST = [ 1 -1; -2 0; 1 0; -2 0; 1 0; 0 1];
E* ST
ans =    0      0
         0      0
         0      0
         0      0
```

All the force vectors given by Eq. (9.127) are said to lie in the *null space* of the equilibrium matrix $[\tilde{E}_f]$, i.e., they are all sets of forces that generate zero (null) applied loads. The basis vectors, or equivalently the matrix $[\tilde{S}]^T$ of these basis vectors, can be said to describe the null space since we can write any vector in this space in terms of them through Eq. (9.127). We can generate the matrix $[\tilde{S}]^T$ directly in MATLAB® without going through all the steps outlined above since there is a function null that does all the work for us:

```
ST =null(E, 'r' )
ST =     1     -1
        -2      0
         1      0
        -2      0
         1      0
         0      1
```

We see that the function null outputs the transpose of the compatibility matrix $[\tilde{S}]$, The 'r' argument in the null function produces a "rational" set of basis vectors for the null space obtained from the row reduced echelon form. What this means is that if the matrix $[\tilde{E}]$ is a small matrix with integer elements, the elements of the basis vectors generated by null(E, 'r') will be ratios of small integers, generating results similar to what we would obtain by hand elimination. Consider now if we call the null function without this rational option. We find:

```
ST2 = null(E)
ST2 =   -0.2302   -0.6862
         0.6136   -0.0670
        -0.3068    0.0335
         0.6136   -0.0670
        -0.3068    0.0335
        -0.0766    0.7197
```

This looks nothing like the previous transpose of the compatibility matrix but it is also a legitimate solution. Recall, we can use combinations of the basis vectors multiplied by arbitrary constants and still have equivalent compatibility equations. In a number of previous examples, we used this flexibility in writing the compatibility equations to produce a more consistent appearance with the equilibrium equations. If we check the null property of these alternative compatibility equations, we find

```
E* ST2
ans =   1.0e-15 *
   0.0278   -0.2220
  -0.1665   -0.0208
   0.0555   -0.0069
  -0.1665    0.0139
```

which is not identically zero (due to numerical round off errors) but is indeed very small. The null function called without the rational 'r' argument present computes the null space from a singular value decomposition method and produces orthonormal basis vectors. More details can be found in linear algebra texts and by looking at the MATLAB® reference page for singular value decomposition (svd), but we will not discuss them here. [Note: when using the null function with a symbolic matrix, as we will do shortly, one should use the function without the 'r' argument.]

9.5 Numerical Solutions – Beam Bending

Axial-load problems provide a simple context in which to examine the method of finite elements. As mentioned previously, the torsion of circular shafts will satisfy the same basic equations as axial loads if we make the replacements $F_x \to M_x$, $EA \to GJ$, $u \to \phi$, $q_x \to t_x$ where M_x is the internal axial torque, GJ is the shear modulus times the polar area moment of the cross-section, ϕ is the angle of twist, and t_x is the distributed torque/unit length. Thus, we will not discuss separately finite elements for torsion problems. Beam-bending problems, however, have a more general structure than axial-load and torsion problems, so we will consider those problems here.

9.5.1 Principles of Virtual Work and Complementary Virtual Work

Figure 9.17a shows an element of a beam and the forces acting on the element. The strain energy for bending was given in Eq. (8.59), which can be written for the element as

$$U = \frac{1}{2}\int_0^l E(\xi)I(\xi)[\kappa(\xi)]^2 d\xi \tag{9.129}$$

where $\kappa = d^2w/d\xi^2$ is the curvature of the beam in the element, $w(\xi)$ is the z-displacement of the neutral axis, and (E, I) are Young's modulus and the area moment of the cross-section about the y-axis (both of which can be functions of ξ). As is customary, the strain energy due to shear stresses is ignored as it is usually much smaller than the bending strain energy of Eq. (9.129). The principle of virtual work says that the virtual change of this strain energy is equal to the virtual work done by the forces acting on the element, so from Figs. 9.17a, b we have

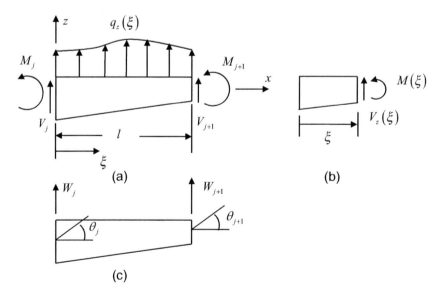

Figure 9.17 (a) A beam element for use in finite element analysis. (b) The internal force and moment in the element. (c) The nodal displacements and slopes at the ends of the element.

$$\delta U = \int_0^l E(\xi)I(\xi)\kappa(\xi)\delta\kappa(\xi)d\xi \qquad (9.130)$$

$$= \delta W_v = \int_0^l q_z(\xi)\delta w(\xi)d\xi + V_j\delta W_j + V_{j+1}\delta W_{j+1} + M_j\delta\theta_j + M_{j+1}\delta\theta_{j+1}$$

where q_z is the force/unit length acting in the z-direction, (V_j, V_{j+1}) are the shear forces acting at the end nodes, and (M_j, M_{j+1}) are the end moments. The displacements (W_j, W_{j+1}) are the end nodal displacements of the beam in the z-direction, and (θ_j, θ_{j+1}) are the nodal angular displacements. For small slopes, as assumed in engineering beam theory, we have

$$\theta_j = \left.\frac{dw}{d\xi}\right|_{\xi=0}, \quad \theta_{j+1} = \left.\frac{dw}{d\xi}\right|_{\xi=l} \qquad (9.131)$$

The positive directions of all the external forces and moments are shown in Fig. 9.17a. The corresponding positive directions of the nodal displacements and nodal angular displacements are given in Fig. 9.17c. Figure 9.17b shows the positive directions assumed for the internal shear force and bending moment within the beam.

The principle of virtual work can also be obtained directly from Eq. (8.56) for the equilibrium of a beam element. Multiplying that equilibrium equation by a virtual displacement $\delta w(\xi)$ and integrating over an element, we have

$$\int_0^{l^{(j)}} \left(\frac{d^2M}{d\xi^2} - q_z\right)\delta w(\xi)d\xi = 0 \qquad (9.132)$$

which we can rewrite as

$$\int_0^l \frac{d}{d\xi}\left(\frac{dM}{d\xi}\delta w\right)d\xi - \int_0^l \left(\frac{dM}{d\xi}\frac{d}{d\xi}(\delta w)\right)d\xi = \int_0^l q_z(\xi)\delta w(\xi)d\xi \qquad (9.133)$$

Carrying out the integration of the perfect differential and using $dM/d\xi = -V(\xi)$ and $dw/d\xi = \theta(\xi)$, Eq. (9.133) can be shown to become

$$-\int_0^l \left(\frac{dM}{d\xi}\delta\theta\right)d\xi = \int_0^l q_z(\xi)\delta w(\xi)d\xi + V(l)\delta w(l) - V(0)\delta w(0) \qquad (9.134)$$

which can again be rewritten as

$$-\int_0^l \frac{d}{d\xi}(M\delta\theta) + \int_0^l \left(M\frac{d}{d\xi}(\delta\theta)\right)d\xi = \int_0^l q_z(\xi)\delta w(\xi)d\xi + V(l)\delta w(l) - V(0)\delta w(0) \qquad (9.135)$$

We have $d\theta/d\xi = d^2w/d\xi^2 = \kappa(\xi)$, where $\kappa(\xi)$ is the curvature of the neutral axis and the perfect differential in Eq. (9.135) can be performed, giving

$$\int_0^l (M\delta\kappa)\,d\xi = \int_0^l q_z(\xi)\delta w(\xi)\,d\xi \qquad (9.136)$$
$$+ V(l)\delta w(l) - V(0)\delta w(0) + M(l)\delta\theta(l) - M(0)\delta\theta(0)$$

However, we can recognize the relations (see Figs. 9.18a, b, which show the forces and moments at the ends):

$$\begin{aligned} V(l) &= V_{j+1}, & V(0) &= -V_j \\ M(l) &= M_{j+1}, & M(0) &= -M_j \\ \delta w(l) &= \delta W_{j+1}, & \delta w(0) &= \delta W_j \\ \delta\theta(l) &= \delta\theta_{j+1}, & \delta\theta(0) &= \delta\theta_j \end{aligned} \qquad (9.137)$$

so that by using Eq. (9.137) in Eq. (9.136) as well as the moment–curvature relation

$$M(\xi) = EI\kappa(\xi) = EI\frac{d^2w(\xi)}{d\xi^2} \qquad (9.138)$$

we obtain

$$\int_0^l (E(\xi)I(\xi)\kappa(\xi)\delta\kappa(\xi))\,d\xi = \int_0^l q_z(\xi)\delta w(\xi)\,d\xi \qquad (9.139)$$
$$+ V_{j+1}\delta W_{j+1} + V_j\delta W_j + M_{j+1}\delta\theta_{j+1} + M_j\delta\theta_j$$

which is identical to Eq. (9.130).

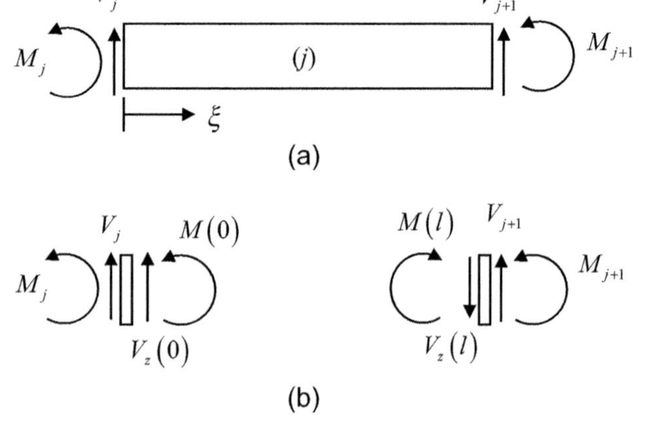

Figure 9.18 (a) A beam element showing the end loads. (b) Small portions of the element at the ends. These portions must be in equilibrium, which gives the relationship between the end loads and the internal force and moment in the beam at the ends.

Now consider the principle of complementary virtual work for a beam. For linear problems the complementary strain energy is the same as the strain energy, which written in terms of the internal moment for an element is

$$U = \frac{1}{2}\int_0^l \frac{1}{E(\xi)I(\xi)}[M(\xi)]^2 d\xi \qquad (9.140)$$

If we consider the principle of complementary virtual work for variations of the internal moment, δM, that satisfy the homogeneous equilibrium equation ($\delta q_z = 0$) then that principle for an element states that the virtual change of the (complementary) strain energy is equal to the complementary work of the external loads, i.e. (see Fig. 9.17)

$$\begin{aligned}\delta U &= \int_0^l \frac{1}{E(\xi)I(\xi)} M(\xi)\delta M(\xi) d\xi \\ &= \delta W_v^c \\ &= W_{j+1}\delta V_{j+1} + \theta_{j+1}\delta M_{j+1} + W_j \delta V_j + \theta \delta M_j \end{aligned} \qquad (9.141)$$

Alternatively, we can use the generalized strain–displacement relation

$$\kappa(\xi) = \frac{d^2 w(\xi)}{d\xi^2} \qquad (9.142)$$

and multiply it by a virtual change of the internal moment that satisfies the homogeneous equilibrium equation to give

$$\int_0^l \left(\kappa(\xi) - \frac{d^2 w(\xi)}{d\xi^2}\right)\delta M(\xi) d\xi = 0 \qquad (9.143)$$

which we can rewrite as

$$\int_0^l \kappa(\xi)\delta M(\xi) d\xi - \int_0^l \frac{d}{d\xi}\left(\frac{dw(\xi)}{d\xi}\delta M(\xi)\right) d\xi + \int_0^l \frac{dw(\xi)}{d\xi}\delta V(\xi) d\xi = 0 \qquad (9.144)$$

where we have used the relationship between the bending moment and the shear force. Carrying out the integral of the perfect differential and rearranging the last integral in Eq. (9.144) as the sum of a perfect differential and an additional integral, we have

$$\begin{aligned}&\int_0^l \kappa(\xi)\delta M(\xi) d\xi - \int_0^l \frac{d}{d\xi}(w(\xi)\delta V(\xi)) \\ &= \frac{dw(l)}{d\xi}\delta M(l) - \frac{dw(0)}{d\xi}\delta M(0)\end{aligned} \qquad (9.145)$$

where no additional integral is present in Eq. (9.145) because the virtual change of the shear force also satisfies a homogeneous equilibrium equation, i.e., $d(\delta V(\xi))/d\xi = 0$. Performing the integration of the perfect differential in Eq. (9.145) gives

$$\int_0^l \kappa(\xi)\delta M(\xi)d\xi = w(l)\delta V(l) - w(0)\delta V(0) + \frac{dw(l)}{d\xi}\delta M(l) - \frac{dw(0)}{d\xi}\delta M(0) \quad (9.146)$$

which we can write in terms of the nodal displacements and angular displacements of Fig. 9.17 and the variations of the end forces and moments as:

$$\int_0^l \kappa(\xi)\delta M(\xi)d\xi = W_{j+1}\delta V_{j+1} + W_j\delta V_j + \theta_{j+1}\delta M_{j+1} + \theta_j\delta M_j \quad (9.147)$$

which is identical to the principle of complementary virtual work expression of Eq. (9.141) if we use the moment–curvature relation in Eq. (9.142).

9.6 Stiffness-Based Finite Elements for Beam Bending

To use the principle of virtual work in a stiffness-based finite element approach, we need to define the shape functions that we will use to approximate the displacement in an element. Since there are four nodal variables $(W_j, \theta_j, W_{j+1}, \theta_{j+1})$ that we use as discrete unknowns, we write the displacement in an element in terms of four constants in a cubic polynomial:

$$w(\xi) = a_1 + a_2\xi + a_3\xi^2 + a_4\xi^3 \quad (9.148)$$

We must relate these constants to the nodal unknowns by using the conditions

$$w(0) = W_j, \quad w(l) = W_{j+1}$$
$$\left.\frac{dw(\xi)}{d\xi}\right|_{\xi=0} = \theta_j, \quad \left.\frac{dw(\xi)}{d\xi}\right|_{\xi=l} = \theta_{j+1} \quad (9.149)$$

which leads to

$$\begin{Bmatrix} W_j \\ \theta_j \\ W_{j+1} \\ \theta_{j+1} \end{Bmatrix} = \begin{bmatrix} 1 & 0 & 0 & 0 \\ 1 & 1 & 0 & 0 \\ 1 & l & l^2 & l^3 \\ 1 & 1 & 2l & 3l^2 \end{bmatrix} \begin{Bmatrix} a_1 \\ a_2 \\ a_3 \\ a_4 \end{Bmatrix} \quad (9.150)$$

Inverting Eq. (9.150) gives

$$\begin{Bmatrix} a_1 \\ a_2 \\ a_3 \\ a_4 \end{Bmatrix} = \frac{1}{l^3}\begin{bmatrix} l^3 & 0 & 0 & 0 \\ 0 & l^3 & 0 & 0 \\ -3l & -2l^2 & 3l & -l^2 \\ 2 & l & -2 & l \end{bmatrix} \begin{Bmatrix} W_j \\ \theta_j \\ W_{j+1} \\ \theta_{j+1} \end{Bmatrix} \quad (9.151)$$

which, when placed into Eq. (9.148) gives

$$w(\xi) = \left(1 - 3\frac{\xi^2}{l^2} + 2\frac{\xi^3}{l^3}\right)W_j + \left(\xi - 2\frac{\xi^2}{l} + \frac{\xi^3}{l^2}\right)\theta_j$$
$$+ \left(3\frac{\xi^2}{l^2} - 2\frac{\xi^3}{l^3}\right)W_{j+1} + \left(-\frac{\xi^2}{l} + \frac{\xi^3}{l^2}\right)\theta_{j+1} \quad (9.152)$$

and which we can write in matrix–vector form as

$$\{w\} \equiv w(\xi) = [H]\{U\}, \quad \{U\} = \begin{Bmatrix} W_j \\ \theta_j \\ W_{j+1} \\ \theta_{j+1} \end{Bmatrix} \quad (9.153)$$

where $\{U\}$ is a vector of generalized nodal displacements. Based on this approximation we have the curvature given by

$$\kappa(\xi) = \frac{d^2 w}{d\xi^2} = \left(-\frac{6}{l^2} + \frac{12\xi}{l^3}\right)W_j + \left(-\frac{4}{l} + \frac{6\xi}{l^2}\right)\theta_j$$
$$+ \left(\frac{6}{l^2} - \frac{12\xi}{l^3}\right)W_{j+1} + \left(-\frac{2}{l} + \frac{6\xi}{l^2}\right)\theta_{j+1} \quad (9.154)$$

which in matrix–vector form is

$$\{\kappa(\xi)\} \equiv \kappa(\xi) = [B]\{U\} \quad (9.155)$$

Now, consider the principle of virtual work, Eq. (9.139), which we write in matrix–vector form as

$$\int_0^l \left(E(\xi)I(\xi)\{\delta\kappa(\xi)\}^T\{\kappa(\xi)\}\right)d\xi = \int_0^l \{\delta w(\xi)\}^T\{q_z(\xi)\}d\xi + \{\delta U\}^T\{P\} \quad (9.156)$$

where $\{P\}$ is a vector of generalized loads, i.e.,

$$\{P\} = \begin{Bmatrix} V_j \\ M_j \\ V_{j+1} \\ M_{j+1} \end{Bmatrix} \quad (9.157)$$

Placing the matrix–vector expressions for the displacement and curvature and their variations into Eq. (9.156), we have

$$\{\delta U\}^T \int_0^l E(\xi)I(\xi)[B(\xi)]^T[B(\xi)]d\xi \, \{U\}$$
$$= \{\delta U\}^T \int_0^l [H(\xi)]^T q_z(\xi)d\xi + \{\delta U\}^T\{P\} \quad (9.158)$$

which must be true for all variations of the generalized nodal displacements so that we find a set of linear equilibrium equations for the generalized displacements:

$$[K]\{U\} = \{Q\} + \{P\} \tag{9.159}$$

where $[K]$ is the stiffness matrix for an element:

$$[K] = \int_0^l E(\xi)I(\xi)[B(\xi)]^T[B(\xi)]d\xi \tag{9.160}$$

and $\{Q\}$ are the equivalent nodal forces from the distributed force q_z:

$$\{Q\} = \int_0^l [H(\xi)]^T q_z(\xi)d\xi$$
$$= \int_0^l [H(\xi)]^T q_z(x_j + \xi)d\xi \tag{9.161}$$

where the second form in Eq. (9.161) should be used when the distributed force expression is given in terms of global x-coordinates instead of the local ξ-coordinate, as discussed for axial load problems. For a constant distributed force acting in the negative z-direction, $q_z = -q_0$, and an element of length l either form can be used, giving

$$\{Q\} \equiv \begin{Bmatrix} V_j^q \\ M_j^q \\ V_{j+1}^q \\ M_{j+1}^q \end{Bmatrix} = \begin{Bmatrix} -q_0 \int_0^l (1 - 3\xi^2/l^2 + 2\xi^3/l^3)d\xi \\ -q_0 \int_0^l (\xi - 2\xi^2/l + \xi^3/l^2)d\xi \\ -q_0 \int_0^l (3\xi^2/l^2 - 2\xi^3/l^3)d\xi \\ -q_0 \int_0^l (-\xi^2/l + \xi^3/l^2)d\xi \end{Bmatrix} = \begin{Bmatrix} -q_0 l/2 \\ -q_0 l^2/12 \\ -q_0 l/2 \\ q_0 l^2/12 \end{Bmatrix} \tag{9.162}$$

where $\left(V_m^q, M_m^q\right)$ $(m = j, j+1)$ are the nodal forces and moments equivalent to the distributed force. Figure 9.19 shows an element of length l carrying a uniform distributed force and the equivalent nodal forces and moments consistent with Eq. (9.162).

The element stiffness matrix $[K]$ is a 4×4 matrix with terms K_{mn} $(m, n = 1, 2, 3, 4)$. We will not give the details of the calculations for this matrix but instead simply show one of the element terms explicitly for a uniform element ($EI = $ constant), namely

9.6 Stiffness-Based Finite Elements for Beam Bending

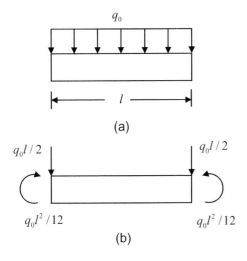

Figure 9.19 (a) A beam element with a uniform distributed force. (b) The equivalent nodal forces and moments.

$$K_{11} = EI \int_0^l \left(\frac{12\xi}{l^3} - \frac{6}{l^2}\right)^2 d\xi = \frac{EI\, l^3}{12} \int_{-6/l^2}^{6/l^2} v^2 dv$$

$$= 12EI/l^3$$

Carrying out the other integrations similarly, we have

$$[K] = \frac{EI}{l^3} \begin{bmatrix} 12 & 6l & -12 & 6l \\ 6l & 4l^2 & -6l & 2l^2 \\ -12 & -6l & 12 & -6l \\ 6l & 2l^2 & -6l & 4l^2 \end{bmatrix} \quad (9.163)$$

which is, as expected, a symmetric matrix.

When the elements of the beam are assembled, the sum of the forces and moments acting at the ends of adjacent elements must be in equilibrium with any applied concentrated applied forces and moments at a node, as shown in Fig, 9.20b, or equivalently the internal forces and moments from the adjacent elements must be in equilibrium with these applied forces and moments, as shown in Fig. 9.20c. At node $j+1$ between element (j) and $(j+1)$ as shown in Fig. 9.20 we have, therefore,

$$\begin{aligned} V_{j+1}^{(j)} + V_{j+1}^{(j+1)} &= V_{j+1}^* \\ M_{j+1}^{(j)} + M_{j+1}^{(j)} &= M_{j+1}^* \\ V_z^{(j)}\left(l^{(j)}\right) - V_z^{(j+1)}(0) &= V_{j+1}^* \\ M^{(j)}\left(l^{(j)}\right) - M^{(j+1)}(0) &= M_{j+1}^* \end{aligned} \quad (9.164)$$

where V_{j+1}^* and M_{j+1}^* are the applied vertical force and moment at node $j+1$ as seen in Fig. 9.20 and we have used the more explicit notation of $M_{j+1}^{(j)}$, $V_{j+1}^{(j)}$, etc. for the end moments and forces. These are very similar to the same relations we obtained in axial-load problems.

Example 9.7 Statically Determinate Beam – Stiffness Method

As a very simple example of a statically determinate problem, consider the cantilever beam of Fig. 9.21, which is acted upon by a uniform force along its entire length. In this case we can use a single element as shown in Fig. 9.21b and from Eq. (9.163) our equilibrium equation is

Figure 9.20 (a) The end moments and forces at node $j + 1$ between element (j) and element $(j+1)$. (b) Forces and moments at node $j + 1$ where an applied force, V^*_{j+1}, and applied moment, M^*_{j+1}, are present. (c) The same node $j + 1$ but where the internal forces and moments in elements (j) and $(j+1)$ are shown instead.

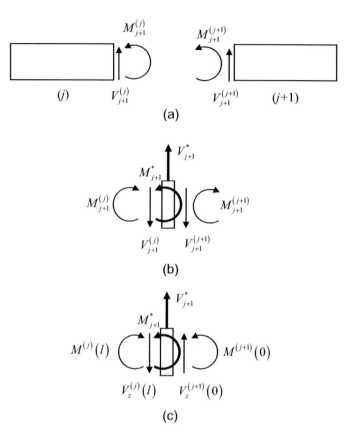

$$\frac{EI}{L^3}\begin{bmatrix} 12 & 6L & -12 & 6L \\ 6L & 4L^2 & -6L & 2L^2 \\ -12 & -6L & 12 & -6L \\ 6L & 2L^2 & -6L & 4L^2 \end{bmatrix}\begin{Bmatrix} W_1 \\ \theta_1 \\ W_2 \\ \theta_2 \end{Bmatrix} = \begin{Bmatrix} V_1^{q;(1)} \\ M_1^{q;(1)} \\ V_2^{q;(1)} \\ M_2^{q;(1)} \end{Bmatrix} + \begin{Bmatrix} V_1^{(1)} \\ M_1^{(1)} \\ V_2^{(1)} \\ M_2^{(1)} \end{Bmatrix}$$

$$= \begin{Bmatrix} -q_0 L/2 \\ -q_0 L^2/12 \\ -q_0 L/2 \\ q_0 L^2/12 \end{Bmatrix} + \begin{Bmatrix} V_{R1} \\ M_{R1} \\ 0 \\ 0 \end{Bmatrix} \quad (9.165)$$

where (V_{R1}, M_{R1}) are the reactions at the fixed support as shown in Fig. 9.21b. To solve this problem we must apply the boundary conditions at the wall which are $W_1 = \theta_1 = 0$

9.6 Stiffness-Based Finite Elements for Beam Bending

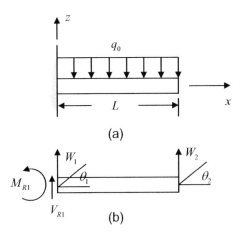

Figure 9.21 (a) A cantilever beam carrying a uniform distributed load. (b) The single finite element used for this problem, showing the nodal displacements and angular displacements and the reactions at the wall.

(no deflection and slope) and which we can do so by eliminating the first two rows in Eq. (9.165) as well as the first two columns of the stiffness matrix. We find

$$\frac{EI}{L^3}\begin{bmatrix} 12 & -6L \\ -6L & 4L^2 \end{bmatrix}\begin{Bmatrix} W_2 \\ \theta_2 \end{Bmatrix} = \begin{Bmatrix} -q_0 L/2 \\ q_0 L^2/12 \end{Bmatrix} \quad (9.166)$$

which we solve symbolically in MATLAB®:

```
syms  E I L q
P = [ -q*L/2;  q*L^2/12];                    % the known loads
Kr = (E*I/L^3)*[ 12   -6*L; -6*L  4*L^2];    % the reduced
% stiffness matrix
inv(Kr)*P   % solve for the generalized displacements (can
% also use Kr\P)
ans =     -(L^4*q)/(8*E*I)
          -(L^3*q)/(6*E*I)
```

This gives the free end deflection and slope $W_2 = -q_0 L^4/8EI$, $\theta_2 = -q_0 L^3/6EI$. You can verify these are actually the exact results. To find the end reactions, we can go back to the original stiffness matrix in Eq. (9.165) and multiply by the known generalized displacements:

```
syms  E I L q
K = [ 12,  6*L ,  -12,  6*L;     % the original stiffness matrix
6*L ,  4*L^2,  -6*L,  2*L^2;     % without the EI/L^3 factor
-12, -6*L  12 ,  -6*L;
6*L,  2*L^2, -6*L,  4*L^2]

K =    [   12,    6*L,   -12,    6*L]
       [  6*L,  4*L^2,  -6*L,  2*L^2]
       [  -12,   -6*L,    12,   -6*L]
       [  6*L,  2*L^2,  -6*L,  4*L^2]

U = [ 0 ; 0 ; -(L^4*q)/(8*E*I); -(L^3*q)/(6*E*I)]      % all
% the generalized displacements
U =                   0
                      0
```

```
            -(L^4*q)/(8*E*I)
            -(L^3*q)/(6*E*I)
(E*I/L^3)*K*U                    % find the loads, including
% reactions
ans =       (L*q)/2
            (5*L^2*q)/12
           -(L*q)/2
            (L^2*q)/12
```

From these results and from Eq. (9.165) we find $V_{R1} - q_0 L/2 = q_0 L/2$, $M_{R1} - q_0 L^2/12 = 5q_0 L^2/12$ so that $V_{R1} = q_0 L$ and $M_{R1} = q_0 L^2/2$, which are obviously the correct reactions by considering equilibrium of the entire beam.

As a slightly more complex statically indeterminate problem, consider for the next example the beam shown in Fig. 9.22a.

Figure 9.22 (a) A statically indeterminate beam. (b) The beam broken into two finite elements, showing the nodal displacements and angular displacements, the reaction force and moment at the fixed wall, and the reaction force at the smooth roller. (c) The end forces and moments acting on each element.

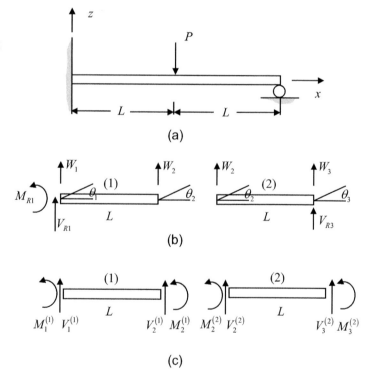

Example 9.8 Statically Indeterminate Beam – Stiffness Method

If we use two finite elements to describe the problem of Fig. 9.22a, the nodal displacements and angular displacements are shown in Fig. 9.22b as well as the end reaction force, F_{R1}, and moment, M_{R1}. The end forces and moments acting on these elements are shown in Fig. 9.22c, where we have $V_1^{(1)} = V_{R1}$, $M_1^{(1)} = M_{R1}$, $V_3^{(2)} = V_{R3}$, $M_3^{(2)} = 0$. Here, when we assemble the elements, the system equilibrium equations will be:

$$\frac{EI}{L^3} \begin{bmatrix} 12 & 6L & -12 & 6L & 0 & 0 \\ 6L & 4L^2 & -6L & 2L^2 & 0 & 0 \\ -12 & -6L & 24 & 0 & -12 & 6L \\ 6L & 2L^2 & 0 & 8L^2 & -6L & 2L^2 \\ 0 & 0 & -12 & -6L & 12 & -6L \\ 0 & 0 & 6L & 2L^2 & -6L & 4L^2 \end{bmatrix} \begin{Bmatrix} W_1 \\ \theta_1 \\ W_2 \\ \theta_2 \\ W_3 \\ \theta_3 \end{Bmatrix} = \begin{Bmatrix} V_1^{(1)} \\ M_1^{(1)} \\ V_2^{(1)} + V_2^{(2)} \\ M_2^{(1)} + M_2^{(2)} \\ V_3^{(2)} \\ M_3^{(2)} \end{Bmatrix} = \begin{Bmatrix} V_{R1} \\ M_{R1} \\ -P \\ 0 \\ V_{R3} \\ 0 \end{Bmatrix}$$

(9.167)

where V_{R3} is the vertical reaction force at the roller. The assembled stiffness matrix comes from the two element contributions shown in Fig. 9.23 which can be seen to add to produce the matrix of Eq. (9.167). The element end forces combine during the assembly to produce any existing external forces or moments, as outlined in Fig. 9.20. To solve this system, we apply the boundary conditions $W_1 = \theta_1 = 0$ at the wall and $W_2 = 0$ at the roller by eliminating the first and second rows and columns of the stiffness matrix, and the fifth row and column leaving

element (1)

$$\frac{EI}{L^3} \begin{bmatrix} 12 & 6L & -12 & 6L & 0 & 0 \\ 6L & 4L^2 & -6L & 2L^2 & 0 & 0 \\ -12 & -6L & 12 & -6L & 0 & 0 \\ 6L & 2L^2 & -6L & 4L^2 & 0 & 0 \\ 0 & 0 & 0 & 0 & 0 & 0 \\ 0 & 0 & 0 & 0 & 0 & 0 \end{bmatrix}$$

$$\frac{EI}{L^3} \begin{bmatrix} 24 & 0 & 6L \\ 0 & 8L^2 & 2L^2 \\ 6L & 2L^2 & 4L^2 \end{bmatrix} \begin{Bmatrix} \theta_2 \\ W_3 \\ \theta_3 \end{Bmatrix} = \begin{Bmatrix} -P \\ 0 \\ 0 \end{Bmatrix} \quad (9.168)$$

element (2)

$$\frac{EI}{L^3} \begin{bmatrix} 0 & 0 & 0 & 0 & 0 & 0 \\ 0 & 0 & 0 & 0 & 0 & 0 \\ 0 & 0 & 12 & 6L & -12 & 6L \\ 0 & 0 & 6L & 4L^2 & -6L & 2L^2 \\ 0 & 0 & -12 & -6L & 12 & -6L \\ 0 & 0 & 6L & 2L^2 & -6L & 4L^2 \end{bmatrix}$$

Figure 9.23 The stiffness matrices for the two elements used in the beam problem of Fig. 9.22a.

which we will solve symbolically in MATLAB®:

```
syms E   I   L   P
K = [ 12    6*L   -12    6*L   0   0;       % form up stiffness matrix
% (without EI/L^3)
6*L   4*L^2   -6*L   2*L^2   0   0;
-12   -6*L   24   0   -12   6*L;
6*L   2*L^2   0   8*L^2   -6*L   2*L^2;
0   0   -12   -6*L   12   -6*L;
0   0   6*L   2*L^2   -6*L   4*L^2];

Kr = K;                 % define reduced matrix where fifth row
% and column are eliminated
Kr(:, 5) = [ ];         % and first and second rows and columns
% are eliminated
Kr(5,:) = [ ];
Kr(:, 1:2) = [ ];
Kr(1:2, :) = [ ]        % (remember to go from highest to lowest
% rows and columns)

Kr = [    24,      0,     6*L]         % show reduced stiffness matrix
     [     0,  8*L^2,   2*L^2]
     [   6*L,  2*L^2,   4*L^2]

F = [-P; 0; 0];                 % define generalized force vector
% of applied loads
U = (L^3/(E*I))*inv(Kr)*F    % solve for generalized
% displacements, putting EI/L^3 factor back   (can also use U =
% (L^3/(E*I))*(Kr\F))
U =   -(7*L^3*P)/(96*E*I)
      -(L^2*P)/(32*E*I)
       (L^2*P)/(8*E*I)

U2 = [ 0; 0; U(1);U(2);0; U(3)]    % define  vector of all
% displacements
U2 =                 0
                     0
     -(7*L^3*P)/(96*E*I)
       -(L^2*P)/(32*E*I)
                     0
        (L^2*P)/(8*E*I)
```

```
(E* I/L^3)*K*U2      % multiply all displacements by original
% stiffness matrix, putting the EI/L^3 factor back
ans = (11*P)/16
      (3*L*P)/8
      -P
      0
      (5*P)/16
      0
```

These results show that the unknown generalized displacements are $W_2 = -7PL^3/96EI$, $\theta_2 = -PL^2/32EI$, $\theta_3 = PL^2/8EI$. Also obtained were the unknown reactions $V_{R1} = 11P/16$, $M_{R1} = 3PL/8$, and $V_{R3} = 5P/16$ which you can verify are in force and moment equilibrium with the applied force P for the entire beam.

9.7 Force-Based Finite Elements for Beam Bending

To apply a force-based finite element approach to our bending problems, we write the principle of virtual work and complementary virtual work in matrix–vector notation for an element as

$$\int_0^l \{\delta\kappa(\xi)\}^T \{M(\xi)\}d\xi = \int_0^l \{\delta w(\xi)\}^T \{q_z(\xi)\}d\xi + \{\delta U\}^T \{P\} \qquad (9.169)$$

and

$$\int_0^l \{\delta M(\xi)\}^T \frac{1}{E(\xi)I(\xi)} \{M(\xi)\}d\xi = \{\delta P\}^T \{U\} \qquad (9.170)$$

where

$$\{U\} = \begin{Bmatrix} W_j \\ \theta_j \\ W_{j+1} \\ \theta_{j+1} \end{Bmatrix}, \quad \{P\} = \begin{Bmatrix} V_j \\ M_j \\ V_{j+1} \\ M_{j+1} \end{Bmatrix}, \quad \{\delta U\} = \begin{Bmatrix} \delta W_j \\ \delta\theta_j \\ \delta W_{j+1} \\ \delta\theta_{j+1} \end{Bmatrix}, \quad \{\delta P\} = \begin{Bmatrix} \delta V_j \\ \delta M_j \\ \delta V_{j+1} \\ \delta M_{j+1} \end{Bmatrix} \qquad (9.171)$$

We will use the same shape functions described previously to write the displacement and curvature as

$$\{w(\xi)\} \equiv w(\xi) = [H]\{U\}$$
$$\{\kappa(\xi)\} \equiv \kappa(\xi) = [B]\{U\} \qquad (9.172)$$

and their variations as

$$\{\delta w(\xi)\} \equiv \delta w(\xi) = [H]\{\delta U\}$$
$$\{\delta \kappa(\xi)\} \equiv \delta \kappa(\xi) = [B]\{\delta U\} \tag{9.173}$$

Here, we will also use shape functions to define the internal moment and its variation as

$$\{M(\xi)\} \equiv M(\xi) = [N]\{F\}$$
$$\{\delta M(\xi)\} \equiv \delta M(\xi) = [N]\{\delta F\} \tag{9.174}$$

where $\{F\}$ is a set of generalized force parameters that represent the internal moment. The nodal displacements and angular displacements are similarly parameters that define the components of the generalized displacement vector $\{U\}$. In the principle of complementary virtual work the variations of the moment must satisfy the homogeneous equilibrium equations so we will choose shape functions such that

$$\frac{d^2}{d\xi^2}(\delta M(\xi)) = 0 \tag{9.175}$$

which can be satisfied if

$$\frac{d^2}{d\xi^2}[N] = 0 \tag{9.176}$$

so that the shape functions must be linear functions. For the (j)th element we will take

$$\{F^{(j)}\} = \left\{\begin{matrix} M_{2j-1} \\ M_{2j} \end{matrix}\right\}, \quad [N] = \left[\left(1 - \frac{\xi}{l}\right) \quad \frac{\xi}{l}\right] \tag{9.177}$$

so that the internal bending moment for the (j)th element, $M^{(j)}(\xi)$, is

$$M^{(j)}(\xi) = M_{2j-1} + (M_{2j} - M_{2j-1})\frac{\xi}{l} \tag{9.178}$$

and (M_{2j-1}, M_{2j}) can be seen to be just the values of the *internal moment* at the beginning and end of the element, i.e., $M_{2j-1} = M^{(j)}(0)$ and $M_{2j} = M^{(j)}(l)$. Thus, there are two internal moment values for each element. Note, however, that these internal moment values are different from the *end moments* acting on an element that were sometimes previously labeled as (M_j, M_{j+1}) (see Fig. 9.18a). To be more precise, those end moments appearing in Fig. 9.18a should be labeled, using our more explicit notation for an element (j), as $\left(M_j^{(j)}, M_{j+1}^{(j)}\right)$. We previously omitted the (j) superscripts to keep the expressions for the end moments and forces less complex-looking during derivations, but now it is necessary to be more precise. Those end moments and these internal moments are related through $M_{2j-1} = -M_j^{(j)}$ and $M_{2j} = M_{j+1}^{(j)}$.

If we place all of these shape functions into the principle of virtual work, we have

9.7 Force-Based Finite Elements for Beam Bending

$$\{\delta U\}^T \int_0^l [B(\xi)]^T [N(\xi)] d\xi \{F\}$$

$$= \{\delta U\}^T \int_0^l [H]^T \{q_z(\xi)\} d\xi + \{\delta U\}^T \{P\} \tag{9.179}$$

which must be satisfied for all variations of the generalized displacements so we arrive at the discrete equilibrium equations

$$[\tilde{E}]\{F\} = \{Q\} + \{P\} \tag{9.180}$$

where

$$[\tilde{E}] = \int_0^l [B(\xi)]^T [N(\xi)] d\xi$$

$$\{Q\} = \int_0^l [H(\xi)]^T \{q_z(\xi)\} d\xi \tag{9.181}$$

These equations are identical in form to those we obtained for axial-load problems (see Eq. (9.61) and Eq. (9.62)), although the shape functions used here are different.

Now, consider the principle of complementary virtual work. Placing the approximations in terms of the shape functions into this principle, we have

$$\{\delta F\}^T \int_0^l [N(\xi)]^T \frac{1}{E(\xi)I(\xi)} [N(\xi)] d\xi \{F\} = \{\delta P\}^T \{U\} \tag{9.182}$$

If we now define a set of discrete generalized strains $\{\Delta\}$ as

$$\{\Delta\} = [G]\{F\} \tag{9.183}$$

where we have the generalized flexibility matrix

$$[G] = \int_0^l [N(\xi)]^T \frac{1}{E(\xi)I(\xi)} [N(\xi)] d\xi \tag{9.184}$$

then Eq. (9.182) becomes

$$\{\delta F\}^T \{\Delta\} = \{\delta P\}^T \{U\} \tag{9.185}$$

However, since the virtual changes of the internal moments must satisfy homogeneous equilibrium equations, we have

$$[\tilde{E}]\{\delta F\} = \{\delta P\} \tag{9.186}$$

so placing the transpose of Eq. (9.186) into Eq. (9.185) and requiring the result be satisfied for all variations of the generalized forces, we obtain

$$\{\Delta\} = [\tilde{E}]^T \{U\} \tag{9.187}$$

which is the generalized strain–displacement relationship, again in the same form as found for axial loads.

To see what the various matrices are explicitly, we need to carry out the evaluations of the integrals. For the generalized flexibility matrix, for example, if we assume a uniform element (EI = constant) of length l we have

$$[G] = \frac{1}{EI} \begin{bmatrix} \int_0^l \left(1-\frac{\xi}{l}\right)^2 d\xi & \int_0^l \left(1-\frac{\xi}{l}\right)\frac{\xi}{l} d\xi \\ \int_0^l \left(1-\frac{\xi}{l}\right)\frac{\xi}{l} d\xi & \int_0^l \frac{\xi^2}{l^2} d\xi \end{bmatrix} = \frac{l}{EI} \begin{bmatrix} \frac{1}{3} & \frac{1}{6} \\ \frac{1}{6} & \frac{1}{3} \end{bmatrix} \tag{9.188}$$

The equilibrium matrix given in Eq. (9.181) is a 4×2 matrix. If we write the shape functions in this matrix as

$$[N(\xi)] = [N_1(\xi), \ N_2(\xi)]$$
$$[B(\xi)] = [B_1(\xi), \ B_2(\xi), \ B_3(\xi), \ B_4(\xi)] \tag{9.189}$$

where

$$N_1(\xi) = \left(1 - \frac{\xi}{l}\right), \quad N_2(\xi) = \frac{\xi}{l}$$

$$B_1(\xi) = \left(-\frac{6}{l^2} + \frac{12\xi}{l^3}\right), \quad B_2(\xi) = \left(-\frac{4}{l} + \frac{6\xi}{l^2}\right) \tag{9.190}$$

$$B_3(\xi) = \left(\frac{6}{l^2} - \frac{12\xi}{l^3}\right), \quad B_4(\xi) = \left(-\frac{2}{l} + \frac{6\xi}{l^2}\right)$$

then the equilibrium matrix $[\tilde{E}]$ has components given by

$$\tilde{E}_{mn} = \int_0^l B_m(\xi) N_n(\xi) d\xi \quad (m = 1,2,3,4) \quad (n = 1,2) \tag{9.191}$$

If we carry out all the integrations of these products of shape functions, we find

$$[\tilde{E}] = \begin{bmatrix} -\frac{1}{l} & \frac{1}{l} \\ -1 & 0 \\ \frac{1}{l} & -\frac{1}{l} \\ 0 & 1 \end{bmatrix} \tag{9.192}$$

We can use these matrices to find force-based beam solutions. However, recall we can also define a pseudostiffness matrix and solve for the displacements. For an element this pseudostiffness matrix, we recall, is given by $[K_{pseudo}] = [\tilde{E}][G]^{-1}[\tilde{E}]^T$. Let us use MATLAB® to obtain the pseudostiffness for a beam element explicitly:

```
syms L E I
Em = [ -1/L , 1/L; -1 , 0; 1/L , -1/L; 0 , 1];    % define the
% equilibrium matrix

G = (L/(E*I))*[ 1/3 , 1/6; 1/6, 1/3];              % define the
% flexibility matrix

Kp = Em*inv(G)*Em.'                                % compute the
% pseudo-stiffness matrix

Kp = [    (12*E*I)/L^3,    (6*E*I)/L^2,  -(12*E*I)/L^3,   (6*E*I)/L^2]
     [    (6*E*I)/L^2,     (4*E*I)/L,    -(6*E*I)/L^2,    (2*E*I)/L]
     [   -(12*E*I)/L^3,   -(6*E*I)/L^2,   (12*E*I)/L^3,  -(6*E*I)/L^2]
     [    (6*E*I)/L^2,     (2*E*I)/L,    -(6*E*I)/L^2,    (4*E*I)/L]

(L^3/(E*I))*Kp           % remove the EI/L^3 factor to make
                         % it easier to read


ans = [   12,     6*L,    -12,    6*L]
      [  6*L,    4*L^2,  -6*L,   2*L^2]
      [  -12,    -6*L,    12,    -6*L]
      [  6*L,    2*L^2,  -6*L,   4*L^2]
```

Comparing these MATLAB® results with Eq. (9.163), which we write here again:

$$[K] = \frac{EI}{l^3}\begin{bmatrix} 12 & 6l & -12 & 6l \\ 6l & 4l^2 & -6l & 2l^2 \\ -12 & -6l & 12 & -6l \\ 6l & 2l^2 & -6l & 4l^2 \end{bmatrix}$$

we see that the pseudostiffness matrix is indeed the same as the stiffness matrix obtained previously. However, we should emphasize again that the pseudostiffness matrix is not in general always identical to the stiffness matrix used in stiffness-based finite elements.

Example 9.9 Statically Determinate Beam – Force Method

Consider now the statically determinate problem of Fig. 9.21a (a cantilever beam carrying a distributed force) with the force-based method. Since there is only one element, the equilibrium equations are just

$$\begin{bmatrix} -\dfrac{1}{L} & \dfrac{1}{L} \\ -1 & 0 \\ \dfrac{1}{L} & -\dfrac{1}{L} \\ 0 & 1 \end{bmatrix} \begin{Bmatrix} M_1 \\ M_2 \end{Bmatrix} = \begin{Bmatrix} V_1^{q;(1)} \\ M_1^{q;(1)} \\ V_2^{q;(1)} \\ M_2^{q;(1)} \end{Bmatrix} + \begin{Bmatrix} V_1^{(1)} \\ M_1^{(1)} \\ V_2^{(1)} \\ M_2^{(1)} \end{Bmatrix} = \begin{Bmatrix} -q_0 L/2 \\ -q_0 L^2/12 \\ -q_0 L/2 \\ q_0 L^2/12 \end{Bmatrix} + \begin{Bmatrix} V_{R1} \\ M_{R1} \\ 0 \\ 0 \end{Bmatrix} \quad (9.193)$$

where (V_{R1}, M_{R1}) are the reactions at the wall. Applying the boundary conditions to this problem we can eliminate the first two rows of these equations and find

$$\begin{bmatrix} 1/L & -1/L \\ 0 & 1 \end{bmatrix} \begin{Bmatrix} M_1 \\ M_2 \end{Bmatrix} = \begin{Bmatrix} -q_0 L/2 \\ q_0 L^2/12 \end{Bmatrix} \quad (9.194)$$

Solving this in MATLAB® we find:

```
syms L q
Eg = [ -1/L , 1/L; -1 , 0; 1/L , -1/L; 0 , 1];   % the
% original equilibrium matrix
Er = Eg;
Er(1:2, :) = [ ]            % generate reduced equilibrium
% matrix and show it
Er = [ 1/L, -1/L]
     [  0,    1]
P = [ -q*L/2 ; q*L^2/12];    % the known loads
F = inv(Er)*P               % solve [Er]{F} ={P} for the
                            % generalized forces (actually
F =  -(5*L^2*q)/12          % moments. Can also use F = Er\P)
      (L^2*q)/12

Eg*F - [ -q*L/2 ; -q*L^2/12; -q*L/2 ; q*L^2/12]    % compute the
% end reactions at the wall (and zeros)
ans =      L*q
        (L^2*q)/2
           0
           0
```

We see the solution is $M_1 = -5q_0 L^2/12$, $M_2 = q_0 L^2/12$. The reactions at the wall are also obtained by using the original equilibrium matrix and subtracting off the distributed force contributions. These reactions are $V_{R1} = q_0 L$, $M_{R1} = q_0 L^2/2$, which agrees with our previous results. The terms in the final MATLAB® result given above are the applied force and moment at the free end of the beam, which are both zero.

9.7 Force-Based Finite Elements for Beam Bending

Consider now the statically indeterminate problem of Fig. 9.22a with the force-based method.

Example 9.10 Statically Indeterminate Beam – Force Method

To solve the problem of Fig. 9.22a, we use two elements as shown in Fig. 9.24 and four unknown generalized internal forces (M_1, M_2, M_3, M_4). Assembling these two elements the principle of virtual work gives

$$\begin{bmatrix} -\dfrac{1}{L} & \dfrac{1}{L} & 0 & 0 \\ -1 & 0 & 0 & 0 \\ \dfrac{1}{L} & -\dfrac{1}{L} & -\dfrac{1}{L} & \dfrac{1}{L} \\ 0 & 1 & -1 & 0 \\ 0 & 0 & \dfrac{1}{L} & -\dfrac{1}{L} \\ 0 & 0 & 0 & 1 \end{bmatrix} \begin{Bmatrix} M_1 \\ M_2 \\ M_3 \\ M_4 \end{Bmatrix} = \begin{Bmatrix} V_1^{(1)} \\ M_1^{(1)} \\ V_2^{(1)} + V_2^{(2)} \\ M_2^{(1)} + M_2^{(2)} \\ V_3^{(2)} \\ M_3^{(2)} \end{Bmatrix} = \begin{Bmatrix} V_{R1} \\ M_{R1} \\ -P \\ 0 \\ V_{R3} \\ 0 \end{Bmatrix} \quad (9.195)$$

where again (V_{R1}, M_{R1}) are the fixed end reactions and V_{R3} is the reaction force at the roller support. If we generate a reduced set of "free" equilibrium equations by eliminating the first two equations and the fifth equation in Eq. (9.195), we have

$$\begin{bmatrix} \dfrac{1}{L} & -\dfrac{1}{L} & -\dfrac{1}{L} & \dfrac{1}{L} \\ 0 & 1 & -1 & 0 \\ 0 & 0 & 0 & 1 \end{bmatrix} \begin{Bmatrix} M_1 \\ M_2 \\ M_3 \\ M_4 \end{Bmatrix} = \begin{Bmatrix} -P \\ 0 \\ 0 \end{Bmatrix} \quad (9.196)$$

which are three equations for the four unknown moments. Thus, we need one compatibility equation. Note that the third equation in Eq. (9.196) gives $M_4 = 0$ so we could eliminate that unknown, but we will still have two equations in three unknowns so that the problem remains statically indeterminate to one degree. We will, however, for generality keep all four equations here, which we will write as $[\tilde{E}_f]\{F\} = \{P\}$. We need to define the compatibility equation $[\tilde{S}]\{\Delta\} = 0$ associated with these equilibrium equations and write that compatibility equation in terms of the generalized forces by using the generalized flexibility matrix, $[G]$,

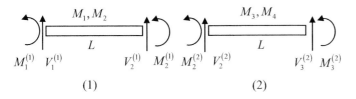

Figure 9.24 The two force-based elements used for the problem of Fig. 9.22a.

i.e., $\{\Delta\} = [G]\{F\}$, where here the flexibility matrix is the diagonal concatenation of the element flexibility matrices:

$$[G] = \frac{L}{EI} \begin{bmatrix} 1/3 & 1/6 & 0 & 0 \\ 1/6 & 1/3 & 0 & 0 \\ 0 & 0 & 1/3 & 1/6 \\ 0 & 0 & 1/6 & 1/3 \end{bmatrix} \qquad (9.197)$$

Appending the compatibility relation $[\tilde{S}][G]\{F\} = 0$ to the free equilibrium equations, we can then solve for the unknown generalized forces. As we have done before, placing these forces back into the original equilibrium equations we can find the unknown reactions in Eq. (9.197). These steps in MATLAB® are:

```
syms  E  I  L  P
Eg = [ -1/L   1/L    0     0;         % define equilibrium matrix
       -1     0      0     0;
       1/L  -1/L   -1/L   1/L;
       0     1     -1     0;
       0     0     1/L   -1/L;
       0     0      0     1] ;

Er = Eg;                  % define reduced, "free" equilibrium matrix
Er(5, :) = [ ];           % and show it
Er(1:2, :) = [ ]

Er = [ 1/L,  -1/L,  -1/L,  1/L]
     [  0,    1,    -1,    0]
     [  0,    0,     0,    1]

null(Er)                  % obtain the compatibility equation for
                          % this statically indeterminate system
ans =  2
       1
       1
       0

Gn = [ 1/3 1/6 0 0;       % define the flexibility matrix with the
       1/6 1/3 0 0;       % L/EI factor removed since it cancels
       0 0 1/3 1/6;       % in the subsequent compatibility term
       0 0 1/6 1/3] ;

SG = (null(Er)).'*Gn      % form up compatibility times the
% flexibility term

SG = [ 5/6,  2/3,  1/3,  1/6]
Es = [ Er ; SG]           % append this term to the free equilibrium
```

```
% equations
Es = [ 1/L,  -1/L,  -1/L,  1/L]
     [   0,     1,    -1,    0]
     [   0,     0,     0,    1]
     [ 5/6,   2/3,   1/3,  1/6]
Pv = [ -P; 0; 0; 0];    % define the vector of known loads
F = inv(Es)*Pv          % solve for the generalized forces (can
% also use F = Es\Pv)
F =    -(3*L*P)/8
        (5*L*P)/16
        (5*L*P)/16
           0

Eg*F                    % multiply generalized forces by the original
                        % equilibrium matrix to get reactions as well
                        % as original loads
ans =     (11*P)/16
          (3*L*P)/8
             -P
              0
           (5*P)/16
              0
```

These results show that the reactions are $V_{R1} = 11P/16$, $M_{R1} = 3PL/8$, and $V_{R3} = 5P/16$, which are same as the previous values obtained with the stiffness-based method.

9.8 Some Extensions of a Force-Based Approach

We have used strength of materials problems as examples of the generation of both stiffness-based and force-based finite elements. In the application of the finite element approach we have solved problems using very few elements so that all the steps and intermediate results are able to be examined. Such solutions, however, may not be possible in all cases. For example, consider the axial-load problem of Fig. 9.25a, where a uniformly loaded bar is fixed at both ends to rigid supports. If we try to use a single element in either a stiffness-based approach or a force-based approach, there are no free displacements to use in the stiffness-based method and no free equilibrium equations in the force-based method since the element is highly constrained. Similarly, in the beam of Fig. 9.25b we cannot use a single element to solve the problem.

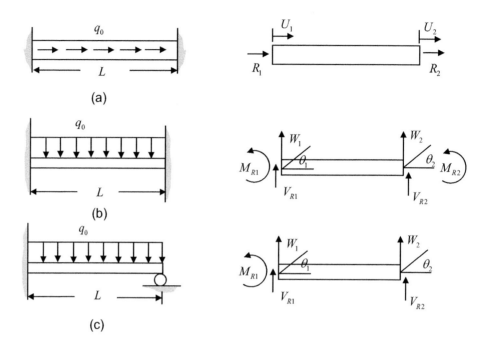

Figure 9.25 (a, b) Two statically indeterminate problems where the use of a single element will not lead to a solution but which can be solved with a force-based method by including the reactions. (c) A statically indeterminate beam problem that will be used to illustrate how reactions are incorporated into the solution. The unknown reactions and the generalized nodal displacements are also shown for these problems.

These difficulties, however, are easily eliminated by simply choosing more elements so that there are sufficient degrees of freedom present in each element to solve the problem. If we want to solve such problems without using more elements there are other options. We will see that one way is to bring the reactions into the problem explicitly as additional unknowns. In fact, this approach will lead to a new way to solve all strength of materials problems, not just the problems such seen in Figs. 9.25a, b. We will demonstrate this new approach with the problem shown in Fig. 9.25c (which can be solved with a single element and an "ordinary" force-based finite element method as well). This statically indeterminate problem is very similar to the statically determinate beam that was solved previously. The equations are, therefore, the same as those in Eq. (9.193) except there is now an unknown reaction force, V_{R2}, at the roller. Thus, in place of Eq. (9.193) we have

$$\begin{bmatrix} -\frac{1}{L} & \frac{1}{L} \\ -1 & 0 \\ \frac{1}{L} & -\frac{1}{L} \\ 0 & 1 \end{bmatrix} \begin{Bmatrix} M_1 \\ M_2 \end{Bmatrix} = \begin{Bmatrix} V_1^{q;(1)} \\ M_1^{q;(1)} \\ V_2^{q;(1)} \\ M_2^{q;(1)} \end{Bmatrix} + \begin{Bmatrix} V_1^{(1)} \\ M_1^{(1)} \\ V_2^{(1)} \\ M_2^{(1)} \end{Bmatrix} = \begin{Bmatrix} -q_0 L/2 \\ -q_0 L^2/12 \\ -q_0 L/2 \\ q_0 L^2/12 \end{Bmatrix} + \begin{Bmatrix} V_{R1} \\ M_{R1} \\ V_{R2} \\ 0 \end{Bmatrix} \quad (9.198)$$

We could write these equations in the more general matrix–vector form

$$[\tilde{E}]\{F_e\} = \{Q\} + \{P\} + \{R\} \tag{9.199}$$

where $\{F_e\}$ is the vector of generalized internal elastic forces, which in this case has as components the moments (M_1, M_2), $\{Q\}$, is the vector of discrete distributed loads, and $\{R\}$ is the vector of generalized reactions, which in this case has the components $(V_{R1}, M_{R1}, V_{R2}, 0)$. We have also for generality included a set of discrete applied loads $\{P\}$, which is of course absent in the present problem.

To bring the reactions into the problems as unknowns, we place them on the left side of the equations in Eq. (9.198), giving

$$\begin{bmatrix} -\dfrac{1}{L} & \dfrac{1}{L} & -1 & 0 & 0 \\ -1 & 0 & 0 & -1 & 0 \\ \dfrac{1}{L} & -\dfrac{1}{L} & 0 & 0 & -1 \\ 0 & 1 & 0 & 0 & 0 \end{bmatrix} \begin{Bmatrix} M_1 \\ M_2 \\ V_{R1} \\ M_{R1} \\ V_{R2} \end{Bmatrix} = \begin{Bmatrix} -q_0 L/2 \\ -q_0 L^2/12 \\ -q_0 L/2 \\ q_0 L^2/12 \end{Bmatrix} \tag{9.200}$$

In the general matrix–vector form, we could write Eq. (9.199) as

$$[\hat{E}]\{\hat{F}\} \equiv [\tilde{E}|\tilde{E}_a] \begin{Bmatrix} F_e \\ - \\ R \end{Bmatrix} = \{Q\} + \{P\} \tag{9.201}$$

where in the present problem $[\hat{E}]$ is the equilibrium matrix appearing in Eq. (9.200) and $\{\hat{F}\}$ is now a vector of generalized forces with components consisting of both generalized internal forces and generalized reactions. The equilibrium matrix $[\hat{E}]$ is our original equilibrium matrix $[\tilde{E}]$ of Eq. (9.198) augmented by a matrix $[\tilde{E}_a]$, which in this case is

$$[\tilde{E}_a] = \begin{bmatrix} -1 & 0 & 0 \\ 0 & -1 & 0 \\ 0 & 0 & -1 \\ 0 & 0 & 0 \end{bmatrix} \tag{9.202}$$

Now, consider how this inclusion of the reactions into the equilibrium equations affects the principle of complementary virtual work. The complementary strain energy can be written as a function of the generalized forces $\{\hat{F}\}$ and we have

$$\delta U^c(\hat{F}) = \delta W_v^c = \{\delta Q\}^T \{U\} + \{\delta P\}^T \{U\} \tag{9.203}$$

The displacements $\{U\}$ contain the generalized displacements associated with the displacement degrees of freedom present in the problem. In the present case there are four displacements (U_1, U_2, U_3, U_4) in $\{U\}$, where U_i is the ith generalized displacement associated with the ith equilibrium equation at a node. The first equilibrium equation, for example, was summation of forces in the vertical direction at the wall so that $U_1 = W_1$ is the vertical displacement at the wall (see Fig. (9.25c). Similarly, $U_2 = \theta_1$, the slope at the wall, and in the same fashion $U_3 = W_2$, $U_4 = \theta_2$ (Fig. 9.25c). The variation of the complementary strain energy can be written as

$$\delta U^c(\hat{F}) = \sum \frac{\partial U^c}{\partial \hat{F}_i} \delta \hat{F}_i \qquad (9.204)$$

We could also write Eq. (9.204) as

$$\delta U^c(\hat{F}) = \{\delta \hat{F}\}^T \left\{ \frac{\partial U^c}{\partial \hat{F}} \right\} \qquad (9.205)$$

where $\{\delta \hat{F}\}$ now contains variations of both the internal elastic forces as well as the reactions, and $\left\{ \frac{\partial U^c}{\partial \hat{F}} \right\}$ contains both derivatives with respect to the internal elastic forces as well as the reactions, i.e.,

$$\left\{ \frac{\partial U^c}{\partial \hat{F}} \right\} = \left\{ \begin{array}{c} \left\{ \frac{\partial U^c}{\partial F_e} \right\} \\ - \\ \left\{ \frac{\partial U^c}{\partial R} \right\} \end{array} \right\} = \left\{ \begin{array}{c} \partial U^c / \partial M_1 \\ \partial U^c / \partial M_2 \\ \partial U^c / \partial V_{R1} \\ \partial U^c / \partial M_{R1} \\ \partial U^c / \partial V_{R2} \end{array} \right\} \qquad (9.206)$$

where the last column vector on the right-hand side of Eq. (9.206) is the explicit expression for the present problem. From the equations of equilibrium, Eq. (9.201), we have

$$[\hat{E}]\{\delta \hat{F}\} = \{\delta Q\} + \{\delta P\} \qquad (9.207)$$

From the complementary strain energy principle, Eq. (9.203) and from Eq. (9.205) and Eq. (9.207), it follows

$$\{\delta \hat{F}\}^T \left\{ \frac{\partial U^c}{\partial \hat{F}} \right\} = \{\delta \hat{F}\}^T [\hat{E}]^T \{U\} \qquad (9.208)$$

which can be satisfied for all $\{\delta \hat{F}\}$ when

$$\left\{ \frac{\partial U^c}{\partial \hat{F}} \right\} = [\hat{E}]^T \{U\} \qquad (9.209)$$

Equation (9.209) is in the form of a strain–displacement relation if we define generalized strains $\{\hat{\Delta}\}$ associated with both the internal forces and reactions as

$$\{\hat{\Delta}\} = \left\{ \frac{\partial U^c}{\partial \hat{F}} \right\} \qquad (9.210)$$

or, more explicitly in component form

$$\Delta_{ei} = \frac{\partial U^c(F_e, R)}{\partial F_{ei}}, \quad \Delta_{Ri} = \frac{\partial U^c(F_e, R)}{\partial R_i} \qquad (9.211)$$

Because the internal forces, reactions, and applied loads are related to each other through the equilibrium equations, we can write the complementary strain energy $U^c(F_e, R)$ in terms of different combinations of the internal forces and reactions. Thus, in a given problem the complementary strain energy will not depend on all the elastic internal forces and reactions, and in Eq. (9.211) values of the subscript i will run only over those generalized forces actually present. For the present problem we have a uniform beam element where

$$U^c = \frac{1}{2EI} \int_0^l [M(\xi)]^2 d\xi \qquad (9.212)$$

We can write the internal moment, for example, solely in terms of the internal elastic moments as

$$M(\xi) = M_1 + (M_2 - M_1)\xi/l \qquad (9.213)$$

so that $U^c = U^c(M_1, M_2)$. The internal moment terms (M_1, M_2) are the same types of generalized internal forces we have considered before and we can relate those moments to corresponding generalized internal strains $(\Delta_{M_1}, \Delta_{M_2})$ through a flexibility matrix, $[G]$. Previously, we calculated this flexibility matrix through its definition in terms of the shape functions (see Eq. (9.68)). However, we can also compute this matrix with the use of Eq. (9.211) which now becomes explicitly

$$\Delta_{M_1} = \frac{\partial U^c}{\partial M_1}, \quad \Delta_{M_2} = \frac{\partial U^c}{\partial M_2}, \quad \Delta_{V_{R1}} = \Delta_{M_{R1}} = \Delta_{V_{R2}} = 0 \qquad (9.214)$$

We can place Eq. (9.213) into Eq. (9.212) and then use Eq. (9.214). Carrying out the integration, it is easy to show that we find the same flexibility matrix as before (see Eq. (9.188)), namely

$$\begin{Bmatrix} \Delta_{M_1} \\ \Delta_{M_2} \end{Bmatrix} = \frac{l}{EI} \begin{bmatrix} 1/3 & 1/6 \\ 1/6 & 1/3 \end{bmatrix} \begin{Bmatrix} M_1 \\ M_2 \end{Bmatrix} \qquad (9.215)$$

for the generalized internal elastic strains $(\Delta_{M_1}, \Delta_{M_2})$ associated with the generalized internal forces (M_1, M_2). Note that the relationships between the elastic generalized strains and the derivatives of the complementary strain energy terms in Eq. (9.214) are similar to the results we obtained in Chapter 8 for the relationship between the internal strains and the derivative of the complementary strain energy density with respect to the stresses (Eq. (8.46a)) which we rewrite here:

$$e_I = \frac{\partial u_0^c(\sigma)}{\partial \sigma_I} \quad (I = 1, 2, \ldots, 6) \qquad (9.216)$$

so that we could consider Eq. (9.214) to be analogous to Eq. (9.216) in terms of generalized strains and generalized forces.

Writing the complementary strain energy in terms of (M_1, M_2) is only one possibility. We could instead write the complementary strain energy in terms of one or more of the

Figure 9.26 (a) Free body diagram involving the reaction force at the roller of Fig. 9.25c. (b) Free body diagram involving the reaction force and moment at the fixed wall of Fig. 9.25c.

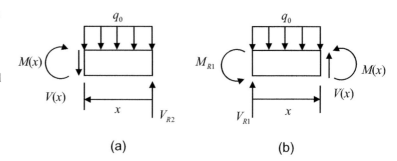

reactions. For example, we can write the internal moment in terms of the reaction force at the roller as

$$M(x) = V_{R2}x - q_0 x^2/2 \qquad (9.217)$$

(see Fig. 9.26a). In this case we have the complementary strain energy given as $U^c = U^c(V_{R2}, q_0)$. Alternatively, we could write the internal moment in terms of the reactions at the fixed wall:

$$M(x) = -M_{R1} + V_{R1}x - q_0 x^2/2 \qquad (9.218)$$

(see Fig. 9.26b) in which case $U^c = U^c(M_{R1}, V_{R1}, q_0)$. Thus, Eq. (9.211) gives us a very general way to relate the generalized strains $\{\hat{\Delta}\}$ to the generalized $\{\hat{F}\}$.

We can determine the compatibility equations for Eq. (9.200) of our present problem by first computing the null space matrix $[\tilde{S}]$ for the equilibrium matrix $[\hat{E}]$. In MATLAB® we find:

```
syms L q E I
Eg = [ -1/L 1/L -1  0   0;           % obtain the equilibrium matrix
       -1    0   0 -1   0;
       1/L -1/L  0  0  -1;
        0    1   0  0   0];
S = (null(Eg)).'                     % generate the null space matrix

S = [ L, 0, -1, -L, 1]
```

which shows that $[\tilde{S}] = [L, 0, -1, -L, 1]$. With this null space matrix and Eq. (9.210), we then can write the compatibility equation(s) in terms of the generalized internal forces.

To summarize: if we form up a set of equilibrium equations $[\hat{E}]$ that includes internal forces and/or reactions and determine the compatibility matrix, $[\tilde{S}]$ from those equilibrium equations, we can obtain the generalized forces by solving a combination of the equilibrium equations and compatibility equations in the form:

$$\begin{aligned}
[\hat{E}]\{\hat{F}\} &= \{P\} + \{Q\} \\
[\tilde{S}]\{\hat{\Delta}\} &= 0 \\
\{\hat{\Delta}\} &= \left\{\frac{\partial U^c}{\partial \hat{F}}\right\}
\end{aligned} \qquad (9.219)$$

where now the generalized forces $\{\hat{F}\}$ can include internal generalized forces, $\{F_e\}$, and the external reactions, $\{R\}$.

Let us now use this formulation to solve the problem of Fig. 9.25c, making several different choices for how we write the complementary strain energy.

Example 9.11 Statically Indeterminate Beam – Use of Complementary Strain Energy

As a first choice for solving the beam of Fig. 9.25c, let us express the internal bending moment in terms of (M_1, M_2). This is the same choice as found in the "ordinary" force-based finite element method. In this case, the first two generalized strains are given by Eq. (9.215) in terms of the flexibility matrix, $[G]$, and the remaining generalized strains are all zero (Eq. (9.214). Thus, the compatibility equation matrix in terms of the generalized forces becomes:

```
G = (L/(E*I))*[ 1/3 1/6; 1/6 1/3]    % define the flexibility
% matrix
Se = [ [L ,0]*[G] , 0, 0, 0]         % obtain the compatibility
% equation matrix in terms of the generalized forces
Se = [ L^2/(3*E*I), L^2/(6*E*I), 0, 0, 0]
```

If we then append this compatibility equation to the original equilibrium equations and also modify the vector of known applied forces appropriately, we can then solve the problem for all the generalized forces:

```
Es = [Eg; Se]    % add compatibility equation to the
% equilibrium matrix to obtain the system equilibrium matrix
Es = [          -1/L,       1/L,      -1,  0,  0]
     [           -1,         0,        0, -1,  0]
     [          1/L,      -1/L,        0,  0, -1]
     [            0,         1,        0,  0,  0]
     [ L^2/(3*E*I), L^2/(6*E*I),       0,  0,  0]
Pv = [ -q*L/2; -q*L^2/12; -q*L/2; q*L^2/12; 0]    % extend the
% applied loads
Pv =      -(L*q)/2
          -(L^2*q)/12
          -(L*q)/2
           (L^2*q)/12
            0
F = inv(Es)*Pv    % solve for the generalized forces
```

Figure 9.27 (a) The reactions, the applied nodal force and moment, and the internal force, V, and moment, M_1, in the beam of Fig. 9.25c at the wall. (b) The reactions, the applied nodal force and moment, and the internal force, V, and moment, M_2, of the beam of Fig. 9.25c at the roller support.

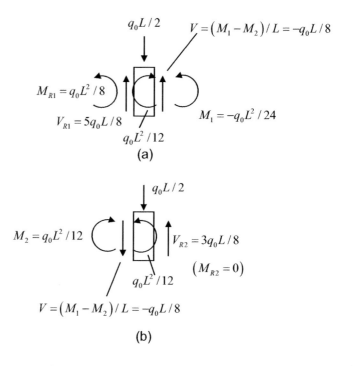

```
F  =   - (L^2* q) /24
       (L^2* q) /12
       (5* L* q) /8
       (L^2* q) /8
       (3* L* q) /8
```

These results give the moments $M_1 = -q_0 L^2/24$, $M_2 = q_0 L^2/12$ with the reactions given by $V_{R1} = 5q_0 L/8$, $M_{R1} = q_0 L^2/8$, and $V_{R2} = 3q_0 L/8$. There are quite a few unknowns here so we have explicitly shown these results together with the applied nodal forces and moments from the distributed force and the internal shear force, V, in Fig. 9.27. Results at both ends of the beam are given in that figure, where it can be seen that all the forces and moments are in equilibrium. Problem P9.1 asks you to also use the strain–displacement equations for this problem to find the slope of the beam at the roller support, which is one of the generalized displacements present. However, you should realize that one cannot always solve the generalized strain–displacement relations for the displacements because the generalized forces in $\{\hat{F}\}$ are not all independent and the displacements may in some cases not have a definite meaning. We will say more about this shortly. However, here the generalized strains $(\Delta_{M_1}, \Delta_{M_2})$ are independent (and all the others are zero) so, as seen in problem P9.1, we can solve the strain–displacement relations for the correct unknown generalized displacement (rotation) in this case.

If instead of using the moments (M_1, M_2), we chose to write the internal moment in terms of the reaction force at the roller, using the free body diagram seen in Fig. 9.26a, we would have (see Eq. (9.217)):

$$\Delta_{V_{R2}} = \frac{\partial U^c}{\partial V_{R2}} = \frac{1}{EI} \int_0^L (V_{R2}x - q_0 x^2/2) x \, dx$$

$$= \frac{V_{R2}L^3}{3EI} - \frac{q_0 L^4}{8EI} \quad (9.220)$$

$$\Delta_{M_1} = \Delta_{M_2} = \Delta_{V_{R1}} = \Delta_{M_{R1}} = 0$$

Since all the other derivatives of the complementary strain energy are zero, the compatibility equation in terms of the generalized forces simply becomes $\partial U^c/\partial V_{R2} = 0$ or, equivalently, $V_{R2} = 3q_0 L/8$, which is the value of the reaction force we obtained previously. If we append this compatibility equation (which we note is now an inhomogeneous equation, so there will be a component of $3q_0 L/8$ added to the vector of loads) to the equilibrium matrix and solve we find

```
Es2 = [ Eg; 0 0 0 0 1]       % add compatibility equation to the
% equilibrium matrix
Es2 = [ -1/L,    1/L,  -1,   0,   0]
      [  -1,     0,    0,   -1,   0]
      [  1/L,  -1/L,   0,    0,  -1]
      [   0,     1,    0,    0,   0]
      [   0,     0,    0,    0,   1]

Pv2 = [ -q*L/2; -q*L^2/12; -q*L/2 ; q*L^2/12; 3*q*L/8]
% extend the vector of loads

Pv2 =    -(L*q)/2
         -(L^2*q)/12
         -(L*q)/2
          (L^2*q)/12
          (3*L*q)/8

F2 = inv(Es2)*Pv2        % solve for the generalized forces

F2 =  -(L^2*q)/24
       (L^2*q)/12
       (5*L*q)/8
       (L^2*q)/8
       (3*L*q)/8
```

which is identical to the solution we obtained before.

Note that the use here of the compatibility equation leads to $\partial U^c/\partial V_{R2} = 0$, which is identical to simply using the principle of least work to solve for the unknown redundant reaction force at the roller, as we did in Chapter 8 (see Fig. 8.17 and Eq. (8.164)). Thus, to solve problems where we want to include unknown redundant reactions explicitly, we could alternatively use the principle of least work directly, as done in Chapter 8, to augment the equations of equilibrium and produce a determinate set of equations we can solve. However, the use of the compatibility equation is more general than using the principle of least work in that we can use unknowns that do not have to be independent redundants. For example, see problem P.9.2 where you are asked to write the complementary strain energy in terms of the reactions (M_{R1}, V_{R1}) to solve for the generalized forces in the current problem.

Both in the present case, where the complementary strain energy is written in terms of V_{R2}, and when we write the complementary strain energy in terms of (M_{R1}, V_{R1}), if we try to solve the strain–displacement relations for the displacements present we may find incorrect values for those displacements since the generalized strains, as defined here, are not all independent. See, for example, problem P9.2 for the case involving (M_{R1}, V_{R1}). This, however, does not affect the solution of Eq. (9.219) for the generalized forces. Also, in some cases the displacements appearing in the strain–displacement relations may not have a clear meaning. For example, consider the axial-load problem of Fig. 9.25a. The equilibrium equation in matrix–vector form is

$$[1 \ 1]\begin{Bmatrix} R_1 \\ R_2 \end{Bmatrix} = -q_0 L$$

and the corresponding generalized strain–displacement relation is

$$\begin{Bmatrix} \Delta_{R_1} \\ \Delta_{R_2} \end{Bmatrix} = \begin{bmatrix} 1 \\ 1 \end{bmatrix} U$$

However, what is the displacement, U, here? All that we can say is that U is an x-displacement degree of freedom associated with the force equation in the x-direction. If we write the complementary strain energy in terms of either R_1 or R_2, then these strain–displacement relations will simply yield $U = 0$. Thus, in general we can solve Eq. (9.219) for the generalized forces but we should normally determine the displacements separately by other means.

Equation (9.219) is an extension of the formulation used previously in force-based finite elements where all the unknown generalized forces were internal generalized elastic forces $\{F_e\}$ that could be related to internal generalized strains through a flexibility matrix, $[G]$, i.e., where Eq. (9.219) becomes

$$[\tilde{E}]\{F_e\} = \{P\} + \{Q\}$$
$$[\tilde{S}]\{\Delta_e\} = 0 \qquad (9.221)$$
$$\{\Delta_e\} = [G]\{F_e\}$$

where $[\tilde{S}]$ is the compatibility matrix associated with the equilibrium matrix $[\tilde{E}]$ and $\{\Delta_e\}$ are the elastic strains associated with the elastic forces. Since the strains play only an intermediate role in either Eq. (9.219) or Eq. (9.221) and can be eliminated, either of these equations can be considered to be purely force-based methods, similar to the Matrix–Vector 3-D Governing Set 3 described in Chapter 6. The strain–displacement relations also do not play a direct role in either of these formulations since the compatibility matrix $[\tilde{S}]$ can be calculated directly from the equilibrium matrix, so successful use of either Eq. (9.219) or Eq. (9.221), as mentioned previously, does not depend on our ability to solve those strain–displacement relations for the displacements. In fact, all of the statically indeterminate problems that commonly appear in strength of materials texts can be solved for the generalized forces present by the purely forced-based approaches of either Eq. (9.219) or Eq. (9.221). Once all the unknown generalized forces are obtained in this fashion, one way we can obtain the generalized displacements is through integration, as is normally done in elementary strength of materials courses. In bending problems, for example, one integrates the moment–curvature relationship twice and applies boundary conditions to obtain the beam slope and vertical deflection. For a comparison of the use of Eq. (9.219) with other methods commonly used in advanced or elementary strength of materials courses, see problem P9.3.

Example 9.12 Specified Displacements at Supports – Force Method

Another situation where it is useful to bring the reactions into the problem formulation is in an ordinary force-based finite element solution when there are nonzero displacements specified at the supports of a body. As a simple example, consider the bar shown in Fig. 9.28a where a force P acts at the center of the bar and the ends of the bar have known displacements $U_1 = \bar{U}_1, U_3 = \bar{U}_3$. The unknown displacement at the force P is U_2. If we let the reaction forces at those supports be (R_A, R_B) and consider two elements as shown in Fig. 9.28b, then the equilibrium equations are

$$\begin{bmatrix} -1 & 0 \\ 1 & -1 \\ 0 & 1 \end{bmatrix} \begin{Bmatrix} F_1 \\ F_2 \end{Bmatrix} = \begin{Bmatrix} R_A \\ P \\ R_B \end{Bmatrix} \qquad (9.222)$$

where (F_1, F_2) are the internal forces in the two elements. We can incorporate the reactions into the equilibrium matrix by rewriting Eq. (9.222) as

$$\begin{bmatrix} -1 & -1 & 0 & 0 \\ 0 & 1 & -1 & 0 \\ 0 & 0 & 1 & -1 \end{bmatrix} \begin{Bmatrix} R_A \\ F_1 \\ F_2 \\ R_B \end{Bmatrix} = \begin{Bmatrix} 0 \\ P \\ 0 \end{Bmatrix} \qquad (9.223)$$

which we can again write symbolically as

$$[\hat{E}]\{\hat{F}\} = \{P\} \qquad (9.224)$$

Figure 9.28 (a) A bar acted upon by a load P at its center and where the ends have specified displacements $(\bar{U}_1, \bar{U}_3)$. (b) The reaction forces, internal forces and nodal displacements.

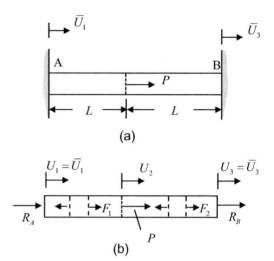

The generalized strains $(\Delta_A, \Delta_1, \Delta_2, \Delta_B)$ associated with the forces (R_A, F_1, F_2, R_B) can then be related to the nodal displacements (U_1, U_2, U_3) through the transpose of the equilibrium matrix in Eq. (9.223) to give the strain–displacement relations as

$$\begin{Bmatrix} \Delta_A \\ \Delta_1 \\ \Delta_2 \\ \Delta_B \end{Bmatrix} = \begin{bmatrix} -1 & 0 & 0 \\ -1 & 1 & 0 \\ 0 & -1 & 1 \\ 0 & 0 & -1 \end{bmatrix} \begin{Bmatrix} U_1 \\ U_2 \\ U_3 \end{Bmatrix} \tag{9.225}$$

which we can also write symbolically as

$$\{\hat{\Delta}\} = [\hat{E}]^T \{U\} \tag{9.226}$$

Now, consider the second and third equations of these strain–displacement relations, which we write as

$$\begin{Bmatrix} \Delta_1 \\ \Delta_2 \end{Bmatrix} = \begin{bmatrix} -1 & 0 \\ 0 & 1 \end{bmatrix} \begin{Bmatrix} \bar{U}_1 \\ \bar{U}_3 \end{Bmatrix} + \begin{bmatrix} 1 \\ -1 \end{bmatrix} \{U_2\} \tag{9.227}$$

We will express Eq. (9.227) symbolically as

$$\{\Delta_e\} = [\tilde{E}_R]^T \{\bar{U}\} + [\tilde{E}_f]^T \{U_f\} \tag{9.228}$$

where $\{\Delta_e\}$ is the vector of generalized internal "elastic" strains for the body, $\{\bar{U}\}$ is a vector of the given displacements, and $\{U_f\}$ are the free displacements, which in this case are just U_2. The matrix $[\tilde{E}_f] = [\,1\ -1\,]$ is the "free" equilibrium matrix (obtained by eliminating the equilibrium equations associated with the reactions from Eq. (9.222)), as seen previously in other problems. We will discuss the meaning of the matrix $[\tilde{E}_R]^T$ shortly. Thus, in general, by first including the reactions (and their associated known displacements), we can always generate a problem with specified nodal displacements having the form

9.8 Some Extensions of a Force-Based Approach

$$[\tilde{E}_f]\{F_e\} = \{Q\} + \{P\}$$
$$\{\Delta_e\} = [\tilde{E}_R]^T\{\bar{U}\} + [\tilde{E}_f]^T\{U_f\} \tag{9.229}$$
$$\{\Delta_e\} = [G]\{F_e\}$$

where now $\{F_e\} = [F_1, F_2]^T$ is the vector of the internal forces only, $[G]$ is the compliance matrix that relates those internal forces to the elastic strains, and $\{P\}$ is a vector of the known applied generalized forces only, which is the vector obtained by eliminating the reaction forces from the right-hand side of Eq. (9.222). For generality, we have also included nodal forces $\{Q\}$ due to any distributed forces that may be present. By introducing the compatibility matrix $[\tilde{S}]$ associated with the free equilibrium matrix $[\tilde{E}_f]$ we can eliminate the free displacements and obtain a set of inhomogeneous compatibility equations (which is a single equation in the present case)

$$[\tilde{S}][G]\{F_e\} = [\tilde{S}][\tilde{E}_R]^T\{\bar{U}\} \tag{9.230}$$

which then leads to the solvable set of equations

$$[\tilde{E}_{system}]\{F_e\} \equiv \begin{bmatrix} [\tilde{E}_f] \\ \overline{[\tilde{S}][G]} \end{bmatrix} \{F_e\} = \left\{ \begin{array}{c} \{Q\} + \{P\} \\ \overline{[\tilde{S}][\tilde{E}_R]^T\{\bar{U}\}} \end{array} \right\} \tag{9.231}$$

In our case $[\tilde{S}] = [1 \quad 1]$. Note that the compatibility equation, Eq. (9.230), written in terms of the elastic strains is $[\tilde{S}]\{\Delta_e\} = [\tilde{S}][\tilde{E}_R]^T\{\bar{U}\}$, which gives, explicitly for our problem,

$$\Delta_1 + \Delta_2 = (\bar{U}_3 - \bar{U}_1) \tag{9.232}$$

This makes sense since Eq. (9.232) says that the sum of the generalized elastic strains (elongations) of the two elements is equal to the overall elongation of the entire bar. If we express Eq. (9.231) for this problem in MATLAB® we can solve it symbolically:

```
syms   E   A   L   P   U1   U3
Es = [ 1 -1 ;  1    1];          % system equilibrium matrix (note
% that the compatibility equation is multiplied by EA/L)

Pv = [  P;    E*A*(U3-U1)/L] ;   % right hand side of known loads
% and displacements
inv(Es)*Pv                       % solve for F1, F2
ans =        P/2 -  (A*E*(U1 - U3))/(2*L)
         -   P/2 -  (A*E*(U1 - U3))/(2*L)
```

In this case we see that the internal forces in the two elements are

$$F_1 = \frac{P}{2} + \frac{AE}{2L}(\bar{U}_3 - \bar{U}_1)$$
$$F_2 = -\frac{P}{2} + \frac{AE}{2L}(\bar{U}_3 - \bar{U}_1) \tag{9.233}$$

We could also set up and solve the extended equilibrium equations of Eq. (9.223) by appending the compatibility equation to those equations:

```
Es2 = [-1 -1 0 0; 0 1 -1 0; 0 0 1 -1; 0 1 1 0];   % use system
% equilibrium matrix with reactions
Pv2= [ 0; P; 0; E*A*(U3-U1)/L]          % new right-hand side with
% known loads and displacements where the compatibility
% equation is again multiplied by EA/L

inv(Es2)*Pv2                            % solve for RA, F1, F2, RB
ans =   (A*E*(U1 - U3))/(2*L) - P/2
        P/2 - (A*E*(U1 - U3))/(2*L)
      - P/2 - (A*E*(U1 - U3))/(2*L)
      - P/2 - (A*E*(U1 - U3))/(2*L)
```

which again leads to the internal forces of Eq. (9.233) as well as the reactions:

$$R_A = -\frac{P}{2} - \frac{EA}{2L}(\bar{U}_3 - \bar{U}_1)$$
$$R_B = -\frac{P}{2} + \frac{EA}{2L}(\bar{U}_3 - \bar{U}_1) \quad (9.234)$$

We can also solve for the free displacements (here the unknown displacement at the force P) by the same process as used previously. Combining the last two equations in Eq. (9.229), we have

$$[G]\{F_e\} = [\tilde{E}_R]^T\{\bar{U}\} + [\tilde{E}_f]^T\{U_f\} \quad (9.235)$$

which can be extended as

$$[G]\{F_e\} = [\tilde{E}_R]^T\{\bar{U}\} + \left[[\tilde{E}_f]^T \mid [G]^T[\tilde{S}]^T\right]\begin{Bmatrix} \{U_f\} \\ - \\ \{0\} \end{Bmatrix}$$

$$= [\tilde{E}_R]^T\{\bar{U}\} + [\tilde{E}_{system}]^T\begin{Bmatrix} \{U_f\} \\ - \\ \{0\} \end{Bmatrix} \quad (9.236)$$

and then solved as

$$\begin{Bmatrix} \{U_f\} \\ - \\ \{0\} \end{Bmatrix} = \left[[\tilde{E}_{system}]^T\right]^{-1}\left([G]\{F_e\} - [\tilde{E}_R]^T\{\bar{U}\}\right) \quad (9.237)$$

9.8 Some Extensions of a Force-Based Approach

In our present problem it is easy to show that Eq. (9.237) gives

$$\left\{ \begin{array}{c} U_2 \\ 0 \end{array} \right\} = \begin{bmatrix} 1 & 1 \\ -1 & 1 \end{bmatrix}^{-1} \left\{ \begin{array}{c} \dfrac{L}{EA}F_1 + \bar{U}_1 \\ \dfrac{L}{EA}F_2 - \bar{U}_3 \end{array} \right\} \qquad (9.238)$$

$$= \begin{bmatrix} 1/2 & -1/2 \\ 1/2 & 1/2 \end{bmatrix} \left\{ \begin{array}{c} \dfrac{L}{EA}F_1 + \bar{U}_1 \\ \dfrac{L}{EA}F_2 - \bar{U}_3 \end{array} \right\} = \left\{ \begin{array}{c} \dfrac{PL}{2EA} + \dfrac{(\bar{U}_1 + \bar{U}_3)}{2} \\ 0 \end{array} \right\}$$

so that we see that the displacement at the applied force P given by

$$U_2 = \frac{PL}{2EA} + \frac{(\bar{U}_1 + \bar{U}_3)}{2} \qquad (9.239)$$

This displacement is composed of two terms. The first term is the displacement at the force P when the ends of the bar are held rigidly fixed. The second term is the displacement at the center of the bar when the ends are displaced and the force P is absent. This center displacement is just the average of the two end displacements.

To summarize, we see that in a problem where there are known displacements $\{\bar{U}\}$ at the support nodes present, the unknown internal forces and free displacements are obtained from solving Eq. (9.231) and Eq. (9.237):

$$[\tilde{E}_{system}]\{F_e\} \equiv \begin{bmatrix} [\tilde{E}_f] \\ - \\ [\tilde{S}][G] \end{bmatrix} \{F_e\} = \left\{ \begin{array}{c} \{Q\} + \{P\} \\ - \\ [\tilde{S}][\tilde{E}_R]^T\{\bar{U}\} \end{array} \right\} \qquad (9.240)$$

$$\left\{ \begin{array}{c} \{U_f\} \\ - \\ \{0\} \end{array} \right\} = \left[[\tilde{E}_{system}]^T \right]^{-1} \left([G]\{F_e\} - [\tilde{E}_R]^T\{\bar{U}\} \right) \qquad (9.241)$$

which is identical in form to the cases considered earlier when the displacements at the supports are zero (see Eq. (9.104) and Eq. (9.111)). Although we obtained Eq. (9.240) and Eq. (9.241) for an axial-load problem, these equations can also be used in a general force-based finite element analysis.

The effect of the support displacements is described by the presence of the $[\tilde{E}_R]^T\{\bar{U}\}$ term, so now let us consider the meaning of $[\tilde{E}_R]^T$. We saw that $[\tilde{E}_R]^T$ appears as the matrix multiplying the known displacements when the strain–displacement relations are reduced to those involving only the free displacements on the right-hand side of those relations. This procedure can always be used to obtain the $[\tilde{E}_R]^T$ matrix but this is a "process" type of definition that does not reveal what this matrix represents. Consider the strain–displacement relations, Eq. (9.226), involving all the generalized strains and displacements. We can write this equation in matrix–vector form as

$$\Delta_k = \sum E_{ik} U_i \qquad (9.242)$$

where k takes on values for all the forces present (internal and reaction forces) and i takes on values for all the all displacements (free and known). We will drop the use of the " ^ " superscripts in Eq. (9.242) and the following equations, and will use different subscript symbols to represent specific parts of the matrix or vector terms. For example, let K represent the values of k for all the internal forces and let I represent the values of i for all the known displacements. Then, if we examine only those values of $k = K$ for the internal strains corresponding to the internal forces, we can write Eq. (9.242) for those strains as

$$\begin{aligned}\Delta_K &= \sum E_{iK} U_i \\ &= \sum E_{IK} U_I + \sum_{i \neq I} E_{iK} U_i \end{aligned} \qquad (9.243)$$

which is nothing more than Eq. (9.228), since the displacements U_i $(i \neq I)$ are just the free displacements and E_{iK} is the transpose of the free equilibrium matrix. However, now let us consider the equilibrium equations involving both the internal forces and reaction forces, Eq. (9.224), which we write as

$$\sum E_{ik} F_k = P_i \qquad (9.244)$$

(For simplicity we will omit the distributed load terms.) Now, let us consider only those equilibrium equations that involve the known displacement degrees of freedom $(i = I)$. They are:

$$\sum E_{Ik} F_k = P_I = 0 \qquad (9.245)$$

since there are no applied loads at the reaction force (known displacement) nodes. We can rewrite Eq. (9.245) as

$$\sum E_{IK} F_K + \sum_{k \neq K} E_{Ik} F_k = 0 \qquad (9.246)$$

However, we have

$$\sum_{k \neq K} E_{Ik} F_k = -R_I \qquad (9.247)$$

since these terms arise simply from including the negative of the reactions R_I in the equilibrium equations for $i = I$ so we have

$$R_I = \sum E_{IK} F_K \qquad (9.248)$$

9.8 Some Extensions of a Force-Based Approach

which we can write as

$$\{R\} = [\tilde{E}_R]\{F_e\} \tag{9.249}$$

where we have again labeled the internal forces (due to elastic deformations) as $\{F_e\}$ to explicitly distinguish them from our original set of forces $\{F\}$ that included the reactions. Thus, we see that $[\tilde{E}_R]$ is just an equilibrium matrix that relates the reactions to the internal forces. (If nodal forces from the distributed loads are present at the reactions, then those forces, which have been neglected in our discussion, will also be present in Eq. (9.249) but will not alter the $[\tilde{E}_R]$ matrix.) Identifying the equilibrium equations from Eq. (9.249) is therefore another way to define the matrix $[\tilde{E}_R]$. The strain–displacement relation of Eq. (9.243) can similarly be written as

$$\{\Delta_e\} = [\tilde{E}_R]^T\{\bar{U}\} + [\tilde{E}_f]^T\{U_f\} \tag{9.250}$$

Equation (9.250) is of course identical to the previous result seen in Eq. (9.228).

One can obtain a different viewpoint of these strain–displacement and compatibility equations by bringing the term involving the known displacements over to the left side of Eq. (9.250) and defining a set of generalized strains, $\{\bar{\Delta}\}$, associated with those displacements as

$$\{\bar{\Delta}\} = -[\tilde{E}_R]^T\{\bar{U}\} \tag{9.251}$$

[Note: these strains $\{\bar{\Delta}\}$ are not the generalized strains (Δ_A, Δ_B) seen earlier.] Then the strain–displacement equation can be written as

$$\{\Delta_t\} = \{\Delta_e\} + \{\bar{\Delta}\} = [\tilde{E}_f]^T\{U_f\} \tag{9.252}$$

where $\{\Delta_t\}$ is the total generalized strains produced by both the specified displacements as well as the elastic behavior of the body. The compatibility equations then are simply

$$\begin{aligned}[\tilde{S}]\{\Delta_t\} &= 0 \\ \rightarrow [\tilde{S}]\{\Delta_e\} &= -[\tilde{S}]\{\bar{\Delta}\}\end{aligned} \tag{9.253}$$

where $[\tilde{S}]$ is the compatibility matrix associated with the free equilibrium matrix $[\tilde{E}_f]$. Equation (9.253) shows that with this definition of the initial strains $\{\bar{\Delta}\}$ the compatibility equations are just the ordinary compatibility conditions for the total strains $\{\Delta_t\}$. In terms of these total strains, the compatibility equations are again homogeneous equations but when expressed in terms of the elastic strains the equations are now inhomogeneous as seen in Eq. (9.253).

To close our discussion of the case where there are nonzero displacements at the supports, we will also further connect the strains $\{\bar{\Delta}\}$ to the strain energy concepts discussed in Chapter 8. First, consider a 1-D stress problem like the axial-load case just considered where there is an linear elastic strain, e_e, as well as a specified strain, $\bar{e}$, due to some specified displacements at the supports. The total strain, e_t, is $e_t = \bar{e} + e_e$. Then the total strain energy density, u_0, is given by

$$u_0 = \int_{\bar{e}}^{e_t} \sigma \, de_t = \int_{\bar{e}}^{e_t} E(e_t - \bar{e}) \, de_t$$

$$= \frac{E}{2}(e_t - \bar{e})^2 \qquad (9.254)$$

which is just the triangular area shown in Fig. 9.29. Similarly, the complementary strain energy density, u_0^c, is given by

$$u_0^c = \int_0^\sigma e_t \, d\sigma = \int_0^\sigma \left(\frac{\sigma}{E} + \bar{e}\right) d\sigma$$

$$= \frac{\sigma^2}{2E} + \bar{e}\sigma \qquad (9.255)$$

which is the other area shown in Fig. 9.29. In terms of generalized strain and the internal force, the total complementary strain energy for a constant area bar under uniaxial load is therefore

$$U^c = \frac{1}{2EA}\int_0^L F^2 \, dx + \frac{\bar{\Delta}}{L}\int_0^L F \, dx \qquad (9.256)$$

which for a constant internal force, F, gives

$$U^c = \frac{F^2 L}{2EA} + \bar{\Delta} F \qquad (9.257)$$

We can extend these 1-D results to a general finite element setting by noting that, if we examine the equations of equilibrium for all the "free" nodes present in a problem, we have

$$[\tilde{E}_f]\{F_e\} = \{P\} + \{Q\} \qquad (9.258)$$

where $[\tilde{E}_f]$ is the free equilibrium matrix and $\{F_e\}$ is a vector of all the internal generalized elastic forces. The principle of complementary strain energy can also be written as

$$\delta U^c(F_e) = \{\delta F_e\}^T \{\Delta_e\}$$
$$= \{\delta R\}^T \{\bar{U}\} + \{\delta Q\}^T \{U_f\} + \{\delta P\}^T \{U_f\} \qquad (9.259)$$

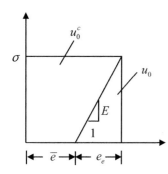

Figure 9.29 The 1-D stresses and strains for an linear elastic solid when a specified strain, $\bar{e}$, is present in addition to the elastic strains, e_e, due to the applied loads. The corresponding strain energy density, u_0, and complementary strain energy density, u_0^c, are shown as the two areas indicated in the figure.

where $\{\Delta_e\}$ are the internal elastic strains associated with the internal forces, $\{\bar{U}\}$ are the known displacements at the support nodes (where the reactions $\{R\}$ are present), and $\{U_f\}$ is the vector of free nodal displacements. Now, suppose we relate the reactions to the internal elastic forces (see Eq. (9.249)) and define the strains $\{\bar{\Delta}\}$ associated with the displacements $\{\bar{U}\}$, as done in Eq. (9.251), then Eq. (9.259) becomes

$$\{\delta F_e\}^T\{\Delta_e\} + \{\delta F_e\}^T\{\bar{\Delta}\} = \{\delta Q\}^T\{U_f\} + \{\delta P\}^T\{U_f\} \qquad (9.260)$$

which we could rewrite as

$$\delta U^c(F_e, \bar{\Delta}) = \delta W_v^c \qquad (9.261)$$

where δW_v^c is the virtual change of the complementary work done by the applied loads and $\delta U^c(F_e, \bar{\Delta})$ is the virtual change of the total complementary strain energy due to both the internal elastic forces and the strains due to the specified displacements. Since the strains $\{\bar{\Delta}\}$ are fixed and hence do not vary, the total complementary strain energy, therefore, is

$$U^c(F_e, \bar{\Delta}) = U^c(F_e) - \{R\}^T\{\bar{\Delta}\} \qquad (9.262)$$

which is the generalization of Eq. (9.257) that we sought.

If we use the definition of the strain $\bar{\Delta}$ in terms of the known displacements (see Eq. (9.251)), and the relationship between the reactions and the elastic internal forces, Eq. (9.249), then we can also write the complementary strain energy of Eq. (9.262) in the form

$$U^c(F_e, R, \bar{U}) = U^c(F_e) - \{R\}^T\{\bar{U}\} \qquad (9.263)$$

But we can also solve for the elastic internal forces in terms of the applied loads and the known displacements (see, for example, Eq. (9.239)) and use that solution in Eq. (9.263) to write the complementary strain energy as

$$U^c(P, Q, R, \bar{U}) = U^c(P, Q, \bar{U}) - \{R\}^T\{\bar{U}\} \qquad (9.264)$$

We can use Eq. (9.257) and Engesser's first theorem of Chapter 8 to determine the displacement at the force P for the axial load problem we have just considered in Example 9.12, where

$$U^c = \frac{F_1^2 L}{2AE} + \bar{\Delta}_1 F_1 + \frac{F_2^2 L}{2AE} + \bar{\Delta}_2 F_2 \qquad (9.265)$$

and (F_1, F_2) were found in terms of P and the known displacements in Eq. (9.233), which we rewrite here as

$$\begin{aligned} F_1 &= \frac{P}{2} + \frac{AE}{2L}(\bar{U}_3 - \bar{U}_1) \\ F_2 &= -\frac{P}{2} + \frac{AE}{2L}(\bar{U}_3 - \bar{U}_1) \end{aligned} \qquad (9.266)$$

The generalized initial strains due to the specified displacements are simply $\bar{\Delta}_1 = \bar{U}_1$, $\bar{\Delta}_2 = -\bar{U}_3$ (see Eq. (9.227)). Thus, from Engesser's first theorem we have

$$U_2 = \frac{\partial U^c}{\partial P} = \frac{L}{2AE}\left[\frac{P}{2} + \frac{AE}{2L}(\bar{U}_3 - \bar{U}_1)\right] + \frac{\bar{\Delta}_1}{2}$$
$$- \frac{L}{2AE}\left[-\frac{P}{2} + \frac{AE}{2L}(\bar{U}_3 - \bar{U}_1)\right] - \frac{\bar{\Delta}_2}{2}$$
(9.267)

Canceling common terms, we find

$$U_2 = \frac{PL}{2AE} + \frac{\bar{U}_1 + \bar{U}_3}{2}$$
(9.268)

which is identical to the results obtained previously (see Eq. (9.239)). Note that if we had used Eq. (9.264) instead to calculate this displacement, we must consider the reactions to be functions of the load P, and we cannot treat the reactions as constants in applying Engesser's first theorem, as was done in Chapter 8 when discussing the principle of least work for the case where the generalized displacements at the supports are zero.

9.9 Finite Elements for General Deformable Bodies

We have seen the use of finite elements for axial loads and bending problems. In this section we want to outline briefly the finite element method more generally for deformable bodies. Both stiffness-based and force-based approaches will be given.

9.9.1 Stiffness-Based Finite Elements

For a general 3-D body, the principle of virtual work gives (see Eq. (8.71b))

$$\int_V \sum_{i=1}^{3}\sum_{j=1}^{3} \sigma_{ij}\delta e_{ij} dV = \int_V \sum_{j=1}^{3} f_j \delta u_j dV + \int_S \sum_{j=1}^{3} T_j^{(\mathbf{n})} \delta u_j dS$$
(9.269)

which can be written in matrix–vector form as

$$\int_V \{\delta e\}^T\{\sigma\} dV = \int_V \{\delta u\}^T\{f\} dV + \int_S \{\delta u\}^T\{T^{(\mathbf{n})}\} dS$$
(9.270)

where $\{\sigma\}$ is a column vector of the six stresses σ_I $(I = 1, 2, \ldots, 6)$ and $\{e\}$ is a similar column vector of the six strains e_I $(I = 1, 2, \ldots, 6)$, as discussed previously. The vector $\{u\}$ has three components u_i $(i = 1, 2, 3)$, and the body force vector $\{f\}$ has the three components f_i $(1 = 1, 2, 3)$. Similarly, the stress vector $\{T^{(\mathbf{n})}\}$ has three components $T_i^{(\mathbf{n})}$ $(i = 1, 2, 3)$. The displacements are represented in terms of a matrix of shape functions $[H]$ and a vector of nodal displacements $\{U\}$ as $\{u\} = [H]\{U\}$. Recall from Chapter 6 that the strains and displacements are related through a matrix of derivatives, i.e., $\{e\} = [\mathcal{L}_E]^T \{u\}$ (see Eq. (6.49)), so we have $\{e\} = [\mathcal{L}_E]^T \{u\} = [\mathcal{L}_E]^T [H]\{U\} = [B]\{U\}$, where $[B]$ is the matrix of differentiated displacement shape functions. From the stress–strain relations we have $\{\sigma\} = [C]\{e\}$, where $[C]$ is the symmetrical 6×6 matrix of elastic constants discussed in Chapter 5. Placing these expressions into Eq. (9.270) gives

$$\{\delta U\}^T \int_V [B]^T [C][B] dV \{U\} = \{\delta U\}^T \int_V [H]^T \{f\} dV + \{\delta U\}^T \int_S [H]^T \{T^{(n)}\} dS \quad (9.271)$$

which must be satisfied for all variations of the nodal displacements so that we find a set of linear equations of the form

$$[K]\{U\} = \{Q\} + \{P\} \quad (9.272)$$

where

$$[K] = \int_V [B]^T [C][B] dV$$

$$\{Q\} = \int_S [H]^T \{f\} dV \quad (9.273)$$

$$\{P\} = \int_S [H]^T \{T^{(n)}\} dS$$

The matrix $[K]$ is the stiffness matrix, $\{Q\}$ is a vector of equivalent nodal body force terms, and $\{P\}$ is a set of nodal forces representing the surface stress vector. If we consider the body here to be an element in a larger body, then we see that Eq. (9.272) is identical to the form we obtained previously for strength of materials problems such as, for example, the axial-load problem. Similarly, the definitions of Eq. (9.273) have analogous terms in those simpler problems. In most cases the terms in Eq. (9.273) must be found through numerical integrations so that, in a general finite element solution process, those integrations become an important part of that process. Needless to say, the assembly process for these general problems is more complex than the simple problems we have considered in this book. For the stiffness matrix, the assembly process is basically placing the element stiffness matrix terms in the appropriate positions in the overall system stiffness matrix for the body. The stress-vector terms are integrations over the surface(s) of the element. Most of those surfaces will be adjacent to other interior element surfaces where the stress-vector terms will cancel, so that the only surface integrations needed for the entire body are for those surface parts of the elements that are at the outer surface of the entire body, where either the stress vectors are known applied values or represent unknown stress vectors corresponding to reactions where the displacements are specified. As with the strength of materials problems we have previously examined, if we take Eq. (9.272) as representing the entire body, the equations involving unknown reactions can be eliminated by applying the displacement boundary conditions and the remaining equations solved for the "free" nodal displacements. The strains and hence the stresses then are also expressible in terms of these nodal displacements. This, in a nutshell, is the stiffness-based finite element method as applied to general 3-D stress analysis problems.

9.9.2 Force-Based Finite Elements

The force-based approach relies on both the principle of virtual work and the principle of complementary virtual work. Thus, in addition to Eq. (9.270) we must examine the complementary virtual work principle. From Eq. (8.96) we found

$$\int_V \sum_{i=1}^{3} \sum_{j=1}^{3} e_{ij} \delta \sigma_{ij} dV - \int_S \sum_{i=1}^{3} \delta T_i^{(n)} u_i dS - \int_V \sum_{i=1}^{3} \delta f_i u_i dV = 0 \qquad (9.274)$$

As in the strength of materials problems we previously examined, we will consider this principle for variations of the stresses that satisfy homogeneous equilibrium equations ($\delta f_i = 0$), so when we write this principle in matrix–vector form we have

$$\int_V \{\delta \sigma\}^T \{e\} dV = \int_S \{\delta T^{(n)}\}^T \{u\} dS \qquad (9.275)$$

Thus, when we write the stresses in terms of a vector of generalized "force" parameters, $\{F\}$ (which are not values defined at nodes, as mentioned previously), as

$$\{\sigma\} = [N]\{F\} \qquad (9.276)$$

we will use shape functions $[N]$ that satisfy the homogeneous equilibrium equations, i.e., $[\mathcal{L}_E][N] = 0$. The principle of virtual work now becomes, with the use of the displacement and strain expressions defined previously,

$$\{\delta U\}^T \int_V [B]^T [N] dV \{F\} = \{\delta U\}^T \int_V [H]^T \{f\} dV + \{\delta U\}^T \int_S [H]^T \{T^{(n)}\} dS \qquad (9.277)$$

which for arbitrary variations of the nodal displacements gives the set of equilibrium equations for the generalized forces

$$\int_V [B]^T [N] dV \{F\} = \int_V [H]^T \{f\} dV + \int_S [H]^T \{T^{(n)}\} dS \qquad (9.278)$$

which we write more succinctly as

$$[\tilde{E}]\{F\} = \{Q\} + \{P\} \qquad (9.279)$$

where $\{Q\}$ and $\{P\}$ have already been defined and we also have the equilibrium matrix, $[\tilde{E}]$, given by

9.9 Finite Elements for General Deformable Bodies

$$[\tilde{E}] = \int_V [B]^T [N] dV \tag{9.280}$$

Now, consider the principle of complementary virtual work, Eq. (9.275). Using the stress–strain law written as $\{e\} = [D]\{\sigma\}$ in terms of the 6×6 symmetric compliance matrix, $[D]$, we find

$$\{\delta F\}^T \int_V [N]^T [D][N] dV \ \{F\} = \int_S \left\{ \delta T^{(n)} \right\}^T [H] dS \ \{U\} \tag{9.281}$$

However, from Eq. (9.278) since the variations of the stresses must satisfy the homogeneous equilibrium equations, we have

$$[\tilde{E}]\{\delta F\} = \int_S [H]^T \left\{ \delta T^{(n)} \right\} dS \tag{9.282}$$

or, taking the transpose of Eq. (9.282),

$$\{\delta F\}^T [\tilde{E}]^T = \int_S \left\{ \delta T^{(n)} \right\}^T [H] dS \tag{9.283}$$

Using Eq. (9.283) in Eq. (9.281) and the fact that the resulting equation must be satisfied for all variations of the unknown forces, we obtain

$$[G]\{F\} = [\tilde{E}]^T \{U\} \tag{9.284}$$

where $[G]$ is the flexibility matrix defined as

$$[G] = \int_V [N]^T [D][N] dV \tag{9.285}$$

We can also define a vector of generalized strains, $\{\Delta\}$, associated with the generalized forces, $\{F\}$, through

$$\{\Delta\} = [G]\{F\} \tag{9.286}$$

in which case we obtain the generalized strain–displacement relations as

$$\{\Delta\} = [\tilde{E}]^T \{U\} \tag{9.287}$$

Thus, the three basic relations we have developed for the force-based method are:

$$\begin{aligned} [\tilde{E}]\{F\} &= \{Q\} + \{P\} \\ \{\Delta\} &= [\tilde{E}]^T \{U\} \\ \{\Delta\} &= [G]\{F\} \end{aligned} \tag{9.288}$$

We can consider Eq. (9.288) as the equations for a single element or, if we assemble all those elements, Eq. (9.288) can represent the relations for the entire body where the flexibility matrix will be the concatenation (along the diagonal of the of the flexibility matrix for the entire body) of the element flexibility matrices, as we have seen in the strength of materials problems. The assembly of the equilibrium matrices of the elements, as we have seen in the strength of materials problems, is different in detail from the assembly process for the stiffness-based method but again it is basically the placing of the element equilibrium matrices in the proper places of the overall equilibrium matrix for the entire body. If we consider Eq. (9.288) as the assembled equations for the entire body that includes reactions as well as the elastic internal generalized forces, then we can eliminate the unknown reaction forces from those equilibrium equations, producing a "free" equilibrium matrix $\left[\tilde{E}_f\right]$ and similarly eliminate the known displacements, leaving the "free" displacements $\{U_f\}$. We then obtain these equations in the form

$$\left[\tilde{E}_f\right]\{F_e\} = \{Q\} + \{P\}$$
$$\{\Delta\} = \left[\tilde{E}_f\right]^T \{U_f\} \qquad (9.289)$$
$$\{\Delta\} = [G]\{F_e\}$$

where $\{P\}$ now will contain only known applied loads and $\{F\}$ contains only the internal forces, which we have previously called $\{F_e\}$. To solve these equations, we obtain the compatibility equations by eliminating the free displacements from the strain–displacement relations to produce the compatibility equations

$$\left[\tilde{S}\right]\{\Delta\} = \left[\tilde{S}\right][G]\{F_e\} = 0 \qquad (9.290)$$

which we append to the "free" equilibrium equations to obtain the final system of equations

$$\left[\tilde{E}_{system}\right]\{F_e\} \equiv \begin{bmatrix} \left[\tilde{E}_f\right] \\ - \\ \left[\tilde{S}\right][G] \end{bmatrix} \{F_e\} = \begin{Bmatrix} Q \\ - \\ 0 \end{Bmatrix} + \begin{Bmatrix} P \\ - \\ 0 \end{Bmatrix} \qquad (9.291)$$

These equations can then be solved for the unknown generalized forces. As demonstrated earlier, we can also obtain the unknown free nodal displacements from these forces through (see Eq. (9.111))

$$\begin{Bmatrix} \{U_f\} \\ - \\ \{0\} \end{Bmatrix} = \left(\left[\tilde{E}_{system}\right]^T\right)^{-1} [G]\{F_e\} \qquad (9.292)$$

Note that the inverse and the transpose in Eq. (9.292) can be interchanged without changing the end result. Also note that we can easily include specified displacements at nodes by including terms similar to those discussed in the previous section but we will not show that generalization here.

We have also shown previously that we can develop a pseudostiffness type of approach with the force-based method. From the equations in Eq. (9.289) we can express the forces in terms of the displacements and arrive at the equations

$$[\tilde{E}_f][G]^{-1}[\tilde{E}_f]^T\{U_f\} = \{Q\} + \{P\} \qquad (9.293)$$

which can be written as

$$[K_{pseudo}]\{U_f\} = \{Q\} + \{P\} \qquad (9.294)$$

where the pseudostiffness matrix, $[K_{pseudo}]$, is

$$[K_{pseudo}] = [\tilde{E}_f][G]^{-1}[\tilde{E}_f]^T \qquad (9.295)$$

Thus, all of the equations for a general deformable body using the force-based finite element approach are identical in form to the equations seen in the strength of materials problems.

One of the crucial parts of the force-based method is to obtain a set of shape functions for the stresses that satisfy the homogeneous equilibrium equations, i.e., $[\mathcal{L}_E][N] = 0$. For the strength of materials problems, this satisfaction was very easy but it is less so for more general problems. Fortunately, one can use stress functions to help define the needed shape functions. For 2-D problems, for example, the Airy stress function can be employed. The details are beyond the scope of this book but the reader can find further information in [1] about the conditions that must be met for acceptable shape functions. Determining the compatibility matrix, $[\tilde{S}]$ is also an important step. As we have seen, this entails determining the null space of the equilibrium matrix, $[\tilde{E}_f]$, or explicitly eliminating the free displacements from the generalized strain–displacement relations $\{\Delta\} = [\tilde{E}]^T\{U_f\}$ to produce the compatibility equations $[\tilde{S}]\{\Delta\} = 0$. Neither the shape function requirements nor the need to generate a compatibility matrix prevent the force-based finite element method from being a very general computational tool capable of solving complex engineering problems (see [4], [5], [6]). It is hoped that the coverage of the force-based method in this book, which is a departure from the numerical stress analysis discussions found in most texts, can help to raise the awareness of this method.

9.10 The Boundary Element Method

Navier's equations are partial differential equations for the displacements that, together with appropriate boundary conditions, form the governing equations of a deformable body. The algebraic equations of stiffness-based or force-based finite elements together with boundary conditions provide an alternate set of (approximate) governing equations. There is, however, another formulation of the governing equations based on integral equations defined over the surface of a deformable body that can be used effectively. If those surface integral equations are solved approximately by breaking the surface of a body into elements and replacing the integral equations by a set of algebraic equations, this method is called the boundary element method.

The starting point for formulating the boundary element method is usually the reciprocity theorem for a linear elastic body developed in Chapter 8 (see Eq. (8.175)) where here we have let $T_j^{(m)} = t_j^{(m)}$), which is a relationship between two different solutions (labeled (1) and (2)) in that body, namely

$$\sum_{j=1}^{3}\int_S t_j^{(1)}u_j^{(2)}dS + \sum_{j=1}^{3}\int_V f_j^{(1)}u_j^{(2)}dV = \sum_{j=1}^{3}\int_S t_j^{(2)}u_j^{(1)}dS + \sum_{j=1}^{3}\int_V f_j^{(2)}u_j^{(1)}dV \qquad (9.296)$$

Now consider taking solution (2) to be a set of body forces given by

$$f_j^{(2)} = e_j\delta(\mathbf{x}-\mathbf{y}) \quad (j=1,2,3) \qquad (9.297)$$

where e_j is a unit vector acting in the x_j direction and $\delta(\mathbf{x}-\mathbf{y})$ is a generalized "function" representing a concentrated body force acting at a point $\mathbf{y}$ in the body. This generalized "function" has the property that, for $\mathbf{x} \neq \mathbf{y}$ $\delta(\mathbf{x}-\mathbf{y}) = 0$ and for any function $g(\mathbf{y})$,

$$\int_V g(\mathbf{y})\delta(\mathbf{x}-\mathbf{y})dV(\mathbf{y}) = \begin{cases} g(\mathbf{x}) & \mathbf{y} \text{ inside } V \\ 0 & \mathbf{y} \text{ outside } V \end{cases} \qquad (9.298)$$

The three body forces will each generate three displacements in the body, producing a 3×3 matrix of displacements, $U_{ij}(\mathbf{x},\mathbf{y})$ (see Fig. 9.30), where the j subscript represents the x_j direction of the displacement at a point $\mathbf{x}$ in the body, and the i subscript represents the direction (along the x_i-axis) of the concentrated body force acting at point $\mathbf{y}$. Similarly, these body forces will produce a 3×3 matrix of stress vectors, T_{ij}, as shown in Fig. 9.31. The total displacement generated by these unit forces is $u_j^{(2)} = \sum_{i=1}^{3}U_{ij}e_i$ and the total stress vector is $t_j^{(2)} = \sum_{i=1}^{3}T_{ij}e_i$. For an isotropic linear elastic solid, one can obtain explicit expressions for these point-source solutions (which are also called *fundamental solutions*) given by [7]:

$$U_{ij}(\mathbf{x},\mathbf{y}) = \frac{1}{16\pi G(1-v)r}\left[(3-4v)\delta_{ij} + \frac{\partial r}{\partial x_i}\frac{\partial r}{\partial x_j}\right]$$

$$T_{ij}(\mathbf{x},\mathbf{y}) = \frac{-1}{8\pi(1-v)r^2}\left[\frac{\partial r}{\partial n}\left[(1-2v)\delta_{ij} + 3\frac{\partial r}{\partial x_i}\frac{\partial r}{\partial x_j}\right] + (1-2v)\left(n_i\frac{\partial r}{\partial x_j} - n_j\frac{\partial r}{\partial x_i}\right)\right] \qquad (9.299)$$

where (G, v) are the shear modulus and Poisson's ratio, respectively, δ_{ij} is the Kronecker delta, n_i are components of a unit normal, and $r = |\mathbf{x}-\mathbf{y}|$ is the distance between points $\mathbf{x}$ and $\mathbf{y}$ (Figs. 9.30 and 9.31). Placing these fundamental solutions into Eq. (9.296) as solution (2) we find

$$\sum_{i=1}^{3}\sum_{j=1}^{3}\left(\int_S U_{ij}(\mathbf{x}_s,\mathbf{y})t_j(\mathbf{x}_s)dS(\mathbf{x}_s)\right)e_i + \sum_{i=1}^{3}\sum_{j=1}^{3}\left(\int_V U_{ij}(\mathbf{x},\mathbf{y})f_j(\mathbf{x})dV(\mathbf{x})\right)e_i$$

$$= \sum_{i=1}^{3}\sum_{j=1}^{3}\left(\int_S T_{ij}(\mathbf{x}_s,\mathbf{y})u_j(\mathbf{x}_s)dS(\mathbf{x}_s)\right)e_i + \sum_{i=1}^{3}\left(\int_V u_i(\mathbf{x})\delta(\mathbf{x}-\mathbf{y})dV(\mathbf{x})\right)e_i \qquad (9.300)$$

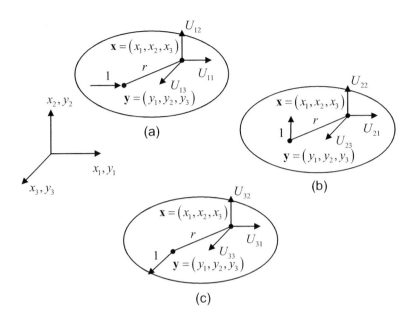

Figure 9.30 The components of the displacements, U_{ij}, in a deformable body at point **x**, when a concentrated body force at point **y** in the body acts (a) in the x_1-direction, (b) in the x_2-direction, and (c) in the x_3-direction.

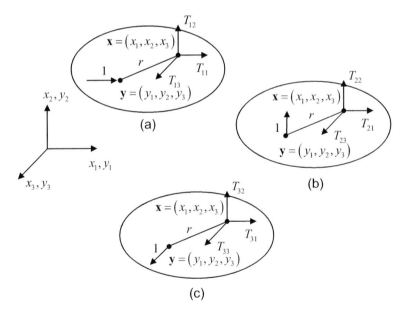

Figure 9.31 The components of the stress vector, T_{ij}, in a deformable body at point **x**, when a concentrated body force at point **y** in the body acts (a) in the x_1-direction, (b) in the x_2-direction, and (c) in the x_3-direction.

where $\mathbf{x}_s$ is an arbitrary point on the surface, S, and **x** is an arbitrary point in the volume, V. We have dropped the (1) superscript on solution (1), which we will take as the problem we wish to solve for the deformable body. Using the sampling properties of the generalized delta function $\delta(\mathbf{x} - \mathbf{y})$ given by Eq. (9.298) and the fact that Eq. (9.300) must be true for each component, e_i ($i = 1, 2, 3$) separately, we find

$$\sum_{j=1}^{3}\int_{S}U_{ij}(\mathbf{x}_s,\mathbf{y})t_j(\mathbf{x}_s)dS(\mathbf{x}_s) + \sum_{j=1}^{3}\int_{V}U_{ij}(\mathbf{x},\mathbf{y})f_j(\mathbf{x})dV(\mathbf{x})$$
$$= \sum_{j=1}^{3}\int_{S}T_{ij}(\mathbf{x}_s,\mathbf{y})u_j(\mathbf{x}_s)dS(\mathbf{x}_s) + u_i(\mathbf{y}) \qquad (i=1,2,3)$$
(9.301)

Equation (9.301) is called *Somigliana's identity*. It can be written more compactly in matrix–vector notation as

$$\{u(\mathbf{y})\} = \int_{S}[U(\mathbf{x}_s,\mathbf{y})]\{t(\mathbf{x}_s)\}dS(\mathbf{x}_s) - \int_{S}[T(\mathbf{x}_s,\mathbf{y})]\{u(\mathbf{x}_s)\}dS(\mathbf{x}_s)$$
$$+ \int_{V}[U(\mathbf{x},\mathbf{y})]\{f(\mathbf{x})\}dV(\mathbf{x})$$
(9.302)

If the stress vector, $\{t\}$, and the displacements, $\{u\}$, are known on the surface (as well as the body force, $\{f\}$, acting in the volume) then Eq. (9.302) is an integral representation that gives the displacement at any point, $\mathbf{y}$, in the body. Of course, in general we do not know both the surface displacements and stress vector on the surface, so before we can use Eq. (9.302) we must solve for whatever displacements of stress-vector components are not known. As it stands, Eq. (9.302) is a mixture of surface and volume integrals. For many body forces, however, we can transform the volume integral into an equivalent surface integral (see [8], for example, for the details) in which case Eq. (9.302) will be a purely surface integral relationship. This will obviously also be the case if there are no body forces. In either case, the body force term will be a known integral so it does not play a major role in the remainder of the solution process. Thus, we will henceforth drop this term to keep our discussion focused on the other important terms.

We can turn the integral relationship of Eq. (9.302) into an integral equation by taking the limit as the point $\mathbf{y}$ in the volume, V, goes to a point, $\mathbf{y}_s$, on the surface. However, this limit must be done with care since the fundamental solution displacements and stress-vector terms are singular when $\mathbf{x}_s = \mathbf{y}_s$ (see the expressions in Eq. (9.299) where the distance, r, appears in the denominator). We will not go into the details of this process but simply write down the end result, namely

$$[C]\{u(\mathbf{y}_s)\} = \int_{S}[U(\mathbf{x}_s,\mathbf{y}_s)]\{t(\mathbf{x}_s)\}dS(\mathbf{x}_s) - \int_{S}[T(\mathbf{x}_s,\mathbf{y}_s)]\{u(\mathbf{x}_s)\}dS(\mathbf{x}_s) \qquad (9.303)$$

where there is a known constant 3×3 matrix, $[C]$, that depends on the geometry of the surface at point $\mathbf{y}_s$. If the surface is smooth at that point $[C]$ is just the scalar value of 1/2. Both integrals in Eq. (9.303) are singular integrals where the integral involving $[U]$ is weakly singular while the integral containing $[T]$ is highly singular and must be interpreted as a Cauchy principal value integral. Again, we will not go into the details of how those integrals are calculated but will refer you to some of the many texts (for example, [7], [8]) that describe the integration procedures explicitly.

Equation (9.303) is a surface (boundary) integral equation that we can use to find whatever displacement or stress-vector components are unknown. For example, if the stress vector is known for the entire surface, then Eq. (9.303) is an integral equation (called an integral equation of the second kind [9]) for the unknown surface displacements. If, instead, the displacements were given for the entire surface, Eq. (9.303) is an integral equation for the unknown stress-vector components (called an integral equation of the first kind [9]). Other combinations of known displacement and stress-vector components similarly lead to integral equations for whatever are the unknown displacements or stress-vector components. To solve for the unknowns in these integral equations, we can break the surface (boundary) up into elements where we assume the displacement and stress-vector components have a simple, known behavior within each element. This is obviously similar to what we do in finite elements so this type of solution is called the boundary element method. For example, we could take both the displacement and stress vectors to be simply constants within a surface element, S_j, that are equal to their values at the centroid of the element. In that simple case then Eq. (9.303) becomes a system of linear equations

$$\left[C^{(i)} \right] \left\{ u^{(i)} \right\} = \sum_{j=1}^{N} \int_{S_j} \left[U\left(\mathbf{x}_s, \mathbf{y}^{(i)}\right) \right] dS(\mathbf{x}_s) \left\{ t^{(j)} \right\} - \sum_{j=1}^{N} \int_{S_j} \left[T\left(\mathbf{x}_s, \mathbf{y}^{(i)}\right) \right] dS(\mathbf{x}_s) \left\{ u^{(j)} \right\} \quad (9.304)$$

for the known and unknown centroidal (nodal) values of displacements and stress-vector components of the N elements. These known and unknown values can be rearranged to produce a set of simultaneous linear equations for the unknowns that can then be solved. Once this solution is available, then all the displacements and stress-vector components in Eq. (9.302) are known and can be used to obtain the displacements at any point within the body. If the displacement expression of Eq. (9.302) is differentiated, then a similar integral relationship for the stresses can be obtained and used to find those interior stress values at any point as well.

This is a very abbreviated description of the boundary element method. Although we have examined only the case when the shape functions were constants for the elements, other shape functions can be used to make the method more efficient. As we have seen, the method basically relies on the representation of the unknowns as a superposition of known solutions (the fundamental solutions) to generate the requisite equations for those unknowns. Thus, the method requires that we know those fundamental solutions and, since it relies on superposition, it typically means that we must be dealing with linear problems. In contrast, the finite element method can be generalized to nonlinear as well as linear problems. One of the main attractive features of the boundary element method is that it requires that we break only the surface of the body into elements, not the entire volume. This usually means that the discretization of the geometry of the body is easier with boundary elements than with finite elements. This feature also makes it easier to handle problems with boundary elements where the geometry is highly irregular, such as in cracked bodies. As we have seen in this short description, the integrations involved in boundary elements are a very important part of the method since one must deal with fundamental solutions and their derivatives, which are singular. Thus, the details of these integrations make the boundary element method inherently more "mathematical" than finite elements, where there are also integrals involved but where those integrals are normally well-behaved.

9.11 PROBLEMS

P9.1 Consider the statically indeterminate beam problem of Fig. 9.25c, which was solved for the generalized internal forces and reactions using the internal moments (M_1, M_2) as the fundamental unknowns in the complementary strain energy expression in Example 9.11 in Section 9.8.

(a) Obtain the unknown slope of the beam at the roller (which is one of the generalized displacements present in this problem) by extending the strain–displacement relations, as done previously, and solving in MATLAB® the set of equations

$$\{\hat{\Delta}\} = \left[\hat{E}_{system}\right]^T \begin{Bmatrix} \{U\} \\ - \\ \{0\} \end{Bmatrix} \tag{P9.1}$$

for the generalized displacements $\{U\}$ where $\left[\hat{E}_{system}\right]$ is the matrix of equilibrium equations that include the internal moments and reactions together with the compatibility equation.

(b) Use the solution for the generalized forces obtained in Section 9.8 for this problem and the unit-load method of Chapter 8 to obtain the slope of the beam at the roller and verify that it agrees with the solution of Eq. (P9.1).

P9.2 Consider the statically indeterminate beam problem of Fig. 9.25c which was solved for the generalized internal forces and reactions using either the internal moments (M_1, M_2) or the reaction force, V_{R2}, as the fundamental unknowns in the complementary strain energy expression in Example 9.11 in Section 9.8.

(a) Write the complementary strain energy instead in terms of the reactions (V_{R1}, M_{R1}) at the wall and use it to obtain the relationship between the generalized strains and generalized forces. Solve the combination of the equilibrium equations obtained in Section 9.8 and the resulting compatibility equation for all the unknown generalized forces in this problem using MATLAB®.

(b) Use the strain–displacement relations to try to solve for the displacements in this problem. Do you obtain a correct solution? Give a physical interpretation of those displacements that are nonzero.

P9.3 Consider the statically indeterminate beam of Fig. P9.1. Write the equations of equilibrium for the entire beam and solve for all the reactions by
(a) using the principle of least work,

(b) integrating the moment–curvature expression $M(x) = EI d^2 w/dx^2$, where w is the vertical displacement, and applying the slope and displacement boundary conditions,

(c) using the force-based approach of Eq. (9.219).

Applying the principle of least work to solve statically indeterminate problems is often found in intermediate and advanced strength of materials texts, while the use of the integration method is normally discussed in elementary texts. The force-based approach of Eq. (9.219) is an alternative that is not commonly covered.

> The problems shown in Fig. P9.2 and Fig. P9.3 are axial-load and beam problems considered in the problems at the end of Chapter 1. There, you were asked to obtain the exact analytical solutions. Here, we want to develop MATLAB® finite element solutions of various types. These are relatively lengthy problems that are probably best treated as class projects. For each case you should write a MATLAB® script that will let you obtain a finite element solution where the number of elements can be varied (but the number of elements must be an even number to match the boundary conditions seen in some of the problems). Note that the displacement boundary conditions are the same in all of these problems (fixed–fixed restraints) except for the problem in Fig. P9.3d so that you only need to change the force boundary conditions in those problems with common restraints. Write these scripts, treating these problems as numerical problems, not symbolic algebra problems.

P9.4 The axial-load problem shown in Fig. P9.2a is the fixed-fixed bar with a distributed load discussed with the stiffness-based method in Example 9.2 and with the force-based method in Example 9.5. Write a MATLAB® script that solves this problem via a stiffness-based finite element solution for any number of elements. (Note, however, you should use an even number of elements, as this is required by the loading in other cases we will consider.) Plot the nodal displacements in the bar. Plot the internal forces (which are constants), as discrete values at the centroids of

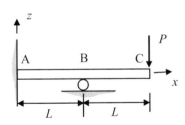

Figure P9.1 A statically indeterminate beam problem.

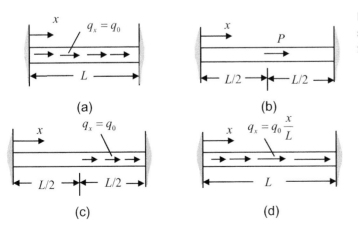

Figure P9.2 Axial-load problems whose solutions are to be obtained by various finite element methods.

Figure P9.3 Beam problems whose solutions are to be obtained by various finite element methods.

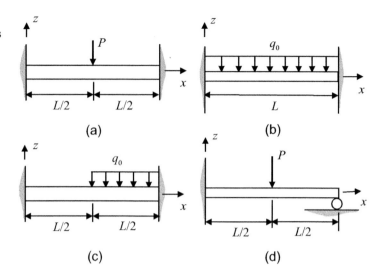

the elements. Obtain the solution and plots for 20 elements. Compare these plots to the exact solution obtained in Chapter 1 for this problem. Let $L = 3000$ mm, $A = 9$ mm^2, $E = 10\,000$ N/mm^2 (10 GPa), $q_0 = 10$ N/mm.

P9.5 Repeat problem P9.4, but use a force-based finite element solution instead.

P9.6 Repeat problem P9.4, but use a pseudostiffness finite element solution instead.

P9.7 Consider the axial load problem in Fig. P9.2b. Solve this problem using 20 elements and a stiffness-based finite element method. Obtain the same types of plots as in problem P9.4 and compare with the exact solution obtained in Chapter 1 for this problem. Let $L = 3000$ mm, $A = 9$ mm^2, $E = 10\,000$ N/mm^2 (10 GPa), $P = 20\,000$ N.

P9.8 Repeat problem P9.7, but use a force-based finite element method instead.

P9.9 Repeat problem P9.7, but use a pseudostiffness finite element method instead.

P9.10 Consider the axial load problem in Fig. P9.2c. Solve this problem using 20 elements and a stiffness-based finite element method. Obtain the same types of plots as in problem P9.4 and compare with the exact solution obtained in Chapter 1 for this problem. Let $L = 3000$ mm, $A = 9$ mm^2, $E = 10\,000$ N/mm^2 (10 GPa), $q_0 = 10$ N/mm.

P9.11 Repeat problem P9.10, but use a force-based finite element solution instead.

P9.12 Repeat problem P9.10, but use a pseudostiffness finite element solution instead.

P9.13 Consider the axial-load problem in Fig. P9.2d. Solve this problem using 20 elements and a stiffness-based finite element method. Obtain the same types of plots as in problem P9.4 and compare with the exact solution obtained in Chapter 1 for this problem. Let $L = 3000$ mm, $A = 9$ mm^2, $E = 10\,000$ N/mm^2 (10 GPa), $q_0 = 10$ N/mm.

P9.14 Repeat problem P9.13, but use a force-based finite element solution instead.

P9.15 Repeat problem P9.13, but use a pseudostiffness finite element solution instead.

P9.16 Consider the fixed–fixed beam with a central concentrated load as shown in Fig. P9.3a. Solve this problem using 20 elements and the stiffness-based finite element method. Obtain solutions and plots for the internal shear force, bending moment, and displacement, and compare with the exact solution obtained in Chapter 1 for this problem. Let $L = 3000$ mm, $E = 2 \times 10^5 \text{N/mm}^2$ (200 GPa), $I = 3 \times 10^6$ mm^4, $P = 20\,000$ N.

P9.17 Repeat problem P9.16, but use a force-based finite element solution instead.

P9.18 Repeat problem P9.16, but use a pseudostiffness finite element solution instead.

P9.19 Consider the fixed–fixed beam with a uniform distributed load shown in Fig. P9.3b. Solve this problem using 20 elements and the stiffness-based finite element method. Obtain solutions and plots for the internal shear force, bending moment, and displacement and compare with the exact solution obtained in Chapter 1 for this problem. Let $L = 3000$ mm, $E = 2 \times 10^5 \text{N/mm}^2$ (200 GPa), $I = 3 \times 10^6$ mm^4, $q_0 = 10$ N/mm.

P9.20 Repeat problem P9.19, but use a force-based finite element solution instead.

P9.21 Repeat problem P9.19, but use a pseudostiffness finite element solution instead.

P9.22 Consider the fixed–fixed beam with a uniform distributed load acting over its right half as shown in Fig. P9.3c. Solve this problem using 20 elements and the stiffness-based finite element method. Obtain solutions and plots for the internal shear force, bending moment, and displacement, and compare with the exact solution obtained in Chapter 1 for this problem. Let $L = 3000$ mm, $E = 2 \times 10^5 \text{N/mm}^2$ (200 GPa), $I = 3 \times 10^6$ mm^4, $q_0 = 10$ N/mm.

P9.23 Repeat problem P9.22, but use a force-based finite element solution instead.

P9.24 Repeat problem P9.22, but use a pseudostiffness finite element solution instead.

P9.25 Consider a beam that is clamped at one end and simply supported at the other, and carries a concentrated load at its center, as shown in Fig. P9.3d. Solve this problem using 20 elements and the stiffness-based finite element method. Obtain solutions and plots for the internal shear force, bending moment, and displacement, and compare with the exact solution obtained in Chapter 1 for this problem. Let $L = 3000$ mm, $E = 2 \times 10^5 \text{N/mm}^2$ (200 GPa), $I = 3 \times 10^6$ mm^4, $P = 20\,000$ N. Note that the displacement boundary conditions are different from the other beam cases considered in Fig. P9.2.

P9.26 Repeat problem P9.25, but use a force-based finite element solution instead.

P9.27 Repeat problem P9.25, but use a pseudostiffness finite element solution instead.

P9.28 Consider the simply supported beam shown in Fig. P9.4 where a beam of length $2L$ with three supports carries a uniform distributed force q_0 (force/unit length). Solve this statically indeterminate problem using the force-based method of Eq. (9.219) where we combine equilibrium and compatibility and use the complementary strain energy to obtain the compatibility equation. Take a single "element" (i.e., the entire beam) and obtain the unknown reactions in terms of q_0 and L.

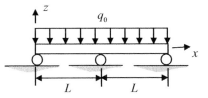

Figure P9.4 A simply supported beam with three supports.

P9.29 Consider the statically indeterminate beam problem of Fig. 9.25c where a beam is fixed at one end and simply supported by a roller at the other end and subjected to a uniform distributed force, q_0, over its entire length, L. In this case, assume that the simply supported end is given a known z-displacement, w_R. Show explicitly the following steps.

(a) Using a single element (i.e., the entire beam), write down the finite element equilibrium equations for the internal moments (M_1, M_2) in terms of the reactions and the equivalent nodal forces and moments coming from the distributed load.

(b) Eliminate those equations that involve the unknown reactions and write the remaining equilibrium equation in terms of a free equilibrium matrix, $[\tilde{E}_f]$. From the equilibrium equations of part (a), identify the $[\tilde{E}_R]$ matrix associated with the known displacement and obtain (1) the free equilibrium equation, (2) the strain–displacement relation (in terms of $[\tilde{E}_f]^T$ and $[\tilde{E}_R]^T$) that relates the generalized strains (Δ_1, Δ_2) associated with (M_1, M_2) to the free generalized displacement, and (3) the relationship between the generalized strains and the generalized internal forces (moments). These are the three sets of equations seen in Eq. (9.229).

(c) Obtain the compatibility matrix associated with the free equilibrium matrix and use it to write the compatibility equation in terms of the internal generalized forces (see Eq. (9.230)). Solve the combination of the free equilibrium equation and the compatibility equation for the internal generalized forces and obtain the value of the reaction force at the roller in terms of q_0, L, EI, and w_R.

(d) Finally, determine the unknown slope at the roller in terms of q_0, L, EI, and w_R using Eq. (9.237). You might want to use the symbolic algebra capabilities of MATLAB® for this part. Can you give a simple explanation for the terms appearing in the expressions for the reaction force and slope at the roller due to the known displacement?

P9.30 In this chapter we showed how known support displacements can be incorporated into the force-based finite element method. Incorporation of known displacements with the stiffness-based method is much easier since we are already working with the displacements. Consider for example, the very simple problem of Fig. 9.28 where a bar of length $2L$ has a concentrated axial load P at its center and the end supports have displacements $(\bar{U}_1, \bar{U}_3)$. Using two elements:

(a) Write down the stiffness-based equations of equilibrium for the bar,

(b) Eliminate the equations involving the end reactions from the equilibrium matrix and move the columns of the matrix associated with the known displacements to the right side with the known load(s). Solve for the free displacement(s) in terms of the known load(s) and displacements. This same process will work for much more general problems.

(c) Determine the internal forces in the two elements. Show all your results agree with the force-based solution obtained previously.

References

1. S. N. Patnaik, D. A. Hopkins, and G. R. Halford, Integrated Force Method Solution to Indeterminate Structural Mechanics Problems (NASA/TP-2004-207430, Washington, D.C., 2004)
2. I. Kaljević, S. N. Patnaik, and D. A. Hopkins, "Three-dimensional structural analysis by the Integrated Force Method," Comp. Struct., **58** (1996), 869–886
3. S. N. Patnaik and D. A. Hopkins, *Strength of Materials: A Unified Theory* (Oxford, UK: Elsevier, 2004)
4. I. Kaljević, S. N. Patnaik, and D. A. Hopkins, Development of Finite Elements for Two-Dimensional Structural Analysis using the Integrated Force Method (NASA Technical Memorandum 4655, 1996)
5. I. Kaljević, S. N. Patnaik, and D. A. Hopkins, Element Library for Three-Dimensional Stress Analysis by the Integrated Force Method (NASA Technical Memorandum 4686, 1996)
6. S. N. Patnaik, L. Berke, and R. H. Gallagher, Integrated Force Method Versus Displacement Method for Finite Element Analysis (NASA Technical Paper 2937, 1990)
7. C. A. Brebbia and J. Dominguez, *Boundary Elements: An Introductory Course*, 2nd edn (New York, NY: McGraw-Hill, 1992)
8. J. T. Katsikadelis, *The Boundary Element Method for Scientists and Engineers* (Boston, MA: Academic Press, 2016)
9. M. A. Jaswon and G. T. Symm, *Integral Equation Methods in Potential Theory and Elastostatics* (Boston, MA: Academic Press, 1977)

10 Unsymmetrical Beam Bending

The engineering beam-bending theory summarized in Chapter 1 assumed that the beam cross-sectional area has a plane of symmetry and that bending moments were acting along a single axis (taken to be the y-axis). In this chapter we want to remove those restrictions and to examine the multiaxis bending of beams with nonsymmetrical cross-sections. This will lead to generalizations of the flexure formula (see Eq. (1.20)). In Chapter 1 we also obtained an expression for the shear stresses induced in symmetrical beams. It is difficult to obtain similar analytical shear stress forms for beams with unsymmetrical cross-sections except in the case when the cross-section is thin, a case we will also consider in this chapter.

10.1 Multiple Axis Bending of Nonsymmetrical Beams

Figure 10.1 shows a cantilever beam with a triangular-area cross-section, a typical example we will use to discuss the extensions needed in elementary beam theory to handle the bending of such unsymmetrical cross-sectional beams. Also shown in Fig. 10.1 are the positive directions assumed for the internal bending moments (M_y, M_z), shear forces (V_y, V_z), and torsional moment, M_x, acting in the beam. These directions as well as the coordinate system shown will be used consistently throughout this chapter and in Chapters 11, 12 when we discuss torsion and combined bending and torsion problems. With these assumed directions, when we look down the negative y-axis at a small element of the beam, we see the free body diagram of Fig. 10.2a and obtain from equilibrium

$$\sum M_o = M_y + V_z dx - \left(M_y + \frac{dM_y}{dx} dx \right) = 0$$
$$\rightarrow \frac{dM_y}{dx} = V_z$$
(10.1)

(Any distributed loads are not shown as they do not affect the moment shear relationships.) Similarly, looking down the z-axis (Fig. 10.2b) we find

$$\sum M_o = -M_z + V_y dx - \left(M_z + \frac{dM_z}{dx} dx \right) = 0$$
$$\rightarrow \frac{dM_z}{dx} = -V_y$$
(10.2)

Equation (10.1) is the same relationship we obtained earlier (see Eq. (1.23)) while Eq. (10.2) is new because in Chapter 1 the internal bending moment was assumed to act only about the y-axis.

10.1 Multiple Axis Bending of Nonsymmetrical Beams

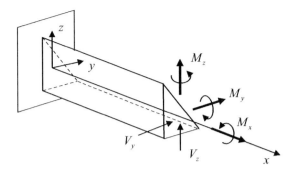

Figure 10.1 A cantilever beam with an unsymmetrical cross-section and the positive directions assumed for the internal moments and shear forces in the beam.

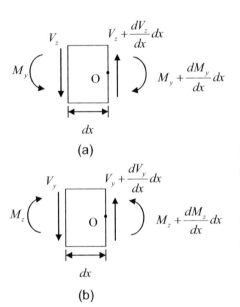

Figure 10.2 (a) The internal moments and shear forces seen looking down the negative y-axis. (b) The internal moments and shear forces seen when looking down the z-axis.

The main difference between bending of an unsymmetrical beam and a symmetrical beam is that, even if the internal moment acts only about the plus or minus y-axis, as assumed in Chapter 1, bending displacements will take place about both the y- and z-axes. Multiaxis bending will also occur when we have internal moments acting about both the y- and z-axes, as assumed here. Consider, as we did in Chapter 1, the pure bending case but where now there are both internal moments (M_y, M_z) acting. If we assume, as we did in Chapter 1, that cross-sections that were originally planar cross-sections normal to the beam axis in the undeformed beam are still planar and normal to the neutral axis in the deformed beam, then a cross-section plane AB has the behavior seen in Figs. 10.3a, b, where the plane experiences small rotations about both the negative y-axis (Fig. 10.3a) and the z-axis (Fig. 10.3b), and these rotations are just equal to the slopes of the neutral axis, dw/dx and dv/dx, respectively, where w is the z-displacement of the neutral axis and v is the y-displacement of the neutral axis. Thus, the beam will experience an x-displacement, u_x, that is given by

$$u_x = -z\frac{dw}{dx} - y\frac{dv}{dx} \tag{10.3}$$

This, in turn, will generate an axial normal strain, e_{xx}, given by

$$e_{xx} = \frac{\partial u_x}{\partial x} = -z\frac{d^2w}{dx^2} - y\frac{d^2v}{dx^2} \tag{10.4}$$

Figure 10.3 The behavior of a cross-section AB (originally normal to the x-axis in the undeformed beam) during bending where it assumed the cross-section remains planar and normal to the neutral axis. (a) The plane AB seen in the x–z plane. (b) The same plane AB as seen in the y–z plane.

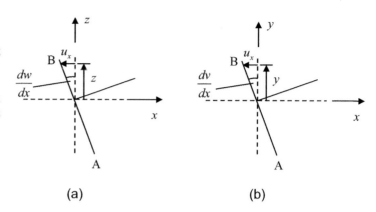

where d^2w/dx^2 and d^2v/dx^2 are curvatures of the beam in the x–z and y–z planes, respectively. If this strain produces only the normal flexure stress, σ_{xx}, we have

$$\sigma_{xx} = E e_{xx} = E\left(-z\frac{d^2w}{dx^2} - y\frac{d^2v}{dx^2}\right) \tag{10.5}$$

where E is Young's modulus. Note that, as in the symmetrical cross-section case, there are also normal strains (e_{yy}, e_{zz}) generated:

$$e_{yy} = e_{zz} = -v e_{xx} = v\left(z\frac{d^2w}{dx^2} + y\frac{d^2v}{dx^2}\right) \tag{10.6}$$

where v is Poisson's ratio.

We can relate the flexure stress to the internal moments via the relations (see Fig. 10.4):

$$\int \sigma_{xx} dA = 0$$

$$\int y\sigma_{xx} dA = -M_z \tag{10.7}$$

$$\int z\sigma_{xx} dA = M_y$$

Placing the normal stress expression of Eq. (10.5) into these relations, we find

$$-E\frac{d^2w}{dx^2}\int z\, dA - E\frac{d^2v}{dx^2}\int y\, dA = 0$$

$$E\frac{d^2w}{dx^2}\int yz\, dA + E\frac{d^2v}{dx^2}\int y^2\, dA = M_z \tag{10.8}$$

$$-E\frac{d^2w}{dx^2}\int z^2\, dA - E\frac{d^2v}{dx^2}\int yz\, dA = M_y$$

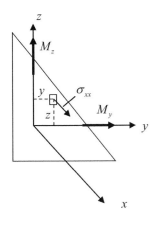

Figure 10.4 Cross-section of an unsymmetrical beam, showing the normal stress and internal bending moments acting in the cross-section.

The first equation in Eq. (10.8) gives

$$\int z dA = \int y dA = 0 \tag{10.9}$$

i.e., the neutral axis is located at the centroid of the cross-section, as it is in the symmetrical cross-section case. The second and third equations in Eq. (10.8) give

$$EI_{yz}\frac{d^2 w}{dx_2} + EI_{zz}\frac{d^2 v}{dx^2} = M_z$$

$$-EI_{yy}\frac{d^2 w}{dx^2} - EI_{yz}\frac{d^2 v}{dx^2} = M_y \tag{10.10}$$

where the second area moments and mixed second area moment of the cross-section are

$$I_{yy} = \int z^2 dA$$

$$I_{zz} = \int y^2 dA \tag{10.11}$$

$$I_{yz} = \int yz dA$$

We can solve the two equations in Eq. (10.10) for the curvatures, obtaining

$$\frac{d^2 v}{dx^2} = \frac{(M_z I_{yy} + M_y I_{yz})}{E(I_{yy}I_{zz} - I_{yz}^2)}$$

$$\frac{d^2 w}{dx^2} = \frac{-(M_y I_{zz} + M_z I_{yz})}{E(I_{yy}I_{zz} - I_{yz}^2)} \tag{10.12}$$

If we place these curvatures into Eq. (10.5), we find the flexure stress is given by

$$\sigma_{xx} = \frac{(M_y I_{zz} + M_z I_{yz})z - (M_z I_{yy} + M_y I_{yz})y}{(I_{yy}I_{zz} - I_{yz}^2)} \tag{10.13}$$

Although we used the case of pure bending to derive Eq. (10.13), as was done in Chapter 1, we can simply let $M_y = M_y(x)$ and $M_z = M_z(x)$ to obtain the equivalent engineering beam theory for the more practical case when bending is due to various applied loads along the beam. This, in turn, will require the existence of shear stresses as well, which we will discuss shortly. If either the y- or z-axis is an axis of symmetry then $I_{yz} = 0$ and we have

$$\sigma_{xx} = \frac{M_y z}{I_{yy}} - \frac{M_z y}{I_{zz}} \tag{10.14}$$

which is just a superposition of the simple flexural stress formula of Chapter 1 for the two internal moments for combined axis bending, and where the difference in sign simply arises because of the assumed directions of the internal moments. Even if we only have a bending moment about the y-axis, as assumed in Chapter 1, for an unsymmetrical cross-section Eq. (10.13) gives

$$\sigma_{xx} = \frac{(M_y I_{zz})z - (M_y I_{yz})y}{\left(I_{yy} I_{zz} - I_{yz}^2\right)} \tag{10.15}$$

which shows that we have bending occurring about both axes, producing a flexure stress that varies linearly along both axes in the cross-section if the cross-section is unsymmetrical.

For multiaxis moments and an unsymmetrical cross-section, Eq. (10.13) shows that the neutral surface in the cross-section (where $\sigma_{xx} = 0$) is along the line $z = \tan\lambda\, y$ (see Fig. 10.5) where

$$\tan\lambda = \frac{M_z I_{yy} + M_y I_{yz}}{M_y I_{zz} + M_z I_{yz}} \tag{10.16}$$

which can also be written as (assuming $M_y \neq 0$, which is usually the case)

$$\tan\lambda = \frac{(M_z/M_y)I_{yy} + I_{yz}}{I_{zz} + (M_z/M_y)I_{yz}} \tag{10.17}$$

Generally, both M_z and M_y are functions of the x-distance in a beam so that the angle of the neutral surface will also be a function of x. However, if the ratio $M_z(x)/M_y(x)$ is a constant, then Eq. (10.17) shows that the angle of the neutral surface will be a constant in the beam. The case where this ratio is a constant is called the case of a *single plane of loading*. Figure 10.6 shows an example of both a single plane of loading case, where all the applied loads lie in a common plane, and a case of general loading. In the latter case the angle neutral surface will vary in the beam.

Equation (10.13) is a general expression for the flexure stress in an unsymmetrical beam under multiaxis bending. There are, however, other forms of this expression that can be obtained. For example, one can use Eq. (10.17) in Eq. (10.13) to eliminate the explicit dependency on M_z in Eq. (10.13) and obtain

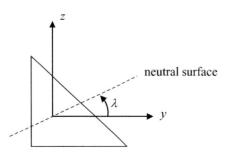

Figure 10.5 The line of the neutral surface in the cross-section of an unsymmetrical beam.

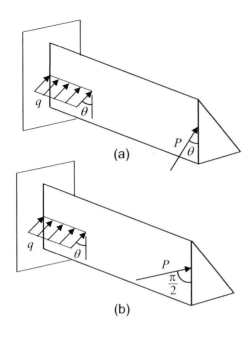

Figure 10.6 (a) Single plane of loading for a beam. (b) A general loading.

$$\sigma_{xx} = \frac{M_y(z - \tan\lambda\, y)}{I_{yy} - I_{yz}\tan\lambda} \tag{10.18}$$

although M_z does not appear explicitly in Eq. (10.18) it is still present implicitly through the $\tan\lambda$ terms. Another choice is to write Eq. (10.13) in terms of a set of principal axes (Y, Z) as shown in Fig. 10.7 where

$$\begin{aligned} Y &= z\sin\theta_p + y\cos\theta_p \\ Z &= z\cos\theta_p - y\sin\theta_p \end{aligned} \tag{10.19}$$

Since principal axes act like axes of symmetry for a general cross-section where $I_{YZ} = 0$ (see Appendix A), we obtain the form of Eq. (10.14) where now

$$\sigma_{xx} = \frac{M_Y Z}{I_{YY}} - \frac{M_Z Y}{I_{ZZ}} \tag{10.20}$$

In this form the moments must be resolved along the principal directions and the second area moments of the cross-section with respect to the principal axes must be calculated, as seen in Appendix A. A fourth and final choice is to use a set of coordinates (η, ζ), which are along and perpendicular to the neutral surface as shown in Fig. 10.8. To obtain the flexure stress in these coordinates note that we can rewrite Eq. (10.13) as

$$\sigma_{xx} = \frac{M_y I_{zz} + M_z I_{yz}}{I_{yy} I_{zz} - I_{yz}^2}(z - \tan\lambda\, y) \tag{10.21}$$

and we have

$$z - \tan\lambda\, y = \frac{z\cos\lambda - y\sin\lambda}{\cos\lambda} = \frac{\zeta}{\cos\lambda} \tag{10.22}$$

so that we obtain

$$\sigma_{xx} = \frac{M_y I_{zz} + M_z I_{yz}}{I_{yy} I_{zz} - I_{yz}^2} \frac{\zeta}{\cos\lambda} \tag{10.23}$$

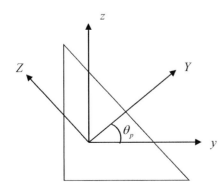

Figure 10.7 A set of principal axes (Y, Z) for an unsymmetrical cross-section.

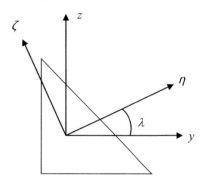

Figure 10.8 A set of (η, ζ) coordinates that lie along and perpendicular to the neutral surface.

All of these different forms can be used and some texts emphasize one form over another. We will simply note that Eq. (10.20) and Eq. (10.23) require extra calculations to use their particular coordinate systems, while Eq. (10.18) can be used directly if one has already calculated the angle, λ, of the neutral surface, as is often the case.

In addition to obtaining the flexure stress, we often want to determine the displacement of the neutral axis. As in the symmetrical cross-section case this means that we must integrate twice the moment–curvature relations, which now are given by Eq. (10.12), and apply the boundary conditions. In the case of a single plane of loading this can be made easier since from Eq. (10.12) we have

$$\frac{d^2v/dx^2}{d^2w/dx^2} = \frac{-(M_z I_{yy} + M_y I_{yz})}{(M_y I_{zz} + M_z I_{yz})} = -\tan\lambda$$

$$\rightarrow \frac{d^2v}{dx^2} = -\tan\lambda \frac{d^2w}{dx^2}$$

(10.24)

and, since the angle λ is a constant, if we let $d^2w/dx^2 = F(x)$ we find, on integration,

$$w = \int\int F(x)dxdx + C_1 x + C_2$$

$$\frac{-v}{\tan\lambda} = \int\int F(x)dxdx + C_3 x + C_4$$

(10.25)

If the boundary conditions are the same for both v and w we have $C_1 = C_3$ and $C_2 = C_4$, and from Eq. (10.25)

$$\frac{-v}{w} = \tan\lambda \qquad (10.26)$$

Geometrically, this condition means that the total displacement of the neutral axis, $\delta = \sqrt{v^2 + w^2}$, is perpendicular to the neutral surface as shown in Fig. 10.9.

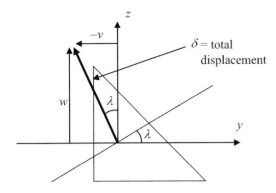

Figure 10.9 The case where the total displacement of the neutral axis is perpendicular to the neutral surface.

Example 10.1 Multiaxis Bending

As a first example of bending under these generalized conditions, consider the cantilever beam shown in Fig. 10.10a. This is a symmetrical beam under multiaxial bending.

The load P is applied to a flange at the bottom of the beam and at a location called the shear center of the beam, which will be defined in a later section. When we discuss shear stresses, we will see that if the load is applied at other locations the beam will twist as well as bend. In this case the load is placed so that the beam experiences bending only, as we have assumed in our present discussions. In calculating the properties of the cross-section, however, this small bottom flange will be neglected (see Fig. 10.10b). In this example we will take $P = 12$ kN, $E = 72$ GPa. From the properties of the cross-section we can determine the area moments are $I_{zz} = 30.73 \times 10^6$ mm^4, $I_{yy} = 39.69 \times 10^6$ mm^4, $I_{yz} = 0$ and the distance to the (y, z) centroid axes (Fig. 10.10b) is $d = 82$ mm (see the MATLAB® function cross_section discussed in Appendix A). In this example we want to find (1) the maximum tensile and compressive stresses and their locations in the cross-section and (2) the total deflection of the beam at the load P.

The maximum bending moments occur at the wall where we have

$$M_y = -3000 P \sin 60° = -31.17 \times 10^6 \text{ N-mm,}$$

$$M_z = 3000 P \cos 60° = 18 \times 10^6 \text{ N-mm,}$$

$$M_z / M_y = -\cot 60° = -0.577$$

Since $I_{yz} = 0$, the expression for the angle the neutral surface makes with respect to the y-axis (see Eq. 10.17) is

$$\tan \lambda = \frac{(M_z/M_y) I_{yy}}{I_{zz}} = (-0.577) \left(\frac{39.69 \times 10^6}{30.73 \times 10^6} \right) = -0.746 \qquad (10.27)$$

which gives $\lambda = -0.640$ rad $= -36.7°$ (see Fig. 10.11b). To compute the flexure stress, we will use Eq. (10.18) with again $I_{yz} = 0$, giving

Figure 10.10 (a) A cantilever beam under multiaxis bending. (b) The cross-section of the beam (the cross-sectional properties of the small flange at the bottom where the load is applied are neglected).

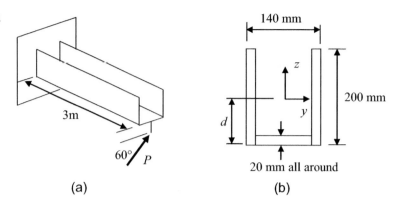

Figure 10.11 (a) The components of the applied load. (b) The orientation of the neutral surface and the points of extreme flexural stress, A and B.

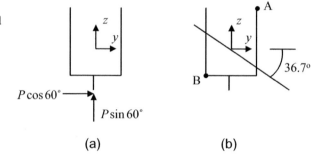

$$\sigma_{xx} = \frac{M_y}{I_{yy}}(z - \tan \lambda y)$$

$$= \frac{-31.17 \times 10^6}{39.69 \times 10^6}(z + 0.746y) \quad (10.28)$$

$$= -0.785z - 0.585y$$

where σ_{xx} is measured in N/mm² or MPa. The extreme flexure stresses in the cross-section at the wall will occur at points A and B (see Fig. 10.11b) that are the farthest points as possible from the neutral axis. At point A $z = 200 - 82 = 118$ mm and $y = 70$ mm, and at point B we have $z = -82$ mm, $y = -70$ mm, so that the stresses at these points are

$$(\sigma_{xx})_A = -(0.785)(118) - (0.585)(70)$$
$$= -133.6 \text{ MPa}$$
$$(\sigma_{xx})_B = -(0.785)(-82) - (0.585)(-70)$$
$$= 105.3 \text{ MPa}$$

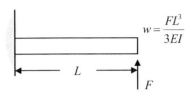

Figure 10.12 A cantilever beam under an applied end load, showing the expression for the vertical deflection, w, at the load.

Alternatively, since the (y, z) axes are principal axes the flexure stress is given by Eq. (10.14), giving

$$\sigma_{xx} = \frac{M_y z}{I_{yy}} - \frac{M_z y}{I_{zz}} = -\frac{31.17 \times 10^6}{39.69 \times 10^6} z - \frac{18 \times 10^6}{30.73 \times 10^6} y \quad (10.29)$$
$$= -0.785 z - 0.585 y$$

which is the same expression as obtained in Eq. (10.28). To compute the deflection of the beam at the load we note that this is a single plane of loading and the beam is rigidly fixed at the wall in both the y- and z-directions, so that we only need to compute the z-deflection, w, and then we can use Eq. (10.26) to find the y-deflection, v. To get the z-deflection, we could integrate the moment–curvature relationship of Eq. (10.12), which in this case, since $I_{yz} = 0$, gives $d^2w/dx^2 = -M_y/EI_{yy}$ but since this a cantilever beam with an end load, most elementary strength of materials books give the end z-deflection explicitly for the geometry seen in Fig. 10.12. Applying that result to our problem we have

$$w = \frac{(P \sin 60°) L^3}{3 E I_{yy}} = \frac{(12,000 \sin 60°)(3000)^3}{3 (72 \times 10^3)(39.69 \times 10^6)} = 32.72 \text{ mm}$$

and, consequently, the y-deflection, v, and the total deflection, δ, are

$$v = -\tan \lambda \, w = 0.745 \, w = 24.38 \text{ mm}$$
$$\delta = \sqrt{w^2 + v^2} = 40.8 \text{ mm}$$

This first example was a case of a single plane of loading, so we now consider a case where the loading is more arbitrary.

Example 10.2 Multiaxis Bending with Loading in Multiple Planes

Let us use the same beam considered in Example 10.1 but change the loading to a combination of a uniform distributed load, $q = 1000$ N/m, and a concentrated end load, $P = 500$ N, as seen in Fig. 10.13. In this example we want to determine the angle the neutral surface makes with respect to the y-axis as a function of the distance, x, from the wall.

Figure 10.13 A cantilever beam subjected to loading in multiple planes.

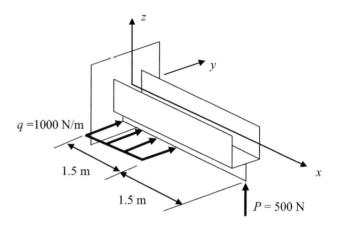

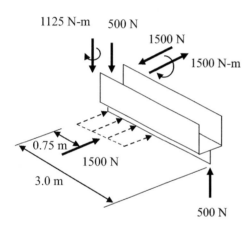

Figure 10.14 The cantilever beam of Fig. 10.13, showing the reactions at the wall.

Figure 10.14 shows a free body diagram of the entire beam and the reactions at the wall calculated from force and moment equilibrium. Also shown in Fig. 10.14 is the resultant force of 1500 N generated by the distributed load and the location of this resultant. It is easy to see that the end reactions are in equilibrium with the applied loads. Figures 10.15a, b show free body diagrams of portions of the beam from the wall to a cutting plane at a distance x taken for (a) $0 < x < 1.5$ m, and (b) $1.5 < x < 3.0$ m. Taking moments at the cutting plane about a point O at the centroid of the cross-section, we find for $0 < x < 1.5$ m:

$$\sum M_{zO} = 0$$
$$M_z + 1500x - 1000x^2/2 - 1125 = 0$$
$$\rightarrow M_z = 500x^2 - 1500x + 1125 \text{ N-m}$$

(10.30)

$$\sum M_{yO} = 0$$
$$M_y - 500x + 1500 = 0$$
$$\rightarrow M_y = -1500 + 500x \text{ N-m}$$

Similarly, for $1.5 < x < 3.0$ m:

10.1 Multiple Axis Bending of Nonsymmetrical Beams

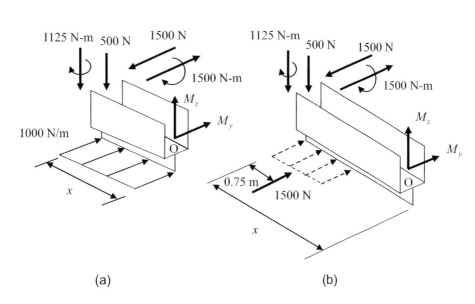

Figure 10.15 Free body diagrams: (a) for $0 < x < 1.5$ m, (b) for $1.5 < x < 3.0$ m.

$$\sum M_{zO} = 0$$
$$M_z + 1500x - 1500(x - 0.75) - 1125 = 0$$
$$\rightarrow M_z = 0$$

(10.31)

$$\sum M_{yO} = 0$$
$$M_y + 1500 - 500x = 0$$
$$\rightarrow M_y = 500x - 1500 \text{ N-m}$$

For the angle that the neutral surface makes, we can again use Eq. (10.27) which gives in this case

$$\tan \lambda = \frac{I_{yy}}{I_{zz}}\left(\frac{M_z}{M_y}\right) = \left(\frac{39.69 \times 10^6}{30.73 \times 10^6}\right) = 1.29\left(\frac{M_z}{M_y}\right)$$

(10.32)

For $0 < x < 1.5$ m we have

$$\tan \lambda = 1.29\left(\frac{500x^2 - 1500x + 1125}{500x - 1500}\right)$$

(10.33)

Figure 10.16 The angle of the neutral surface with respect to the *y*-axis for the beam of Fig. 10.13.

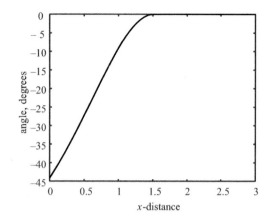

and for $1.5 < x < 3.0$ m

$$\tan \lambda = 1.29 \left(\frac{0}{500x - 1500} \right) = 0 \tag{10.34}$$
$$\rightarrow \lambda = 0$$

The result of Eq. (10.34) is as expected since in that portion of the beam only the bending moment M_y exists, as is the case in elementary bending, as discussed in Chapter 1. If we solve for the angle λ from Eq.(10.33) and plot the results over the entire beam we see (Fig. 10.16) that the neutral axis makes an angle with respect to the *y*-axis of nearly $-45°$ at the wall and gradually becomes less negative, reaching $0°$ at the end of the distributed load.

10.2 Shear Stresses in Thin, Open Cross-Section Beams

In engineering beam theory for symmetrical cross-sections, it was shown that when the flexural stress varies along the *x*-axis of the beam, shear stresses must also be present. By using the flexure stress expression and equilibrium of an element of the beam, an expression was obtained for those shear stresses. For multiaxis bending of unsymmetrical beams, one cannot obtain a similar analytical result except in the case when the beam has a thin cross-section, such as the beam shown in Fig. 10.17a. Commonly used beams such as I-beams and channel sections are composed of thin sections, so analytical results for such thin cross-sections are of significant practical interest. In this section, in addition to assuming the beam is thin, we will assume the beam cross-section is also an open cross-section, i.e., there are no closed loops formed in the cross-section. The beam shown in Fig. 10.17a is such a thin, open cross-section. To analyze the shear stresses, we cut a small element from the beam, as shown in Fig. 10.17a. This element is also shown in Fig. 10.17b. Since the beam is assumed to be

10.2 Shear Stresses in Thin, Open Cross-Section Beams

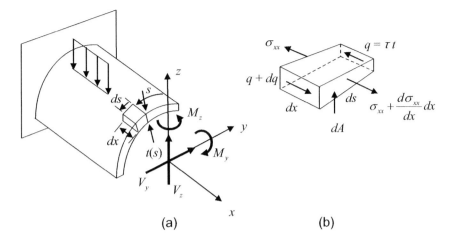

Figure 10.17 (a) An unsymmetrical beam with a thin, open cross-section. (b) A small element of the beam showing the flexure stresses and shear flows acting on it.

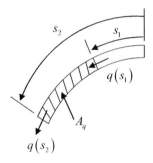

Figure 10.18 An element of a thin beam showing the shear flows on a section.

thin, we expect the shear stresses will be nearly uniform across the thickness of a section and parallel to the adjacent free surfaces. Thus, in thin beam problems, it is common to multiply this uniform stress by the local thickness and examine instead the *shear flow*, q, in the beam defined as $q(s) = \tau(s)t(s)$, where s is the distance measured along the cross-section from a free end of the section, as shown in Fig. 10.17a. The changing flexure stress and the shear flows on a small element of the beam are shown in Fig. 10.17b. If we sum forces in the x-direction for this element we find

$$\sum F_x = 0$$
$$(q + dq)dx - qdx + \left(\sigma_{xx} + \frac{d\sigma_{xx}}{dx}dx\right)dA - \sigma_{xx}dA = 0 \qquad (10.35)$$
$$\rightarrow dq = -\frac{d\sigma_{xx}}{dx}dA$$

If we then integrate Eq. (10.35) between two locations s_1 and s_2, as shown in Fig. 10.18, then we obtain

$$\Delta q = q(s_2) - q(s_1) = -\int_{A_q} \frac{d\sigma_{xx}}{dx}dA \qquad (10.36)$$

where A_q is the area between s_1 and s_2, as shown in Fig. 10.18. For an unsymmetrical section, the normal stress was given by Eq. (10.13) as

Unsymmetrical Beam Bending

$$\sigma_{xx} = \frac{(M_y I_{zz} + M_z I_{yz})z - (M_z I_{yy} + M_y I_{yz})y}{\left(I_{yy} I_{zz} - I_{yz}^2\right)} \quad (10.37)$$

so that if we differentiate this stress on x and use the moment–shear relations of Eq. (10.1) and Eq. (10.2), we obtain

$$\frac{d\sigma_{xx}}{dx} = \frac{(V_z I_{zz} - V_y I_{yz})z - (V_z I_{yz} - V_y I_{yy})y}{\left(I_{yy} I_{zz} - I_{yz}^2\right)} \quad (10.38)$$

Placing Eq. (10.38) into the shear flow expression of Eq. (10.36) gives

$$\Delta q = \frac{(V_z I_{yz} - V_y I_{yy})Q_z + (V_y I_{yz} - V_z I_{zz})Q_y}{D} \quad (10.39)$$

where

$$D = I_{yy} I_{zz} - I_{yz}^2$$

$$Q_z = \int_{A_q} y \, dA \quad (10.40)$$

$$Q_y = \int_{A_q} z \, dA$$

Note that in obtaining our expression for Δq in Eq. (10.39), we showed the shear flow q flowing out from the end of the section A under consideration and the shear flow flowing into A at the beginning of the section (see Fig. 10.18). As shown in Fig. 10.19 for an I-beam, this same convention is followed regardless of the section taken. For all the sections in Fig. 10.19, therefore we have

$$\Delta q = q(s) - q(0) = \frac{(V_z I_{yz} - V_y I_{yy})Q_z + (V_y I_{yz} - V_z I_{zz})Q_y}{D} \quad (10.41)$$

Figure 10.19 An I-beam where various sections are taken, showing the directions of the shear flows.

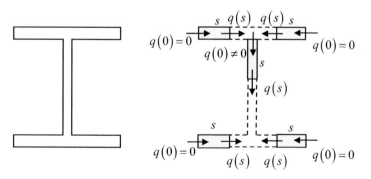

10.2 Shear Stresses in Thin, Open Cross-Section Beams

In a given problem, the shear flow into, or the shear flow out of, a section can be positive or negative, so by associating the positive signs with a figure such as Fig. 10.19 it is easy to see the actual direction in the cross-section

If we have a symmetrical cross-section and if $V_y = 0$, then we are considering the elementary bending theory case considered in Chapter 1. Furthermore, if we take $q(s_1) = 0$ so we are starting at a free end of the open cross-section then $\Delta q = \tau(s_2)t(s_2)$ and Eq. (10.39) becomes simply

$$\tau = -\frac{V_z Q_y}{I_{yy} t} \tag{10.42}$$

which is of very similar form to Eq. (1.25) for the average vertical shear stress in a symmetrical beam. A minus sign is present in Eq. (10.42) but not in Eq. (1.25) because of the difference between the direction of the shear stress in Eq. (1.25) and direction of the shear stress (shear flow) in Eq. (10.42). These differences are shown explicitly in Fig. 10.20 for a rectangular section.

Another fact you should be aware of is that the shear flows must be conserved at a junction between sections such as the junction between the web and flange of an I-beam, for example. This follows directly from equilibrium considerations in Fig. 10.21, which shows the shear flows (q_1, q_2, q_3) at a T-junction. These shear flows must be in equilibrium with the changing flexure stress so that, by summing forces in the x-direction, we obtain from that figure:

$$\sum F_x = 0$$

$$(q_1 + q_2 + q_3)dx + \left(\sigma_{xx} + \frac{d\sigma_{xx}}{dx}dx\right)dA - \sigma_{xx}dA = 0 \tag{10.43}$$

$$\rightarrow (q_1 + q_2 + q_3) + \frac{d\sigma_{xx}}{dx}dA = 0$$

Thus, as we shrink $dA \rightarrow 0$ from Eq. (10.43) we find

$$q_1 + q_2 + q_3 = 0 \tag{10.44}$$

i.e., the net shear flow out of (or into) a junction must be zero. [Note: this result was shown for a T-junction but the same result is obtained at a general junction.]

Equation (10.39) can be used to obtain the shear-stress distribution in a thin, unsymmetrical beam under multiaxis loading, and we will show an example of such explicit calculations

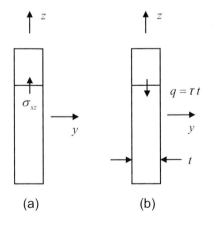

Figure 10.20 (a) The average shear stress, σ_{xz}, in a rectangular section appearing in Eq. (1.25). (b) The shear flow, q, and corresponding stress, τ, appearing in Eq. (10.42).

Figure 10.21 Shear flows at a junction where equilibrium shows that the net shear flow into or out of the junction must be zero.

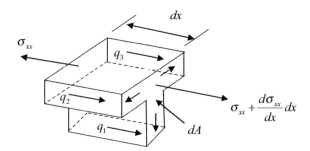

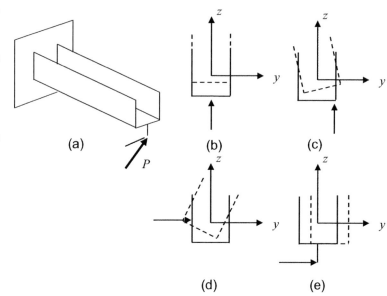

Figure 10.22 (a) A beam loaded so as to produce bending only (no twisting). (b) Only bending is produced when a vertical load acts through the z-axis of symmetry (also a centroidal axis). (c) Bending and twisting occur when the vertical load is not located along the symmetry axis. (d) If the load acts horizontally through the y-centroidal axis, both bending and twisting occur. (e) If the horizontal load acts at a point below the beam on the flange called the shear center, only bending will occur.

shortly. However, before we apply that equation, we will discuss in the next section the important concept of the shear center.

10.3 The Shear Center for Thin, Open Cross-Section Beams

Figure 10.22a shows the thin, open cross-section cantilever beam we previously used to discuss unsymmetrical bending (see Fig. 10.12). There, we placed the load P on a flange beneath the cross-section so that we could assume the beam was only in multiaxial bending. Let us examine that assertion more closely. This beam does have an axis of symmetry (the z-axis) so that if we placed a vertical load anywhere along that axis (which is also a centroidal axis), we will produce bending only, as shown in Fig. 10.22b. However, if the vertical load acted to the right or left of the z-axis, the load will produce twisting (torsion) as well as bending, as shown in Fig. 10.22c. When we discuss torsion in Chapter 11 we will see that, in thin sections, there are also shear stresses due to torsion as well as bending so that shear stresses obtained from our shear flow expression, Eq. (10.39), which assumed bending deformations only are present, is not a correct expression for the total shear stresses when both bending and torsion are present. If we apply a horizontal load, as shown in Fig. 10.22d, through the centroidal y-axis (which is not an axis of symmetry), we will find that both bending and twisting are produced. There must be some vertical location, however, where the horizontal load can be applied and not produce twisting because if the load is applied at a very large positive z-coordinate, it will obviously produce a large counterclockwise twist, similar to that seen in Fig. 10.22d, while if the load acts at very large negative z-values, we would expect to see a twist in the opposite sense. There is some point, therefore, (which in this case turns out to be below the cross-section and along the z-axis) where bending only is produced (see Fig. 10.22e) about both the y- and z-directions. This point is called the shear center and it is the point we placed the load (without specifying its location specifically) when solving this problem previously. Thus, to use our bending shear-flow expression, Eq. (10.39), we must locate the shear center and ensure that the lines of action of the shear force components act through that shear center. There are some special cross-sections where we can locate the shear center by inspection. The shear center, S, must lie on an axis of symmetry of a cross-section, as seen in Fig. 10.23a, since the symmetrical shear flows produced by a shear force acting along that axis will not produce any net twisting moment. We must still calculate, however, where the shear center lies along that symmetry axis. The shear center must lie at the corner of an L-section, as shown in Fig. 10.23b, since the shear flows all have directions that act through that point. For the same reason the shear centers must be located at the points shown in Figs. 10.23c, d for those cross-sections.

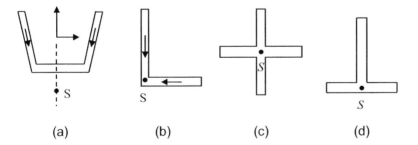

Figure 10.23 Cases where the location of the shear center, S, can be partially or totally specified by inspection. (a) If the cross-section has an axis of symmetry, the shear center lies along that axis since any shear force whose line of action is along that symmetry axis will produce symmetrical shear flows, as shown, that have no tendency to twist the section. (b) The shear center must lie at the corner of an L-section since all the shear flows must have directions which act through that point. The same reasoning applies to the cross-sections seen in (c) and (d).

390 Unsymmetrical Beam Bending

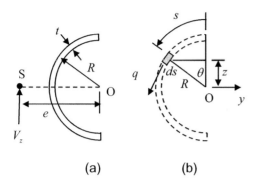

Figure 10.24 (a) A thin semicircular cross-section of radius R and thickness t, where the shear center is at a distance, e, from the center, O. (b) Geometry for calculating the shear flow, where z is measured from the centroidal y-axis of symmetry.

There are basically two ways to calculate the location of the shear center. The first way is based on the fact that we can calculate the moments produced by the shear flows about a reference point in terms of V_y and V_z individually by using the shear-flow expression in Eq. (10.39), which assumes only bending. Equating moments generated by those shear flows to the moments of V_y and V_z when acting at some unknown distances from that reference point then allows us to solve for those distances (the V_y and V_z values cancel out) and hence determine the location of the shear center. For this method to be viable by hand calculations, the integrals of the shear flow distributions needed to calculate the moments must be known in analytical forms. As a simple example, consider the semicircular cross-section shown in Fig. 10.24a. The shear center must lie somewhere along the axis of symmetry. To locate it with respect to the center O, note that for a vertical shear force, V_z, the shear flow expression, Eq. (10.39) becomes

$$q = -\frac{V_z}{I_{yy}} \int_{A_q} z dA = -\frac{V_z}{I_{yy}} \int_0^\theta R \cos\theta \, t R d\theta \qquad (10.45)$$

$$= -\frac{V_z}{I_{yy}} R^2 t \sin\theta$$

For the section we find $I_{yy} = \pi R^3 t / 2$, which is just one fourth the value of the polar area moment for a thin full circular section. Calculating the moment generated by the shear flow about point O and setting it equal to the moment of the shear force in Fig. 10.24a gives

$$M_{xO} = \int_0^\pi Rq ds = R^2 \int_0^\pi q d\theta$$

$$= -\frac{2V_z}{\pi R^3 t} R^4 t \int_0^\pi \sin\theta d\theta = -\frac{4RV_z}{\pi} \qquad (10.46)$$

$$= -V_z e$$

$$\rightarrow e = 4R/\pi$$

Thus, the shear center is outside the cross-section as shown in Fig. 10.24a. For a general unsymmetrical section we can do similar calculations individually for the shear forces V_z and V_y and locate the shear center in both the y- and z-directions.

10.3 The Shear Center for Thin, Open Cross-Section Beams

Another way to find the location of the shear center is to recognize that the shear center is, like the centroid, a purely geometric quantity that depends only on the shape of the cross-section. Thus, if we can determine the geometric parameters that control the location of the shear center, we can determine the shear center directly from the geometry. As we will see, one important geometric parameter is called the *sectorial area function*. Figure 10.25 shows a thin, open cross-section where the shear forces act through the shear center. If we locate the shear center relative to a reference point, B, then the moment generated about B by the shear forces must equal the moment produced by the shear flow under bending only (no torsion). Equating moments about the x-axis we find

$$V_z e_y - V_z e_z = \int_I^E r_\perp q \, ds \tag{10.47}$$

where the integral is over the entire area, A, of the cross-section. Since the shear flow is zero at the starting point, I, from Eq. (10.39) we obtain

$$V_z e_y - V_y e_z = \frac{(V_z I_{yz} - V_y I_{yy})}{D} \int_I^E r_\perp Q_z \, ds + \frac{(V_y I_{yz} - V_z I_{zz})}{D} \int_I^E r_\perp Q_y \, ds \tag{10.48}$$

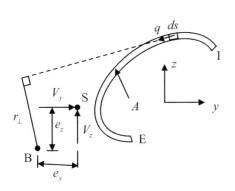

But $r_\perp ds = 2 dA_s = d\omega_B$ is twice the area, dA_s, swept out from point B as we go through a distance ds along the cross-section, as shown in Fig. 10.26, a quantity which we will call the differential of the *sectorial area function*, $d\omega_B$, as measured from point B (do not confuse dA_s with $dA = t \, ds$, which is a differential of the cross-sectional

Figure 10.25 Geometry of an open, unsymmetrical cross-section showing the shear center, S, as measured relative to a reference point, B, and the shear flow, which generates the shear forces.

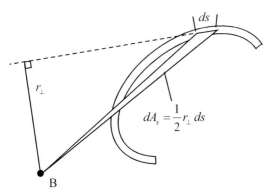

Figure 10.26 Definition of an area measured from the point B swept out by a length ds along the centerline of the cross-section.

area, A). If we do an integration by parts on the integrals in Eq. (10.48), we find

$$\int_I^E Q_z d\omega_B = Q_z\omega_R\Big|_I^E - \int_I^E \omega_B dQ_z = -\int_I^E \omega_B dQ_z$$

$$\int_I^E Q_y d\omega_B = Q_y\omega_B\Big|_I^E - \int_I^E \omega_B dQ_y = -\int_I^E \omega_B dQ_y$$

(10.49)

since both Q_z and Q_y vanish at the starting point I and they are also zero at the end point E because they are the first moments of the total cross-sectional area as measured from the centroid. However, from their definitions we also have $dQ_z = ydA, dQ_y = zdA$ (see Eq. (10.40)) so

$$\int_I^E Q_z d\omega_B = -\int_A y\omega_B dA \equiv -I_{y\omega_B}$$

$$\int_I^E Q_y d\omega_B = -\int_A z\omega_B dA \equiv -I_{z\omega_B}$$

(10.50)

where $I_{y\omega_B}$ and $I_{z\omega_B}$ are called the *sectorial products of areas* for the sectorial area function, ω_B. [See Appendix C, which gives a detailed discussion of the sectorial area function and its properties. You should study that appendix to better understand this function. In torsion problems we will see the sectorial area function also plays an important role.]

From Eq. (10.47), therefore,

$$V_z e_y - V_y e_z = V_z \frac{(I_{zz}I_{z\omega_B} - I_{yz}I_{y\omega_B})}{D} - V_y \frac{(I_{yz}I_{z\omega_B} - I_{yy}I_{y\omega_B})}{D}$$

(10.51)

which must be true for arbitrary V_z and V_y, giving the location of the shear center relative to point B finally as

$$e_y = \frac{(I_{zz}I_{z\omega_B} - I_{yz}I_{y\omega_B})}{I_{yy}I_{zz} - I_{yz}^2}$$

$$e_z = \frac{(I_{yz}I_{z\omega_B} - I_{yy}I_{y\omega_B})}{I_{yy}I_{zz} - I_{yz}^2}$$

(10.52)

The location of the shear center given by Eq. (10.52) is only a function of the geometry of the cross-section, as shown in Appendix C. The sectorial area function can be calculated with respect to any convenient point B. The distances e_y and e_z are then given with respect to that point. The distances y and z appearing in the first moments of the sectorial area function are measured from the centroid of the cross-section to a point on the centerline of the cross-section. To use Eq. (10.52) directly requires that we calculate a large number of parameters.

However, as pointed out in Appendix C, if we construct a sectorial area function about a point B where

$$\int_A y\omega_B dA = 0$$
$$\int_A z\omega_B dA = 0 \tag{10.53}$$

then e_y and e_z are both zero; i.e., the point B is at the shear center. Thus, one can use Eq. (10.53) directly to calculate the location of the shear center without having to calculate all the area moments appearing in Eq. (10.52). Note that we can start the integration from a point where the sectorial area function has a nonzero starting value, ω_0. However, we can always set that constant value equal to zero when using Eq. (10.53) since if $\omega' = \omega + \omega_0$ we have

$$\int_A y\omega' dA = \int_A y\omega dA + \omega_0 \int_A y dA = \int_A y\omega dA$$
$$\int_A z\omega' dA = \int_A z\omega dA + \omega_0 \int_A z dA = \int_A z\omega dA \tag{10.54}$$

where we have used the fact that y and z are measured from the centroid of the total area A of the cross-section to eliminate the terms involving the constant. Thus, these sectorial products of area $(I_{z\omega}, I_{y\omega})$ are not affected by a constant sectorial area value. This is also discussed in Appendix C. We can also remove the restriction that the (y,z) coordinates in Eq. (10.53) be measured from the centroid by adding the requirement that the integral of the sectorial area function be zero, i.e., requiring instead that we satisfy the three conditions:

$$\int_A \omega_B dA = 0$$
$$\int_A y' \omega_B dA = 0 \tag{10.55}$$
$$\int_A z' \omega_B dA = 0$$

where now (y', z') can be measured from any origin in the cross-section we choose. However, if we start out the sectorial area function at a location where the sectorial area function has a constant value, that constant, along with the location of the shear center, are all fixed by Eq. (10.55). A sectorial area function that satisfies Eq. (10.55) is called a *principal sectorial area function* (see Appendix C).

Example 10.3 Calculating the Shear Center

As an example of determining the location of the shear center, consider the C-section considered previously when discussing multiaxis bending. We will assume the thicknesses (t_b, t_h) of the cross-section are thin. For such thin sections it is customary to measure the distances of the parts of the cross-section along the centerlines of the section, as shown in Fig. 10.27a. It is also customary to neglect terms involving higher powers of these thickness. For example, consider calculating the area moment I_{zz}. We have from the parallel axis theorem (see Appendix A)

$$I_{zz} = \frac{t_b b^3}{12} + 2\left[\frac{h t_h^3}{12}\right] + 2\left[h t_h \left(\frac{b}{2}\right)^2\right]$$

$$\cong \frac{t_b b^3}{12} + \frac{h t_h b^2}{2}$$

(10.56)

where the middle term in the first line of Eq. (10.56) is neglected since it is much smaller than the other terms. To determine the shear center, we only need to evaluate its location on the z-axis since it lies on that axis of symmetry as shown in Fig. 10.27b so that $e_y = b/2$. Since $I_{yz} = 0$ from Eq. (10.52), we only have to evaluate

$$e_z = -\frac{I_{y\omega_B}}{I_{zz}}$$

(10.57)

The sectorial area function ω_B calculated using point B as the pole and point I as the initial integration point (origin) is shown as the shaded area in Fig. 10.27b. Specifically, we have $\omega_B = bs$ in the right-hand side and $\omega_B = 0$ otherwise, since there is no projected area from B for the other two sides. The distance y from the centroid to the right-hand side is just $y = b/2$ so

$$I_{y\omega_B} = \int_A y\omega_B dA = \int_{s=0}^{s=h} (b/2)(sb)t_h ds = t_h b^2 h^2/4$$

(10.58)

Figure 10.27 (a) Geometry of a C-section. (b) Parameters for calculating the location of the shear center.

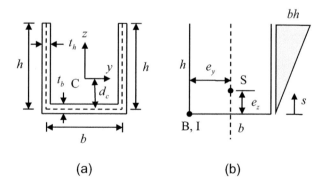

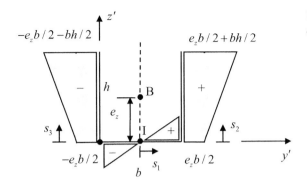

Figure 10.28 The sectorial area function used in determining the shear center.

From this value and Eq. (10.56) and Eq. (10.57) we find, finally,

$$e_z = -\frac{3t_h h^2}{(t_b b + 6t_h h)} \tag{10.59}$$

Thus, the shear center is below the bottom of the C-section, as we assumed in the previous example where we considered multiaxis bending for a beam with this cross-section. Alternatively, we can use, say, the (y', z') coordinates and locate the pole B at a distance along the center axis as seen in Fig. 10.28 and choose the initial point (origin) I as shown. [Note: we could continue to use the centroidal coordinates (y, z) and obtain the same result. We have changed the coordinates to emphasize they can be noncentroidal.] The sectorial area function is then

$$\omega_B = \begin{cases} e_z s_1 & -b/2 \leq s_1 \leq b/2 \\ e_z b/2 + b s_2/2 & 0 \leq s_2 \leq h \\ -e_z b/2 - b s_3/2 & 0 \leq s_3 \leq h \end{cases} \tag{10.60}$$

whose graphs are shown as the shaded areas in Fig. 10.28. From Fig. 10.28 we can see that the conditions

$$\int_A \omega_B dA = 0$$
$$\int_A z' \omega_B dA = 0 \tag{10.61}$$

are automatically satisfied by this sectorial area function so we only need to examine the remaining condition in Eq. (10.55) which gives

$$\int_A y' \omega_B dA = 0 = \int_{-b/2}^{b/2} (s_1 + b/2)(e_z s_1) t_b ds_1 + \int_0^h (b)(e_z b/2 + b s_2/2) t_h ds_2 \tag{10.62}$$

Carrying out the integrals we find

$$e_z t_b b^3/12 + 0 + e_z b^2 h t_h/2 + b^2 h^2 t_h/4 = 0$$

$$\rightarrow e_z = -\frac{3 t_h h^2}{(t_b b + 6 t_h h)} \qquad (10.63)$$

which agrees with our previous result.

We will now return to Eq. (10.41) and consider an example that describes the shear flow in a thin, open section under bending and examine in detail the calculations needed to (1) obtain the shear flow explicitly and (2) locate the shear center. Two key quantities in obtaining the shear flows are the (Q_y, Q_z) terms. These quantities are defined as first moment integrals of a varying cross-sectional area, A_q, as we move in the cross-section (see Eq. (10.40)). The coordinates (y, z) in these first moment integrals are measured from point C, the centroid of the entire cross-section, to a small-area element $dA = t ds$ in A_q and where the (Q_y, Q_z) terms can be calculated as

$$Q_y = \int_{A_q} z dA = \bar{z}(s) A_q(s)$$

$$Q_z = \int_{A_q} y dA = \bar{y}(s) A_q(s) \qquad (10.64)$$

Here, $(\bar{y}(s), \bar{z}(s))$ are measured from the centroid of the entire cross-section, C, to the centroid of the area $A_q(s)$ (see Fig. 10.29). In cases where we can calculate the $(\bar{y}, \bar{z}, A_q)$ terms directly we can, therefore, find the (Q_y, Q_z) terms without performing the underlying integrations.

Figure 10.29 Geometry for determining Q_y and Q_z in the shear-flow expression, Eq. (10.41).

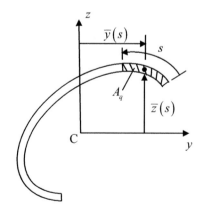

Example 10.4 Shear Flow and Shear Center in a Thin, Open Section Beam

Consider the thin, open section beam cross-section of Fig. 10.30. It is assumed that a 1000 lb vertical shear force acts through the shear center at this section, producing bending but no twisting. We want to determine the bending shear-flow distributions in this section and the location of the shear center. The thickness of all parts of the cross-section is 0.1 in. and all the dimensions (in inches) in Fig. 10.30 are measured to the centerlines. In this case $V_y = 0$ and $I_{yz} = 0$ so the shear-flow expression of Eq. (10.39) becomes much simpler:

$$\Delta q = \frac{-V_z Q_y}{I_{yy}} \qquad (10.65)$$

If we break the cross-section up into three areas, we can calculate the area moment I_{yy} with the parallel axis theorem as

$$I_{yy} = 2 \times \left[\frac{1}{12}(8)(0.1)^3 + (8)(0.1)(5)^2 \right] + \frac{1}{12}(0.1)(10)^3 \qquad (10.66)$$
$$= 48.33 \text{ in.}^4$$

where the first term which involves the thickness cubed is neglected in the end result. Note that the centroid C is along the horizontal centerline of the entire cross-section at a distance y_c as shown in Fig. 10.31, which we could also calculate with these areas but which will not be needed. First, consider the shear flow in a portion of the section AB as shown in Fig. 10.32a. We have

$$\Delta q = q(s_1) - 0 = \frac{-1000}{48.33} Q_y = -20.69 Q_y = -20.69 \bar{z} A_q$$
$$= -20.69[(5)(0.1)(s_1)] \qquad (10.67)$$
$$\rightarrow q(s_1) = -10.35 s_1 \text{lb/in.}$$

The shear flow is a linear function in AB and the value at B, q_B^1, is given in Fig. 10.32b. Since the shear flow is negative throughout AB the direction of the shear flow is as shown. In section BD similar calculations (see Fig. 10.33) give

$$\Delta q = q(s_2) - 0 = -20.69 \bar{z} A_q$$
$$= -20.69[(5)(0.1)(s_2)] \qquad (10.68)$$
$$\rightarrow q(s_2) = -10.35 s_2 \text{lb/in.}$$

and the value at B, q_B^2, is shown in Fig. 10.33b along with the actual shear-flow distribution in BD. Now consider the vertical

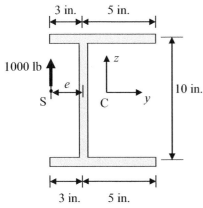

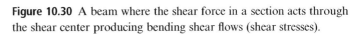

Figure 10.30 A beam where the shear force in a section acts through the shear center producing bending shear flows (shear stresses).

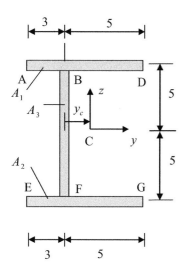

Figure 10.31 Decomposition of the cross-section into three areas $A_m (m = 1, 2, 3)$.

section BF. A portion of this section is shown in Fig. 10.34a. The shear-flow calculations are

$$\Delta q = q(s_3) - q_B^3 = -20.69 \bar{z} A_q$$
$$= -20.69[(5 - s_3/2)(0.1)(s_3)] \quad (10.69)$$
$$\rightarrow q(s_3) = q_B^3 - 10.35 s_3 + 1.035 s_3^2 \text{ lb/in.}$$

To determine the starting value q_B^3 we must use conservation of shear flow at the B junction, as shown in Fig. 10.34b, which gives

$$q_B^3 + 51.75 + 31.03 = 0$$
$$\rightarrow q_B^3 = -82.78 \text{ lb/in.} \quad (10.70)$$

and the shear flow in BF is explicitly

$$q(s_3) = -82.78 - 10.35 s_3 + 1.035 s_3^2 \text{ lb/in.} \quad (10.71)$$

At the middle of BF the shear flow is $q(5) = -109$ lb/in. and the entire distribution is sketched in Fig. 10.34c. For the lower flange Q_y has the same magnitude as for the upper flange but with the opposite sign so that shear-flow distributions in EF and FG are as shown in Fig. 10.35. Figure 10.36a shows the complete shear-flow distribution in the cross-section. These shear flows generate the forces shown in Fig. 10.36b where

Figure 10.32 (a) Geometry for determining the shear flow in section AB. (b) The shear-flow distribution.

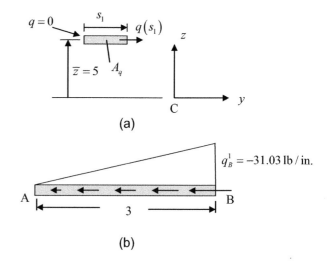

10.3 The Shear Center for Thin, Open Cross-Section Beams

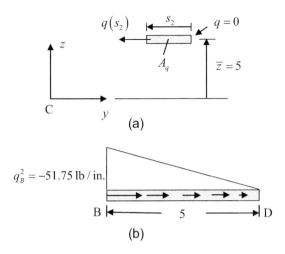

Figure 10.33 (a) Geometry for determining the shear flow in section BD. (b) The shear-flow distribution.

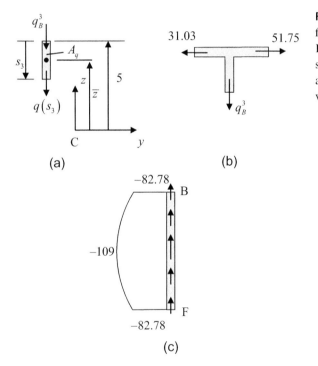

Figure 10.34 (a) Geometry for determining the shear flow in section BF. (b) The shear flows at junction B. (c) The shear-flow distribution in BF (note the shear-flow values are negative so the shear flow is actually up, as shown, with corresponding positive values).

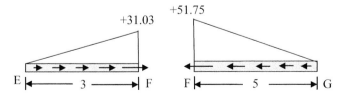

Figure 10.35 The shear-flow distributions in the lower flange.

$$R_1 = \int_0^3 10.35 s_1 \, ds_1 = 46.58 \text{ lb}$$

$$R_2 = \int_0^5 10.35 s_2 \, ds_2 = 129.38 \text{ lb} \qquad (10.72)$$

$$R_3 = \int_0^{10} (82.78 + 10.35 s_3 - 1.035 s_3^2) \, ds_3 = 1000 \text{ lb}$$

[Note: we have removed all the minus signs appearing in the shear flows in Eq. (10.72) so that the forces are positive in the directions shown in Fig. 10.36b.] Since we obtained these shear-flow distributions under the assumption of bending only, the moment of the 1000 lb shear force about any point must be the same as the moment generated by the shear flows, or equivalently the forces seen in Fig. 10.36b. Taking moments about O (Fig. 10.37), we find

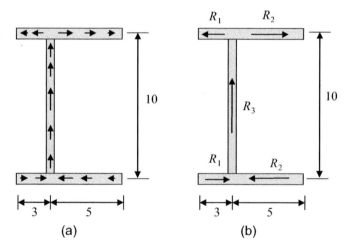

Figure 10.36 (a) The shear-flow distribution in the cross-section. (b) The forces carried in the various parts of the cross-section.

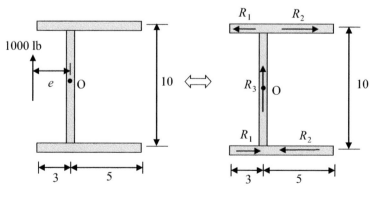

Figure 10.37 The moment about point O of the 1000 lb force acting through the shear center must be equal to the moment about O of the forces generated by the shear flows.

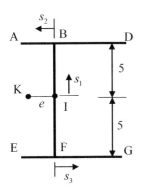

Figure 10.38 The parameters used to locate the shear center by use of the sectorial area function.

$$\sum M_O = 1000e = 10R_2 - 10R_1$$
$$= (10)(129.38 - 46.58) \quad (10.73)$$
$$\rightarrow \quad e = 0.828 \text{ in.}$$

which gives us the shear center location relative to point O.

Another way to locate the shear center directly from the geometry is to determine the principal pole location of the principal sectorial area function. As discussed previously, and as shown in Appendix C, the principal sectorial area function can be found by evaluating the sectorial area function about a pole where Eq. (10.55) is satisfied. We will call that pole here pole K (to avoid confusion with the other labels being used) as shown in Fig. 10.38, and we will take the initial point for the integration to be point I. Then with the coordinates along the various sections, as shown in Fig. 10.38, we find the sectorial area function ω_K to be:

$$\omega_K = \begin{cases} es_1 & -5 \leq s_1 \leq 5 \\ 5e + 5s_2 & -5 \leq s_2 \leq 3 \\ -5e + 5s_3 & -3 \leq s_3 \leq 5 \end{cases} \quad (10.74)$$

A sketch of this function is shown in Fig. 10.39. It is easy to see that this sectorial area function satisfies

$$\int_A \omega_K dA = 0$$
$$\int_A y\omega_K dA = 0 \quad (10.75)$$

so that the only remaining condition is

$$\int_A z\omega_K dA = \int_{-5}^{5} (s_1)(es_1)tds_1 + \int_{-5}^{3} (5)(5e + 5s_2)tds_2$$
$$+ \int_{-3}^{5} (-5)(-5e + 5s_3)tds_3 = 0 \quad (10.76)$$

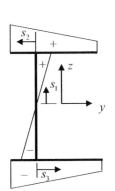

Figure 10.39 The sectorial area function.

where here the distance z is measured from the centroid so that $z = s_1$ in the vertical web and $z = 5$, $z = -5$ for the top and bottom flanges, respectively. Carrying out the integrations and dividing by the common thickness, t, gives

$$\frac{250}{3}e + 400e - 400 = 0$$

$$\rightarrow e = \frac{400}{400 + 250/3} = 0.828 \text{ in.} \tag{10.77}$$

which agrees with our previous result. Note that with pole K located at the distance e given by Eq. (10.77), the sectorial area function is a principal sectorial area function since the three equations of Eq. (10.55) are satisfied, and we rewrite them here:

$$\int_A \omega_K dA = 0$$

$$\int_A y' \omega_K dA = 0 \tag{10.78}$$

$$\int_A z' \omega_K dA = 0$$

With our choice of the pole K and the starting point I, the first two equations were automatically satisfied so that only the third equation was needed. As demonstrated earlier, it was not necessary to choose the (y', z') coordinates to be measured from the centroid, even though that was the choice we made here. However, if in our choice of K and I the first equation in Eq. (10.78) was not satisfied, then we would be forced to use coordinates as measured from the centroid in the other two equations in Eq. (10.78). Those two equations would then locate the principal pole (shear center) even though ω_K is not a principal sectorial area function. The most general way to proceed is to write the starting value of the sectorial area function ω_K in terms of an unknown constant, ω_0, at the starting point I so that $\omega_K = \omega_K(\omega_0, e_y, e_z)$ where (e_y, e_z) are some unknown locations of pole K from an arbitrary fixed point in the cross-section. Then the equations in Eq. (10.78) are three equations for the three unknowns (ω_0, e_y, e_z). Problem P10.3 gives you a chance to examine this general approach for a C-section.

10.4 Thin, Closed Cross-Section Beams

All of the examples considered previously were thin, open cross-sections. One can also examine the bending of thin, closed cross-sections containing one or more closed "cells."

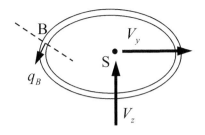

Figure 10.40 Bending of a single-cell closed section where the shear forces act through the shear center. The shear flow has an unknown value, q_B, at a point B where an imaginary cut is taken through the thickness.

The main difference from the thin, open cross-section case is that the shear flows in closed cross-sections contain one or more unknown constant values since there are no edges in closed cross-sections at which the shear flows must be zero. Additional conditions must be considered to evaluate those unknowns. Those conditions will be described in detail in Chapter 12 when we consider the combined bending and torsion of thin open and closed cross-sections. Here, we will just summarize the case of a single closed cell where the shear forces act through the shear center so that the member is in bending only (Fig. 10.40). We can make an imaginary cut of the cross-section at any point B and call the value of the shear flow at that location q_B. Then the shear flow in the cell is given (Eq. (10.41)) as

$$q(s) = q_B + \frac{(V_z I_{yz} - V_y I_{yy})Q_z + (V_y I_{yz} - V_z I_{zz})Q_y}{D} \quad (10.79)$$

$$= q_B + q_0(s)$$

where $q_0(s)$ is the known shear flow for an open section that has a real cut at B. As shown in Chapter 11, the twist/unit length of the section, ϕ', induced by shear flow in a thin section is given by:

$$\phi' = \frac{1}{G\Omega} \oint_C \frac{q(s)}{t(s)} ds \quad (10.80)$$

where t is the thickness of the section, G is the shear modulus, and Ω is twice the cross-sectional area contained within the center line of the cross-section. When the shear forces act through the shear center no twisting is induced, so we can find the constant q_B from

$$\frac{1}{G\Omega} \oint \frac{q}{t} ds = \frac{1}{G\Omega} \oint \frac{(q_B + q_0)}{t} ds = 0$$

$$\rightarrow q_B = -\frac{\oint \frac{q_0}{t} ds}{\oint \frac{ds}{t}} \quad (10.81)$$

and once this constant is known the shear flow is also known explicitly. The location of the shear center can then be determined by equating moments of the shear flow to the moments of the shear forces, as done in Example 10.4. Chapter 12 gives more details when discussing combined bending and torsion and also considers the case of multiple closed sections, so we will not consider those extensions here.

404 Unsymmetrical Beam Bending

10.5 PROBLEMS

P10.1 An I-beam ($I_{yy} = 937 \times 10^6$ mm^4, $I_{zz} = 18.7 \times 10^6$ mm^4) is subjected to a pure bending moment, M, at a very small angle ($\phi = 1°$) with respect to the y-axis, as shown in Fig. P10.1.
(a) Determine the orientation of the neutral axis.
(b) Determine the ratio of the maximum tensile stress in the beam for the moment at this angle to the maximum tensile stress for the symmetrical bending case where $\phi = 0°$.

P10.2 Two 10 mm thick steel plates are welded together to form the 120 mm by 80 mm angle beam shown in Fig. P10.2. The beam, which is simply supported at its ends in both the y- and z-directions, is subjected to a concentrated load $P = 4$ kN acting at point E (which is the shear center) in the y–z plane at an angle of 60° (see Fig. P10.2).
(a) Determine maximum tensile and compressive bending stresses at the section of the beam where the load is applied.
(b) Determine the total deflection (magnitude and direction) at the section of the beam where the load is applied.

For the properties of the cross-section, see Example A.1 in Appendix A.

P10.3 As discussed at the end of Section 10.3, the most general way to use the sectorial area function to find the location of the shear center of a thin open cross-section is to obtain a sectorial area function $\omega_K = \omega_K(\omega_0, e_y, e_z)$ relative to an arbitrary pole K where we solve the three equations of Eq. (10.78), rewritten here as:

$$\int_A \omega_K dA = 0$$
$$\int_A y\omega_K dA = 0 \quad \text{(P10.1)}$$
$$\int_A z\omega_K dA = 0$$

for the values (ω_0, e_y, e_z), which then gives us a principal sectorial area function and the location of the shear center. Note that the location of the origin of the (y, z) coordinates in Eq. (P10.1) is arbitrary so that (y, z) here do not have to be centroidal. Consider the C-section of Fig. P10.3 where the sectorial area function value at the starting point I is

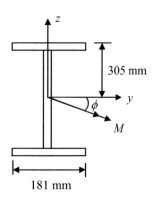

Figure P10.1 An I-beam subjected to a bending moment at a small angle with respect to the y-axis.

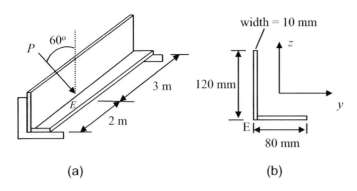

Figure P10.2 (a) Two 10 mm thick steel plates are welded together to form an angle beam that is simply supported at its ends in both the y- and z-directions (details of the supports are not shown). The force P acts in a plane parallel to the y–z plane. (b) The cross-section of the beam (dimensions are measured to the ends of the cross-section).

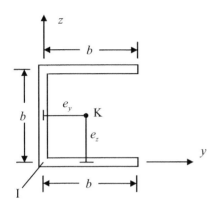

Figure P10.3 A thin C-section of uniform thickness, t. The point, I, is at the origin of the (y, z) coordinates. All dimensions of the cross-section are measured along the center lines.

ω_0 and the location of the pole K is taken to be at the point (e_y, e_z) in the (y, z) coordinates shown. The cross-section has a constant thickness, t. Use Eq. (P10.1) and the (y, z) coordinates of Fig. P10.3 to obtain the location of the shear center and the constant ω_0 without explicitly using the symmetry of the geometry. Sketch the sectorial area function for the cross-section. [Hint: it is advantageous to use the symbolic integration capabilities of MATLAB®.]

P10.4 One general way to find the shear center is by use of the sectorial area function (see problem P10.3). Another general way, as discussed in this chapter, is to find the shear flow due to bending only and then equate the moments of that shear flow to the moments of the shear forces to locate where those shear forces must act. Consider for example, a split circular section of radius R and thickness t, as shown in Fig. P10.4. The shear center must lie along the horizontal axis of symmetry through the center of the section. Determine:

(a) the approximate value of the area moment, I_{yy}, for this thin section,

(b) the shear flow distribution due to a vertical shear force, P, acting through the shear center, in terms of P, θ, and R,

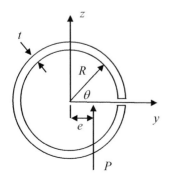

Figure P10.4 A thin, split circular cylinder, of uniform thickness, t, where the shear force P for the section acts through the shear center.

(c) the location, e, of the shear center along the horizontal axis by using the shear flow distribution,

(d) the angle, θ, at which the shear flow is a maximum and that maximum value.

P10.5 The box beam shown in Fig. P10.5 $\left(I_{yy} = 688 \times 10^6 \text{ mm}^4\right)$ supports a vertical shear force $V_z = 10$ kN acting through the shear center. Determine the location of the shear center (the distance, e, as shown) and show on a sketch the complete shear flow distribution. All distances in Fig. P10.5 are measured from the center lines of the cross-section.

P10.6 For the thin J-section shown in Fig. P10.6a, do the following.

(a) Determine the coordinates (c_y, c_z) of the centroid relative to the cross-section center lines (see Fig. P10.6b) and the area moments and mixed area moment with respect to the centroid C. Use the thin-section approximations where the cross-section is decomposed into three rectangle having lengths equal to the center line lengths shown in Fig. P10.6b and terms involving the thickness cubed are neglected when computing the area moments and mixed area moment.

(b) Repeat part (a) where exact expressions for the centroid location, area moments and mixed area moments are obtained with the MATLAB® script sections (see Appendix A),

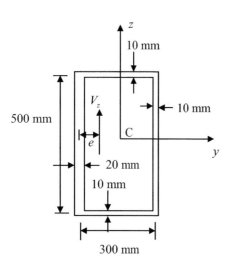

Figure P10.5 A box beam in bending.

Figure P10.6 (a) A thin J-section where all distances are measured along the center lines. (b) The center lines of the section and the location (c_y, c_z) of the centroid relative to those center lines.

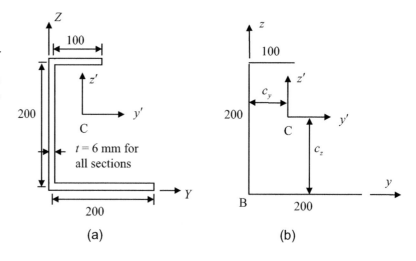

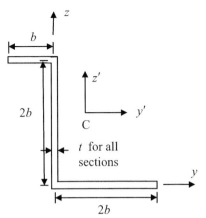

Figure P10.7 A thin, unsymmetrical Z-section. All distances are measured along the center lines.

using the (X, Y, Z) axes shown in Fig. P10.6a. Show that your answers agree with those of part (a).

(c) Determine the location of the shear center (e_y, e_z) relative to the (y, z) axes shown in Fig. 10.6b. Use Eq. (10.52) and the origin of the (y, z) coordinates, point B, as the pole for calculating the sectorial area function.

P10.7 In the bending of unsymmetrical sections, the quantity $D = I_{yy}I_{zz} - I_{yz}^2$ appears in the denominator of the expression for the flexure stress (see Eq. (10.13)). Using the fact that the smallest principal area moment for the cross-section is always positive (see Appendix A), show that D can never be zero.

P10.8 For the thin unsymmetrical Z-section shown in Fig. P10.7, obtain explicit expressions for the location of the centroid and the area moments and mixed area moment using the approximation for thin cross-sections, where terms involving the thickness cubed are neglected. Determine the accuracy of these thin section results by comparing them to the more exact values obtained with the MATLAB® script sections for $b = 10$ in., $t = 2$ in. (see Appendix A). All the dimensions and the (y, z) axes shown in Fig. P10.7 are measured along the center lines.

P10.9 An L-section with $b = 3$ in., $h = 5$ in., $t_1 = t_2 = 0.25$ in. (see Fig. A.2 in Appendix A) is subjected to a bending moment $M = 1000$ in.-lb about the y-axis. Using centroidal axes (y, z) parallel to the axes shown in Fig. A.2:

(a) determine the angle that the neutral axis makes with respect to the y-axis,
(b) find the maximum tensile and compressive flexural stresses and their locations in the cross-section,
(c) repeat (a) and (b) using the flexure formula written in principal (Y, Z) axes (see Eq. (10.20)). [Note: these are not the (Y, Z) axes shown in Fig. A.2.]

P10.10 Consider the triangular cross-section shown in Fig. P10.8. From tables we can find that the area moments and mixed area moment are given by

$$I_{yy} = \frac{bh^3}{36}, I_{zz} = \frac{hb^3}{36}, I_{yz} = \frac{b^2h^2}{72}$$

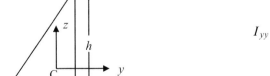

Figure P10.8 A triangular cross-section.

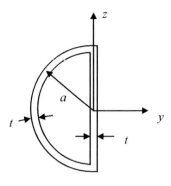

Figure P10.9 A thin, closed semicircular cross-section of constant thickness, t, where $t \ll a$.

If a bending moment is applied only along the y-axis, show that neutral axis intersects the midpoint of the vertical side AC of the triangle.

P10.11 Consider the thin, closed semicircular cross-section shown in Fig. P10.9. A shear force V_z acts on this section through the shear center. Determine the location of the shear center and obtain the complete distribution of the shear flow in the section in terms of V_z and the radius a.

11 Uniform and Nonuniform Torsion

The torsion of a solid or hollow bar with a circular cross-section is one of the important problems considered in elementary strength of material texts. In this chapter we consider the torsion of bars having more general cross-sections, where the axial warping deformations produced require that one develops a much more complex solution procedure. Like most advanced strength of materials discussions, we will first consider the idealized case of uniform torsion (also called Saint-Venant torsion) where the bar is completely free to warp. Solutions of uniform torsion problems are obtained using both a warping function and a Prandtl stress function approach. The case of nonuniform torsion, where the rate of twist varies along the length of the bar and/or the warping of the bar is restrained, is also considered using a warping function approach. Nonuniform torsion for general cross-sections is not normally treated in advanced strength of materials texts. This omission is unfortunate as many practical torsion problems likely fall in this category. This chapter will give a complete description of the governing equations for nonuniform torsion of members with general cross-sections.

Nonuniform torsion problems for general cross-sections typically require a numerical solution. However, in the case of thin members, one can obtain more direct solutions for both open and closed cross-sections. It will be shown that the sectorial area function plays a major role in the solution of both uniform and nonuniform torsion problems for thin cross-sections.

11.1 Torsion of Circular Cross-Sections – A Summary

In Chapter 1 we outlined the major elements of the strength of materials solution for the torsion of circular shafts. To set the stage for dealing with more complex torsion problems in this section, we want to summarize those results for circular sections so that we can contrast them with more general cases. For a circular cross-section, we assumed that the displacements were due to a pure rotational twisting, $\phi(x)$, of the cross-section with no displacement in the axial direction, so that the displacements were given as

$$\begin{aligned} u_x &= 0 \\ u_y &= -z\phi(x) \\ u_z &= y\phi(x) \end{aligned} \qquad (11.1)$$

We also assumed that the rate of twist $d\phi/dx = \phi'$ was a constant. In that case the only strains are (e_{xy}, e_{xz}) and the only nonzero stresses are likewise

Uniform and Nonuniform Torsion

$$\sigma_{xy} = -G\phi'z$$
$$\sigma_{xz} = G\phi'y \tag{11.2}$$

where G is the shear modulus. These stresses were then related to the internal torque, T, in the cross-section given by

$$T = GJ\phi' \tag{11.3}$$

where

$$J = \int_A (y^2 + z^2)dA = \int_A r^2 dA \tag{11.4}$$

is the polar area moment of the cross-section. If the rate of twist, ϕ', is a constant, Eq. (11.3) shows that the internal torque is also a constant, a condition we will call *uniform torsion*. Figure 11.1a shows a bar, fixed at $x = 0$ and subjected to a torque, T_0, at $x = L$. In this case the internal torque, $T = T_0$ is a constant throughout the bar, as is the rate of twist, so this is a case of uniform torsion. Since $\phi' = d\phi/dx$ is a constant $\phi(x) = \phi_0 x/L$, where ϕ_0 is the twist at $x = L$, and we have from Eq. (11.3) $\phi_0 = T_0 L/GJ$, which is the familiar torque–twist relationship used in strength of materials. The shear stresses in Eq. (11.2) are components of the total shear stress, τ, in the cross-section, which varies radially from the center of cross-section (see Fig. 11.1b) as

$$\tau = \frac{Tr}{J} \tag{11.5}$$

so that the maximum stress in the cross-section occurs at the outer radius $r = c$ and is given by

$$\tau_{max} = \frac{Tc}{J} \tag{11.6}$$

For a circular bar, in the case of *nonuniform torsion*, where the rate of twist, ϕ', and the internal torque, T, are functions of x, the basic assumptions made on the deformations for

Figure 11.1 (a) A bar of circular cross-section under uniform torsion. (b) The total shear stress, τ, in the cross-section.

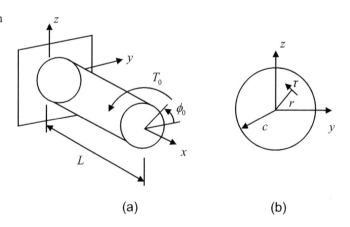

the uniform case are not violated so that nonuniform torsion problems of circular cross-section members involving loading and boundary conditions different from that seen in Fig. 11.1a can be solved directly without difficulty. As we will see, this is not the case for noncircular sections. We will begin by considering the uniform torsion case for those more general sections.

11.2 Uniform Torsion of Noncircular Cross-Sections – Warping function

The strength of materials solution described in the previous section cannot be applied to the torsion of noncircular sections because the basic assumption of the deformation, Eq. (11.1), requires that each planar cross-section of the bar remains planar and merely rotates (twists) about the central axis. If one draws a grid on the surface of a circular rubber rod and then twists it, as shown in Fig. 11.2a, then one can clearly see the twisting deformation of the horizontal lines (which are transformed by the twisting from straight lines along the x-axis into helixes) and that there is no distortion of the cross-section in the horizontal x-direction, as assumed. However, if one takes a noncircular rubber rod and twists it, then the deformation is as shown in Fig. 11.2b. There is again a twisting deformation of the horizontal lines but there is also significant distortion of the cross-section in the x-direction, i.e., the cross-sections both twist and warp. Warping significantly affects the strains and their corresponding stresses in the bar. To include this warping, we will replace Eq. (11.1) by

$$\begin{aligned} u_x &= \phi' \psi(y,z) \\ u_y &= -z\phi(x) \\ u_z &= y\phi(x) \end{aligned} \quad (11.7)$$

where $\psi(y,z)$ is the *warping function*, which represents the out-of-plane deformation (that has for convenience been multiplied by the rate of twist, ϕ'), and we will assume that ϕ' again is a constant so that we have the bar under uniform torsion. In this case the strains are given by

$$\begin{aligned} e_{xx} &= \frac{\partial u_x}{\partial x} = 0, \; e_{yy} = \frac{\partial u_y}{\partial y} = 0, \; e_{zz} = \frac{\partial u_z}{\partial z} = 0 \\ \gamma_{xy} &= \frac{\partial u_x}{\partial y} + \frac{\partial u_y}{\partial x} = \phi' \left(\frac{\partial \psi}{\partial y} - z \right) \\ \gamma_{xz} &= \frac{\partial u_x}{\partial z} + \frac{\partial u_z}{\partial x} = \phi' \left(\frac{\partial \psi}{\partial z} + y \right) \\ \gamma_{yz} &= \frac{\partial u_z}{\partial y} + \frac{\partial u_y}{\partial z} = \phi(x) - \phi(x) = 0 \end{aligned} \quad (11.8)$$

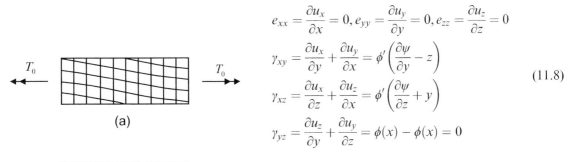
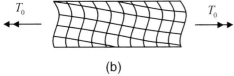

Figure 11.2 (a) Uniform torsion of a circular bar with no out-of-plane warping. (b) A noncircular bar showing out-of-plane warping.

so that the only nonzero stresses again are

$$\sigma_{xy} = G\phi'\left(\frac{\partial \psi}{\partial y} - z\right)$$
$$\sigma_{xz} = G\phi'\left(\frac{\partial \psi}{\partial z} + y\right)$$
(11.9)

These stresses must satisfy the equilibrium equations, which are

$$\frac{\partial \sigma_{xy}}{\partial y} + \frac{\partial \sigma_{xz}}{\partial z} = 0$$
$$\frac{\partial \sigma_{xy}}{\partial x} = 0$$
$$\frac{\partial \sigma_{xz}}{\partial x} = 0$$
(11.10)

The second and third equations are satisfied identically while the first equation gives

$$\frac{\partial^2 \psi}{\partial y^2} + \frac{\partial^2 \psi}{\partial z^2} = 0$$
(11.11)

so that the warping function must satisfy Laplace's equation. To solve this equation also requires that we specify the behavior of ψ on the boundary line that defines the edge of cross-section. We can obtain this boundary condition by examining the stresses on the boundary. Figure 11.3a shows the shear-stress components $(\sigma_{xn}, \sigma_{xt})$ at the edge of the cross-section acting in the **n** and **t** directions, where **n** is a unit outward normal vector to the curved edge and **t** is a unit tangent vector to that edge. The corresponding stress components $(\sigma_{xy}, \sigma_{xz})$ acting in the y- and z-directions are shown in Fig. 11.3b, together with the angle α that the normal and tangent vectors make with respect to the y- and z-axes. Relating these components, we find

$$\sigma_{xn} = \sigma_{xy} \cos \alpha + \sigma_{xz} \sin \alpha$$
$$= \sigma_{xy} n_y + \sigma_{xz} n_z$$
$$\sigma_{xt} = -\sigma_{xy} \sin \alpha + \sigma_{xz} \cos \alpha$$
$$= -\sigma_{xy} n_z + \sigma_{xz} n_y$$
(11.12)

in terms of the components (n_y, n_z) of the unit normal. The shear-stress component σ_{xn} must vanish on the boundary because $\sigma_{xn} = \sigma_{nx}$ where σ_{nx} is the corresponding shear stress acting on the lateral surface (n = constant) of the bar in the x-direction and there are no such axial

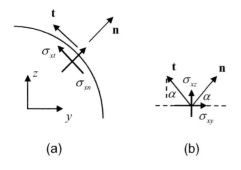

Figure 11.3 (a) The stresses at the edge of the cross-section in the directions normal and tangential to that edge. (b) The angles that the unit normal and tangent vectors make with respect to the y- and z-axes.

surface stresses in our torsion problem. From the boundary condition $\sigma_{xn} = 0$ and Eq. (11.12) and Eq. (11.9), we find

$$\frac{\partial \psi}{\partial y} n_y + \frac{\partial \psi}{\partial z} n_z - z n_y + y n_z = 0 \tag{11.13}$$

or, equivalently,

$$\frac{\partial \psi}{\partial n} = z n_y - y n_z \tag{11.14}$$

where $\partial \psi / \partial n$ is the normal derivative of the warping function. We can express this boundary condition in another equivalent form if we let s be the arc length along the boundary. Figure 11.4a then shows a small change of this arc length (in the t-direction) and Fig. 11.4b shows the relationship between ds and changes in the y- and z-direction along the boundary. It follows that

$$\begin{aligned} n_y &= \cos \alpha = \frac{dz}{ds} \\ n_z &= \sin \alpha = -\frac{dy}{ds} \end{aligned} \tag{11.15}$$

so from Eq. (11.15) and Eq. (11.14) we can also write the boundary condition as

$$\begin{aligned} \frac{\partial \psi}{\partial n} &= z \frac{dz}{ds} + y \frac{dy}{ds} \\ &= \frac{1}{2} \frac{d}{ds} (y^2 + z^2) \end{aligned} \tag{11.16}$$

However, it is usually more convenient to use the boundary condition in the form of Eq. (11.14). Once the warping function is obtained by solving Laplace's equation subject to the boundary condition, we can obtain the relationship between the torque and rate of twist. In this case we have

$$\begin{aligned} T &= \int_A (y \sigma_{xz} - z \sigma_{xy}) dA \\ \sigma_{xy} &= G \phi' \left(\frac{\partial \psi}{\partial y} - z \right) \\ \sigma_{xz} &= G \phi' \left(\frac{\partial \psi}{\partial z} + y \right) \end{aligned} \tag{11.17}$$

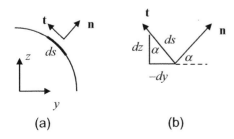

Figure 11.4 (a) An element of length ds along the boundary. (b) The geometry relating ds to changes in the y- and z-coordinates.

so that

$$T = G\phi' \int_A \left[(y^2 + z^2) + \left(y\frac{\partial \psi}{\partial z} - z\frac{\partial \psi}{\partial y}\right)\right] dA \qquad (11.18)$$
$$= G\phi' J_{eff}$$

where J_{eff} is the *effective polar area moment* for the noncircular cross-section. Equation (11.18) shows that we can write the torque-rate of twist relation in the same form as for the circular cross-section. Note that when the warping is absent $\psi = 0$, and J_{eff} just becomes the ordinary polar area moment. The form seen in Eq. (11.18) is not the only form possible for J_{eff}. If we multiply the boundary condition of Eq. (11.13) by ψ and use Gauss's theorem, we have

$$\int_C \left[\psi\left(\frac{\partial \psi}{\partial y} - z\right)n_y + \psi\left(\frac{\partial \psi}{\partial z} + y\right)n_z\right] ds$$
$$= \int_A \left[\frac{\partial}{\partial y}\left\{\psi\left(\frac{\partial \psi}{\partial y} - z\right)\right\} + \frac{\partial}{\partial z}\left\{\psi\left(\frac{\partial \psi}{\partial z} + y\right)\right\}\right] dA \qquad (11.19)$$
$$= \int_A \left[\left(\frac{\partial \psi}{\partial y}\right)^2 - z\frac{\partial \psi}{\partial y} + \left(\frac{\partial \psi}{\partial z}\right)^2 + y\frac{\partial \psi}{\partial z}\right] dA = 0$$

where C is the boundary of the cross-sectional area A and we have used the fact that ψ satisfies Laplace's equation in obtaining the final form seen in Eq. (11.19). If we add the integral in Eq. (11.19) to the integral appearing in Eq. (11.18) we can find an alternate expression for J_{eff} given by

$$J_{eff} = \int_A \left[\left(y + \frac{\partial \psi}{\partial z}\right)^2 + \left(z - \frac{\partial \psi}{\partial y}\right)^2\right] dA \qquad (11.20)$$

which is a form we will use shortly. It can be shown that the largest total shear stress, τ_{max}, occurs at some point on the boundary of the cross-section [1]. Since on the boundary $\sigma_{xn} = 0$ the shear stress is tangent to the boundary and we have

$$\tau_{max} = \{\sigma_{xt}\}_{max} = \{-\sigma_{xy}n_z + \sigma_{xz}n_y\}_{max}$$
$$= G\phi'\left\{\left(z - \frac{\partial \psi}{\partial y}\right)n_z + \left(y + \frac{\partial \psi}{\partial z}\right)n_y\right\}_{max} \qquad (11.21)$$

We can define an *effective maximum radius*, c_{eff}, as

$$c_{eff} = \left\{\left(z - \frac{\partial \psi}{\partial y}\right)n_z + \left(y + \frac{\partial \psi}{\partial z}\right)n_y\right\}_{max} \qquad (11.22)$$

so from Eq. (11.21) and Eq. (11.18) it follows that the maximum shear stress can also be expressed in a form entirely similar to that of the circular cross-section, namely

$$\tau_{\max} = \frac{T c_{\text{eff}}}{J_{\text{eff}}} \qquad (11.23)$$

but here the J_{eff} and c_{eff} values are typically obtained from a numerical solution.

We have shown that the shear stresses of Eq. (11.9) produce a twisting moment in the member. To ensure they generate a purely torsional moment, we also need to show that they do not produce shear forces. Thus, consider the force generated in the y-direction given by

$$V_y = \int_A \sigma_{xy} dA = G\phi' \int_A \left(\frac{\partial \psi}{\partial y} - z \right) dA \qquad (11.24)$$

Using the fact that the warping function satisfies Laplace's equation, we can rewrite Eq. (11.24) in terms of an equivalent area integral and then use Gauss's theorem to transform that area integral into an integral around the boundary of the cross-section:

$$\begin{aligned}
V_y &= G\phi' \int_A \left\{ \frac{\partial}{\partial y} \left[y \left(\frac{\partial \psi}{\partial y} - z \right) \right] + \frac{\partial}{\partial z} \left[y \left(\frac{\partial \psi}{\partial z} \right) + y \right] \right\} dA \\
&= G\phi' \int_C y \left[\frac{\partial \psi}{\partial n} - z n_y + y n_z \right] ds = 0
\end{aligned} \qquad (11.25)$$

The boundary integral in Eq. (11.25) is zero because of the boundary condition, Eq. (11.14), showing that $V_y = 0$. An entirely similar set of steps shows that the shear force $V_z = 0$ also.

This formulation we have been analyzing for the uniform torsion case is also called the *Saint-Venant torsional theory*. Generally, we need to numerically solve Laplace's equation, Eq. (11.11), subject to the boundary conditions, Eq. (11.13), a task that can be done with finite elements or with boundary elements. We can, however, obtain an approximate analytical expression for the torsion of a thin rectangular section, as we will show in the following example.

Example 11.1 Torsion of a Thin Rectangle

Consider the uniform torsion of the thin rectangular section of length b and thickness t shown in Fig. 11.5a, where $t \ll b$. On the two long faces of the section, the warping function must satisfy the boundary conditions

$$\left. \frac{\partial \psi}{\partial y} \right|_{y=\pm t/2} = z \qquad (11.26)$$

Since the other two faces of the cross-section at $z = \pm b/2$ are small, we will assume we can treat the cross-section as if it were infinitely long and simply ignore the boundary conditions at those faces. In that case we see that a warping function given by

$$\psi = yz \qquad (11.27)$$

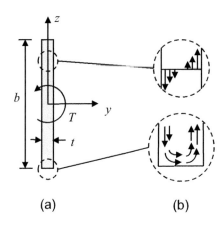

Figure 11.5 (a) Torsion of a thin rectangular section. (b) The shear-stress distributions across the thickness and near an end of the cross-section.

satisfies both Laplace's equation and the boundary conditions of Eq. (11.26). The shear stresses, Eq. (11.9), are then

$$\sigma_{xz} = 2G\phi' y$$
$$\sigma_{xy} = 0 \qquad (11.28)$$

which produces a linear distribution of stress across the thickness as shown in Fig. 11.5b. This distribution does represent well the actual stress distribution over most of the cross-section, although near the ends it is obviously in error since the shear stress must change direction and "flow" around the ends so as to remain tangent to the boundary at the edge of the cross-section (see Fig. 11.5b). If we compute the effective polar area moment from Eq. (11.20), we find

$$J_{eff} = 4\int_A y^2 dA = 4I_{zz} = \frac{bt^3}{3} \qquad (11.29)$$

Since the maximum shear stress occurs at $y = \pm t/2$, from Eq. (11.22) we also find

$$c_{eff} = \pm 2y|_{y=\pm t/2} = t \qquad (11.30)$$

which gives

$$\tau_{max} = \frac{Tt}{J_{eff}} \qquad (11.31)$$

We should note that if we had calculated J_{eff} with the integral expression given in Eq. (11.18), which came directly from the equation

$$T = \int_A (y\sigma_{xz} - z\sigma_{xy}) dA \qquad (11.32)$$

we would obtain a value for J_{eff} that is smaller by a factor of two from the correct value given by Eq. (11.29)! This discrepancy occurs because we have ignored the σ_{xy} stress in the cross-section and those stresses near the ends cannot be neglected when using either Eq. (11.32) or, equivalently, Eq. (11.18). In contrast, using Eq. (11.20) the end effects are negligible. Thus, when using approximate expressions in calculating such torsional constants, all forms are not equally good for such approximations (although all forms are equivalent for exact solutions). We will discuss this point again later.

11.3 Uniform Torsion of Noncircular Cross-Sections – Prandtl Stress function

In the Saint-Venant theory of torsion, the displacements are assumed to be

$$u_x = \frac{d\phi}{dx}\psi(y,z)$$
$$u_y = -z\phi(x) \qquad (11.33)$$
$$u_z = y\phi(x)$$

and for uniform torsion, the rate of twist $d\phi/dx = \phi'$ is a constant. As we have seen, this assumed deformation produces only shear stresses in the cross-section that we can write in terms of the displacements as

$$\sigma_{xz} = G\left(\frac{\partial u_x}{\partial z} + \frac{\partial u_z}{\partial x}\right) = G\left(\frac{\partial u_x}{\partial z} + \phi' y\right)$$
$$\sigma_{xy} = G\left(\frac{\partial u_x}{\partial y} + \frac{\partial u_y}{\partial x}\right) = G\left(\frac{\partial u_x}{\partial y} - \phi' z\right) \qquad (11.34)$$

All the equilibrium equations are satisfied if

$$\frac{\partial \sigma_{xy}}{\partial y} + \frac{\partial \sigma_{xz}}{\partial z} = 0 \qquad (11.35)$$

This equation can be satisfied automatically by writing the stresses in terms of a function, Φ, called the *Prandtl stress function*, where

$$\sigma_{xy} = \frac{\partial \Phi}{\partial z}, \quad \sigma_{xz} = -\frac{\partial \Phi}{\partial y} \qquad (11.36)$$

However, from Eq. (11.34) we have

$$G\frac{\partial u_x}{\partial z} = -\frac{\partial \Phi}{\partial y} - G\phi' y$$
$$G\frac{\partial u_x}{\partial y} = \frac{\partial \Phi}{\partial z} + G\phi' z \qquad (11.37)$$

which also implies

$$G\frac{\partial^2 u_x}{\partial z \partial y} = -\frac{\partial^2 \Phi}{\partial y^2} - G\phi'$$
$$G\frac{\partial^2 u_x}{\partial y \partial z} = \frac{\partial^2 \Phi}{\partial z^2} + G\phi' \qquad (11.38)$$

These mixed derivatives of the displacement u_x must be equal if we are to be able to integrate the strains to find this displacement and this compatibility condition requires that the stress function satisfy

$$\frac{\partial^2 \Phi}{\partial y^2} + \frac{\partial^2 \Phi}{\partial z^2} = -2G\phi' \tag{11.39a}$$

which is also often written as

$$\nabla^2 \Phi = -2G\phi' \tag{11.39b}$$

in terms of the Laplacian operator $\nabla^2 = \partial^2/\partial y^2 + \partial^2/\partial z^2$. Equation (11.39a) is called *Poisson's equation*. On the outer surface of the bar, there is no component of the stress vector acting in the x-direction so that

$$T_x^{(n)} = \sigma_{nx} = \sigma_{xy} n_y + \sigma_{xz} n_z = 0 \tag{11.40}$$

and the y- and z-components of the stress vector are identically zero. In terms of the Prandtl stress function Eq. (11.40) becomes

$$\frac{\partial \Phi}{\partial z} n_y - \frac{\partial \Phi}{\partial y} n_z = 0 \tag{11.41}$$

By examining a small element along the boundary (Fig. 11.6) we see that Eq. (11.41) also implies

$$\frac{d\Phi}{ds} = \frac{\partial \Phi}{\partial z}\frac{dz}{ds} + \frac{\partial \Phi}{\partial y}\frac{dy}{ds} = 0 \tag{11.42}$$

so that Φ must be a constant on the boundary. For a simply connected cross-section (a cross-section with no holes), we can take that constant to be zero so the boundary condition becomes simply

$$\Phi = 0 \quad \text{on the boundary } C \tag{11.43}$$

To find the internal torque, T, in terms of the stress function we start from the relation between the torque and the shear stresses and express that relationship in terms of the stress function, yielding

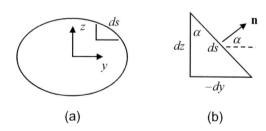

Figure 11.6 (a) A small element at the boundary. (b) The geometry relating that element to the unit normal.

$$T = \int_A (\sigma_{xz}y - \sigma_{xy}z)dA$$

$$= -\int_A \left(\frac{\partial \Phi}{\partial y}y + \frac{\partial \Phi}{\partial z}z\right)dA$$

$$= -\int_A \left[\frac{\partial}{\partial y}(\Phi y) + \frac{\partial}{\partial z}(\Phi z) - 2\Phi\right]dA \quad (11.44)$$

$$= -\int_C \left[\Phi y n_y + \Phi z n_z\right]dA + 2\int_A \Phi dA$$

$$= 2\int_A \Phi dA$$

where we have used Gauss's theorem and where the integral over the boundary C vanishes since $\Phi = 0$. Thus, the torque is just twice the integral of the $\Phi(y,z)$ function over the cross-section.

As mentioned previously, the largest shear stress in the cross-section occurs at some point on the boundary. Since the total shear stress τ must be tangent to the boundary (Fig. 11.7), we have

$$\sigma_{xz} = \tau \cos \alpha$$
$$\sigma_{xy} = -\tau \sin \alpha \quad (11.45)$$

so that

$$\frac{\partial \Phi}{\partial n} = \frac{\partial \Phi}{\partial y}n_y + \frac{\partial \Phi}{\partial z}n_z$$
$$= -\sigma_{xz}\cos \alpha + \sigma_{xy}\sin \alpha \quad (11.46)$$
$$= -\tau$$

which gives

$$\tau_{max} = \left(-\frac{\partial \Phi}{\partial n}\right)_{max \text{ on } C} \quad (11.47)$$

As with the warping function, we can make these results appear similar to the familiar formulas for the torsion of a circular cross-section if we define a normalized stress function $\bar{\Phi} = \Phi/G\phi'$ where

$$\nabla^2 \bar{\Phi} = -2 \quad \text{in the cross-section}$$
$$\bar{\Phi} = 0 \quad \text{on the boundary} \quad (11.48)$$

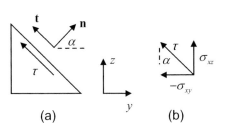

Figure 11.7 (a) The total shear stress on the boundary. (b) The relationship of the total shear stress to the stress components in the (y, z) directions.

If we let $T = GJ_{eff}\phi'$ then

$$J_{eff} = 2\int_A \bar{\Phi}dA \qquad (11.49)$$

and

$$\tau_{max} = \frac{Tc_{eff}}{J_{eff}} \qquad (11.50)$$

where

$$c_{eff} = \left(-\frac{\partial\bar{\Phi}}{\partial n}\right)_{max \ on \ C} \qquad (11.51)$$

Prandtl Stress Function Solutions

There are a number of cases where we can obtain explicit results using stress functions. Here we discuss some of the important examples.

(1) Circular cross-section of radius c

In this case, the normalized Prandtl stress function is

$$\bar{\Phi} = -\frac{1}{2}\left(y^2 + z^2 - c^2\right) = -\frac{1}{2}\left(r^2 - c^2\right) \qquad (11.52)$$

The effective polar area moment is just the ordinary polar area moment where

$$J_{eff} = J = 2\int_A \bar{\Phi}dA = 2\pi \int_{r=0}^{r=c} (c^2 - r^2)r\,dr = \frac{\pi c^4}{2} \qquad (11.53)$$

For the maximum shear stress, we have

$$\tau_{max} = \frac{T}{J_{eff}}\left(-\frac{\partial\bar{\Phi}}{\partial r}\right)_{r=c} = \frac{Tc}{J} \qquad (11.54)$$

note that in this case

$$\sigma_{xz} = -\frac{\partial\Phi}{\partial y} = G\phi'y = G\left(\frac{\partial u_x}{\partial z} + \phi'y\right)$$

$$\sigma_{xy} = \frac{\partial\Phi}{\partial z} = -G\phi'z = G\left(\frac{\partial u_x}{\partial y} - \phi'z\right) \qquad (11.55)$$

which implies that

$$\frac{\partial u_x}{\partial y} = \frac{\partial u_x}{\partial z} = 0 \qquad (11.56)$$

11.3 Uniform Torsion of Noncircular Cross-Sections

so that u_x is a constant that we can take equal to zero. Thus, there is no warping, as expected, in this case.

(2) Elliptical cross-section (with semimajor axes of lengths a and b along the y- and z-axes, respectively)

Here, we have

$$\bar{\Phi} = \frac{-a^2 b^2}{(a^2 + b^2)} \left(\frac{y^2}{a^2} + \frac{z^2}{b^2} - 1 \right) \tag{11.57}$$

and the effective polar area moment is

$$J_{eff} = \frac{\pi a^3 b^3}{(a^2 + b^2)} \tag{11.58}$$

The maximum shear stress is

$$(\tau_{max})_{z=\pm b} = \frac{2T}{\pi a b^2} \quad \text{for } b < a \tag{11.59}$$

If we integrate the shear-stress expressions to find the warping displacement, u_x, we find

$$u_x = \frac{T(b^2 - a^2)}{\pi a^3 b^3 G} yz \tag{11.60}$$

(3) Thin rectangular section

Consider the same case examined with the warping function where the cross-section is of length b in the z-direction and has a small thickness t in the y-direction (Fig. 11.5a). In this case we might expect that $\bar{\Phi} = f(y)$ so that

$$\nabla^2 \bar{\Phi} = \frac{d^2 f}{dy^2} = -2 \tag{11.61}$$
$$f = 0 \text{ on } y = \pm t/2$$

which gives

$$\bar{\Phi} = \left(\frac{t^2}{4} - y^2 \right) \tag{11.62}$$

Using this stress function, the effective polar area moment is

$$J_{eff} = 2b \int_{-t/2}^{t/2} \bar{\Phi} dy = \frac{1}{3} b t^3 \tag{11.63}$$

which agrees with our previous result. The maximum shear stress is

$$\tau_{max} = \frac{T}{J_{eff}}\left(\mp\frac{\partial\bar{\Phi}}{\partial y}\right)_{y=\pm t/2} = \frac{3T}{bt^2} \qquad (11.64)$$

This value of J_{eff}, which was obtained with approximate solutions for both the warping function and the Prandtl stress function, can be compared with the first term in an infinite series solution that is good to several percent for a rectangle of arbitrary size given by [2]

$$J_{eff} = \frac{bt^3}{3}\left[1 - \frac{192}{\pi^5}\frac{t}{b}\tan\left(\frac{\pi b}{2t}\right)\right] \qquad (11.65)$$

Consider a thin rectangle where $t/b = 0.2$, for example. In that case Eq. (11.65) gives

$$J_{eff} = \frac{bt^3}{3}(1.017) \qquad (11.66)$$

so that the approximate value of $bt^3/3$ is quite good.

If we use the relationship between the torque and the stresses to obtain J_{eff}, we find

$$T = \int_A (\sigma_{xz}y - \sigma_{xy}z)dA$$

$$= G\phi'\int_{-t/2}^{t/2} 2y^2 b\,dy \qquad (11.67)$$

$$= G\phi'\frac{bt^3}{6}$$

so that

$$J_{eff} = \frac{bt^3}{6} \qquad (11.68)$$

which is an incorrect value that is only one half the value obtained from either the warping function and Eq. (11.20) or the Prandtl stress function and Eq. (11.49). As mentioned previously, the reason for this discrepancy is that we neglected in our approximation the shear stress σ_{xy} that develops near the end of the cross-section. While those shear stresses may be localized in their importance, they do have a large moment arm (on the order of length $b/2$) as opposed to the maximum shear stresses σ_{xz}, whose moment arm is on the order of the thickness, $t/2$. To estimate the size of the contribution from the σ_{xy} stress, we will assume that the stress function near the ends of the cross-section varies linearly from zero to its maximum value over a parabolic-shaped area, A_{end}, of length h (Fig. 11.8), i.e.,

$$\bar{\Phi} \cong \frac{t^2}{4}\frac{z}{h} \qquad (11.69)$$

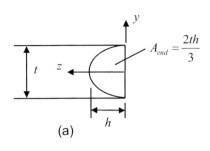
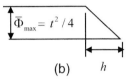

Figure 11.8 (a) A parabolic area of height h over which the shear stress σ_{xy} is assumed to be significant, and over which Φ is assumed to vary linearly from zero to its maximum value, as shown in (b).

and

$$\sigma_{xy} \cong G\phi' \frac{t^2}{4h} \tag{11.70}$$

We can estimate the torque produced by this stress (from both ends) as

$$T \cong 2\left[\sigma_{xy}\frac{b}{2}\int_{A_{end}} dA\right] = G\phi' \frac{t^2}{4h} b\left(\frac{2th}{3}\right) \tag{11.71}$$

$$= G\phi' \frac{bt^3}{6}$$

Thus, we obtain a contribution that is equal to that calculated from only σ_{xz} and the total torque from both stresses gives the correct result.

To evaluate the warping of the thin rectangle, note that for our approximate solution

$$\sigma_{xy} = \frac{\partial \Phi}{\partial z} = G\left(\frac{\partial u_x}{\partial y} - \phi'z\right) = 0$$

$$\sigma_{xz} = -\frac{\partial \Phi}{\partial y} = G\left(\frac{\partial u_x}{\partial z} + \phi'y\right) = 2G\phi'y \tag{11.72}$$

so that

$$\frac{\partial u_x}{\partial y} = \phi'z, \quad \frac{\partial u_x}{\partial z} = \phi'y \tag{11.73}$$

Integrating both of these equations we find

$$u_x = \phi'yz + f(z)$$
$$u_x = \phi'yz + g(y) \tag{11.74}$$

where f and g are both arbitrary functions. Equating these two expressions, we find that we must have $f = g = $ constant, where we can take this constant value to be zero (since it just produces a translation in the x-direction), giving the same warping function obtained earlier in Eq. (11.27).

The Membrane Analogy

One of the reasons the Prandtl stress function has been more widely used to describe torsion problems than the warping function approach is that there is a correspondence between the

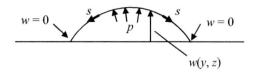

Figure 11.9 A thin membrane under a pressure, p. The membrane is stretched with a tension, s. The behavior of the displacement, w, of the membrane is analogous to that of the normalized Prandtl stress function.

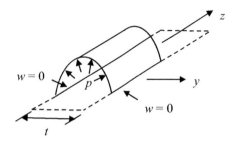

Figure 11.10 The (exaggerated) shape of the deflection of a membrane for a thin rectangular cross-section (away from the ends).

behavior of the stress function and the deflection of a thin membrane under pressure. For example, Fig. 11.9 shows the side view of an inflated membrane under an internal pressure, p, where w is the vertical deflection of the membrane and s is the constant tension in the membrane. From equilibrium of a membrane element one can show [3] that the membrane satisfies

$$\frac{\partial^2 w}{\partial y^2} + \frac{\partial^2 w}{\partial z^2} = -\frac{p}{s} \tag{11.75}$$

$w = 0$ on the boundary

which is identical to the solution of the torsion problem (for a simply connected cross-section) in terms of the normalized Prandtl stress function if we make the substitution

$$w = \frac{p}{s} \bar{\Phi} \tag{11.76}$$

The slopes of the membrane are proportional to the shear stresses because of Eq. (11.36). A similar analogy can be used for multiply connected cross-sections [4]. By examining the deflection and slopes of the membrane at various locations one can obtain a sense of the distribution of the shear stresses in a given cross-section. Figure 11.10, for instance, shows the shape of a membrane for thin rectangular section, where the deflected shape of the membrane follows the same parabolic distribution as the normalized Prandtl function, Eq. (11.62), over most of the cross-section. Near the ends (not shown) the changing slopes of the membrane similarly show the development of the σ_{xy} shear stress.

Torsion of Multiply Connected Cross-Sections

We previously showed that the stress function had to be a constant on any boundary of the cross-section. This is true for all boundaries of a multiply connected cross-section, i.e., the outer boundary and the holes. When we have a simply connected cross-section, we can always take the single constant on the outer boundary to be zero since we always can add or

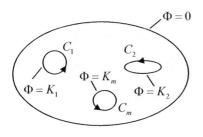

Figure 11.11 A multiply connected cross-section where the Prandtl stress function takes on different constant values at each hole.

subtract a constant from the Prandtl stress function without affecting the stresses. However, if there is one or more holes in a cross-section, the constant values on those boundaries cannot all be taken to be zero. Generally, one takes $\Phi = 0$ on the outer boundary of the cross-section and $\Phi = K_j$ ($j = 1, \ldots, m$) on m holes (Fig. 11.11), where the constants K_j are not zero. To find these m constants we need to specify m additional conditions on our problem. These conditions can be obtained by requiring that the u_x displacement be single-valued. A single-valued displacement can be assured if we require

$$\oint_{C_j} du_x = 0 \quad (j = 1, \ldots, m) \tag{11.77}$$

where C_j ($j = 1, \ldots, m$) are the closed lines defining the holes. But recall

$$u_x = u_x(y, z) = \phi' \psi(y, z), \quad \phi' = d\phi/dx = \text{constant}$$

$$\sigma_{xy} = G\left(\frac{\partial u_x}{\partial y} - z\phi'\right) \tag{11.78}$$

$$\sigma_{xz} = G\left(\frac{\partial u_x}{\partial z} + y\phi'\right)$$

so that placing these relations into

$$\oint_{C_j} du_x = \oint_{C_j} \left(\frac{\partial u_x}{\partial y} dy + \frac{\partial u_x}{\partial z} dz\right) = 0 \tag{11.79}$$

we obtain

$$\oint_{C_j} (\sigma_{xy} dy + \sigma_{xz} dz) + G\phi' \oint_{C_j} (z dy - y dz) = 0 \tag{11.80}$$

where the closed line integrals are taken in a counterclockwise sense. We can write these integrands in terms of the components of the outward unit normal vector (see Eq. (11.15)) and the Prandtl stress function (Eq. (11.36)) as

$$-\oint_{C_j} \left(\frac{\partial \Phi}{\partial z} n_z + \frac{\partial \Phi}{\partial y} n_y\right) ds - G\phi' \oint_{C_j} (zn_z + yn_y) ds = 0 \tag{11.81}$$

or equivalently, using Gauss's theorem

$$\oint_{C_j} \left(-\frac{\partial \Phi}{\partial n}\right) ds = G\phi' \int_{A_j} \left[\frac{\partial}{\partial z}(z) + \frac{\partial}{\partial y}(y)\right] dA$$

$$= 2G\phi' \int_{A_j} dA \qquad (11.82)$$

$$= 2G\phi' A_j$$

where A_j is the area that is enclosed by the contour C_j. Thus, to find the m constants we have to satisfy the m conditions

$$\oint_{C_j} \left(-\frac{\partial \Phi}{\partial n}\right) ds = 2G\phi' A_j \quad (j = 1, \ldots, m) \qquad (11.83)$$

or, in terms of the total shear stress, τ, on these boundaries,

$$\oint_{C_j} \tau\, ds = 2G\phi' A_j \quad (j = 1, \ldots, m) \qquad (11.84)$$

Historically, because of the membrane analogy most texts have emphasized the Prandtl stress function approach over the formulation based on the warping function. However, when solving these torsion problems numerically for multiply connected cross-sections the use of the warping function is easier in that there is only one boundary condition (see Eq. (11.14)) that is applied to all boundaries and there are no additional conditions such as Eq. (11.84) that need to be considered.

11.4 Nonuniform Torsion

The case of uniform torsion is a simple condition, much like the case of pure bending for beams. In Chapter 1, we extended the conditions of pure bending to more practical cases where the bending moment is not a constant along the axis of the beam. By letting the bending moment (and stress σ_{xx}) vary along the beam, however, we found that shear stresses are induced that also must be considered. In Chapter 10 we saw how to model such shear stresses explicitly for the bending of unsymmetrical thin cross-section beams. For the case of uniform torsion, the rate of the angle of twist and the internal torque are constants and the warping of the cross-section is allowed to occur unimpeded. In real applications, however, there typically are some constraints on the warping at boundaries. At a rigid boundary, for example, the warping would be zero. Such restraint of warping will cause the warping displacement to vary along the length of the member, i.e., we will have $u_x = u_x(x)$, and that variation in turn will produce a varying axial stress, $\sigma_{xx}(x)$. We can try to include variable warping by extending the basic assumptions on the deformation (see Eq. (11.33)) to be

$$u_x = \phi'(x)\psi(y,z)$$
$$u_y = -z\phi(x) \qquad (11.85)$$
$$u_z = y\phi(x)$$

i.e., the in-plane displacements still reflect the same basic twisting deformation and we again have out-of-plane warping but we will let the rate of twist be a function of the x-coordinate. Under these conditions we say that the member is under *nonuniform torsion* since, if the rate of twist is not a constant, the internal torque will vary also. Note that we can also have nonuniform torsion when there are applied distributed torsional loads along the length of the member since those loads will cause the internal torque to vary along the length of the member and so the rate of twist will vary also. Based on the displacements assumed in Eq. (11.85) we have the strains

$$
\begin{aligned}
e_{xx} &= \frac{\partial u_x}{\partial x} = \phi''(x)\psi(y,z) \\
\gamma_{xy} &= \frac{\partial u_x}{\partial y} + \frac{\partial u_y}{\partial x} = \phi'(x)\left(\frac{\partial \psi}{\partial y} - z\right) \\
\gamma_{xz} &= \frac{\partial u_x}{\partial z} + \frac{\partial u_z}{\partial x} = \phi'(x)\left(\frac{\partial \psi}{\partial z} + y\right) \\
\gamma_{yz} &= \frac{\partial u_y}{\partial z} + \frac{\partial u_z}{\partial y} = -\phi(x) + \phi(x) = 0
\end{aligned}
\qquad (11.86)
$$

where $\phi'' = d^2\phi/dx^2$. If we assume that σ_{xx} is the only significant normal stress in the member then for an isotropic elastic material the only nonzero stresses are

$$
\begin{aligned}
\sigma_{xx} &= E\phi''(x)\psi(y,z) \\
\sigma_{xy} &= G\gamma_{xy} = G\phi'(x)\left(\frac{\partial \psi}{\partial y} - z\right) \\
\sigma_{xz} &= G\gamma_{xz} = G\phi'(x)\left(\frac{\partial \psi}{\partial z} + y\right)
\end{aligned}
\qquad (11.87)
$$

Placing these stresses into the local equilibrium equations (see Eq. (3.11)) with zero body forces gives

$$
\begin{aligned}
&\frac{\partial}{\partial x}\left[E\phi''(x)\psi(y,z)\right] + \frac{\partial}{\partial y}\left[G\phi'(x)\left(\frac{\partial \psi}{\partial y} - z\right)\right] \\
&\quad + \frac{\partial}{\partial z}\left[G\phi'(x)\left(\frac{\partial \psi}{\partial z} + y\right)\right] = 0 \\
&G\phi''(x)\left(\frac{\partial \psi}{\partial y} - z\right) = 0 \\
&G\phi''(x)\left(\frac{\partial \psi}{\partial z} + y\right) = 0
\end{aligned}
\qquad (11.88)
$$

Equation (11.88) shows that, based on our original deformation assumption, we cannot satisfy the second and third equilibrium equations unless either $\phi'' = 0$ (in which case we are back to the uniform torsion case) or the torsional stresses vanish and we have no torsion. The first equation gives

$$\frac{\partial^2 \psi}{\partial y^2} + \frac{\partial^2 \psi}{\partial z^2} = -\frac{E\phi''(x)}{G\phi'(x)}\psi \tag{11.89}$$

which would require that unless $\phi'' = 0$ we must have $\psi = \psi(x, y, z)$, which contradicts our original assumption that $\psi = \psi(y, z)$. Thus, we need to modify our assumptions about the behavior in this problem. From beam-bending theory, as mentioned previously, we know that a normal stress that varies along the beam axis requires the development of shear stresses in the cross-section. In our torsion case a varying normal stress will produce shear stresses that can cause an additional displacement u_x of the cross-section, u_x^S, called *secondary warping*. Let $\psi^S(x, y, z)$ represent this secondary warping, which will vary in the x-direction as well as the y- and z-directions. This secondary warping will be associated with *secondary stresses* $\left(\sigma_{xy}^S, \sigma_{xz}^S\right)$, where

$$\begin{aligned} u_x^S &= \psi^S(x, y, z) \\ \sigma_{xy}^S &= G\frac{\partial u_x^S}{\partial y} = G\frac{\partial \psi^S}{\partial y} \\ \sigma_{xz}^S &= G\frac{\partial u_x^S}{\partial z} = G\frac{\partial \psi^S}{\partial z} \end{aligned} \tag{11.90}$$

This secondary warping and these secondary stresses will be incorporated into the case of nonuniform torsion in the following manner. Let us again modify the uniform torsion case where we simply let the rate of twist vary, i.e., we assume

$$\begin{aligned} u_x &= \phi'(x)\psi^P(y, z) \\ u_y &= -z\phi(x) \\ u_z &= y\phi(x) \end{aligned} \tag{11.91}$$

Here the in-plane displacements represent the twisting of the cross-section and there is warping present that we will call the *primary warping* of the cross-section, represented by $\psi^P(y, z)$. These are the same expressions as used in uniform torsion theory except the rate of twist is now allowed to vary. This change from the uniform torsion case means that the primary warping will generate a normal stress that we will call the *warping normal stress*, σ_{xx}^w, where

$$\sigma_{xx}^w(x, y, z) = E\frac{\partial u_x}{\partial x} = E\phi''(x)\psi^P(y, z) \tag{11.92}$$

in addition to the primary shear stresses, $\left(\sigma_{xy}^P, \sigma_{xz}^P\right)$, given by

$$\sigma_{xy}^P = G\phi'(x)\left(\frac{\partial \psi^P}{\partial y} - z\right)$$
$$\sigma_{xz}^P = G\phi'(x)\left(\frac{\partial \psi^P}{\partial z} + y\right) \tag{11.93}$$

Since, as we previously indicated, the warping normal stress produces additional secondary shear stresses the total shear stresses will be the sum of the primary and secondary shear stresses, i.e.,

$$\sigma_{xy} = \sigma_{xy}^P + \sigma_{xy}^S$$
$$\sigma_{xz} = \sigma_{xz}^P + \sigma_{xz}^S \tag{11.94}$$

Assuming the warping normal stress and the total shear stresses are in equilibrium (see the first equilibrium equation of Eq. (3.11), in the absence of body forces) we have

$$\frac{\partial \sigma_{xx}^w}{\partial x} + \frac{\partial \sigma_{xy}^P}{\partial y} + \frac{\partial \sigma_{xz}^P}{\partial z} + \frac{\partial \sigma_{xy}^S}{\partial y} + \frac{\partial \sigma_{xz}^S}{\partial z} = 0 \tag{11.95}$$

In order to satisfy Eq. (11.95), we will further require that the primary shear stresses are in equilibrium with themselves while the secondary shear stresses are in equilibrium with the warping normal stress, leading to the conditions

$$\frac{\partial \sigma_{xy}^P}{\partial y} + \frac{\partial \sigma_{xz}^P}{\partial z} = 0$$
$$\frac{\partial \sigma_{xx}^w}{\partial x} + \frac{\partial \sigma_{xy}^S}{\partial y} + \frac{\partial \sigma_{xz}^S}{\partial z} = 0 \tag{11.96}$$

This decomposition makes sense since the primary shear stresses are the same as those in the uniform torsion case where there is no normal stress, while the secondary shear stresses are present because of the varying normal stress. It is true that we cannot satisfy the two remaining equilibrium equations, so this nonuniform torsion theory lacks a completeness that one should be aware of. This is not unlike engineering beam theory for bending, however, where in general all the local equilibrium equations are also not satisfied.

If we place the stresses of Eq. (11.90) and Eq. (11.93) into Eq. (11.96), we find the governing equations for both the primary and secondary warping functions:

$$\nabla^2 \psi^P = \frac{\partial^2 \psi^P}{\partial y^2} + \frac{\partial^2 \psi^P}{\partial z^2} = 0$$
$$\nabla^2 \psi^S = \frac{\partial^2 \psi^S}{\partial y^2} + \frac{\partial^2 \psi^S}{\partial z^2} = -\frac{E\phi'''(x)}{G}\psi^P \tag{11.97}$$

where $\phi''' = d^3\phi/dx^3$. These governing equations are now consistent with our original assumptions that $\psi^P = \psi^P(y,z)$ and $\psi^S = \psi^S(x,y,z)$ so by the inclusion of the secondary warping and secondary shear stresses into the nonuniform torsion case, we have removed the problems associated with Eq. (11.89) described earlier. To solve the equations in Eq. (11.97) one also needs to specify boundary conditions. Recall from Eq. (11.12)

$$\sigma_{xn} = \sigma_{xy}n_y + \sigma_{xz}n_z$$
$$\sigma_{xt} = -\sigma_{xy}n_z + \sigma_{xz}n_y$$
(11.98)

so with the decomposition into primary and secondary shear stresses we also have

$$\sigma_{xn}^P = \sigma_{xy}^P n_y + \sigma_{xz}^P n_z$$
$$\sigma_{xt}^P = -\sigma_{xy}^P n_z + \sigma_{xz}^P n_y$$
$$\sigma_{xn}^S = \sigma_{xy}^S n_y + \sigma_{xz}^S n_z$$
$$\sigma_{xt}^S = -\sigma_{xy}^S n_z + \sigma_{xz}^S n_y$$
(11.99)

Substituting Eq. (11.90) and Eq. (11.93) into Eq. (11.99) we find

$$\sigma_{xn}^P = G\phi'(x)\left(\frac{\partial \psi^P}{\partial n} - zn_y + yn_z\right)$$
$$\sigma_{xt}^P = G\phi'(x)\left(\frac{\partial \psi^P}{\partial t} + yn_y + zn_z\right)$$
$$\sigma_{xn}^S = G\frac{\partial \psi^S}{\partial n}$$
$$\sigma_{xt}^S = G\frac{\partial \psi^S}{\partial t}$$
(11.100)

At the boundary we have seen previously that $\sigma_{xn} = 0$, so requiring that both the primary and secondary shear stresses satisfy this condition, the boundary conditions for the primary and secondary warping functions are

$$\frac{\partial \psi^P}{\partial n} = zn_y - yn_z, \quad \frac{\partial \psi^S}{\partial n} = 0$$
(11.101)

Thus, the case of nonuniform torsion requires that we solve a boundary value problem for the primary warping given as

$$\frac{\partial^2 \psi^P}{\partial y^2} + \frac{\partial^2 \psi^P}{\partial z^2} = 0 \quad \text{in the cross-section } A$$
$$\frac{\partial \psi^P}{\partial n} = zn_y - yn_z \quad \text{on the boundary } C$$
(11.102)

Once this solution is known, then we must solve a second boundary value problem for the secondary warping where

$$\frac{\partial^2 \psi^S}{\partial y^2} + \frac{\partial^2 \psi^S}{\partial z^2} = -\frac{E\phi'''(x)}{G}\psi^P \quad \text{in the cross-section } A \tag{11.103}$$

$$\frac{\partial \psi^S}{\partial n} = 0 \quad \text{on the boundary } C$$

The first boundary value problem is the same problem found earlier in the uniform torsion case. We see that, because of the presence of ψ^P in the governing equation, the secondary boundary value problem requires the solution of the first boundary value problem. A knowledge of the twist $\phi(x)$ is also required because of the $\phi'''(x)$ term. Both boundary value problems must typically be solved numerically. We will say more about their solutions later. Once these solutions are obtained, then we can find the warping normal stress and the shear stresses from Eq. (11.90) and Eq. (11.93). The shear stresses must produce the total twisting moment, $M_x(x)$, which we can also decompose into a primary twisting moment, $M_x^P(x)$, and a secondary twisting moment, $M_x^S(x)$, where

$$\begin{aligned}
M_x^P(x) &= \int_A \left(y\sigma_{xz}^P - z\sigma_{xy}^P \right) dA \\
&= G\phi'(x) \int_A \left[(y^2 + z^2) + y\frac{\partial \psi^P}{\partial z} - z\frac{\partial \psi^P}{\partial y} \right] dA \\
M_x^S(x) &= \int_A \left(y\sigma_{xz}^S - z\sigma_{xy}^S \right) dA \\
&= G \int_A \left[y\frac{\partial \psi^S}{\partial z} - z\frac{\partial \psi^S}{\partial y} \right] dA
\end{aligned} \tag{11.104}$$

We recognize that the expression for the primary twisting moment is the same as the one given by the uniform torsion theory where

$$M_x^P(x) = GJ_{eff}\phi'(x) \tag{11.105}$$

and

$$J_{eff} = \int_A \left[(y^2 + z^2) + y\frac{\partial \psi^P}{\partial z} - z\frac{\partial \psi^P}{\partial y} \right] dA \tag{11.106}$$

is the effective polar area moment of the cross-section (see Eq. (11.18). We can also use the alternate expression given in Eq. (11.20). For the secondary twisting moment, we note that by placing the warping functions (ψ^P, ψ^S) in the Gauss–Green theorem we have

$$\int_A \left(\psi^P \nabla^2 \psi^S - \psi^S \nabla^2 \psi^P \right) dA = \int_C \left(\psi^P \partial \psi^S / \partial n - \psi^S \partial \psi^P \partial n \right) ds \tag{11.107}$$

But using the boundary value problem relations of Eq. (11.102) and Eq. (11.103) we have

$$-\frac{E\phi'''(x)}{G}\int_A (\psi^P)^2 dA = -\int_C \psi^S(zn_y - yn_z)ds \tag{11.108}$$

and we can use Gauss's theorem to transform the boundary integral to an area integral, giving

$$-\frac{E\phi'''(x)}{G}\int_A (\psi^P)^2 dA = -\int_C \psi^S(zn_y - yn_z)ds$$

$$= \int_A \left[-z\frac{\partial \psi^S}{\partial y} + y\frac{\partial \psi^S}{\partial z}\right] dA \tag{11.109}$$

Comparing this result with the secondary twisting moment expression in Eq. (11.104), we obtain

$$M_x^S(x) = -EJ_\psi \phi'''(x) \tag{11.110}$$

where the *warping constant*, J_ψ, is given by (see Appendix C where a similar function is defined in terms of the sectorial area function for thin sections)

$$J_\psi = \int_A (\psi^P)^2 dA \tag{11.111}$$

Like the primary shear stresses, the secondary shear stresses should not generate any net shear forces. Thus, we must show that

$$V_y = \int_A \sigma_{xy}^S dA = G\int_A \frac{\partial \psi^S}{\partial y} dA = 0$$

$$V_z = \int_A \sigma_{xz}^S dA = G\int_A \frac{\partial \psi^S}{\partial z} dA = 0 \tag{11.112}$$

Consider the V_y term. We have

$$\int_A \frac{\partial \psi^S}{\partial y} dA = \int_A \left[\frac{\partial}{\partial y}\left(y\frac{\partial \psi^S}{\partial y}\right) + \frac{\partial}{\partial z}\left(y\frac{\partial \psi^S}{\partial z}\right)\right] dA - \int_A \left[y\left(\frac{\partial^2 \psi^S}{\partial y^2} + \frac{\partial^2 \psi^S}{\partial z^2}\right)\right] dA \tag{11.113}$$

The first integral on the right-hand side of Eq. (11.113) can be transformed by Gauss's theorem into a line integral on the boundary of the cross-section, while in the second integral we can use the governing differential equation in Eq. (11.103) to find

$$\int_A \frac{\partial \psi^S}{\partial y} dA = \int_C y\left(\frac{\partial \psi^S}{\partial y}n_y + \frac{\partial \psi^S}{\partial z}n_z\right) dC + \frac{E\phi'''(x)}{G}\int_A y\psi^P dA \tag{11.114}$$

The first integral in Eq. (11.114) vanishes by the boundary conditions on the secondary warping function while the second integral vanishes if

$$\int_A y\psi^P dA = 0 \tag{11.115}$$

so that under these conditions $V_y = 0$. In an entirely similar fashion, we can show that $V_z = 0$ if

$$\int_A z\psi^P dA = 0 \tag{11.116}$$

We will see shortly that the primary warping function must satisfy Eq. (11.115) and Eq. (11.116) so that the secondary shear stresses indeed produce only a secondary twisting moment. Note that if we integrate the differential equation in Eq. (11.103) over the cross-section and apply the boundary conditions, we also have

$$\int_A \left(\frac{\partial^2 \psi^S}{\partial y^2} + \frac{\partial^2 \psi^S}{\partial z^2} \right) dA = -\frac{E\phi'''(x)}{G} \int_A \psi^P dA$$

$$= \int_C \left(\frac{\partial \psi^S}{\partial y} n_y + \frac{\partial \psi^S}{\partial z} n_z \right) dC = 0 \tag{11.117}$$

which gives an additional condition on the primary warping function, namely

$$\int_A \psi^P dA = 0 \tag{11.118}$$

The physical meaning of these three conditions on the primary warping function is that in the case of nonuniform torsion, this primary warping function must not generate any axial stresses or bending stresses since from Eq. (11.92) we see that Eq. (11.115), Eq. (11.116), and Eq. (11.118) can also be written in terms of the warping normal stress, σ_{xx}^w, i.e.,

$$\int_A y\psi^P dA = \int_A y\sigma_{xx}^w dA = 0$$

$$\int_A z\psi^P dA = \int_A z\sigma_{xx}^w dA = 0 \tag{11.119}$$

$$\int_A \psi^P dA = \int_A \sigma_{xx}^w dA = 0$$

We will show that the three conditions of Eq. (11.119) locate the *center of twist* of the cross-section, which is the point about which the cross-section rotates, and fix the value of an arbitrary constant that appears as part of the solution.

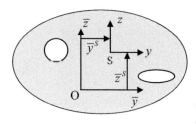

Figure 11.12 The center of twist, S, of a general cross-section as measured with respect to a set of $(\bar{y}, \bar{z})$ coordinates with origin O.

In our original assumptions on the deformation, Eq. (11.85), the distances y and z appearing in that equation are measured from the center of twist, which we will call point S (see Fig. 11.12), but we did not specify where that point is located. In solving for the primary warping function, our solution ψ^P is dependent on having the origin of the (y, z) coordinates be at S so we will label that desired function to be $\psi_S^P(y, z)$. Now, choose a set of coordinates $(\bar{y}, \bar{z})$ relative to a different arbitrary origin, O, and let $(\bar{y}^S, \bar{z}^S)$ be the location of S as measured in those coordinates (Fig. 11.12). From the geometry we have

$$y = \bar{y} - \bar{y}^S$$
$$z = \bar{z} - \bar{z}^S \tag{11.120}$$

Recall that we had

$$\frac{\partial^2 \psi_S^P}{\partial y^2} + \frac{\partial^2 \psi_S^P}{\partial z^2} = 0$$
$$\frac{\partial \psi_S^P}{\partial n} = z n_y - y n_z \tag{11.121}$$

so that by a change of coordinates from (y, z) to $(\bar{y}, \bar{z})$ we have

$$\frac{\partial^2 \psi_S^P}{\partial \bar{y}^2} + \frac{\partial^2 \psi_S^P}{\partial \bar{z}^2} = 0$$
$$\frac{\partial \psi_S^P}{\partial n} = (\bar{z} - \bar{z}^S) n_{\bar{y}} - (\bar{y} - \bar{y}^S) n_{\bar{z}} \tag{11.122}$$

or, equivalently,

$$\frac{\partial^2 \psi_S^P}{\partial \bar{y}^2} + \frac{\partial^2 \psi_S^P}{\partial \bar{z}^2} = 0$$
$$\frac{\partial}{\partial n}\left(\psi_S^P + \bar{y}\bar{z}^S - \bar{z}\bar{y}^S\right) = \bar{z} n_{\bar{y}} - \bar{y} n_{\bar{z}} \tag{11.123}$$

If we let $\psi_O^P = \psi_S^P + \bar{y}\bar{z}^S - \bar{z}\bar{y}^S$ then we see that ψ_O^P satisfies the same boundary value problem as ψ_S^P, namely

$$\frac{\partial^2 \psi_O^P}{\partial \bar{y}^2} + \frac{\partial^2 \psi_O^P}{\partial \bar{z}^2} = 0$$
$$\frac{\partial \psi_O^P}{\partial n} = \bar{z} n_{\bar{y}} - \bar{y} n_{\bar{z}} \tag{11.124}$$

This boundary value problem is called a Neumann problem [4], which has a unique solution (to within an arbitrary constant) so that we can include such a constant, $\bar{c}$, and write

$$\psi_S^P(y, z) = \psi_O^P(\bar{y}, \bar{z}) - \bar{y}z^S + \bar{z}\bar{y}^S + \bar{c} \qquad (11.125)$$

[We should point out that by using Eq. (11.125), it is easy to show [1] that the primary shear stresses generated by either $\psi_S^P(y, z)$ or by $\psi_O^P(\bar{y}, \bar{z})$ are the same so at least for these stresses, the choice of coordinates is not essential.] Now consider Eq. (11.125) and the conditions of Eq. (11.119) which because (y, z) and $(\bar{y}, \bar{z})$ just differ by the constants $(\bar{y}^S, \bar{z}^S)$ can be written equivalently as

$$\int_A \bar{y}\psi_S^P dA = 0$$

$$\int_A \bar{z}\psi_S^P dA = 0 \qquad (11.126)$$

$$\int_A \psi_S^P dA = 0$$

We have

$$\int_A \bar{y}\psi_S^P dA = \int_A \bar{y}\psi_O^P dA - \bar{z}^S \int_A \bar{y}^2 dA + \bar{y}^S \int_A \bar{y}\bar{z} dA + \bar{c} \int_A \bar{y} dA = 0$$

$$\int_A \bar{z}\psi_S^P dA = \int_A \bar{z}\psi_O^P dA - \bar{z}^S \int_A \bar{y}\bar{z} dA + \bar{y}^S \int_A \bar{z}^2 dA + \bar{c} \int_A \bar{z} dA = 0$$

$$\int_A \psi_S^P dA = \int_A \psi_O^P dA - \bar{z}^S \int_A \bar{y} dA + \bar{y}^S \int_A \bar{z} dA + \bar{c}A = 0 \qquad (11.127)$$

Equation (11.127) is a set of three equations for the three unknowns $(\bar{y}^S, \bar{z}^S, \bar{c})$. We can solve these equations directly but they can be simplified by choosing the origin O of the $(\bar{y}, \bar{z})$ coordinates to be at the centroid, C, of the cross-sectional area. In this case we find the set of equations

$$-\bar{y}^S I_{\bar{y}\bar{z}} + \bar{z}^S I_{\bar{z}\bar{z}} = I_{\bar{y}\psi}^P$$

$$-\bar{y}^S I_{\bar{y}\bar{y}} + \bar{z}^S I_{\bar{y}\bar{z}} = I_{\bar{z}\psi}^P$$

$$\bar{c} = -I_\psi^P / A \qquad (11.128)$$

where

$$I_\psi^P = \int_A \psi_C^P dA$$

$$I_{\bar{y}\psi}^P = \int_A \bar{y}\psi_C^P dA \qquad (11.129)$$

$$I_{\bar{z}\psi}^P = \int_A \bar{z}\psi_C^P dA$$

The solution of Eq. (11.128) for the coordinates of the center of twist (*as measured relative to the cross-section centroid*) is then

$$\bar{y}^S = \frac{I_{\bar{y}\bar{z}}I^P_{\bar{y}\psi} - I_{\bar{z}\bar{z}}I^P_{\bar{z}\psi}}{I_{\bar{y}\bar{y}}I_{\bar{z}\bar{z}} - I^2_{\bar{y}\bar{z}}}$$

$$\bar{z}^S = \frac{I_{\bar{y}\bar{y}}I^P_{\bar{y}\psi} - I_{\bar{y}\bar{z}}I^P_{\bar{z}\psi}}{I_{\bar{y}\bar{y}}I_{\bar{z}\bar{z}} - I^2_{\bar{y}\bar{z}}} \quad (11.130)$$

You might have noticed the similarity between the coordinates of the center of twist expression, Eq. (11.130), and the location of the shear center found for thin, open cross-section beams. As we will show later in this chapter, for thin, open cross-sections the primary warping function is simply the negative of the principal sectorial area function. As shown in Appendix C (see Eq. (C.20)), the location (y_P, z_P) of the principal pole (which was shown to also be the shear center) as measured in *centroidal* coordinates (y, z) (where $y_C = z_C = 0$) is given by

$$y_P = \frac{I_{z\omega_C}I_{zz} - I_{y\omega_C}I_{yz}}{I_{yy}I_{zz} - I^2_{yz}}$$

$$z_P = \frac{I_{z\omega_C}I_{yz} - I_{y\omega_C}I_{yy}}{I_{yy}I_{zz} - I^2_{yz}} \quad (11.131)$$

which is the same as Eq. (11.130) with the replacement of the primary warping function calculated with a coordinate system whose origin is at the centroid by the negative of the sectorial area function calculated with a pole at the centroid, i.e., $\psi^P_C \to -\omega_C$. Thus, for thin sections $(y_P, z_P) = (\bar{y}^S, \bar{z}^S)$. [Note: the P subscript in Eq. (11.131) indicates that (y_P, z_P) locates the principal pole in centroidal coordinates while the P superscript in warping functions such as ψ^P_C refers to the fact that this is a primary warping function, so one should be aware of this notational difference.]

As discussed in Chapter 10, texts typically determine the shear center location (for thin beams) through use of either the sectorial area function or by locating where the shear forces must be applied in the cross-section for the bending shear-stress expressions to be valid. However, the center of twist location expression, Eq. (11.130), is generally not derived so the fact that the two locations are identical, as shown here (at least for thin, open cross-sections), is omitted. One reason for this omission is that many texts emphasize only the uniform torsion of general cross-sections and, as described above, it is necessary to consider the case of nonuniform torsion in detail to identify the center of twist location explicitly.

In the case of nonuniform torsion, it is also necessary to determine the twist of the bar as a function of the axial coordinate, x. The primary twisting moment, M^P_x, was given in Eq. (11.105), and the secondary twisting moment, M^S_x, was given by Eq. (11.110). Thus, the total internal twisting moment, M_x, is

$$M_x(x) = M^P_x(x) + M^S_x(x)$$
$$= GJ_{eff}\phi'(x) - EJ_\psi\phi'''(x) \quad (11.132)$$

If the bar carries a distributed applied toque/unit length, $t_x(x)$, then from equilibrium we have

$$\frac{dM_x}{dx} = -t_x \tag{11.133}$$

so we have the differential equation for the twist, $\phi(x)$, given by

$$GJ_{\textit{eff}}\phi''(x) - EJ_\psi \phi^{iv}(x) = -t_x(x) \tag{11.134}$$

where $\phi^{iv} = d^4\phi/dx^4$. We will rewrite Eq. (11.134) as

$$\frac{d^4\phi}{dx^4} - k^2 \frac{d^2\phi}{dx^2} = \frac{t_x(x)}{EJ_\psi} \tag{11.135}$$

$$\text{where } k^2 = \frac{GJ_{\textit{eff}}}{EJ_\psi} \tag{11.136}$$

To solve Eq. (11.135) we need to specify boundary conditions. The boundary conditions most frequently occurring are fixed, simple, and free conditions. At a fixed support there is no twisting or warping allowed ($u_x = 0$) so that

$$\text{fixed support} : \phi = 0, \quad \phi' = 0 \tag{11.137}$$

At a simple support twisting is prevented and the support is free from normal stress so that

$$\text{simple support} : \phi = 0, \quad \phi'' = 0 \tag{11.138}$$

At a free end there is no normal stress and no shear stresses so that the total internal torque, M_x, is zero. These requirements give, since $M_x = EJ_\psi \left(k^2\phi' - \phi'''\right)$, the boundary conditions

$$\text{free end} : \phi'' = 0, \quad k^2\phi' - \phi''' = 0 \tag{11.139}$$

The general solution of Eq. (11.135) is the sum of a homogeneous solution and a particular solution, which can be written as [5]

$$\phi = C_1 + C_2 x + C_3 \cosh(kx) + C_4 \sinh(kx)$$
$$- \frac{1}{kGJ_{\textit{eff}}} \int_0^x [k(x-\xi) - \sinh(k(x-\xi))] \, t_x(\xi) d\xi \tag{11.140}$$

where C_m ($m = 1, \ldots, 4$) are constants and one end of the member is assumed to be at $x = 0$. From Eq. (11.132) the total internal twisting moment (torque) is

$$M_x(x) = GJ_{\textit{eff}} C_2 - \int_0^x t_x(\xi) d\xi \tag{11.141}$$

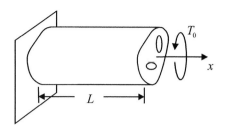

Figure 11.13 Torsion of a noncircular bar of length L that is restrained from twisting and warping at one end and where a torque, T_0, is applied at the other end.

As a simple example, consider a bar (Fig. 11.13) that is fixed at $x = 0$ and where an applied torque, T_0, is applied at the end $x = L$. In this case $t_x = 0$, and at $x = L$ we have from Eq. (11.141)

$$GJ_{eff} C_2 = T_0 \tag{11.142a}$$

The other boundary condition at $x = L$ is that the normal stress is zero, so we have

$$\phi''(L) = k^2 [C_3 \cosh(kL) + C_4 \sinh(kL)] = 0 \tag{11.142b}$$

At the fixed end $x = 0$ we have

$$\begin{aligned} \phi(0) &= C_1 + C_3 = 0 \\ \phi'(0) &= C_2 + kC_4 = 0 \end{aligned} \tag{11.142c, d}$$

These four boundary conditions determine the constants. In MATLAB® we have the solution

```
syms   G J T k L
M = [ 0 1 0 0; 1 0 1 0; 0 1 0 k; 0 0 cosh(k*L) sinh(k*L)];
C = inv(M)*[ T/(G*J); 0; 0; 0]
C =
 -(T*sinh(L*k))/(G*J*k*cosh(L*k))
                          T/(G*J)
  (T*sinh(L*k))/(G*J*k*cosh(L*k))
                       -T/(G*J*k)
```

which can be written as

$$\begin{aligned} C_3 &= -C_1 = \frac{T_0}{GJ_{eff} k} \tanh(kL) \\ C_2 &= \frac{T_0}{GJ_{eff}}, \quad C_4 = \frac{-T_0}{GJ_{eff} k} \end{aligned} \tag{11.143}$$

and the final solution is therefore

$$\phi(x) = \frac{T_0}{GJ_{eff} k} [kx - \sinh(kx) - \tanh(kL)(1 - \cosh(kx))] \tag{11.144}$$

and the twist at $x = L$ is

$$\phi(L) = \frac{T_0 L}{GJ_{eff}} \left[1 - \frac{\tanh(kL)}{kL} \right] \tag{11.145}$$

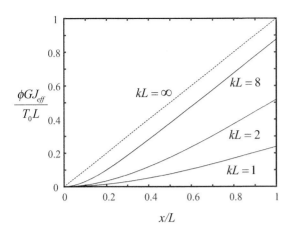

Figure 11.14 Normalized twist versus distance for different values of restraint of warping at $x = 0$, where $kL = \infty$ corresponds to the case where there is no restraint of warping.

The first term in Eq. (11.145) is the twist at $x = L$ when the twist is zero at $x = 0$ but the bar is free to warp. The second term, which represents the effect of restraint of warping at $x = 0$, reduces the twist at $x = L$; i.e., it makes the bar appear stiffer. As $kL \to \infty$ the second term goes to zero. The twist along the entire bar is shown in Fig. 11.14 for different kL values.

To complete our description of nonuniform torsion, we note that in solving for the secondary warping function, ψ^S, the solution of Eq. (11.103) is determined only to within an arbitrary constant. We see from that equation that this secondary warping is proportional to $\phi'''(x)$ so that we could write the secondary warping displacement and normal stress as

$$u_x^S = \psi^S(x, y, z) = \phi'''(x)\tilde{\psi}(y, z)$$
$$\sigma_{xx}^S = E\frac{\partial u_x^S}{\partial x} = E\phi^{iv}(x)\tilde{\psi}(y, z)$$
(11.146)

But this secondary warping must not produce any net axial force so we must have

$$\int_A \sigma_{xx}^S dA = E\phi^{iv}(x)\int_A \tilde{\psi}(y, z)dA = 0$$
(11.147)

which in turn implies that ψ^S must satisfy

$$\int_A \psi^S dA = 0$$
(11.148)

Equation (11.148) is a constraint on the secondary warping function that fixes the value of the constant since, if we solve the boundary value problem of Eq. (11.103) and call that solution $\widehat{\psi}$ and let a constant $\widehat{C}$ be given as

$$\widehat{C} = \frac{1}{A}\int_A \widehat{\psi}\, dA$$
(11.149)

then the secondary warping function

$$\psi^S = \widehat{\psi} - \bar{c} \tag{11.150}$$

is a solution of Eq. (11.103), which also satisfies Eq. (11.148).

11.5 General Solution of the Nonuniform Torsion Problem

We now have a complete formulation of the nonuniform torsion problem where we have the governing equations that determine the primary and secondary warping functions and the shear stresses, the location of the center of twist, and the angle of twist in the member. The solution procedure has a number of steps that we want to outline here explicitly. First, we can locate the centroid of the cross-section and determine the solution for the primary warping function, ψ_C^P, using the centroid as the origin of the coordinates. This warping function satisfies the boundary value problem

$$\frac{\partial^2 \psi_C^P}{\partial \bar{y}^2} + \frac{\partial^2 \psi_C^P}{\partial \bar{z}^2} = 0 \quad \text{in the cross-section}$$

$$\frac{\partial \psi_C^P}{\partial n} = \bar{z} n_{\bar{y}} - \bar{y} n_{\bar{z}} \quad \text{on the boundary} \tag{11.151}$$

From this solution we can then locate the center of twist explicitly and a constant, $\bar{c}$, as (see Eqs. (11.128)–(11.130)):

$$\bar{y}^S = \frac{I_{\bar{y}\bar{z}} I_{\bar{y}\psi}^P - I_{\bar{z}\bar{z}} I_{\bar{z}\psi}^P}{I_{\bar{y}\bar{y}} I_{\bar{z}\bar{z}} - I_{\bar{y}\bar{z}}^2}$$

$$\bar{z}^S = \frac{I_{\bar{y}\bar{y}} I_{\bar{y}\psi}^P - I_{\bar{y}\bar{z}} I_{\bar{z}\psi}^P}{I_{\bar{y}\bar{y}} I_{\bar{z}\bar{z}} - I_{\bar{y}\bar{z}}^2}$$

$$\bar{c} = -I_\psi^P / A \tag{11.152}$$

The primary warping function, which uses the center of twist, S, as the origin of the (y,z) coordinates, is then given by Eq. (11.125):

$$\psi_S^P(y,z) = \psi_C^P(\bar{y}, \bar{z}) - \bar{y}\bar{z}^S + \bar{z}\bar{y}^S + \bar{c} \tag{11.153}$$

[Note: we can use coordinates with an arbitrary origin O rather than the centroid C in the above steps to determine $(\bar{y}^S, \bar{z}^S)$ and $\bar{c}$ if we solve the set of more general equations given by Eq. (11.127).] With this primary warping function known, we can determine the effective polar area moment and warping constant (see Eq. (11.106) and Eq. (11.111)):

11.5 General Solution of the Nonuniform Torsion Problem

$$J_{eff} = \int_A \left[(y^2 + z^2) + y\frac{\partial \psi_S^P}{\partial z} - z\frac{\partial \psi_S^P}{\partial y} \right] dA$$

$$J_\psi = \int_A (\psi_S^P)^2 \, dA$$
(11.154)

Having these constants, we can obtain the twist in the member (see Eq. (11.135) and Eq. (11.136)) as the solution of the equation

$$\frac{d^4\phi}{dx^4} - k^2 \frac{d^2\phi}{dx^2} = \frac{t_x(x)}{EJ_\psi}$$

$$k^2 = \frac{GJ_{eff}}{EJ_\psi}$$
(11.155)

under appropriate specified loads and boundary conditions. This solution also gives us the primary and secondary internal moments as

$$M_x^P(x) = GJ_{eff} \phi'(x)$$

$$M_x^S(x) = -EJ_\psi \phi'''(x)$$
(11.156)

and the corresponding total internal moment $M_x(x) = M_x^P(x) + M_x^S(x)$. The primary shear stresses are then also given explicitly as

$$\sigma_{xy}^P = G\phi'(x) \left(\frac{\partial \psi_S^P}{\partial y} - z \right)$$

$$\sigma_{xz}^P = G\phi'(x) \left(\frac{\partial \psi_S^P}{\partial z} + y \right)$$
(11.157)

The secondary warping function, ψ^S, can be found from the solution of Eq. (11.103):

$$\frac{\partial^2 \widehat{\psi}}{\partial y^2} + \frac{\partial^2 \widehat{\psi}}{\partial z^2} = -\frac{E\phi'''(x)}{G} \psi_S^P \quad \text{in the cross-section } A$$

$$\frac{\partial \widehat{\psi}}{\partial n} = 0 \quad \text{on the boundary } C$$
(11.158)

where

$$\psi^S = \widehat{\psi} - \frac{1}{A} \int_A \widehat{\psi} \, dA$$
(11.159)

which gives the secondary shear stresses as

$$\sigma_{xy}^S = G\frac{\partial \psi^S}{\partial y}$$

$$\sigma_{xz}^S = G\frac{\partial \psi^S}{\partial z}$$
(11.160)

As part of the nonuniform torsion problem solution, one can also examine the warping normal stresses, σ_{xx}^w, generated, where (Eq. (11.92)):

$$\sigma_{xx}^w(x, y, z) = E\phi''(x)\psi_S^P(y, z) \qquad (11.161)$$

Needless to say, a numerical solution is in general needed to implement these steps and solve the general nonuniform torsion problem. Recently, work by Sapountzakis and coworkers has given a very efficient solution using the boundary element method [6], which is a nice illustration of a case where boundary elements is an attractive alternative to the finite element method. We will not go through the details of such numerical solutions here.

11.6 Torsion of Thin, Open Cross-Sections

If the cross-section is thin, then the torsion problem can be treated more explicitly. Consider first the case of thin, open cross-sections such as the geometries shown in Fig. 11.15. Except near ends or junctions, we expect that the primary stress results for a thin rectangle still apply to these cross-sections, namely that the shear stress varies linearly across the thickness. Thus, we have (Fig. 11.16)

$$\sigma_{xs} = 2G\phi'n$$
$$\sigma_{xn} = 0 \qquad (11.162)$$

where n is the distance measured from the centerline in the outward normal direction. This result can be justified with the membrane analogy, since the shape of the membrane will still be approximately parabolic in shape across the thickness for most of the cross-section, as seen in the thin rectangle case (Fig. 11.10). Thus, if we ignore the deviations from this behavior at ends or junctions, then for the cross-sections seen in Figs. 11.15a, b, c the effective polar area moments for these sections can be calculated with the thin rectangle results as

Figure 11.15 Thin, open cross-sections.

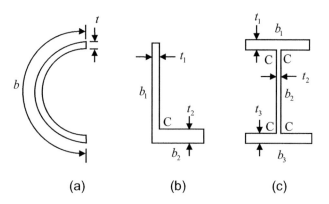

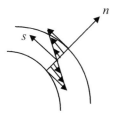

Figure 11.16 Primary shear stress across the cross-section in a thin, open section where s is directed along the tangent to the center line and n is directed normal to the center line.

$$\begin{aligned}\text{(a)} \quad & J_{eff} = \frac{1}{3}bt^3 \\ \text{(b)} \quad & J_{eff} = \frac{1}{3}b_1 t_1^3 + \frac{1}{3}b_2 t_2^3 \\ \text{(c)} \quad & J_{eff} = \frac{1}{3}b_1 t_1^3 + \frac{1}{3}b_2 t_2^3 + \frac{1}{3}b_3 t_3^3\end{aligned} \qquad (11.163)$$

When the thickness is slowly varying function along the length, b, of a cross-section, then we can again use a modified result for the thin rectangle given by

$$J_{eff} = \frac{1}{3} \int_0^b [t(s)]^3 ds \qquad (11.164)$$

which reduces to the thin rectangle result when the thickness is a constant. If we let t_{max} be the largest thickness in a cross-section, then the maximum primary shear stress based on Eq. (11.162) would be

$$(\sigma_{xn})_{max} \equiv \tau_{max} = \frac{T t_{max}}{J_{eff}} \qquad (11.165)$$

However, at reentrant corners such as the points C shown in Fig. 11.15, the changing geometry produces stress concentrations that can produce larger stress than given by Eq. (11.165). Figure 11.17 shows a typical case where (assuming that $t_2 > t_1$ so that $\tau = T t_2 / J_{max}$ is the larger of the two stresses in the sides adjacent to the corner) the stress distribution near the corner is no longer linear and has a maximum value given by

$$\tau_{max} = K \frac{T t_2}{J_{eff}} \qquad (11.166)$$

where K is the stress concentration factor. The stress concentration factor is a function of the radius, r, of the fillet in the corner (Fig. 11.17). Such factors are often determined experimentally or with detailed numerical calculations. Similar concentrations can also appear locally near notches or keyways in the cross-sections.

Unlike a thin rectangular section where the warping is zero along the centerline of the rectangle, a general thin, open section can experience significant warping. You can experience this tendency of thin, open sections to warp by simply twisting a thin rolled-up piece of paper and see what happens. To obtain an expression for this centerline warping, consider the tangential displacement along the centerline, u_s, as seen in Fig. 11.18. Relating this displacement to the components along the y- and z-axes and using the Saint-Venant assumptions for those y- and z-displacements (i.e., the displacements due to a rotation about the center of twist, S, as seen in Eq. (11.7)), we have

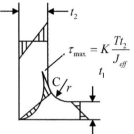

Figure 11.17 Stress concentration at a reentrant corner where it is assumed $t_2 > t_1$.

$$\begin{aligned} u_s &= -u_y \sin\alpha + u_z \cos\alpha \\ &= z\phi(x)\sin\alpha + y\phi(x)\cos\alpha \\ &= \phi(x)\mathbf{r}\cdot\mathbf{e}_n \\ &= r_\perp \phi(x) \end{aligned} \quad (11.167)$$

where $\mathbf{r} = y\mathbf{e}_y + z\mathbf{e}_z$, $\mathbf{e}_n = \cos\alpha\,\mathbf{e}_y + \sin\alpha\,\mathbf{e}_z$, and $r_\perp = \mathbf{r}\cdot\mathbf{e}_n$ is the perpendicular distance from the center of twist, S, to the line of action of u_s. The primary engineering shear strain, γ^P_{xs}, at the centerline is then given by

$$\begin{aligned} \gamma^P_{xs} &= \frac{\partial u_s}{\partial x} + \frac{\partial u_x}{\partial s} \\ &= r_\perp \phi' + \frac{\partial u_x}{\partial s} \end{aligned} \quad (11.168)$$

where s is the distance measured along the centerline (in the s-direction shown in Fig. 11.18) and the corresponding primary shear stress, σ^P_{xs}, is

$$\sigma^P_{xs} = G\gamma^P_{xs} = Gr_\perp \phi' + G\frac{\partial u_x}{\partial s} \quad (11.169)$$

However, just as in a rectangular section, we expect this centerline stress and strain to be zero, so that by setting $\sigma^P_{xs} = 0$ and integrating Eq. (11.169), we find the primary warping displacement, u_x, along the centerline to be given as

$$\begin{aligned} u_x &= -\phi'\left(\int_0^s r_\perp\,ds + \omega_0\right) \\ &= -\phi'\omega(s) \end{aligned} \quad (11.170)$$

where $\omega(s)$ is a sectorial area function, a function we have previously seen in the bending of thin sections, and which is examined in Appendix C. The ω_0 is a constant of integration, which depends on the choice of the starting point for the integration ($s = 0$). This sectorial area function depends on the choice of the constant of integration (starting point), as well as the location of the origin, S, of the coordinates (called the pole of the function) which here is at the center of twist. If we compare Eq. (11.170) with the original Saint-Venant assumption on the deformation (Eq. (11.33)), we see that the primary warping function, ψ^P, along the centerline is just the negative of the sectorial area function, i.e., $\psi^P = -\omega$. Thus, instead of having to solve a differential equation to find the primary warping function, for thin, open sections we can obtain the centerline value directly from the geometry by determining the sectorial area function. As in the case of general cross-sections, to determine explicitly the location of the center of twist, S (which is also the pole of the sectorial area function in Eq. (11.170)), as well as the constant of integration, one must consider the secondary warping and stress present for the case of nonuniform

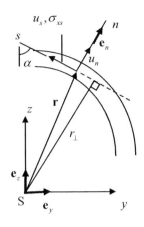

Figure 11.18 Geometry for a thin section, showing the displacement and stress in the cross-section along the s-direction (tangential to the centerline). The displacement also has a component in the n-direction, as shown.

torsion of the thin member. This secondary warping is generated because of the primary warping normal stress produced, i.e., in this case

$$\sigma_{xx}^w(x, y, z) = -E\phi''(x)\omega(y, z) \qquad (11.171)$$

Requiring that this warping stress produces no net axial force or bending moments, we have

$$\int_A \sigma_{xx}^w dA = -E\phi'' \int_A \omega dA = 0$$

$$\int_A y\sigma_{xx}^w dA = -E\phi'' \int_A y\omega dA = 0 \qquad (11.172)$$

$$\int_A z\sigma_{xx}^w dA = -E\phi'' \int_A z\omega dA = 0$$

which are the conditions that require the sectorial area function to be a principal sectorial area function $\omega = \omega_P(s, \omega_0)$, where the principal pole P is also the shear center S and the constant ω_0 defines a principal origin (see Appendix C). Thus, the primary warping is determined explicitly by this principal sectorial area function. The changing warping normal stress produces a secondary shear stress, σ_{xt}^S along the centerline in the thin section. We will assume this secondary shear stress is uniformly distributed over the thickness, t, of the section so that it can be expressed in terms of a secondary shear flow, $q^S(s)$, as $q^S(s) = t(s)\sigma_{xs}^S(s)$. By examining equilibrium of a small element of a thin section (Fig. 11.19) we obtain

$$\left(\frac{\partial \sigma_{xx}^w}{\partial x} dx\right) tds + \left(\frac{\partial q^S}{\partial s} ds\right) dx = 0$$

$$\rightarrow \frac{\partial q^S}{\partial s} = -t\frac{\partial \sigma_{xx}^w}{\partial x} \qquad (11.173)$$

Writing the warping normal stress in terms of the principal sectorial area function and integrating we have

$$q^S(s) = E\phi'''(x) \int_0^s \omega_P(s, \omega_0) t(s) ds \qquad (11.174)$$

where the constant of integration is zero since we have assumed $s = 0$ is taken to be at the end of a thin, open section where $q^S(0) = 0$. Recall that for the secondary warping function, ψ^S, we had (Eq. (11.160))

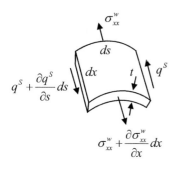

Figure 11.19 A small element of a thin section showing the relationship between the secondary shear flow and the warping normal stress.

$$\sigma_{xy}^S = G\frac{\partial \psi^S}{\partial y}$$

$$\sigma_{xz}^S = G\frac{\partial \psi^S}{\partial z}$$

Since (see Fig. 11.18)

$$\sigma_{xy}^S = G\frac{\partial \psi^S}{\partial y} = -\sigma_{xs}^S \sin \alpha = \frac{q^S}{t}\frac{dy}{ds}$$

$$\sigma_{xz}^S = G\frac{\partial \psi^S}{\partial z} = \sigma_{xs}^S \cos \alpha = \frac{q^S}{t}\frac{dz}{ds}$$
(11.175)

we find

$$\frac{\partial \psi^S}{\partial s} = \frac{\partial \psi^S}{\partial y}\frac{dy}{ds} + \frac{\partial \psi^S}{\partial z}\frac{dz}{ds} = \frac{q^S}{Gt}$$
(11.176)

so, using Eq. (11.174), the secondary warping can be obtained by integration as

$$\psi^S(x,s) = \frac{1}{G}\left[\int_0^s \frac{q^S(u)}{t(u)}du + K\right]$$
(11.177)

where the constant of integration, K, can be obtained, as before, from the condition (see Eq. (11.148))

$$\int_A \psi^S dA = 0$$
(11.178)

In summary, for thin, open sections we have both the primary shear stresses and the secondary shear flow given explicitly as

$$\sigma_{xs}^P = 2G\phi'n$$
$$\sigma_{xn}^P = 0$$
$$q^S = E\phi'''\int_0^s \omega_P t ds$$
(11.179)

We also have the primary and secondary warping functions given by

$$\psi^P = -\omega_P$$
$$\psi^S = \frac{1}{G}\left[\int_0^s \frac{q^S}{t}ds + K\right]$$
(11.180)

The effective polar area moment, J_{eff}, is the thin section result obtained from the thin rectangular section, as discussed previously. This is related to the primary torque, M_x^P, by $M_x^P = GJ_{eff}\phi'$. For the secondary torque, M_x^S, computed about the center of twist (shear center) S, we have

$$M_x^S = \int_0^l r_\perp q^S ds = \int_0^l q^S d\omega_P \qquad (11.181)$$

where l is the length of the open section and we can take $r_\perp ds = d\omega_P$ in terms of the principal sectorial area function since $r_\perp$ is measured from the shear center. If we integrate Eq. (11.181) by parts, we find

$$\begin{aligned} M_x^S &= q_S \omega_P |_0^l - \int_0^l \omega_P \frac{\partial q^S}{\partial s} ds \\ &= -\int_0^l \omega_P \frac{\partial q^S}{\partial s} ds \end{aligned} \qquad (11.182)$$

since $q^S(0) = q^S(l) = 0$ at the free ends of an open section. Placing the expression for the secondary shear flow, Eq. (11.174), into Eq. (11.182) gives

$$\begin{aligned} M_x^S &= -E\phi'''(x) \int_A \omega_P^2 dA \\ &= -E\phi'''(x) J_\omega \end{aligned} \qquad (11.183)$$

(where $dA = tds$ is the element area), which is the same expression for the warping constant, J_ψ, obtained previously for a general cross-section but where here the secondary warping function (see Eq. (11.111)) is replaced by the principal sectorial area function for the thin open section. We can again relate the primary and secondary torques to a distributed torque/unit length, t_x, acting on the bar, and the twist for the thin open section is thus governed by the same differential equation (see Eq. (11.134)) as before:

$$GJ_{eff}\phi''(x) - EJ_\omega \phi^{iv}(x) = -t_x(x) \qquad (11.184)$$

and appropriate boundary conditions.

Example 11.2 Nonuniform Torsion of an I-Beam

We will use the I-beam cross-section of Fig. 11.20a to give an explicit example of the use of these thin, open section formulas. Many texts use an I-beam to discuss nonuniform torsion and restraint of warping since the warping and stress distributions are relatively simple to derive and understand. We will assume this beam has a length L and is fixed at $x = 0$, and a

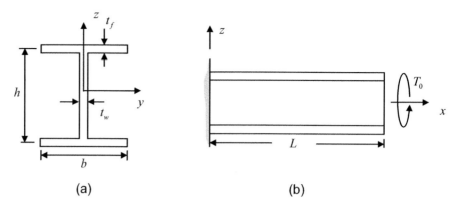

Figure 11.20 (a) Cross-section of an I-beam. (b) An example of restrained warping and nonuniform torsion where the I-beam is loaded in torsion at one end and rigidly fixed at the other end.

has an applied torque, T_0, acting at its other end at $x = L$, as shown in Fig. 11.20b. The effective polar area moment for this section is given by the thin section result as

$$J_{eff} = \frac{2bt_f^3 + ht_w^3}{3} \qquad (11.185)$$

The principal sectorial area function for this cross-section has its pole, P, located at the centroid (which is also the shear center) and the initial point of integration, I, can also be taken at P. The principal sectorial area function is then given by (shown as the shaded areas in Fig. 11.21a)

$$\omega_P = \frac{h}{2} s \qquad (11.186)$$

which is valid for both flanges when the distance, s, is taken as shown in Fig. 11.21. The sectorial area function is zero in the web. The largest and smallest values for this function are $\pm bh/4$. The warping normal stress induced by the restraint of warping is therefore given by

$$\sigma_{xx}^w = -E\phi''(x)\omega_P = -E\phi''(x)\frac{h}{2} s \qquad (11.187)$$

and has the linear distributions in the flanges shown in Fig. 11.21b. [As we will see, ϕ'' is positive so the stress distribution is opposite in sign to the primary warping function.] In the beam cross-section, the primary shear stress distribution is a linear distribution across the thickness

$$\sigma_{xs}^P = 2G\phi'n = 2\frac{M_x^P}{J_{eff}} n \qquad (11.188)$$

where M_x^P is the primary part of the torsional moment and the distance n is normal to the centerline. The primary shear-stress distribution along the cross-section sides is shown in Fig. 11.22a. The secondary shear-stress distribution is described by the shear flow, q^S, where in each flange

11.6 Torsion of Thin, Open Cross-Sections

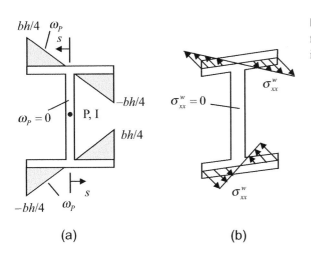

Figure 11.21 (a) The principal sectoral area function for the I-beam. (b) The resulting warping stress induced when the beam is in nonuniform torsion.

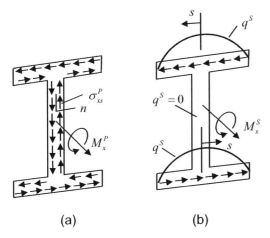

Figure 11.22 (a) The primary torsional moment and associated primary shear stress distribution. (b) The secondary torsional moment and associated secondary shear-flow distribution.

$$q^S = E\phi'''(x)\left[\int_0^s \omega_P t\, ds + C\right]$$
$$= E\phi'''(x)\left[\frac{ht_f s^2}{4} + C\right] \quad (11.189)$$

and where we have included a constant of integration, C, because $s = 0$ is not at a free end. Since we must have $q^S = 0$ at $s = \pm b/2$ we have $C = -h t_f b^2/16$ and

$$q^S = \frac{E\phi'''(x) h t_f}{4}\left(s^2 - \frac{b^2}{4}\right)$$
$$= \frac{M_x^S}{J_\omega}\frac{h t_f}{4}\left(\frac{b^2}{4} - s^2\right) \quad (11.190)$$

which has the parabolic shear flow distribution in the flanges shown in Fig. 11.22b. The warping constant is

$$J_\omega = \int_A \omega_P^2 dA = 2t_f \int_{-b/2}^{b/2} \frac{h^2}{4} s^2 ds \qquad (11.191)$$

$$= \frac{t_f h^2 b^3}{24}$$

Since the boundary conditions on the I-beam are the same as considered in Section 11.4, we have from Eq. (11.144)

$$\phi(x) = \frac{T_0}{GJ_{eff}k}[kx - \sinh(kx) - \tanh(kL)(1 - \cosh(kx))]$$

$$= \frac{T_0}{GJ_{eff}k}\left[kx - \tanh(kL) + \frac{-\sinh(kx)\cosh(kL) + \cosh(kx)\sinh(kL)}{\cosh(kL)}\right]$$

$$= \frac{T_0}{GJ_{eff}k}\left[kx - \tanh(kL) - \frac{\sinh(k(x-L))}{\cosh(kL)}\right]$$

$$(11.192)$$

which gives

$$\phi'(x) = \frac{T_0}{GJ_{eff}}\left[1 - \frac{\cosh(k(x-L))}{\cosh(kL)}\right]$$

$$\phi''(x) = \frac{-T_0 k}{GJ_{eff}}\left[\frac{\sinh(k(x-L))}{\cosh(kL)}\right] \qquad (11.193)$$

$$\phi'''(x) = \frac{-T_0 k^2}{GJ_{eff}}\left[\frac{\cosh(k(x-L))}{\cosh(kL)}\right]$$

Note that throughout the beam we have $\phi' > 0$, $\phi'' > 0$, and $\phi''' < 0$. The primary and secondary torques are then given by

$$M_x^P = GJ_{eff}\phi' = T_0\left[1 - \frac{\cosh(k(x-L))}{\cosh(kL)}\right]$$

$$M_x^S = -EJ_\omega \phi''' = T_0 \frac{\cosh(k(x-L))}{\cosh(kL)}$$

$$(11.194)$$

11.6 Torsion of Thin, Open Cross-Sections

The primary and secondary warping functions in each flange are given by Eq. (11.180):

$$\psi^P = -\omega_P = \frac{-hs}{2}$$

$$\psi^S = \frac{1}{G}\left[\int_{-b/2}^{s} \frac{q^S}{t}ds + K_1\right] \qquad (11.195)$$

$$= \frac{E\phi'''}{G}\left[\frac{h}{4}\int_{-b/2}^{s}\left(s^2 - \frac{b^2}{4}\right)ds + K_2\right] = \frac{E\phi'''}{G}\left[\frac{h}{4}\left(\frac{s^3}{3} - \frac{b^2 s}{4} - \frac{b^3}{12}\right) + K_2\right]$$

where K_1 and K_2 are "constants" (i.e., they not functions of s) and where $E\phi'' K_2 = K_1$. Requiring that there be no net axial force from this secondary warping, we have

$$\int_A \psi^S dA = 2 \times \int_{-b/2}^{b/2} \psi^S t\, ds = 2 \times \frac{E\phi''' t_f}{G}\left[-\frac{b^4 h}{48} + K_2 b\right] = 0 \qquad (11.196)$$

so that $K_2 = b^3 h/48$ and we have in each flange (see Fig. 11.23 for both the primary and secondary warping distributions):

$$\psi^S = \frac{E\phi''' h}{4G}\left(\frac{s^3}{3} - \frac{b^2 s}{4}\right) \qquad (11.197)$$

which attains its maximum magnitude at $x = 0$, $s = \pm b/2$, where from Eq. (11.193),

$$|\psi^S|_{max} = \frac{E\phi'''(0) b^3 h}{48 G} = \frac{T_0 b^3 h}{48 G J_\omega} \qquad (11.198)$$

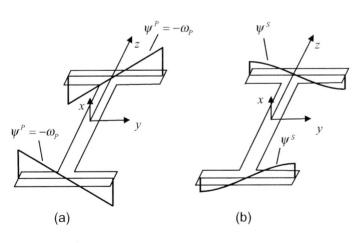

Figure 11.23 (a) The primary centerline warping function. (b) The secondary warping function of the I-beam according to the theory for thin, open sections.

while the maximum magnitude of the primary warping function is at $s = \pm b/2$ where

$$|\psi^P|_{max} = \frac{bh}{4} \tag{11.199}$$

The maximum primary stress in the cross-section (ignoring any stress concentrations) occurs at $x = L$ and at the outer edge of the cross-section given as

$$|\tau^P|_{max} \equiv |\sigma^P_{xs}|_{max} = \frac{M^P_x t}{J_{eff}} = \frac{T_0 t}{J_{eff}}\left(1 - \frac{1}{\cosh(kL)}\right) \tag{11.200}$$

while the maximum secondary shear stress occurs at $x = 0$ and $s = 0$ in the flanges and is

$$|\tau^S|_{max} = \left|\frac{q^S}{t}\right|_{max} = \frac{M^S_x b^2 h}{J_\omega \, 16} = \frac{T_0 b^2 h}{16 J_\omega} \tag{11.201}$$

We want to compare these explicit results for this I-beam with results obtained by numerically solving the nonuniform torsion theory of Section 11.4 (and summarized in Section 11.5), using the boundary element method [6]. The specific parameters are for a cantilever steel I-beam of length $L = 5.0$ m, where Young's modulus is $E = 2.1 \times 10^8$ kN/m², and Poisson's ratio is $\nu = 0.3$. The dimensions of the cross-section are $h = 14/15$ m, $b = 2/3$ m, $t_f = t_w = 1/15$ m, and the applied torque at $x = L$ is $T_0 = 1/3$ kN-m. To compare the thin section results with the more general theory, we will examine six quantities: (1) the effective polar area moment of the I-beam, J_{eff}, (2) the warping constant, J_ω, (3) the maximum value in the cross-section of the principal warping function, (4) the maximum value in the cross-section of the secondary warping function, (5) the maximum value of the primary shear stress in the beam, and (6) the maximum value of the secondary shear stress in the beam.

For this beam we will first examine the split of the total torque, T_0, into the primary and secondary torques, Eq. (11.194). To obtain the distributions of these torques, we need the value of the parameter

$$kL = \sqrt{\frac{GJ_{eff}}{EJ_\omega}} L \tag{11.202}$$

so we will first find the underlying parameters in Eq. (11.202). From the values given for this problem we have the results:

$$G = \frac{E}{2(1+\nu)} = 8.0769 \times 10^8 \text{ kN/m}^2$$

$$J_{eff} = \frac{2bt_f^3 + ht_w^3}{3} = 2.2387 \times 10^{-4} \text{ m}^4 \tag{11.203}$$

$$J_\omega = \frac{t_f h^2 b^3}{24} = 7.1696 \times 10^{-4} \text{ m}^6$$

$$kL = \sqrt{\frac{GJ_{eff}}{EJ_\omega}} L = 1.7327$$

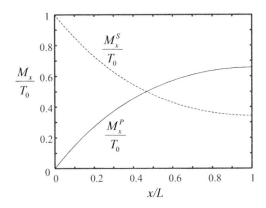

Figure 11.24 The primary twisting moment (solid line) and secondary moment (dashed line) distributions in the cantilever I-beam.

With these values we then can obtain and plot the primary and secondary torsional moments:

$$M_x^P = T_0 \left[1 - \frac{\cosh(kL(x/L - 1))}{\cosh(kL)} \right]$$
$$M_x^S = T_0 \frac{\cosh(kL(x/L - 1))}{\cosh(kL)}$$
(11.204)

Figure 11.24 shows that at the free end, the secondary twisting moment is still about 35 percent of the total torque so that the effects of warping restraint extend over the full length of the member. Next, we will compute the maximum values of the primary and secondary warping functions and the primary and secondary shear stresses:

$$\begin{aligned}
\left| \psi^P \right|_{\max} &= \frac{bh}{4} = 0.1556 \text{ m}^2 \\
\left| \psi^S \right|_{\max} &= \frac{T_0 b^3 h}{48 G J_\omega} = 3.3163 \times 10^{-8} \text{ m} \\
\left| \tau^P \right|_{\max} &= \frac{T_0 t}{J_{eff}} \left(1 - \frac{1}{\cosh(kL)} \right) = 65.2284 \text{ kN/m}^2 \\
\left| \tau^S \right|_{\max} &= \frac{T_0 b^2 h}{16 J_\omega} = 12.0536 \text{ kN/m}^2
\end{aligned}$$
(11.205)

In Table 11.1 these results from the theory for thin, open sections are compared to the results obtained in [6] for the more general theory for arbitrary cross-sections as calculated numerically by the boundary element method (BEM). The agreement in general is good for both the warping functions and the stresses. Although the thin cross-section results predict the warping at the centerline and the boundary element method solves for the warping along the edges, the results from the boundary element method also show that that there is little variation across the thickness [6]. Note that while the secondary warping function is the actual warping displacement, u_x^S, the primary warping displacement, u_x^P, is given by $u_x^P = \phi'(x) \psi^P = -\phi'(x) \omega_P$, which has its maximum value at $x = L$ given by the thin cross-section theory as

Table 11.1 Comparison of the results of the torsion theory for thin sections with the more general theory for arbitrary cross-sections as calculated with the boundary element method (BEM)

	Thin cross-section theory	BEM
$J_{eff}\,(\text{m}^4)$	2.24 E–4	2.47 E–4
$J_\omega\,(\text{m}^6)$	7.17 E–4	7.13 E–4
$\left|\psi^P\right|_{max}\,(\text{m}^2)$	1.56 E–1	1.64 E–1
$\left|\psi^S\right|_{max}\,(\text{m})$	3.32 E–8	3.21 E–8
$\left|\tau^P\right|_{max}\,(\text{kPa})$	65.23	71.29
$\left|\tau^S\right|_{max}\,(\text{kPa})$	12.05	10.85

$$\begin{aligned}\left|u_x^P\right|_{max} &= \phi'(L)\left|\psi^P\right|_{max}\\ &= \frac{T_0}{GJ_{eff}}\left[1-\frac{1}{\cosh(kL)}\right]\left|\psi^P\right|_{max} \qquad (11.206)\\ &= 1.8844\times 10^{-6}\ \text{m}\end{aligned}$$

This is small but it is still significantly larger than the secondary warping displacement. However, the maximum secondary shear stress is not negligibly small in comparison to the maximum primary shear stress.

11.7 Torsion of Thin, Closed Cross-Sections

When a thin cross-section is closed, the analysis is more complex than the open section case. Figure 11.25 shows a closed cross-section with two holes. The section is said to be composed of two "cells." Since the shear stresses must be parallel to the sides of a given section, the stresses must act as shown at three positions labeled a–a, b–b, and c–c. We can use the Prandtl stress function approach to obtain the distribution of these stresses. As discussed previously, the Prandtl function must be zero on the outer boundary and have constant values K_1, K_2 at the inner holes. Since the cross-section is thin, we can approximate the Prandtl function as just a linear function across the thickness of the sections, as shown in Fig. 11.26a. This also makes sense from the membrane analogy because, to first order, we expect that the membrane displacement across the thickness will also be linear. If we let the thickness of the cross-section at a–a, b–b, and c–c be t_a, t_b, and t_c, respectively, then since the shear stress is just proportional to the slope of the Prandtl stress function, the stresses at these three locations are

$$\tau_a = -\left(\frac{\partial \Phi}{\partial n}\right)_a = \frac{K_1}{t_a},\quad \tau_b = -\left(\frac{\partial \Phi}{\partial n}\right)_b = \frac{K_1-K_2}{t_b},\quad \tau_c = -\left(\frac{\partial \Phi}{\partial n}\right)_c = \frac{K_2}{t_c} \qquad (11.207)$$

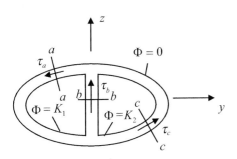

Figure 11.25 Torsion of a closed cross-section with two "cells," showing the shear stresses in the section at three locations and the value of the Prandtl stress function at the outer and inner boundaries of the cells.

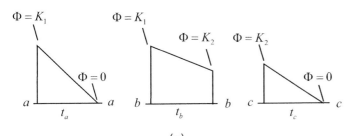

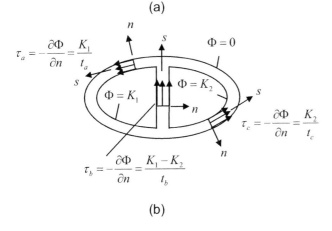

Figure 11.26 (a) Approximating the Prandtl stress function (and corresponding membrane analogy displacement) as linear functions across the thickness of the sections. (b) The shear stresses generated in the different sections.

(see Eq. (11.46)), as shown in Fig. 11.26b. These shear stresses are uniform across the thickness in each section but if the thickness changes, these stresses can vary as we go along the center line of each cell. However, if as in the bending of thin sections, we introduce the shear flow $q = \tau t$, Eq. (11.207) shows that these shear flows depend only on the constants K_1, K_2, so the shear flows in a closed section are constant both across the thickness and along the centerline. For the two-cell closed cross-section, therefore, we can describe the shear flows as two constant shear flows q_i ($i = 1, 2$), where q_i is the shear flow for the ith cell (Fig. 11.27). Note that while the shear flow in each cell is a constant, in sections that share the sides of two cells, such as the vertical section in Fig. 11.27, the total shear flow is a different constant from the remainder of the cells.

We have also used shear flows to describe the bending of unsymmetrical sections in Chapter 10. As in the bending case, we can show that these torsional shear flows must be

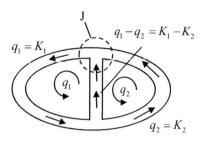

Figure 11.27 The constant shear flows (q_1, q_2) in the two cells and the total shear flows flowing in the various sections, where at a junction, such as J, the sum of the shear flows into or out of the junction must be zero.

conserved at a junction in the sense that the sum of the shear flows flowing into or out of a junction must be zero. This can easily be seen to be the case, for example, for the shear flows in Fig. 11.27 at junction J. We will take a positive shear flow to be flowing in a counterclockwise sense around any given cell, as shown in Fig. 11.27. Such flows will then generate internal torques which are positive along the positive x-axis of the member, as we have already assumed in our previous discussions of torsion. To obtain the torque generated by the shear flows in a closed cross-section, consider the ith cell, as shown in Fig. 11.28. The torsion moment, M_{xi}, generated about a point P by the shear flow, q_i, in the ith cell is

$$M_{xi} = \oint q_i r_\perp ds$$
$$= q_i \left(\int_{ABC} r_\perp ds + \int_{CDA} r_\perp ds \right) \quad (11.208)$$
$$= q_i (\omega_{ABC} + \omega_{CDA})$$

In Eq. (11.208) ω_{ABC} and ω_{CDA} are the sectorial areas obtained by integrating along the curved side ABC and the straight side CDA of the cell, respectively. However, the sectorial area is positive when the area is swept out in a counterclockwise direction and negative when sept out in a clockwise direction (see Appendix C) so that ω_{ABC} is positive and ω_{CDA} is negative, where $\omega_{ABC} = 2A_{ABC}$, $\omega_{CDA} = -2A_{CDA}$; i.e., these sectorial areas are just twice the areas A_{ABC}, A_{CDA} being swept out as shown in Fig. 11.28b. It then follows that

$$\omega_{ABC} + \omega_{CDA} = 2A_{ABC} - 2A_{CDA} \\ = \Omega_i \quad (11.209)$$

where $\Omega_i = 2A_i$ is just twice the area contained with the centerline of the ith cell, as shown in Fig. 11.28b. Thus, the torque being carried by the ith cell is just

$$M_{xi} = q_i \Omega_i \quad (11.210)$$

The total primary (Saint-Venant) torque for a closed section with N cells, M_x^P, is therefore given by

$$M_x^P = \sum_{i=1}^{N} M_{xi} = \sum_{i=1}^{N} q_i \Omega_i \quad (11.211)$$

For a single cell if the primary torque is known then the primary shear flow, q^P, in the cell is simply given as

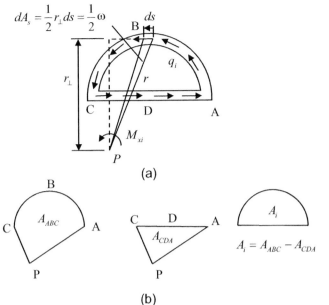

Figure 11.28 (a) The geometry of the ith cell for calculating the torque produced by the shear flow. (b) The areas A_{ABC}, A_{CDA} swept out from point P and the area, A_i, within the centerline of the cell.

$$q^P = \frac{M_x^P}{\Omega} \tag{11.212}$$

where $\Omega = 2A$, i.e., it is twice the area, A, contained within the centerline of the single cell. For multiple cells we need to obtain additional relations to solve for the shear flows. These relations are obtained by requiring that the warping displacement, u_x, must be single valued. Recall, from Eq. (11.169) we found that the primary shear stress on the centerline of a section can be expressed as

$$\sigma_{xs}^P = G\gamma_{xs}^P = Gr_\perp \phi' + G\frac{\partial u_x}{\partial s} \tag{11.213}$$

For a thin, open section, we set this shear stress to be zero but for a closed section we have

$$\sigma_{xs}^P = \frac{q_t}{t} \tag{11.214}$$

where q_t is the shear flow. [Note: q_t here is *total shear flow anywhere in a cell* so that it may be q_i in some parts of the ith cell but combinations of q_i and other shear flows q_j ($j \neq i$) in other parts of the ith cell. Thus, we have placed the t subscript on this shear flow to remind you that it is not just q_i in the ith cell.]. Thus, we have

$$\frac{\partial u_x}{\partial s} = \frac{q_t}{Gt} - \phi' r_\perp \tag{11.215}$$

Integrating Eq. (11.215) around each cell and requiring the warping displacement to be single valued, we find

$$\oint_{ith\ cell} \frac{\partial u_x}{\partial s} ds = 0 = \frac{1}{G} \oint_{ith\ cell} \frac{q_t}{t} ds - \phi' \oint_{ith\ cell} r_\perp ds \qquad (11.216)$$

where ϕ' is the same for each cell and the second integral on the right-hand side of Eq. (11.216) is just twice the area contained within the ith cell, so we have finally the conditions

$$\frac{1}{G\Omega_i} \oint_{ith\ cell} \frac{q_t}{t} ds = \phi' \quad (i = 1, \ldots, N) \qquad (11.217)$$

For a single cell Eq. (11.217) and Eq. (11.212) simply give

$$\phi' = \frac{M_x^P}{G\Omega^2} \oint \frac{ds}{t} \qquad (11.218)$$

which shows that for the single-cell closed section we again can write the relationship between the primary torque and rate of twist as

$$M_x^P = GJ_{eff} \phi' \qquad (11.219)$$

where the effective polar area moment is

$$J_{eff} = \frac{\Omega^2}{\oint \frac{ds}{t}} \qquad (11.220)$$

For multiple cells we cannot write down a closed form expression for J_{eff} but we can write Eq. (11.211) in the form of Eq. (11.219) where

$$J_{eff} = \sum_{i=1}^{N} f_i \Omega_i \qquad (11.221)$$

and where $f_i = q_i/G\phi'$. The normalized total shear flow, $\hat{q}_t = q_t/G\phi'$, is just a linear function of the f_i so that the equations

$$\oint_{ith\ cell} \frac{\hat{q}_t}{t} ds = \Omega_i \quad (i = 1, \ldots, N) \qquad (11.222)$$

are just N linear equations that can be solved for the N unknown f_i values and those results placed into Eq. (11.221) to obtain J_{eff}.

For a general closed section with multiple cells, we can also obtain the primary warping displacement by returning to Eq. (11.215) and integrating. We find

$$u_x(s) = -\phi' \int_{s_P}^{s} r_\perp \, ds + \frac{1}{G} \int_{s_P}^{s} \frac{q_t}{t} \, ds$$

$$= -\phi' \left[\int_{s_P}^{s} r_\perp \, ds - \int_{s_P}^{s} \frac{\hat{q}_t}{t} \, ds \right] \quad (11.223)$$

$$= -\phi' \widehat{\omega}_P (s, s_P)$$

in terms of a function, $\widehat{\omega}_P(s, s_P)$, given by

$$\widehat{\omega}_P(s, s_P) = \int_{s_P}^{s} r_\perp \, ds - \int_{s_P}^{s} \frac{\hat{q}_t}{t} \, ds \quad (11.224)$$

which acts as a *modified sectorial area function* for the closed section. For a single closed cell Eq. (11.224) has the more explicit form

$$\widehat{\omega}_P(s, s_P) = \int_{s_P}^{s} r_\perp \, ds - \frac{\Omega}{\oint \frac{ds}{t}} \int_{s_P}^{s} \frac{ds}{t} \quad (11.225)$$

Like the ordinary sectorial area function, the modified sectorial area function depends on the pole P (which is the origin from which the $r_\perp$ distance is measured) and the constant s_P (which is a origin location where this modified function is zero). The modified function is a function of $r_\perp$ and $\hat{q}_t$ (which is obtained from Eq. (11.221)) so, like the ordinary sectorial area function, it depends only on the geometry. For the nonuniform torsion of open sections we found that in order to ensure that the warping does not generate any net axial force or bending moments, the sectorial area function appearing in the primary warping displacement had to be a principal sectorial area function. Here, the modified function $\widehat{\omega}_P(s, s_P)$ must satisfy the same conditions so that it is a principal modified sectorial area function where the pole P and the origin s_P can be found from the three conditions (see Eq. (11.172)):

$$\int_A \widehat{\omega} \, dA = 0$$

$$\int_A y \, \widehat{\omega} \, dA = 0 \quad (11.226)$$

$$\int_A z \, \widehat{\omega} \, dA = 0$$

where $\widehat{\omega}$ is a modified principal sectorial area function calculated with respect to an arbitrary pole and origin. In Appendix C we derive a wide range of properties of the ordinary sectorial area function. It is easy to show that the modified sectorial area function defined in

Eq. (11.223) also satisfies the following properties found in Appendix C for the ordinary sectorial area function.

(1) Relationship for changing the pole and origin of the modified sectorial area function (see Eq. (C.12)):

$$\widehat{\omega}_A(s, s_0) = \widehat{\omega}_B(s, s_1) - \widehat{\omega}_B(s_0, s_1) \\ + (y_B - y_A)(z(s) - z_0) - (z_B - z_A)(y(s) - y_0) \quad (11.227)$$

(2) Relationship between the principal modified sectorial area function, $\widehat{\omega}_P(s, s_P)$, and a modified function, $\widehat{\omega}_B(s, s_1)$, computed with an arbitrary pole B and origin s_1 (see Eq. (C.43)):

$$\widehat{\omega}_P(s, s_P) = \widehat{\omega}_B(s, s_1) - \frac{Q_{\widehat{\omega}_B}}{A} + (z_P - z_B)y - (y_P - y_B)z \quad (11.228)$$

where

$$Q_{\widehat{\omega}_B} = \int_A \widehat{\omega}_B \, dA \quad (11.229)$$

(3) Location (y_P, z_P) of the principal pole P (center of twist or shear center) (see Eq. (C.21)) as calculated from a modified sectorial area function, $\widehat{\omega}_B(s, s_1)$, with an arbitrary pole B and origin s_1:

$$e_y = y_P - y_B = \frac{I_{z\widehat{\omega}_B} I_{zz} - I_{y\widehat{\omega}_B} I_{yz}}{I_{yy} I_{zz} - I_{yz}^2}$$

$$e_z = z_P - z_B = \frac{I_{z\widehat{\omega}_B} I_{yz} - I_{y\widehat{\omega}_B} I_{yy}}{I_{yy} I_{zz} - I_{yz}^2} \quad (11.230)$$

where, as shown in Appendix C, y and z are centroidal axes. In Eq. (11.230) the distances (e_y, e_z) are the distances of the principal pole P from the pole B in the y- and z-directions.

(4) The warping constant, $J_{\widehat{\omega}_P}$, calculated with the principal modified sectorial area function, as calculated from the same constant using a modified sectorial area function, $\widehat{\omega}_B$, with arbitrary pole and origin (see Eq. (C.46)):

$$J_{\widehat{\omega}_P} = \frac{J_{\widehat{\omega}_B} - Q_{\widehat{\omega}_B}^2}{A} - (y_P - y_B)^2 I_{yy} - (z_P - z_B)^2 I_{zz} + 2(y_P - y_B)(z_P - z_B)I_{yz} \quad (11.231)$$

Of course, the warping constant can also be calculated directly by integration using its definition

$$J_{\widehat{\omega}_P} = \int_A \widehat{\omega}_P^2 \, dA \quad (11.232)$$

11.7 Torsion of Thin, Closed Cross-Sections

In the case of nonuniform torsion there will again be a warping normal stress generated where

$$\sigma_{xx}^w(s,x) = -E\phi''(x)\widehat{\omega}_P(s, s_P) \tag{11.233}$$

As in the thin, open section case, equilibrium requires that this normal stress produce a secondary shear flow, $q^S(s)$, where

$$\frac{\partial q^S}{\partial s} = -t\frac{\partial \sigma_{xx}^w}{\partial x} = tE\phi'''(x)\widehat{\omega}_P(s, s_P) \tag{11.234}$$

As with the case of uniform torsion for a closed cross-section with multiple cells, it is difficult to obtain explicit results for this secondary shear flow and the secondary warping. Later, we will outline the steps needed for the multiple cell case. Here, however, we will obtain explicit results for the case of a single-cell closed section. If we integrate Eq. (11.234) for a single cell, we obtain

$$q^S(s,x) = q_0 + E\phi'''(x)\int_0^s \widehat{\omega}_P \, dA$$
$$= q_0 + E\phi'''(x)S_{\widehat{\omega}_P}(s) \tag{11.235}$$

where q_0 is a constant and we have defined $S_{\widehat{\omega}_P}$ as

$$S_{\widehat{\omega}_P}(s) = \int_0^s \widehat{\omega}_P \, dA \tag{11.236}$$

[Note: in the case of thin, open sections we had $q^S(s,x) = E\phi'''(x)\int^s \omega_P dA$, which we could also have written as $q^S(s,x) = E\phi'''(x)S_{\omega_P}(s)$.] To obtain the constant q_0 we use the fact that the secondary warping displacement, u_x^S, must be single valued so that the following closed integral around the cell is zero:

$$\oint \frac{\partial u_x}{\partial s} ds = \oint \frac{\partial \psi^S}{\partial s} ds = 0 \tag{11.237}$$

But from Eq. (11.176) this is equivalent to

$$\frac{1}{G}\oint \frac{q^S}{t} ds = 0 \tag{11.238}$$

Instead of writing Eq. (11.238) in terms of q_0 we will write it instead in terms of the secondary torsional moment, M_x^S. We have

$$M_x^S = \oint r_\perp q^S ds$$
$$= q_0 \oint r_\perp ds + E\phi''' \oint S_{\widehat{\omega}_P} r_\perp ds \tag{11.239}$$

which we can solve for q_0 as

$$q_0 = \frac{M_x^S}{\Omega} - \frac{E\phi'''}{\Omega}\oint S_{\widehat{\omega}_P} r_\perp \, ds \tag{11.240}$$

which, when placed back into Eq. (11.235) gives

$$\begin{aligned} q^S(s,x) &= \frac{M_x^S}{\Omega} - \frac{E\phi'''}{\Omega}\oint S_{\widehat{\omega}_P} r_\perp \, ds + E\phi''' S_{\widehat{\omega}_P}(s) \\ &= E\phi'''\left[\frac{\hat{M}_x^S}{\Omega} - \frac{1}{\Omega}\oint S_{\widehat{\omega}_P} r_\perp \, ds + S_{\widehat{\omega}_P}(s)\right] \end{aligned} \tag{11.241}$$

where $\hat{M}_x^S = M_x^S/E\phi'''$ is a normalized moment. Placing Eq. (11.241) into Eq. (11.238) then gives (after canceling common terms)

$$\frac{\hat{M}_x^S}{\Omega}\oint \frac{ds}{t} - \frac{\oint \frac{ds}{t}}{\Omega}\oint S_{\widehat{\omega}_P} r_\perp \, ds + \oint \frac{S_{\widehat{\omega}_P}}{t} \, ds = 0 \tag{11.242}$$

From Eq. (11.225) we have

$$d\widehat{\omega}_P(s,s_P) = r_\perp \, ds - \frac{\Omega}{\oint \frac{ds}{t}}\frac{ds}{t} \tag{11.243}$$

so that

$$\oint S_{\widehat{\omega}_P} r_\perp \, ds = \oint S_{\widehat{\omega}_P} d\widehat{\omega}_P + \frac{\Omega}{\oint \frac{ds}{t}}\oint \frac{S_{\widehat{\omega}_P}}{t} \, ds \tag{11.244}$$

Placing Eq. (11.244) into Eq. (11.242) then gives

$$\begin{aligned} \frac{\hat{M}_x^S}{\Omega}\oint \frac{ds}{t} - \frac{\oint \frac{ds}{t}}{\Omega}\oint S_{\widehat{\omega}_P} d\widehat{\omega}_P &= 0 \\ \rightarrow \hat{M}_x^S &= \oint S_{\widehat{\omega}_P} d\widehat{\omega}_P \end{aligned} \tag{11.245}$$

However, by integrating by parts we have

$$\begin{aligned} \oint S_{\widehat{\omega}_P} d\widehat{\omega}_P &= S_{\widehat{\omega}_P}\widehat{\omega}_P\Big|_0^{s_f} - \oint (\widehat{\omega}_P)^2 \, dA \\ &= -J_{\widehat{\omega}_P} \end{aligned} \tag{11.246}$$

where $s = s_f$ is at the same point in the cross-section as $s = 0$ and the first term vanishes since both $S_{\widehat{\omega}_P}$ and $\widehat{\omega}_P$ are single-valued functions. We have also defined the *modified warping constant*, $J_{\widehat{\omega}_P}$, as

$$J_{\hat{\omega}_P} = \oint (\hat{\omega}_P)^2 dA \qquad (11.247)$$

(written here as a line integral rather than an area integral since $dA = tds$). Thus, from Eq. (11.245) we have the secondary torsional moment given by

$$M_x^S = -EJ_{\hat{\omega}_P} \phi'''(x) \qquad (11.248)$$

and the secondary shear flow becomes

$$\begin{aligned} q^S(s,x) &= E\phi'''(x) \left[\frac{\hat{M}_x^S}{\Omega} - \frac{1}{\Omega} \oint S_{\hat{\omega}_P} r_\perp ds + S_{\hat{\omega}_P}(s) \right] \\ &= E\phi'''(x) \left[-\frac{J_{\hat{\omega}_P}}{\Omega} - \frac{1}{\Omega} \left(-J_{\hat{\omega}_P} + \frac{\Omega}{\oint \frac{ds}{t}} \oint \frac{S_{\hat{\omega}_P}}{t} ds \right) + S_{\hat{\omega}_P}(s) \right] \\ &= E\phi'''(x) \bar{S}_{\hat{\omega}_P}(s) \end{aligned} \qquad (11.249)$$

where

$$\bar{S}_{\hat{\omega}_P}(s) = S_{\hat{\omega}_P}(s) - \frac{1}{\oint \frac{ds}{t}} \oint \frac{S_{\hat{\omega}_P}}{t} ds \qquad (11.250)$$

The total twisting moment is

$$\begin{aligned} M_x &= M_x^P + M_x^S \\ &= GJ_{eff} \phi' - EJ_{\hat{\omega}_P} \phi''' \end{aligned} \qquad (11.251)$$

so that, from equilibrium of moments, we again have the differential equation for the twist given in the same form as found before for both general sections and thin open sections, namely

$$GJ_{eff} \phi''(x) - EJ_{\hat{\omega}_P} \phi^{iv}(x) = -t_x(x) \qquad (11.252)$$

where t_x is the distributed torque/unit length along the member.

Finally, the secondary warping displacement for the single closed cell is obtained by integrating

$$\frac{\partial \psi^S}{\partial s} = q^S = E\phi''' \bar{S}_{\hat{\omega}_P}(s) \qquad (11.253)$$

giving

$$\psi^S(s) = E\phi''' \int_0^s \bar{S}_{\hat{\omega}_P} ds + K \qquad (11.254)$$

where the constant K can again be found from the condition

$$\oint \psi^S dA = 0 \tag{11.255}$$

For a multicell closed cross-section with N cells, the steps are very similar but we cannot write them down as explicit expressions. First, we can integrate Eq. (11.235) and define a set of shear flows flowing around each cell as

$$\begin{aligned} q_i^S(s,x) &= q_i + E\phi'''(x) \int_0^s \widehat{\omega}_p dA \\ &= q_i + E\phi'''(x) S_{\widehat{\omega}_p}(s) \quad (i=1,\ldots,N) \end{aligned} \tag{11.256}$$

where q_i are constants at the starting integration points in each cell (Fig. 11.29). We can also define the twisting moments for each cell as

$$M_{xi} = \oint_{\text{ith cell}} q_i^S r_\perp ds \quad (i=1,\ldots,N) \tag{11.257}$$

and, as in the single-cell case, write shear-flow constants q_i in terms of the normalized moments, $\hat{M}_{xi} = M_{xi}/E\phi'''$. The total secondary moment, M_x^S, is then

$$M_x^S = E\phi''' \sum_{i=1}^N \hat{M}_{xi} \tag{11.258}$$

which we can put in the form used previously

$$M_x^S = -E\phi''' J_{\widehat{\omega}}^{\text{eff}} \tag{11.259}$$

where

$$J_{\widehat{\omega}}^{\text{eff}} = -\sum_{i=1}^N \hat{M}_{xi} \tag{11.260}$$

To obtain the normalized moments we use the fact that the warping displacement must be single valued so that now we must have (see Eq. (11.238))

$$\frac{1}{G} \oint_{\text{ith cell}} \frac{q_i^S}{t} ds = 0 \quad (i=1,2,\ldots,N) \tag{11.261}$$

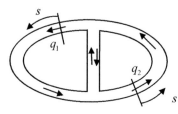

which are N linear equations for the N $\hat{M}_{xi}$ values. As in the primary shear-flow case we have called the total secondary shear flow in each

Figure 11.29 A two-cell cross-section showing the starting shear flow values (q_1, q_2) at the $s = 0$ positions in the cells.

cell q_i^S since certain walls may contain several $q_i^S(s, x)$ shear-flow functions (Fig. 11.29). Solving Eq. (11.261) for the $\hat{M}_{xi}$ values then gives us the individual secondary shear flows, $q_i^S(s, x)$ and the effective warping constant, $J_{\hat{\omega}}^{eff}$. The secondary warping function can then be obtained by integration of the total secondary shear flow:

$$\psi^S(s, x) = \int_0^s q_i^S ds + K \qquad (11.262)$$

where again the constant K can be found from the condition

$$\oint \psi^S dA = 0 \qquad (11.263)$$

Obviously, this is a rather complex process where we have not delineated many of the explicit details but instead listed the basic steps involved, following our previous approach in describing the primary shear flows.

Example 11.3 Primary Warping of a Thin Rectangular Closed Cell

Consider a thin rectangular closed cross-section as shown in Fig. 11.30. We will examine the primary warping of this section, using the expression for the principal modified sectorial area function given by Eq. (11.225). For a rectangular section with constant thickness, that equation can be written along each straight side as

$$\widehat{\omega}_P(s) = \int_0^s r_\perp ds - \frac{ab}{(a+b)} s + \widehat{\omega}_P(0) \qquad (11.264)$$

Because of the symmetry of the rectangular section, the shear center is located at the center, O, of the rectangle so we can take the principal pole at O. If we take the starting point of the

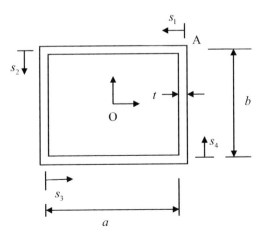

Figure 11.30 A thin rectangular cross-section of constant thickness, t. The lengths (a, b) are measured from the centerlines of the cross-sections and the distances (s_1, s_2, s_3, s_4) are also measured along the centerlines.

integrations to be at point A in Fig. 11.30 and let the value of the modified sectorial area function be ω_0 at that point then in the four parts of the section we have, using Eq. (11.264),

$$\widehat{\omega}_1 = \frac{b(b-a)}{2(a+b)} s_1 + \omega_0$$

$$\widehat{\omega}_2 = \frac{a(a-b)}{2(a+b)} s_2 + \omega_0 + \frac{ab(b-a)}{2(a+b)}$$

$$\widehat{\omega}_3 = \frac{b(b-a)}{2(a+b)} s_3 + \omega_0$$

$$\widehat{\omega}_4 = \frac{a(a-b)}{2(a+b)} s_4 + \omega_0 + \frac{ab(b-a)}{2(a+b)} \quad (11.265)$$

To obtain the value of ω_0 we must satisfy (see Eq. (11.226))

$$I = \int \widehat{\omega}_P dA = 0 \quad (11.266)$$

We can do the integration in MATLAB®, giving:

```
syms a b w0 w1 w2 w3 w4 x
w1 = b*(b-a)*x/(2*(a+b)) +w0;
w2 = a*(a -b)*x/(2*(a+b)) + w0 + a*b*(b -a)/(2*(a+b));
w3 = b*(b-a)*x/(2*(a+b)) +w0;
w4 = a*(a-b)*x/(2*(a+b))+ w0 +a*b*(b-a)/(2*(a+b));
I = int(w1,x, 0,a) +int(w2,x,0,b) +int(w3,x,0,a)...
+int(w4,x,0,b);
simplify(I)
ans = 2*a*w0 + 2*b*w0 + (a*b^2)/2 - (a^2*b)/2
```

Setting this integral I found in MATLAB® equal to zero, we find

$$\omega_0 = \frac{ab(a-b)}{4(a+b)} \quad (11.267)$$

We can write the principal modified sectorial area function succinctly in terms of this constant as

$$\widehat{\omega}_1 = -2\omega_0 s_1/a + \omega_0$$
$$\widehat{\omega}_2 = 2\omega_0 s_2/b - \omega_0$$
$$\widehat{\omega}_3 = -2\omega_0 s_3/a + \omega_0$$
$$\widehat{\omega}_4 = 2\omega_0 s_4/b - \omega_0 \quad (11.268)$$

which is shown in Fig. 11.31. For $a = b$ the constant ω_0 vanishes so that a square section is free of warping. Having the principal modified sectorial area function, one can calculate the warping constant as

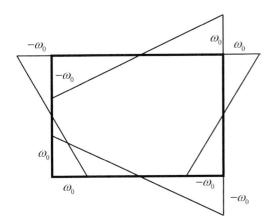

Figure 11.31 The principal modified sectorial area function for a rectangular section.

$$J_{\bar{\omega}_P} = \frac{ta^2b^2(b-a)^2}{24(a+b)} \tag{11.269}$$

Note that it is also easy to evaluate the area moments approximately when the rectangular section is thin as

$$I_{yy} = \frac{tb^2(b+3a)}{6}$$
$$I_{zz} = \frac{ta^2(a+3b)}{6} \tag{11.270}$$

11.8 PROBLEMS

P11.1 For a solid bar whose cross-section is an equilateral triangle of height, h, (see Fig. P11.1), the normalized Prandtl stress can be obtained exactly as:

$$\bar{\Phi} = \frac{1}{2h}\left(y - \sqrt{3}z - \frac{2h}{3}\right)\left(y + \sqrt{3}z - \frac{2h}{3}\right)\left(y + \frac{h}{3}\right)$$

Note that each of the quantities in parentheses in the above expression vanish on an edge of the triangle so that $\bar{\Phi} = 0$ on the boundary, as required. For this shape one can show by integration that

$$J_{eff} = 2\iint \bar{\Phi} dA = \frac{\sqrt{3}h^4}{45}$$

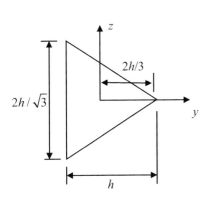

Figure P11.1 A cross-section in the form of an equilateral triangle.

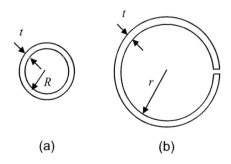

Figure P11.2 (a) A thin, closed circular cross-section. (b) A thin, open circular cross-section.

(a) For this cross-section, determine and plot the shear stress along the y-axis.
(b) Find the maximum shear stress, τ_{max}, in terms of the torque, T, carried by the section, and h. Where is τ_{max} found?

P11.2 Consider a thin, closed circular cross-section of radius R and thickness t, (see Fig. P11.2a), and a thin open section circular cross-section of radius r and thickness t, as shown in Fig. P11.2b. In these sections the effective polar area moment, J_{eff}, determines the stiffness of the cross-section with respect to torsion. If the effective polar area moments of these sections are equal, determine the radius, r, of the open section in terms of R and t for the closed section. What value of r do you obtain in this case if $t = 1$ mm, $R = 10$ mm?

P11.3 The thin extruded cross-section shown in Fig. P11.3 is subjected to a uniform torque, T. Determine the maximum allowable value for T if the maximum permissible shear stress is 75 MPa. Neglect stress concentrations. All dimensions are in millimeters and measured with respect to the cross-section centerlines.

P11.4 The thin aluminum ($G = 26.7$ GPa) C-section torsional member shown in Fig. P11.4 is subjected to a uniform torque $T = 1356$ N-m.

(a) Determine the angle of twist per unit length.
(b) Determine the maximum shear stress and the location of this maximum stress in the cross-section. Neglect stress concentrations. All dimensions are in millimeters and measured with respect to the cross-section-centerlines.

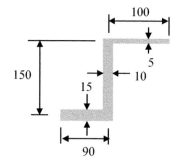

Figure P11.3 A thin, open cross-section. All dimensions are given in millimeters as measured along the centerlines.

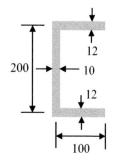

Figure P11.4 A thin aluminum C-section. All dimensions are given in millimeters as measured along the centerlines.

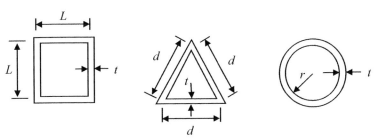

Figure P11.5 Three thin cross-sections in the form of a square, an equilateral triangle, and a circle. All sections have the same uniform thickness, t, and the same length along their center line perimeters (i.e., the lengths L, d, r are related).

P11.5 Compare the shear stress and twist per unit length of a thin square tube and a thin tube in the shape of an equilateral triangle with the corresponding shear stress and twist per unit length, respectively, for a thin circular tube (Fig. P11.5). The three sections have equal wall thicknesses and equal perimeters. The dimensions are measured with respect to the cross-section centerlines.

P11.6 Consider a single closed cell that has "fins" as seen in Fig. P11.6a. In the fins we might expect the stress distributions to be those we have found for thin, open sections while in the closed cell we might expect the stress to be nearly uniform across the thickness. We will still keep these assumptions on the behavior of the stresses but will take the presence of the fins into account in the torque–twist relationship by writing the effective polar area moment, J_{eff}, as the sum of the closed cell and open cell values, i.e.,

$$J_{eff} = \frac{\Omega^2}{\oint \frac{ds}{t}} + \sum_i \frac{1}{3} \int_0^{b_i} t_i(s)^3 ds$$

where the summation is over all the fins present and we have allowed the fins to have variable thicknesses (see Eq. (11.164)). Generally, these fin contributions are small.

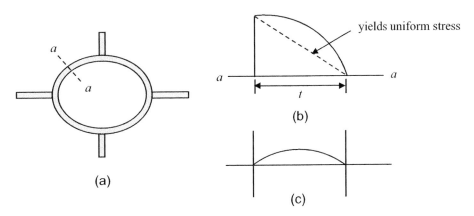

Figure P11.6 (a) A single closed cell section with fins. (b) The shape of a membrane across the thickness of the closed cell at a–a when using the membrane analogy, showing the difference of the membrane shape from the assumed straight-line behavior, an assumption which yields a uniform shear stress across the thickness, t. (c) A plot of a membrane having the shape of the difference seen in (b).

However, there is also a small correction to the closed cell value that we should include that is of the same order. To see where this correction comes from, consider the membrane analogy we discussed previously. Figure P11.6b shows the shape of the membrane across the thickness, t, at some location a–a in the closed cell, as seen in Fig. P11.6a. We had assumed to first order the membrane has the straight-line shape seen in Fig. P11.6b. It was this assumption that led to the shear stress being taken as uniform across the thickness. However, there will be a difference between the actual membrane shape and the assumed straight-line behavior and that difference is shown by itself in Fig. P11.6c. The membrane shape of this difference looks much like that seen in a thin, open section so we will add a similar open section contribution for the closed cell and write

$$J_{eff} = \frac{\Omega^2}{\oint \frac{ds}{t}} + \frac{1}{3}\oint t(s)^3 ds + \sum_i \frac{1}{3}\int_0^{b_i} t_i(s)^3 ds$$

where the middle term is a line integral around the closed cell.

Now, let us apply this result to the cross-section seen in Fig. P11.7a where a torque $T = 6000$ N-mm is applied to the section and the shear modulus is $G = 80$ GPa.

(a) Compute the angle of twist/unit length and the maximum shearing stress (neglecting stress concentrations) in the closed section and the fin.

(b) Now, consider the same geometry but where the cross-section is split at A to produce the open section seen in Fig. P11.7b. Again, determine the twist/unit length and the maximum shearing stress for this open section. How does the slitting change these values from those calculated in part (a)?

P11.7 A member with the two-cell cross-section shown in Fig. P11.8a carries a uniform torque, T. The shear modulus is $G = 80$ GPa.

(a) If the maximum allowable shear stress is 10 MPa, determine the maximum allowable torque for this closed cross-section. Ignore any stress concentrations.

Figure P11.7 (a) A box beam with an extension. (b) The same cross-section, which is split at point A. All dimensions are outside measurements.

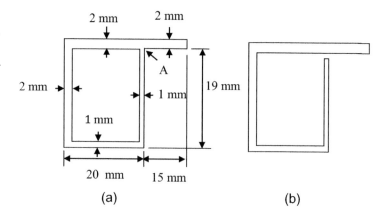

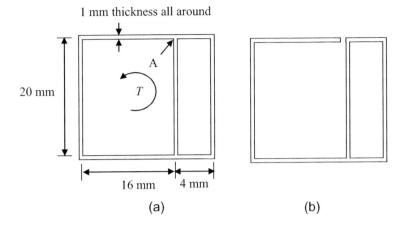

Figure P11.8 (a) A two-cell closed cross-section. (b) The same cross-section when it is slit at point A. All dimensions are measured from the center lines.

(b) If the cross-section is split at point A, producing the cross-section shown in Fig. P11.8b, determine the maximum allowable torque under the same conditions as in (a). How much do your answers change if you include the small correction to J_{eff} for the closed part as described in problem P11.6?

P11.8 Two cover plates (shown as dashed lines in Fig. P11.9a) are welded onto the outside of an I-beam. Both the cover plates and the I-beam have the same uniform thickness, t. What is the ratio of the torsional stiffness of the I-beam with the plates to the beam without the plates?

P11.9 Given a thin rectangular section of length b and thickness, t (Fig. P11.9b) what should the radius, R, be (in terms of b and t) of two thin, closed circular tubes, also of thickness t, that we weld onto the rectangular section if we wish to double the torsional stiffness?

P11.10 The two-cell thin-walled box beam shown in Fig. P11.10 is subjected to a pure torque, T. If the maximum allowable shear stress in the beam is 40 MPa (neglecting stress concentrations), determine the applied torque, T, and the rate of twist, ϕ' ($G = 70$ GPa). All wall thicknesses are 0.5 mm and the dimensions shown in Fig. P11.10 are all measured from the centerlines of the cross-section.

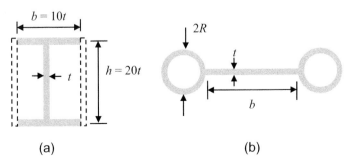

Figure P11.9 (a) An I-beam with and without cover plates. The thickness of all components is t. (b) A thin rectangular section of length b augmented by two thin circular tubes of radius R. The thickness of all components is t.

Figure P11.10 A two-cell box beam.

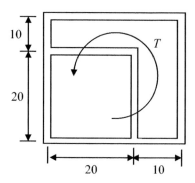

P11.11 The two-cell box beam shown in Fig. P11.11 carries a pure torque, T. If the maximum shearing stress in the section is limited to 5 ksi, determine the corresponding allowable torque and twist/unit length. Neglect any effects of stress concentrations. Take the box beam to be made of steel $(G = 12 \times 10^6 \text{ psi})$ and let the thickness $t = 0.5$ in. and the dimension $a = 6$ in. All the dimensions are measured from the centerlines of the beam walls.

Figure P11.11 A two cell box beam where $a = 6$ in., $t = 0.5$ in.

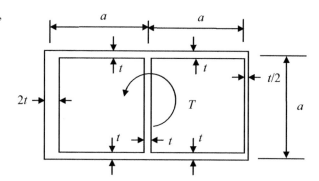

P11.12 Consider a member with the thin, split (open) circular cross-section shown in Fig. P11.12a. We will assume that this member is subjected to a torque, T_0, at one end and is fixed at the other end, as shown (for a general section) in Fig. 11.13, so that it is a state of nonuniform torsion. Since nonuniform torsion involves quite a few parameters and functions, we want to describe those functions and parameters in general terms for this geometry.

(a) In Chapter 10 we showed how to use the shear flow in bending to determine the shear center. We also have shown in Appendix C how we can use the sectorial area function to obtain explicit expressions for the shear center location (see Eq. (C.21)) or instead calculate the location by using the properties of the principal sectorial area function (see Eq. (C.32)). Here, we can take advantage of the relationship obtained in Appendix C for how the sectorial area function changes

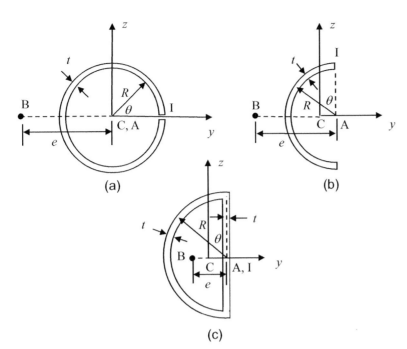

Figure P11.12 (a) A thin, open circular section that is split at point I. (b) A thin, open semicircular section. (c) A thin, closed semicircular section.

when we change the pole location for the sectorial function (Eq. (C.14)) which we rewrite here as

$$\omega_B(s, s_0) = \omega_A(s, s_0) + (y_A - y_B)(z(s) - z_0) - (z_A - z_B)(y(s) - y_0) \quad \text{(P11.1)}$$

to determine the principal sectorial function for the cross-section of Fig. P11.12a. Recall that the (y, z) coordinates in this relationship are measured from the centroid.

First, determine the sectorial area function with the pole located at point A (see Fig. P11.12a) and with an initial starting point (origin) at point I (this is a very simple function). Calculate the sectorial area function, in terms of the distance, e, from Eq. (P11.1) when the pole is located at point B instead. Then change the origin, adding an unknown constant to the sectorial area function. Finally, use the criteria of Eq. (C.32) to obtain the principal sectorial area function, ω_P, and the value of e that defines the shear center location (and the unknown constant). Note that the primary warping function for the cross-section, ψ^P, is then $\psi^P = -\omega_P$.

(b) Determine the sectorial second moment (warping constant), J_{ω_P}, the effective polar area moment, J_{eff}, and the constant, kL (see Eq. (11.136)), where L is the length of the member. Note that expressions for the twist and derivatives of the twist $\left(\phi(x), \phi'(x), \phi''(x), \phi'''(x)\right)$ are then given by Eq. (11.192) and Eq. (11.193).

(c) Determine the secondary warping function, ψ^S, and expressions for the primary and secondary shear stresses in the cross-section as well as the warping normal stress, σ_{xx}^w.

P11.13 Repeat the steps of problem P11.12 for the cross-section shown in Fig. P11.12b.

P11.14 Repeat part (a) of problem P11.12 for the closed cross-section of Fig. P11.12c. Specifically, determine (1) the principal shear flow, (2) the location of the shear center, and (3) the principal modified sectorial area function, which gives us the primary warping function of the cross-section. Equation (P11.1) can also be used for the modified sectorial area function. Note that in a closed cross-section the warping is normally much smaller than for open sections and the shear center is usually near the centroid.

References

1. I. S. Sokolnikoff, *Mathematical Theory of Elasticity* (New York, NY: McGraw-Hill, 1956)
2. W. B. Bickford, *Advanced Mechanics of Materials* (Menlo Park, CA: Addison-Wesley, 1998)
3. R. D. Cook and W. C. Young, *Advanced Mechanics of Materials*, 2nd edn (Upper Saddle River, NJ: Prentice-Hall, 1999)
4. C. E. Pearson, *Theoretical Elasticity* (Cambridge, MA: Harvard University Press, 1959)
5. W.D. Pilkey and L. Kitis, Notes on the Linear Analysis of Thin-walled Beams, https://vdocuments.mx/bmbk.html (1996)
6. E. J. Sapounzakis and V. G. Mokos, "Warping shear stresses in non-uniform torsion by BEM," Comp. Mech., **30** (2003), 131–142

12 Combined Deformations

The bending of unsymmetrical beams was considered in Chapter 10 where we showed that, for thin sections, we could obtain explicit expressions for both the flexure stresses and the shear flows (shear stresses). In Chapter 11 similar explicit expressions were found for the shear stresses or shear flows for the torsion of thin, open and closed cross-sections. In this chapter we will examine cases where a thin member is in a combination of bending and torsion. We will also consider the case where axial loads acting on a thin section can induce torsional deformations. A new quantity called a *bimoment* will be shown to be a resultant of the axial stresses in that case. Although we will deal only with combined bending and torsional and combined axial extension/compression and torsion in this chapter, we should note that parts of our discussion are elements of a theory called Vlasov theory [1], [2], which examines the more general case where there can be any combinations of bending, torsion, and axial extension/compression of thin, open sections.

12.1 Bending and Torsion of Thin, Open Cross-Sections

Consider a thin, open section where the shear forces act on a general point in the cross-section, as defined by the distances (d_y, d_z) from a fixed point, O, as shown in Fig. 12.1. These shear forces will produce both bending and torsion of the section. If the shear forces acted through the shear center, then bending only would be produced so we can imagine decomposing the case of Fig. 12.1 into pure bending and pure torsional problems as shown in Fig. 12.2 where (e_y, e_z) are measured from point O to the shear center and the torque T is given by

$$T = V_z(d_y - e_y) - V_y(d_z - e_z) \tag{12.1}$$

We know from our discussions in Chapter 10 and Chapter 11 how to solve these bending and torsional problems separately, so by superposition we can obtain the total shear stresses in the general case. This process is best illustrated with an example so we will use the thin, open section considered in Example 10.4 in Chapter 10.

Example 12.1 Combined Bending and Torsion of a Thin, Open Cross-Section

In Example 10.4 considered in Chapter 10, there was a vertical 1000 lb shear force acting through the shear center, producing bending only of the section. In this case we want to

476 Combined Deformations

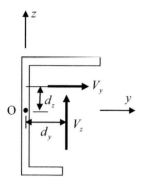

Figure 12.1 Shear forces acting on an unsymmetrical cross-section.

obtain the maximum shear stress and its location in the cross-section for the problem where the shear force acts through the center point O on the web (Fig. 12.3) so that both bending and torsion are produced. For the torsional shear stress, we will only consider the primary shear stress for uniform torsion. The shear center was found to be 0.828 in. to the left of point O in the web so that the pure bending and torsional problems are as shown in Fig. 12.3. The effective polar area moment, J_{eff}, and the maximum primary shear stress, $(\tau_{max})_T$ for the pure torsion case (neglecting any stress concentrations) are

$$J_{eff} = 2 \times \frac{1}{3}(8)(0.1)^3 + \frac{1}{3}(10)(0.1)^3$$
$$= 0.008666 \text{ in.}^4$$

$$(\tau_{max})_T = \frac{Tt_{max}}{J_{eff}} = \frac{(828.2)(0.1)}{0.008666}$$
$$= 9556 \text{ psi}$$

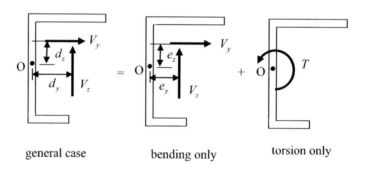

Figure 12.2 A general loading case decomposed into pure bending and pure torsion problems.

Figure 12.3 The beam of Example 10.4 in Chapter 10 where the shear force acts through the center point O in the web, producing the bending and torsional loadings shown. The beam has a constant thickness $t = 0.1$ in.

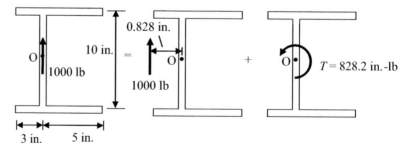

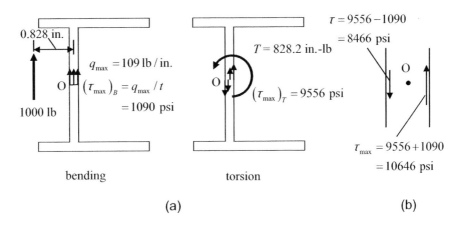

Figure 12.4 (a) The maximum bending and torsional shear stresses. (b) The stresses at point O.

This maximum shear stress acts at the outer boundaries of the cross-section and is the same in all sections since the thickness is a constant 0.1 in. Thus, we need only to find the location where this maximum torsional stress adds to the maximum bending stress. In the bending case considered in Chapter 10, we found the maximum shear flow to be 109 lb/in. flowing upwards at point O so that the maximum bending shear stress (which is uniform across the cross-section), $(\tau_{max})_B$, is given by $(\tau_{max})_B = q_{max}/t = 1090$ psi. Both the maximum bending and torsional shear stresses at O are shown in Fig. 12.4a. Since the primary torsional shear stress varies linearly across the thickness, the sum of the maximum bending and torsional stresses are as shown in Fig. 12.4b where the largest total shear stress $\tau_{max} = 10\,646$ psi will occur on the right-hand side of the section at point O.

12.2 Bending and Torsion of Thin, Closed Cross-Sections

12.2.1 Single-Cell Cross-Sections

Consider first the case of a single-cell closed section as shown in Fig. 12.5, where shear force components (V_y, V_z) are located with respect to an arbitrary point P. The torsion that the shear forces will induce a constant shear flow, q_T, as shown. However, in the bending part the shear flow, q_B, is known only up to a constant since we have, from Eq. (10.41),

$$q_B(s) = q_B(0) + \frac{(V_z I_{yz} - V_y I_{yy})Q_z + (V_y I_{yz} - V_z I_{zz})Q_y}{D} \quad (12.2)$$

and, unlike open sections, we cannot find an end where $q_B = 0$. Thus, the total shear flow will be determined only to within a constant value (that we must determine) and that constant is partly due to bending and partly due to torsion. If we sum moments about P we have

Figure 12.5 Bending and torsion of a thin, single-cell closed section decomposed into its bending and torsional components.

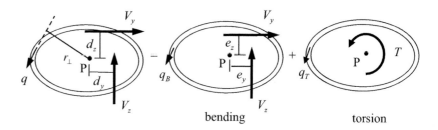

$$\sum M_P = V_z d_y - V_y d_z = \oint_C q r_\perp \, ds \qquad (12.3)$$

and the total shear flow in the section can be written as

$$q(s) = q_c + V_y f(s) + V_z g(s) \qquad (12.4)$$

where q_c is a constant shear flow, part of which comes from bending and part of which comes from the primary shear stress due to uniform torsion. (We will not include the secondary torsional shear flows in this discussion.) From Eq. (12.2) we have the remaining bending shear-flow terms

$$f(s) = \frac{I_{yz} Q_y - I_{yy} Q_z}{I_{yy} I_{zz} - I_{yz}^2}$$

$$g(s) = \frac{I_{yz} Q_z - I_{zz} Q_y}{I_{yy} I_{zz} - I_{yz}^2} \qquad (12.5)$$

Placing Eq. (12.4) and Eq. (12.5) into Eq. (12.3) gives

$$V_z d_y - V_y d_z = \Omega q_c + \oint_C [V_y f(s) + V_z g(s)] r_\perp \, ds \qquad (12.6)$$

where Ω is twice the area contained within the centerline of the cross-section. Also recall that we have a relationship between the shear flow and the twist/unit length induced by that shear flow given as (Eq. (11.217)):

$$\phi' = \frac{1}{G\Omega} \oint_C \frac{q}{t} \, ds \qquad (12.7)$$

We will examine two types of problems.

(1) The shear forces and their positions are known.

In this case q_c can be found directly from Eq. (12.6) since the left-hand side of that equation is known explicitly as are all the other terms on the right-hand side of that equation

12.2 Bending and Torsion of Thin, Closed Cross-Sections

except for q_c. Once q_c is known, then the total shear flow q is known completely so we can use it in Eq. (12.7) to obtain the twist/unit length ϕ'.

(2) The shear forces are known and assumed to act through the shear center, whose position (e_y, e_z) relative to point P is unknown.

In this case there is no twisting of the section so that this is not a combined deformation problem and we have $\phi' = 0$. We can also set $V_y = 0$ and solve Eq. (12.7) for an unknown constant shear flow q_{c1}. Then Eq. (12.6) gives the location of the shear center $d_y = e_y$ since

$$V_z e_y = q_{c1}\Omega + \oint V_z g(s) r_\perp ds \qquad (12.8)$$

We can repeat this process by setting $\phi' = 0$ and $V_z = 0$ and solving Eq. (12.7) for a new constant shear-flow value q_{c2}. Then Eq. (12.6) gives the location of the shear center $d_z = e_z$ since

$$-V_y e_z = q_{c2}\Omega + \oint V_y f(s) r_\perp ds \qquad (12.9)$$

We can then combine these two cases and obtain the total shear flow and corresponding shear stress when both V_y and V_z are present.

12.2.2 Multiple-Cell Cross-Sections

Figure 12.6a shows a cross-section with two cells. We know that in the case of torsion only there are two constant shear flows $((q_1)_T, (q_2)_T)$ in each cell (not shown). In the case of bending only we can imagine breaking the cells at two locations and writing the bending

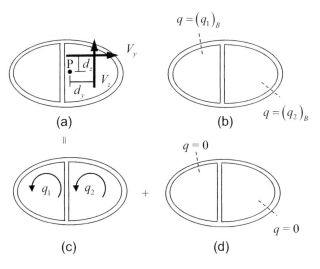

Figure 12.6 (a) A two-cell closed section under bending and torsion. (b) The two shear-flow constants appearing in the bending part of the problem. (c, d) The decomposition of the original problem into two parts where (c) constant shear flows (due to both bending and torsion) are flowing in the cells, and (d) an equivalent open section where the shear flows are zero at imaginary cuts in the cross-section.

shear stresses in terms of two unknown "starting" shear flows $((q_1)_B, (q_2)_B)$, as seen in Fig. 12.6b. Thus, we can combine the two sets of unknown constant shear flows and consider the original problem of Fig. 12.6a as the sum of a problem where there are constant shear flows (q_1, q_2) in the cross-section, as shown in Fig. 12.6c (where $q_i = (q_i)_T + (q_i)_B (i = 1, 2)$) and a problem where there are bending shear flows in an open-like cross-section where the starting shear flows at the imaginary breaks in the cross-section are both zero, as shown in Fig. 12.6d. This means that if we sum moments about point P we will find

$$V_z d_y - V_y d_z = \Omega_1 q_1 + \Omega_2 q_2 + \sum_{m=1}^{2} \oint_{C_m} [V_y f(s) + V_z g(s)] r_\perp ds \qquad (12.10)$$

where $\Omega_i (i = 1, 2)$ are twice the areas contained inside the centerlines of the two cells and the two line integrals are taken around the two closed cells. In addition, we have, as shown in Chapter 11, the angle of twist/unit length related to the shear flows in each cell as

$$\begin{aligned} \phi' &= \frac{1}{G\Omega_1} \oint_{C_1} \frac{q}{t} ds \\ \phi' &= \frac{1}{G\Omega_2} \oint_{C_2} \frac{q}{t} ds \end{aligned} \qquad (12.11)$$

As in the single cell problem we can examine two types of problems.

(1) The shear forces and their positions are known.

In this case we can solve the two equations in Eq. (12.11) for (q_1, q_2) in terms of the unknown ϕ' and place these values into Eq. (12.10), which is then an equation that can be solved for ϕ'. Once ϕ' is obtained in this manner then the previously obtained solutions of Eq. (12.11) yield the values of (q_1, q_2).

(2) The shear forces are known and assumed to act through the shear center, whose position is unknown.

There is no torsion in this case so this is not a combined deformation problem and we can set $\phi' = 0$, $V_y = 0$ and solve the two equations in Eq. (12.11) for the unknown (q_1, q_2) values. Then Eq. (12.10) becomes

$$V_z e_y = \Omega_1 q_1 + \Omega_2 q_2 + \sum_{m=1}^{2} \oint_{C_m} [V_z g(s)] r_\perp ds \qquad (12.12)$$

which can be solved for the shear center location e_y. We can repeat this process by setting $\phi' = 0$ and $V_z = 0$, solving for new values (q_1, q_2), and then obtain the shear center location e_z from

$$-V_y e_z = \Omega_1 q_1 + \Omega_2 q_2 + \sum_{m=1}^{2} \oint_{C_m} [V_y f(s)] r_\perp \, ds \tag{12.13}$$

In this chapter we have emphasized obtaining the bending and primary torsional shear stresses in the combined bending and torsion of open and closed sections. There is also warping present in these combined problems as well as secondary torsional shear stresses for nonuniform torsion cases that can be obtained with the methods outlined in Chapter 11. We will now consider in some detail an example of a single-cell closed section to illustrate the steps outlined previously in an explicit case.

Example 12.2 Combined Bending and Torsion of a Single-Cell Closed Cross-Section

In this problem we will calculate the shear flows and determine the shear center for the box beam shown in Fig. 12.7. The beam has a uniform thickness of $t = 0.1$ in., and all the dimensions shown in Fig. 12.7 are measured in inches. A 1000 lb vertical force acts along the centerline of the left-hand vertical section, producing bending and torsion. We have labeled the four sides of the section as sides (1)–(4). The location of the centroid has been calculated as $Y_C = 5.21$ in., $Z_C = 3.5$ in. and the area moments and mixed area moment of the four sides have been calculated as

$$\begin{aligned}
I_{yy}^{(1)} &= 1.433 & I_{zz}^{(1)} &= 18.44 & I_{yz}^{(1)} &= -4.074 \\
I_{yy}^{(2)} &= 14.41 & I_{zz}^{(2)} &= 16.41 & I_{yz}^{(2)} &= -3.419 \\
I_{yy}^{(3)} &= 6.975 & I_{zz}^{(3)} &= 24.43 & I_{yz}^{(3)} &= -4.689 \\
I_{yy}^{(4)} &= 14.70 & I_{zz}^{(4)} &= 15.15 & I_{yz}^{(1)} &= -3.318
\end{aligned} \tag{12.14}$$

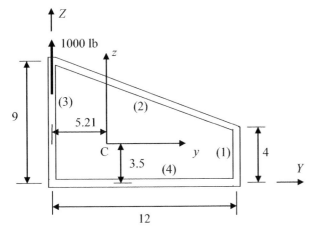

Figure 12.7 Bending and torsion of a closed single-cell section of uniform thickness $t = 0.1$ in. All the dimensions listed are in inches.

482 Combined Deformations

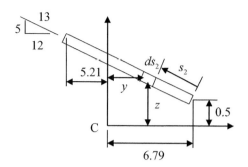

Figure 12.8 Geometry of the inclined side as seen in centroidal coordinates.

all measured in units of in.^{4}. For sides (1), (3), and (4) these values follow directly from the parallel axis theorem (see Appendix A) but calculations for the inclined side (2) are a bit more complex. We could use a combination of the parallel axis theorem and the change of the area moment under rotations but it is easier to simply do the integrations directly, using the geometry of the inclined side relative to the cross-section centroid as shown in Fig. 12.8. It follows that the (y,z) locations of a small element ds_2 are given by

$$z = 0.5 + \frac{5}{13}s_2$$
$$y = 6.79 - \frac{12}{13}s_2$$
(12.15)

The area moment, $I_{yy}^{(2)}$ is then just an integral that can easily be calculated:

$$I_{yy}^{(2)} = \int_0^{13} z^2 dA$$
$$= \int_0^{13} (0.5 + 5s_2/13)^2 (0.1) ds_2$$
$$= 14.41 \text{ in.}^4$$
(12.16)

and the others can be calculated similarly, yielding

$$I_{zz}^{(2)} = \int_0^{13} y^2 dA = 16.41 \text{ in.}^4$$
$$I_{yz}^{(2)} = \int_0^{13} yz dA = -3.419 \text{ in.}^4$$
(12.17)

From all these values for the individual parts we then have for the entire cross-section

12.2 Bending and Torsion of Thin, Closed Cross-Sections

$$I_{yy} = \sum_{m=1}^{4} I_{yy}^{(m)} = 37.52 \text{ in.}^4$$

$$I_{zz} = \sum_{m=1}^{4} I_{zz}^{(m)} = 74.43 \text{ in.}^4 \quad (12.18)$$

$$I_{yz} = \sum_{m=1}^{4} I_{yz}^{(m)} = -15.50 \text{ in.}^4$$

In our problem we have $V_z = 1000$ lb, $V_y = 0$ so the equation for the bending shear flows (see Eq. (10.41)) is

$$\Delta q = \frac{-\left(I_{zz}Q_y - I_{yz}Q_z\right)V_z}{I_{yy}I_{zz} - I_{yz}^2}$$

$$= \frac{\left(-74.43 Q_y - 15.5 Q_z\right)(1000)}{(37.52)(74.43) - (15.5)^2} \quad (12.19)$$

$$\rightarrow \Delta q = -29.16 Q_y - 6.073\, Q_z$$

where Δq is the difference between the shear flow at a given location and its constant value at some starting location. However, we can also use Eq. (12.19) for the total shear flow since the torsional part simply adds an additional constant value, which we could combine with the constant in the bending expression, as discussed previously. Thus, we will now use Eq. (12.19) for the total shear flow in each part of the section. Consider part (1) as shown in Fig. 12.9. The quantities (Q_y, Q_z) are calculated for an area A_s as shown in Fig. 12.9 as

$$Q_y = \int_{A_s} z dA = \bar{z} A_s$$

$$Q_z = \int_{A_s} y dA = \bar{y} A_s \quad (12.20)$$

where $(\bar{y}, \bar{z})$ are the distances from the centroid of the entire cross-section to the centroid of the area A_s as shown in Fig, 12.9 for part (1). Thus, we have for part (1)

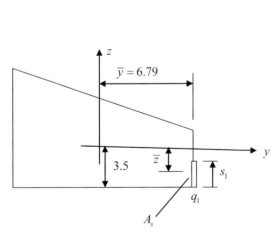

Figure 12.9 The geometry for part (1) of the cross-section.

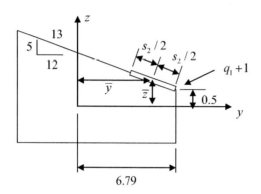

Figure 12.10 The geometry for part (2) of the cross-section.

$$Q_y^{(1)} = \int z \, dA = -(3.5 - s_1/2)(0.1)(s_1)$$
$$= -0.35 s_1 + 0.05 s_1^2$$
$$Q_z^{(1)} = \int y \, dA = (6.79)(0.1)(s_1)$$
$$= 0.679 s_1$$
(12.21)

which gives the shear flow in part (1) as

$$q^{(1)}(s_1) = q_1 - 29.16\left[-0.35 s_1 + 0.05 s_1^2\right] - 6.073[0.679 s_1]$$
$$= q_1 + 6.082 s_1 - 1.458 s_1^2 \text{ lb/in.}$$
(12.22)

where q_1 is the starting value in the part, as shown in Fig. 12.9. At the end of part (1) we have $q^{(1)}(4) = q_1 + 1$ lb/in., which is also the starting value for the shear flow in part (2) as shown in Fig. 12.10. In this part we have

$$\bar{y} = 6.79 - 6 s_2/13$$
$$\bar{z} = 0.5 + 5 s_2/26$$
$$Q_y^{(2)} = \bar{z} A_s = (0.5 + 5 s_2/26)(0.1)(s_2) = 0.05 s_2 + 0.1923 s_2^2$$
$$Q_z^{(2)} = \bar{y} A_s = (6.79 - 6 s_2/13)(0.1)(s_2) = 0.679 s_2 - 0.04615 s_2^2$$
(12.23)

so that the shear flow in this part is

$$q^{(2)}(s_2) = q_1 + 1 - 29.16\left[0.05 s_2 + 0.01923 s_2^2\right] - 6.073\left[0.679 s_2 - 0.04615 s_2^2\right]$$
$$= q_1 + 1 - 5.58 s_2 - 0.2802 s_2^2 \text{ lb/in.}$$
(12.24)

and at the end of part (2) we have $q^{(2)}(13) = q_1 - 118.89$ lb/in. For part (3), as shown in Fig. 12.11, the same types of calculations give

Figure 12.11 The geometry for part (3) of the cross-section.

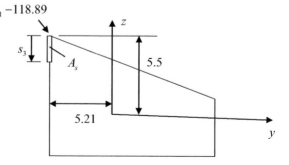

$$\bar{y} = -5.21$$

$$\bar{z} = 5.5 - s_3/2$$

$$Q_y^{(3)} = (5.5 - s_3/2)(0.1)(s_3) = 0.55s_3 - 0.05s_3^2$$

$$Q_z^{(3)} = (-5.21)(0.1)(s_3) = -0.521s_3$$

$$q^{(3)}(s_3) = q_1 - 118.89 - 29.16\left[0.55s_3 - 0.05s_3^2\right] - 6.073[-0.521s_3]$$

$$= q_1 - 118.89 - 12.87s_3 + 1.458s_3^2 \text{ lb/in.}$$

(12.25)

and at the end of part (3) we have $q^{(3)}(9) = q_1 - 116.62$ lb/in. Finally, for part (4), as shown in Fig. 12.12, we find

$$\bar{y} = -5.21 + s_4/2$$

$$\bar{z} = -3.5$$

$$Q_y^{(4)} = (-3.5)(0.1)(s_4) = -0.35s_4$$

$$Q_z^{(4)} = (-5.21 + s_4/2)(0.1)(s_4) = -0.521s_4 + 0.05s_4^2$$

$$q^{(4)}(s_4) = q_1 - 116.62 - 29.16[-0.35s_4] - 6.073\left[-0.521s_4 + 0.05s_4^2\right]$$

$$= q_1 - 116.62 - 13.37s_4 + 0.3036s_4^2 \text{ lb/in.}$$

(12.26)

We can check the consistency of these many calculations by examining the end value of the shear flow in part (4), where we find $q^{(4)}(12) \cong q_1$, which was our original starting value. Since we now have all the shear flows in the parts, we can calculate the total forces generated in each part approximately as

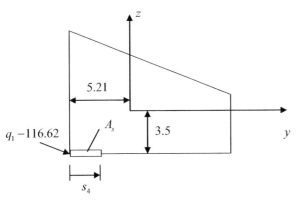

Figure 12.12 The geometry for part (4) of the cross-section.

$$F^{(1)} = \int_0^4 (q_1 + 6.082s_1 - 1.458s_1^2)ds_1 = 4q_1 + 17.55 \text{ lb}$$

$$F^{(2)} = \int_0^{13} (q_1 + 1 - 5.58s_2 - 0.2802s_2^2)ds_2 = 13q_1 - 663.7 \text{ lb}$$

$$F^{(3)} = \int_0^9 (q_1 - 118.89 - 12.87s_3 + 1.458s_3^2)ds_3 = 9q_1 - 1237 \text{ lb}$$

$$F^{(4)} = \int_0^{12} (q_1 - 116.62 + 13.37s_4 - 0.3036s_4^2)ds_4 = 12q_1 + 611.4 \text{ lb}$$

(12.27)

These forces are shown in Fig. 12.13 along with the original 1000 lb force they must produce. These force systems are equivalent so they must generate the same moment about any point (and the same force). Point A is a convenient point to take moments about so we find

$$(12q_1 - 611.4)(9) + (4q_1 + 17.55)(12) = (1000)(0)$$
$$\rightarrow q_1 = 33.92 \text{ lb/in.}$$

(12.28)

Then the shear flows are explicitly

$$q^{(1)}(s_1) = 33.9 + 6.08s_1 - 1.46s_1^2 \text{ lb/in.}$$
$$q^{(2)}(s_2) = 34.9 - 5.58s_2 - 0.208s_2^2 \text{ lb/in.}$$
$$q^{(3)}(s_3) = -85.0 - 12.9s_3 + 1.46s_3^2 \text{ lb/in.}$$
$$q^{(4)}(s_4) = -82.7 + 13.4s_4 - 0.304s_4^2 \text{ lb/in.}$$

(12.29)

These distributions are shown explicitly in Fig. 12.14.

Figure 12.13 The applied 1000 lb force and the forces generated from the shear flows.

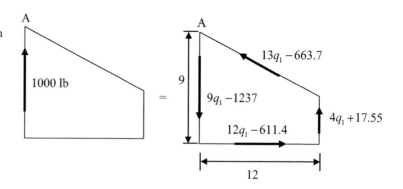

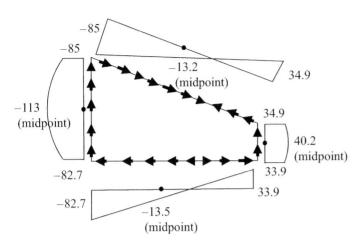

Figure 12.14 The shear-flow distributions in the box beam.

Now, to find the location of the shear center for the vertical force we need to set $\phi' = 0$, which gives

$$\phi' = \frac{1}{2G\Omega}\oint \frac{q}{t}ds = 0 \qquad (12.30)$$

but, since the thickness t is a constant, we have equivalently

$$\oint q\, ds = 0 \qquad (12.31)$$

which is just the sum of the forces we have already computed, giving

$$(4q_1 + 17.55) + (13q_1 - 663.7) + (9q_1 - 1237) + (12q_1 - 611.4) = 0$$
$$\rightarrow q_1 = 65.65 \text{ lb.in.} \qquad (12.32)$$

With this value of q_1 then in place of Fig. 12.13 we have the equivalent systems of Fig. 12.15. Again, equating moments about point A of these two systems we find

$$(4q_1 + 17.55)(12) + (12q_1 - 611.4)(9) = 1000 e_y$$
$$\rightarrow e_y = 4.95 \text{ in.} \qquad (12.33)$$

Having this result, we can then decompose our original combined bending and torsion problem into a purely bending problem where the shear force acts through the shear center plus the torsion produced by the moment of the force about the shear center. This decomposition is shown in Fig. 12.16 where we see that the clockwise torque about the shear center is just $T = 4950$ in.-lb. This torque will generate a constant primary shear flow throughout the cross-section given by (see Eq. (11.212)):

$$q = \frac{T}{\Omega} = \frac{-4950}{2[(4)(12) + (5)(12)/2]} = -31.7 \text{ lb/in.} \qquad (12.34)$$

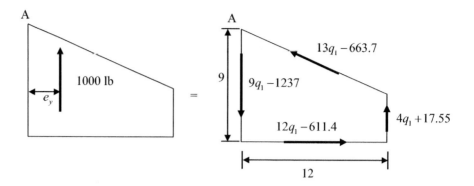

Figure 12.15 The applied 1000 lb force acting through the shear center and the forces generated in the cross-section by the shear flows.

Figure 12.16 The original loading which produces bending when the force acts through the shear center plus torsion of the section generated by the moment of the force acting about the shear center. The shear flows at the starting point of the integrations are also shown.

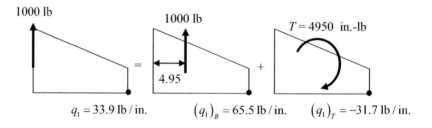

where the minus sign is present since we have been taking positive torques and shear flows as acting counterclockwise. This value of the shear flow is also the value of q_1 at the starting point of the integration due to torsion only, so we can see that the original constant value of q_1 for combined bending and torsion is the sum (approximately) of its value $(q_1)_B$ due to bending and a value $(q_1)_T$ due to torsion as shown in Fig. 12.16, in agreement with our previous discussions.

12.3 Twisting Induced by Axial Stresses

In Chapter 11 we saw that, for nonuniform torsion problems, the torsion of the section induced axial normal stresses. Is it possible to similarly induce torsion by the application of axial normal stresses? The answer is yes. Consider, for example, the cantilever I-beam shown in Fig. 12.17a. Because of the restraint of warping, we saw in Chapter 11 that normal stresses are induced in the flanges (Fig. 12.17b). These normal stresses in turn are statically

12.3 Twisting Induced by Axial Stresses

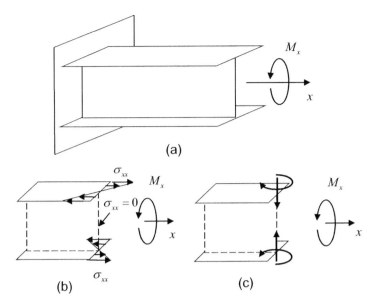

Figure 12.17 (a) Nonuniform torsion of a cantilever I-beam. (b) Normal stress distributions in the flanges. (c) The normal stresses are equivalent to a pair of moments called bimoments.

equivalent to a pair of moments as shown in Fig. 12.17c called, for obvious reasons, *bimoments*. Thus, if torsion can coexist with bimoments (which are statically equivalent to a set of self-equilibrated axial stress distributions), it follows that we could view this problem in reverse and consider the axial stress distributions which generate such bimoments as the applied loads that in turn induce the torsion present. The challenge remains, however, to describe in more general terms such self-equilibrated stresses, which we will now do.

To model such twisting induced by axial stresses we will consider a general restrained thin, open section, as shown in Fig. 12.18, subjected to axial stresses that we will assume produce both an axial displacement of the section, $u_0(x)$, and nonuniform warping. Then the axial displacement will be given by

$$u_x = -\omega_p(y,z)\phi'(x) + u_o(x) \tag{12.35}$$

where ω_P is the principal sectorial area function and ϕ' is the twist/unit length. Recall that the principal sectorial area function satisfies

$$\int_A \omega_p dA = 0$$

$$\int_A y\omega_p dA = 0 \tag{12.36}$$

$$\int_A z\omega_p dA = 0$$

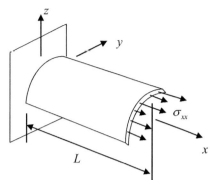

Figure 12.18 A restrained thin, open section subjected to an axial stress distribution.

If we assume σ_{xx} is the only significant normal stress, then

$$\sigma_{xx} = E\frac{du_x}{dx} = -E\omega_p\frac{d^2\phi}{dx^2} + E\frac{du_0}{dx} \quad (12.37)$$

where E is Young's modulus. If F_x is the axial load generated by the stress distribution, then

$$F_x = \int_A \sigma_{xx} dA = -E\frac{d^2\phi}{dx^2}\int_A \omega_p dA + EA\frac{du_0}{dx}$$

$$= EA\frac{du_0}{dx} \quad (12.38)$$

because the integral of the principal sectorial area function over the cross-section is zero (the first condition in Eq. (12.36)). Thus, Eq. (12.38) is just the ordinary relationship seen between the axial force and the axial displacement. Also note that the bending moments vanish since

$$M_y = \int_A y\sigma_{xx} dA = -E\frac{d^2\phi}{dx^2}\int_A y\omega_p dA + E\frac{du_0}{dx}\int_A y dA = 0$$

$$-M_z = \int_A z\sigma_{xx} dA = -E\frac{d^2\phi}{dx^2}\int_A z\omega_p dA + E\frac{du_0}{dx}\int_A z dA = 0 \quad (12.39)$$

if (y, z) are measured from the centroid of the cross-section. Thus, the axial stress of Eq. (12.37) does not generate any bending about the centroid, only axial extension/compression and warping.

The force, F_x, is the resultant that describes axial extension/compression of the cross-section and the moments (M_y, M_z) describe bending, so the question now is what type of resultant describes the warping and associated torsion of the cross-section? We will define such a resultant, $M_\Gamma(x)$, as

$$M_\Gamma = \int_A \sigma_{xx}\omega_p dA \quad (12.40)$$

which, because of the behavior seen in the I-beam discussed previously, we will call the *bimoment*. This definition makes sense because the bimoment is directly related to the twisting of the section, i.e.,

$$M_\Gamma(x) = -E\frac{d^2\phi}{dx^2}\int_A \omega_p^2 dA + E\frac{du_0}{dx}\int_A \omega_p dA \qquad (12.41)$$

$$= -EJ_\omega \frac{d^2\phi}{dx^2}$$

where J_ω is the warping constant. If we place Eq. (12.41) and the axial force relation of Eq. (12.38) into the expression for the axial stress, Eq. (12.37), we find

$$\sigma_{xx} = \frac{M_\Gamma(x)\omega_p(y,z)}{J_\omega} + \frac{F_x}{A} \qquad (12.42)$$

If the applied stress is purely uniform across the cross-section, as assumed in axial-load problems, then no twisting will be induced since

$$M_\Gamma = \int_A \sigma_{xx}\omega_p dA = \sigma_{xx}\int_A \omega_p dA = 0 \qquad (12.43)$$

However, other distributions that generate a bimoment as well as an axial force can produce both twisting and extension/compression. If a given stress distribution does generate a bimoment, then the equation for the twist becomes (see Eq. (11.135))

$$\frac{d^4\phi}{dx^4} - k^2 \frac{d\phi}{dx} = 0 \qquad (12.44)$$

since there is no applied moment, M_x, and where, recall (Eq. (11.136)),

$$k^2 = \frac{GJ_{eff}}{EJ_\psi} \qquad (12.45)$$

To solve Eq. (12.44) for the twist ϕ we need three boundary conditions. For the cantilever beam problem of Fig. 12.18, for example, we have at the wall ($x = 0$) the conditions

$$\phi(0) = 0, \quad \phi'(0) = 0 \qquad (12.46)$$

The other boundary condition comes from the loaded end ($x = L$), where we assume the stress distribution (and hence the bimoment) is known, and from Eq. (12.41) we have

$$\left.\frac{d^2\phi}{dx^2}\right|_{x=L} = -\frac{M_\Gamma(L)}{EJ_\omega} \qquad (12.47)$$

Example 12.3 Twisting of an Axially Loaded Member

Consider the cantilever unsymmetrical beam shown in Fig. 12.19, which is loaded by a concentrated force, $P = 100$ kN, acting at the centroid. We want to determine the twist,

Figure 12.19 (a) An unsymmetrical beam loaded by an axial load acting through the centroid of the cross-section. (b) the geometry of the cross-section where the thickness $t = 5$ mm is constant though out the cross-section.

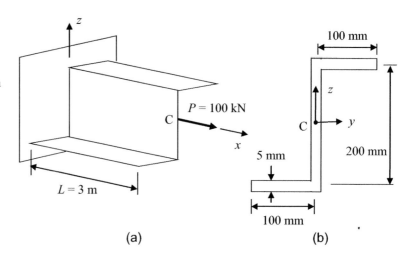

$\phi(x)$, in the beam and the axial stress distribution at the wall. In this case the centroid C is also at the shear center. We will take $E = 200$ GPa $= 2 \times 10^5$ N/mm², $G/E = 0.36$. To determine the principal sectorial area function, we only need to satisfy the first condition on that function in Eq. (12.36) if we take the pole at the shear center, C. Thus, let us take the initial integration point at A, where we let $\omega = \omega_0$ (see Fig. 12.20), and sweep out the sectorial area going from A to D as shown. Then in the three sections of the cross-section, we have

$$\omega = \omega_0 + hs_1/2 \qquad (0 \leq s_1 \leq d)$$
$$\omega = \omega_0 + hd/2 \qquad (0 \leq s_2 \leq h) \qquad (12.48)$$
$$\omega = (\omega_0 + hd/2) - hs_3/2 \quad (0 \leq s_3 \leq d)$$

Setting $\int_A \omega dA = 0$ gives

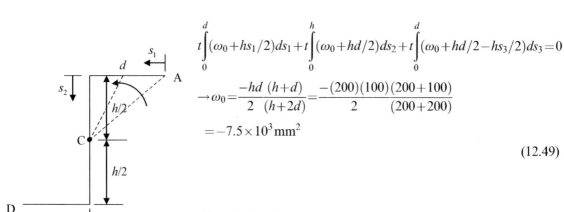

$$t\int_0^d (\omega_0 + hs_1/2)ds_1 + t\int_0^h (\omega_0 + hd/2)ds_2 + t\int_0^d (\omega_0 + hd/2 - hs_3/2)ds_3 = 0$$

$$\rightarrow \omega_0 = \frac{-hd\,(h+d)}{2\,(h+2d)} = \frac{-(200)(100)(200+100)}{2\,(200+200)}$$

$$= -7.5 \times 10^3 \text{ mm}^2$$

(12.49)

Figure 12.20 The geometry for determining the principal sectorial area function, where the centroid C is the shear center.

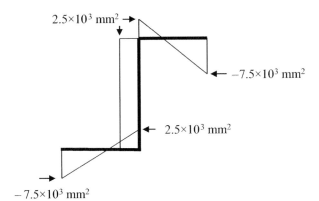

Figure 12.21 The principal sectorial area function.

which gives the principal sectorial area function shown in Fig. 12.21. It can then be shown that

$$J_{eff} = 2 \times \frac{1}{3} d\, t^3 + \frac{1}{3} h t^3 = 0.17 \times 10^5 \text{ mm}^4$$

$$J_\omega = \int_A \omega_p^2 dA = 2.08 \times 10^{10} \text{ mm}^6 \qquad (12.50)$$

$$k = \sqrt{\frac{G J_{eff}}{E J_\omega}} = 5.4 \times 10^{-4} \text{ mm}^{-1}$$

We now are in a position to calculate the twist in the beam. General solutions of the homogeneous equation seen in Eq. (12.44) and its derivatives are

$$\phi = A \cosh(kx) + B \sinh(kx) + C$$
$$\phi' = Ak \sinh(kx) + Bk \cosh(kx) \qquad (12.51)$$
$$\phi'' = Ak^2 \cosh(kx) + Bk^2 \sinh(kx)$$

where A, B, and C are constants. From $\phi'(0) = 0$ we find $B = 0$. From $\phi''(L) = -M_\Gamma(L)/EJ_\omega$ we then find $A = -M_\Gamma(L)/Ek^2 J_\omega \cosh(kL)$. finally, from $\phi(0) = 0$ we have $C = -A$, giving

$$\phi(x) = \frac{M_\Gamma(L)}{Ek^2 J_\omega \cosh(kL)} [1 - \cosh(kx)] \qquad (12.52)$$

In this case it is easy to calculate $M_\Gamma(L)$ since the force at $x = L$ is a concentrated force at the origin of the centroidal coordinates so that

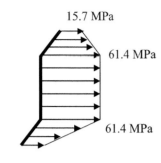

Figure 12.22 Axial stress distribution at the wall.

$$M_\Gamma(L) = \int_A \sigma_{xx}(y,z)\omega_p(y,z)\,dA$$

$$= \omega_p(0,0)\int_{A(0,0)} \sigma_{xx}\,dA \qquad (12.53)$$

$$= P\omega_p(0,0) = 2.5 \times 10^3 P \text{ N-mm}$$

Placing the value of P and the other values into Eq. (12.52) gives approximately

$$\phi(x) = 0.08\left[1 - \cosh\left(5.4 \times 10^{-4} x\right)\right] \text{rad} \qquad (12.54)$$

so at $x = L$ we have $\phi(L) = -0.13$ rad. At the fixed wall $x = 0$ we have

$$M_\Gamma(0) = -\frac{d^2\phi}{dx^2}(0) E J_\omega$$

$$= 95 \times 10^6 \text{ N-mm}^2 \qquad (12.55)$$

and the axial stress is, in terms of the sectorial area function,

$$\sigma_{xx} = \frac{P}{A} + \frac{M_\Gamma(0)\omega_p}{J_\omega} = \frac{10^5}{(400)(5)} + \frac{(95 \times 10^6)\omega_p}{2.08 \times 10^{10}} \qquad (12.56)$$

$$= 50 + 4.57 \times 10^{-3}\omega_p \text{ MPa}$$

which has the distribution at the wall as shown in Fig. 12.22. It can be seen that the torsional warping makes the stress distribution quite different from the constant $P/A = 50$ MPa value computed from ordinary axial-load theory even though the load acted through the centroid of the cross-section.

12.4 PROBLEMS

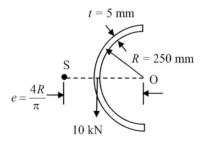

P12.1 A 5mm thick plate is formed into the semicircular shape of radius $R = 250$ mm as shown in Fig. P12.1. The shear center location for this section was obtained in

Figure P12.1 A 10 kN shear force acting on a thin, open semicircular cross-section.

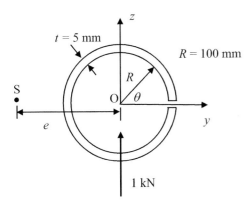

Figure P12.2 A 1 kN shear force acting on a thin, open circular cross-section.

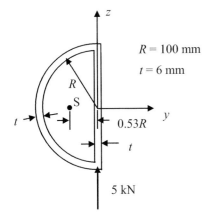

Figure P12.3 A 5 kN shear force acting on a closed semicircular section.

Chapter 10 (see Eq. (10.46)) to be at a distance $e = 4R/\pi$ from point O, as shown in the figure. Determine the maximum shear stress and its location in the cross-section if a 10 kN vertical shear force acts as shown in Fig. P12.1.

P12.2 Consider the thin, open circular section of Fig. P12.2. The thickness is $t = 5$ mm and the radius of the section is $R = 100$ mm. A 1 kN shear force acts through the centroid, O, of the cross-section. Determine the maximum shear stress and its location in the cross-section.

P12.3 The 100 mm radius closed semicircular section of Fig. P12.3 carries a 5 kN shear force acting on the straight vertical part of the section. The thickness, $t = 6$ mm, of the section is constant throughout the section. The shear center, S, is located as shown in Fig. P12.3. Determine the maximum shear stress and its location in the cross-section (neglect stress concentrations). Also, plot the shear stresses separately in the straight vertical part and the curved part.

References

1. V. Z. Vlasov, *Thin-Walled Elastic Beams*, 2nd edn (translated from Russian. Washington, DC: Clearinghouse for Federal and Technical Information, TT61–11400, 1959)
2. D. V. Wallerstein, *A Variational Approach to Structural Analysis* (New York, NY: John Wiley and Sons, 2002)

13 Material Failure and Stability

One of the primary reasons for doing stress analysis of structures is to ensure that those structures remain safe. Structures can fail in different ways, so one needs to examine a variety of failure modes. In this chapter we will consider (1) a number of the commonly used static failure theories, (2) fatigue failure under alternating loads, and (3) fracture theory. We will also briefly discuss how nondestructive inspections can be used in conjunction with crack growth laws to keep structures safe while in use. Another way that structures can fail is through a loss of stability. The sudden buckling of columns, also called a bifurcation type of instability, will be described as well as other types of instabilities such as limit-load instabilities and snap-through buckling instabilities.

13.1 Theories of Static Failure

The basic idea behind theories of static failure is that if nontime-varying stresses are too large, then a material will fail. To examine static failure, let us consider first the special case of plane stress (Fig. 13.1a). We can define what we mean by "too large" by plotting combinations of the two principal stresses and defining a safe region where any of those combinations do not lead to failure (Fig. 13.1b). If we can determine the boundary of such a safe region through particular failure theories and experiments, then we have a rational way to guarantee the safety of the structure. Let us now examine a number of static failure theories.

13.1.1 Maximum Normal-Stress Theory

Materials are often classified into two types: ductile and brittle. A ductile material under large loads exhibits significant permanent strains, called "slipping," when unloaded. Brittle materials, however, do not have this behavior over the full range of stresses all the way to ultimate fracture. Many steels, for example, exhibit ductile behavior, whereas materials such as cast iron are brittle. The *maximum normal-stress theory* for a brittle material says that fracture occurs if either $|\sigma_{p1}| = \sigma_f$ or $|\sigma_{p2}| = \sigma_f$, where $(\sigma_{p1}, \sigma_{p2})$ are the principal stresses and σ_f is a fracture stress as determined by pulling a specimen to failure in a uniaxial tension test (see Fig. 13.2a). A plot of the safe region for this theory is shown in Fig. 13.2b.

One of the limitations of this theory is that it does not distinguish between tensile or compressive stress behavior and materials do typically show different failure criterion in tension or compression. There is another theory, called the Coulomb–Mohr theory, that

13.1 Theories of Static Failure   497

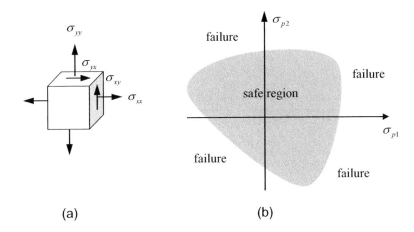

Figure 13.1 (a) A plane state of stress. (b) The definition of failure in terms of a region in the plane of principal stresses.

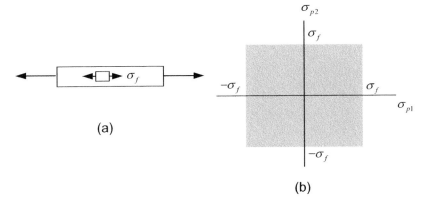

Figure 13.2 (a) A uniaxial tension test where the bar is loaded to the fracture stress, σ_f. (b) The safe region (shaded area) for the maximum normal-stress theory.

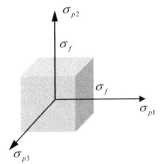

Figure 13.3 The safe region for a general state of stress according to the maximum normal-stress theory.

does take such tensile/compressive differences into account but we will not cover it here. This theory also does not depend on the orientation of the principal planes so it is only applicable to isotropic materials. The generalization of this theory to a 3-D state of stress is easy, where the corresponding safe region is the cube shown in Fig. 13.3.

13.1.2 Maximum Shearing-Stress Theory

For ductile materials the *maximum shearing-stress theory* (also called the *Tresca theory*) says that failure by slip (also called yielding) occurs when the maximum shearing stress, τ_{max}, exceeds the *yield shear stress*, τ_y, as determined experimentally in a uniaxial tension test, as shown in Fig. 13.4a. In uniaxial tension, this yield shear stress is $\tau_y = \sigma_y/2$, where σ_y is the

Figure 13.4 (a) A uniaxial tension test where the bar is loaded to the normal yield stress, σ_y, and the corresponding shear stress $\tau_y = \sigma_y/2$. (b) A stress–strain curve, where the yield stress, σ_y, occurs when, upon unloading, the material is left with a specified permanent strain, e_s, at zero stress.

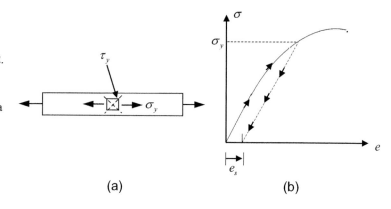

Figure 13.5 The safe region (shaded area) for no failure by slip according to the maximum shearing stress theory.

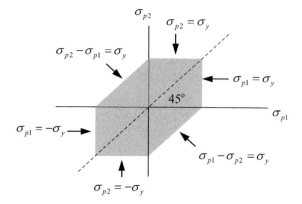

normal stress at yielding, and this yield shear stress occurs on planes at $\pm 45°$ to the x-axis, as shown in Fig. 13.4a. The yield stress, σ_y, however, is not the stress at which the bar fractures. Instead it defined as the normal stress that, when the bar is completely unloaded, results in a specified residual strain, e_s (also called a permanent slip), as shown in Fig. 13.4b. This permanent slip is why this type of failure is called failure by slip.

In a plane state of stress there are three principal stresses $(\sigma_{p1}, \sigma_{p2}, 0)$, and depending on their values, the maximum shear stress is the largest of the three extreme shear stresses given by

$$\tau_{max} = \max \begin{cases} |\sigma_{p1} - \sigma_{p2}|/2 \\ |\sigma_{p1} - 0|/2 \\ |\sigma_{p2} - 0|/2 \end{cases} = \max \begin{cases} |\sigma_{p1} - \sigma_{p2}|/2 \\ |\sigma_{p1}|/2 \\ |\sigma_{p2}|/2 \end{cases} \qquad (13.1)$$

The safe region $\tau_{max} \leq \tau_y = \sigma_y/2$ is thus the shaded hexagonal area shown in Fig. 13.5. Note that this theory also does not depend on the orientations of the planes of extreme shear so that it is only strictly applicable to isotropic materials.

13.1.3 Maximum Distortional Strain Energy Theory

Intuitively, we expect a material will fail if we place too much energy into it. Thus, we might expect that we can use the size of the strain energy density, u_0, as a failure criterion. However, as discussed in Chapter 8, experiments have shown that a pure hydrostatic pressure will not cause yielding, so failure due to slip must be independent of the hydrostatic part of u_0. By removing that hydrostatic part (see Chapter 8), we can define the distortional stain energy density, u_d, as

$$u_d = \frac{(1+v)}{6E}\left[(\sigma_{p1} - \sigma_{p2})^2 + (\sigma_{p1} - \sigma_{p3})^2 + (\sigma_{p2} - \sigma_{p3})^2\right] \tag{13.2}$$

The *maximum distortional strain energy theory* (also called the *von Mises failure theory*) predicts that failure with respect to slip (yielding) occurs when the distortional strain energy density is equal to its value, u_y, as determined in a uniaxial tension test where there is only one principal stress and that is equal to the normal stress at yield, σ_y, so that

$$u_y = \frac{(1+v)}{6E}\left[2\sigma_y^2\right] \tag{13.3}$$

Thus, the failure criterion $u_d = u_y$ gives

$$(\sigma_{p1} - \sigma_{p2})^2 + (\sigma_{p1} - \sigma_{p3})^2 + (\sigma_{p2} - \sigma_{p3})^2 = 2\sigma_y^2 \tag{13.4}$$

which for a state of plane stress reduces to

$$\sigma_{p1}^2 - \sigma_{p1}\sigma_{p2} + \sigma_{p2}^2 = \sigma_y^2 \tag{13.5}$$

Equation (13.5) represents the equation of a rotated ellipse, as shown in Fig. 13.6. The corresponding hexagonal shape from the maximum shearing stress theory is also shown in Fig. 13.6 so we see that there is little difference between the two theories.

In a general 3-D state of stress, we can also interpret the failure criterion in two other equivalent ways. Recall, in Chapter 2 we obtained the magnitude of the total shear stress on the octahedral plane as (see Eq. (2.82)):

$$|\tau_s|_{oct} = \frac{1}{3}\sqrt{(\sigma_{p1} - \sigma_{p2})^2 + (\sigma_{p1} - \sigma_{p3})^2 + (\sigma_{p3} - \sigma_{p2})^2} \tag{13.6}$$

In the uniaxial load case at the yield stress, σ_y, we have this shear stress on the octahedral plane, τ_y^{oct}, given by $\tau_y^{oct} = \sqrt{2}\sigma_y/3$. The failure criterion $|\tau_s|_{oct} = \tau_y^{oct}$ for the stress on the octahedral plane gives Eq. (13.4) so that it is

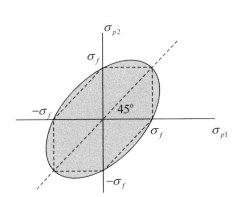

Figure 13.6 The safe region (shaded area) for no failure by slip according to the maximum distortional strain energy theory.

equivalent to the distortional strain energy criterion $u_d = u_y$. Alternatively, if we define the von Mises or effective stress, σ_v, as $\sigma_v = 3|\tau_s|_{oct}/\sqrt{2}$ (see Eq. (2.84)), then at the yield stress this effective stress is just σ_y, so the failure criterion $\sigma_v = \sigma_y$ is again equivalent to $u_d = u_y$. Thus, we can view the distortional strain energy criterion as equivalent to a criterion on either the octahedral stress or the effective stress.

In formulating this failure theory, we used generalized Hooke's law so that this theory as presented is applicable only to isotropic materials, but it can be generalized to anisotropic materials. The maximum distortional strain energy theory is less conservative than the maximum shear-stress theory, but generally there is little difference in their predictions of failure. Most experimental results tend to fall on or between these two theories.

For 3-D states of stress, the maximum distortional strain energy theory is given by Eq. (13.4) while the maximum shear-stress theory becomes

$$\tau_{\max} = \tau_y = \sigma_y/2 = \max \begin{cases} |\sigma_{p1} - \sigma_{p2}|/2 \\ |\sigma_{p1} - \sigma_{p3}|/2 \\ |\sigma_{p2} - \sigma_{p3}|/2 \end{cases} \tag{13.7}$$

Both Eq. (13.4) and Eq. (13.7) remain unchanged if we add equal stresses to all principal stress components, i.e., if we let

$$\sigma'_{p1} = \sigma_{p1} + \sigma$$
$$\sigma'_{p2} = \sigma_{p2} + \sigma$$
$$\sigma'_{p3} = \sigma_{p3} + \sigma$$

This is because both theories are independent of adding or subtracting a hydrostatic component. What this means is that in the 3-D case the failure (yield) surfaces are cylinders whose sides are parallel to a line that makes equal axes with all three principal stress directions, as seen in Fig. 13.7.

Figure 13.7 Failure surface for a 3-D state of stress using the maximum shearing stress (Tresca) theory or the maximum distortional strain energy (von Mises) theory.

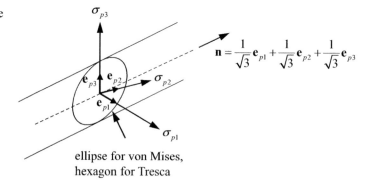

Example 13.1 Static Failure Criteria

A solid circular shaft having a 60 mm radius is loaded by both a 10 kN-m bending moment and a 12 kN-m torque. The isotropic material of the shaft has a measured yield stress of 220 MPa in uniaxial tension. What is the safety factor for failure by slip according to the maximum shearing-stress theory and the maximum distortional strain energy theory?

The largest flexure normal stress and torsional shear stress occur at the outer radius, c, of the shaft and are given by

$$\sigma_{xx} = \frac{Mc}{I} = \frac{(10 \times 10^6)(60)}{\pi(60)^4/4} = 58.9 \text{ MPa}$$

$$\tau = \frac{Tc}{J} = \frac{(12 \times 10^6)(60)}{\pi(60)^4/2} = 35.4 \text{ MPa}$$

At this outer radius we have a state of plane stress so principal stresses are given by (see Eq. (2.67))

$$\sigma_{p1}, \sigma_{p2} = \frac{\sigma_{xx} + \sigma_{yy}}{2} \pm \sqrt{\left(\frac{\sigma_{xx} - \sigma_{yy}}{2}\right)^2 + \sigma_{xy}^2} = \frac{58.9}{2} \pm \sqrt{\left(\frac{58.9}{2}\right)^2 + (35.4)^2}$$

$$\sigma_{p3} = 0$$

which gives

$$\sigma_{p1} = 75.5 \text{ MPa}, \sigma_{p2} = -16.6 \text{ MPa}, \sigma_{p3} = 0$$

If we examine the Mohr circles, we see the maximum shear stress is just

$$\tau_{max} = \frac{\sigma_{p1} - \sigma_{p2}}{2} = 46.05 \text{ MPa}$$

Thus, according to the maximum shear-stress theory the safety factor, SF, is just

$$SF = \frac{\tau_y}{\tau_{max}} = \frac{\sigma_y/2}{\tau_{max}} = \frac{110}{46.05} = 2.4$$

In contrast, for the distortional strain energy theory, written in terms of the effective stress:

$$SF = \frac{\sigma_y}{\sigma_{eff}} = \frac{\sqrt{2}\sigma_y}{\sqrt{(\sigma_{p2} - \sigma_{p1})^2 + (\sigma_{p2} - \sigma_{p3})^2 + (\sigma_{p1} - \sigma_{p3})^2}}$$

$$= \frac{\sqrt{2}(220)}{\sqrt{(92.1)^2 + (16.6)^2 + (75.5)^2}} = 2.6$$

13.2 Fatigue Failure

The failure theories of the previous section are appropriate for structures under static loads. However, many structures are subjected to significant time-varying loads so we must develop equivalent failure theories for those cases. In cases such as reciprocating machinery, for example, the time varying loads are sinusoidal. Under such repetitive loading, small imperfections in the material or on the surface, for example, develop into microscopic cracks, which grow over time into larger cracks until eventually a crack of "critical" size is produced that grows rapidly, causing the material to fracture. This type of process is called *fatigue failure*. As you might expect, fatigue failure is in detail a very complex process so that one cannot develop relatively simple failure criterion as found in the case of static failure. Thus, in fatigue one relies heavily on the use of experiments. Figure 13.8a shows a typical experimental setup that is used to generate fatigue failure data. A cantilever beam of circular cross section is forced to rotate while being loaded by an end force, P through a fixed surrounding sleeve. In this arrangement, the extreme flexure stresses at the top and bottom of the beam go through alternating (i.e., completely reversed) tensile and compressive values as a function of time, as seen in Fig. 13.8b, where σ_a is the magnitude of this alternating stress. The beam is allowed to remain in this setup until the beam ultimately fails at a specific number of cycles, N_1. This whole process is then repeated with a new sample at a new value of P (and, hence, σ_a) resulting in a different number of cycles to failure, N_2. A collection of such experiments then allows us to generate an *S–N curve*, where the alternating stress, σ_a (which is normally labeled as S), is plotted versus the number of cycles to failure, N_f, as seen in Fig. 13.9. Some materials exhibit a stress level, σ_e, called the endurance limit, below which the life of the part is essentially infinite. If such asymptotic behavior is not present, then the endurance limit is usually defined as the stress at a very large number of cycles such as $N_f = 10^8$. The S–N curve is normally generated with a purely reversed alternating stress, σ_a, but in practice a component will often also see a static (mean) stress, σ_m, as seen in Fig. 13.10, where the mean stress, alternating stress, and stress range, $\Delta\sigma$, are defined in terms of the maximum and minimum stresses experienced. Combinations of mean and alternating stresses are often defined in terms of the stress ratio, $R = \sigma_{min}/\sigma_{max}$, and the amplitude ratio, $A = \sigma_a/\sigma_m$. Fig. 13.11b shows a few combinations of (R, A) values that are common. If one measures the number of cycles to failure at various combinations of such

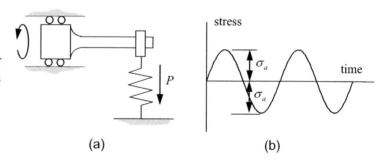

Figure 13.8 (a) Generating a sinusoidal state of stress versus time by rotating a cantilever specimen with a static end load, P. (b) A plot of the stress versus time where the stress is an alternating (i.e., completely reversed) stress, σ_a.

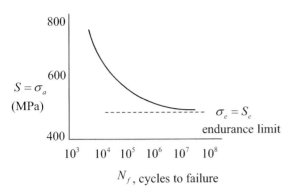

Figure 13.9 A typical S–N curve for fatigue failure.

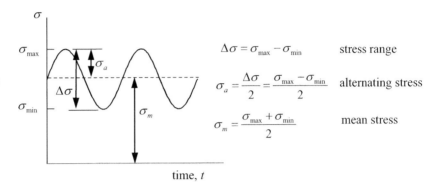

Figure 13.10 A more general time-varying harmonic stress that includes a mean (static) stress component as well as an alternating component.

$\Delta\sigma = \sigma_{max} - \sigma_{min}$ stress range

$\sigma_a = \dfrac{\Delta\sigma}{2} = \dfrac{\sigma_{max} - \sigma_{min}}{2}$ alternating stress

$\sigma_m = \dfrac{\sigma_{max} + \sigma_{min}}{2}$ mean stress

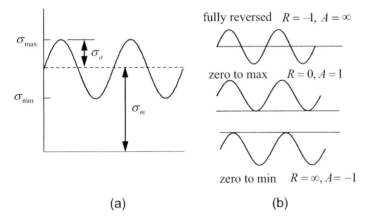

Figure 13.11 (a) A loading with mean stress, σ_m, and alternating stress, σ_a, values. (b) Some specific combinations of (σ_m, σ_a) as defined in terms of the stress ratio, $R = (\sigma_{min}/\sigma_{max})$, and amplitude ratio, $A = \sigma_a/\sigma_m$.

mean and alternating stresses then a sequence of S–N curves can be generated, as shown in Fig. 13.12a. At a given constant life (i.e., constant number of cycles to failure) then one can generate the *constant life curves* of Fig. 13.12b from the data of Fig. 13.12a. Generating such constant life curves experimentally in this fashion is very time consuming and costly so instead one normally uses an empirical constant life curve that has the general behavior seen in Fig. 13.12b. There are a number of different empirical curves that have been used.

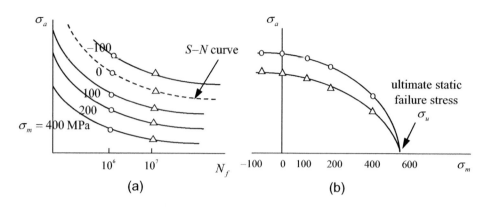

Figure 13.12 (a) S–N curves generated for different mean stresses. (b) Constant life curves of alternating versus mean stress values.

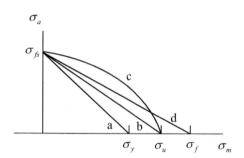

Figure 13.13 Empirical constant life curves: (a) Soderberg, (b) Goodman, (c) Gerber, and (d) Morrow.

The most common of those curves are shown in Fig. 13.13. The names of their authors (as well as the country of origin and the year they were introduced) and the behavior of these curves are summarized as follows:

$$\text{Soderberg (USA, 1930)} \quad \frac{\sigma_a}{\sigma_{fs}} + \frac{\sigma_m}{\sigma_y} = 1 \qquad (13.8a)$$

$$\text{Goodman (England, 1899)} \quad \frac{\sigma_a}{\sigma_{fs}} + \frac{\sigma_m}{\sigma_u} = 1 \qquad (13.8b)$$

$$\text{Gerber (Germany, 1874)} \quad \frac{\sigma_a}{\sigma_{fs}} + \left(\frac{\sigma_m}{\sigma_u}\right)^2 = 1 \qquad (13.8c)$$

$$\text{Morrow (USA, 1960s)} \quad \frac{\sigma_a}{\sigma_{fs}} + \frac{\sigma_m}{\sigma_f} = 1 \qquad (13.8d)$$

As can be seen from Fig. 13.13, all these curves are defined by their limiting values at $\sigma_m = 0$ and $\sigma_a = 0$. The stress σ_{fs}, which is called the *fatigue strength*, is common to all these curves. It is the purely alternating stress at failure for a specified lifetime of N_f cycles. The stress σ_y is the static yield stress while the stress σ_u is the ultimate static stress at fracture. The stress σ_f is the so-called true stress at fracture. Unlike the ultimate stress, which is given by the ratio of the failure load at fracture, $P_{failure}$, to the original cross-sectional area, A_0, of the specimen, i.e., $\sigma_u = P_{failure}/A_0$, the true fracture stress, σ_f, is the ratio of the failure load to the actual cross-sectional area, A, at failure, i.e., $\sigma_f = P_{failure}/A$, so that it accounts for any significant

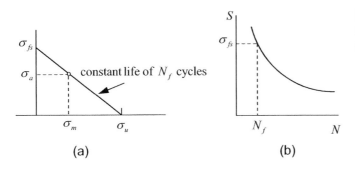

Figure 13.14 A finite life design based on (a) the Goodman line and (b) the S–N curve for the material.

reduction in that area at large strains. Generally, most test data tend to fall between the Goodman line and the Gerber curve. The Soderberg line is very conservative. If the peak stresses used in these fatigue calculations are based on stresses at locations where a stress concentration factor, K_t, is present, one should be aware that the use of such stress concentration factors produces overly conservative results so that in cyclic loading the K_t value is often replaced by a modified fatigue stress concentration factor, $K_f < K_t$ that is based on experimental data.

If one wants to do finite fatigue life design based on these fatigue concepts, there are basically three steps.

(1) A given combination of mean and alternating stresses is taken to lie on a constant life curve such as, for example, the Goodman line (Fig. 13.14a).
(2) The equation of that line is then used to solve for the fatigue strength, σ_{fs}. In the Goodman case we can have

$$\frac{\sigma_a}{\sigma_{fs}} + \frac{\sigma_m}{\sigma_u} = 1 \tag{13.9}$$

which yields

$$\sigma_{fs} = \frac{\sigma_a}{1 - \sigma_m/\sigma_u} \tag{13.10}$$

(3) Using this fatigue strength, the lifetime, N_f, for this combination of mean and alternating stresses is found from the S–N diagram for the given material (Fig. 13.14b).

Even though these three steps are straightforward, note that we must get values for (σ_a, σ_m) and it is not at all clear how to do this if a component is in a general 3-D state of stress. There are a number of possibilities, such as using the largest tensile principal stress behavior or using the effective (i.e., von Mises) stress to calculate these mean and alternating stress values, but there is no fundamental reason for making those choices. One can examine a number of the possible ways to calculate these (σ_a, σ_m) values and evaluate the sensitivity of the fatigue life results to those choices, as we will show in the next example. We will not go into further details here but there is a wide literature on fatigue that can be consulted to help make judgements in specific cases (see [1]–[3]).

Example 13.2 Fatigue Life Design

A closed cylindrical pressure vessel has a mean radius $r_m = 60$ mm and a wall thickness of $t = 4$ mm. The vessel is subjected to an internal pressure that repeatedly cycles between zero and 25 MPa. How many pressure cycles can be applied before fatigue failure is expected? For this multiaxial fatigue problem, use a Goodman line for the constant life curve and compute the mean and alternating stresses using the behavior of the largest tensile principal stress. Take the yield stress (in tension) of the material of the cylinder to be 320 MPa and the ultimate stress (in tension) to be 476 MPa. Assume the S–N curve can be written as

$$S = AN^{-B} \tag{13.11}$$

where S is the alternating stress in MPa, N is the number of cycles to failure, and $A = 839$, $B = 0.102$ for the material of the vessel. How do your answers change if you use the effective (von Mises) stress to calculate mean and alternating stress values?

The hoop stress in the cylinder is $\sigma_h = pr_m/t$ and the axial stress is $\sigma_x = pr_m/2t$ (see Chapter 7). Both of these are principal stresses, so the hoop stress is the largest, varying from zero to a maximum value $\sigma_h = (25)(60)/4 = 375$ MPa. This is an alternating load with $R = 0$, $A = 1$ (see Fig. 13.11b), so we have $\sigma_m = \sigma_a = 187.5$ MPa. Placing these values and the ultimate stress into the Goodman line, we have the fatigue strength

$$\sigma_{fs} = \frac{\sigma_a}{1 - \sigma_m/\sigma_u} = \frac{187.5}{1 - 187.5/476} = 309.4 \text{ MPa}$$

If we place this value in the S–N curve, we can solve for the number of cycles as

$$N_f = \left(\frac{\sigma_{fs}}{A}\right)^{-1/B} = \left(\frac{309.4}{839}\right)^{-9.804} = 1.77 \times 10^4 \text{ cycles}$$

If we had used the effective stress to calculate the mean and alternating values, then letting $\sigma_{p1} = \sigma_x, \sigma_{p2} = \sigma_h = 2\sigma_x, \sigma_{p3} = 0$ we have for the effective stress at the largest pressure

$$\sigma_{eff} = \frac{1}{\sqrt{2}}\sqrt{(\sigma_{p2} - \sigma_{p1})^2 + (\sigma_{p2} - \sigma_{p3})^2 + (\sigma_{p1} - \sigma_{p3})^2}$$

$$= \frac{1}{\sqrt{2}}\sqrt{\sigma_x^2 + 4\sigma_x^2 + \sigma_x^2} = \sqrt{3}\sigma_x = \sqrt{3}(25)(60)/8$$

$$= 324.8 \text{ MPa}$$

so that $\sigma_m = \sigma_a = 162.4$ MPa, which gives a fatigue strength of

$$\sigma_{fs} = \frac{\sigma_a}{1 - \sigma_m/\sigma_u} = \frac{162.4}{1 - 162.4/476} = 246.5 \text{ MPa}$$

and from the S–N curve

$$N_f = \left(\frac{\sigma_{fs}}{A}\right)^{-1/B} = \left(\frac{246.5}{839}\right)^{-9.804} = 1.64 \times 10^5 \text{ cycles}$$

which is about nine times the number of cycles based on the largest principal stress.

In practice, a structure typically will see different numbers of cycles at varying stress levels over its lifetime. To handle this case, consider the following general model where we suppose that a body can tolerate only a certain amount of total damage, D, before failure. If that total damage comes from damages, $D_i = (i = 1, 2, \ldots, N)$ arising from M sources, then we might expect that failure will occur if

$$\sum_{i=1}^{M} D_i = D \tag{13.12}$$

or, equivalently,

$$\sum_{i=1}^{M} \frac{D_i}{D} = 1 \tag{13.13}$$

defines failure, where D_i/D is the fractional damage from the ith source. We can use this linear damage concept in a fatigue setting by considering the situation where a component is subjected to n_1 cycles at alternating stress σ_1, n_2 cycles at stress σ_2, ..., n_M cycles at σ_M. From the S–N curve for this material, we can find the number of cycles to failure, N_1, at σ_1, N_2 at σ_2, ..., N_M at σ_M. It is reasonable in this case to let the fractional damage at stress level σ_i be $D_i = n_i/N_i$. This leads to the *Palmgren–Miner rule*, which says that fatigue failure occurs when

$$\sum_{i=1}^{M} \frac{n_i}{N_i} = 1 \tag{13.14}$$

Example 13.3 Palmgren–Miner Rule

Consider a part in a fatigue environment where 10 percent of its life is spent at an alternating stress level, σ_1, 30 percent is spent at level σ_2, and 60 percent at a level σ_3. How many cycles, n, can the part undergo before failure?

If, from the S–N diagram for this material, the number of cycles to failure at σ_i is N_i ($i = 1, 2, 3$), then from the Palmgren–Miner rule failure occurs when

$$\frac{0.1n}{N_1} + \frac{0.3n}{N_2} + \frac{0.6n}{N_3} = 1 \tag{13.15}$$

so solving for n gives

$$n = \frac{1}{\left(\dfrac{0.1}{N_1} + \dfrac{0.3}{N_2} + \dfrac{0.6}{N_3}\right)} \tag{13.16}$$

In evaluating the Palmgren–Miner rule experimentally, in "High–Low" fatigue tests where testing occurs sequentially at two stress levels (σ_1, σ_2), and where $\sigma_1 > \sigma_2$, it is found that failure occurs when

$$\sum_{i=1}^{2} \frac{n_i}{N_i} = c$$

where c normally is less than one, i.e., the Palmgren–Miner rule is nonconservative. For "Low–High" tests, on the other hand, c values are generally greater than one. For tests with random loading histories at several stress levels (rather than the cyclical histories normally considered), the Palmgren–Miner rule is very good.

13.3 Fracture Mechanics

In fatigue failure the ultimate cause of failure is the growth of cracks until a crack of a critical size is reached that leads to rapid fracturing of the component. Thus, an analysis of the behavior of cracks is essential for a more complete understanding of failure. *Fracture mechanics* (see [4]–[9]) deals with the stresses that surround cracks and the stress criteria for failure due to catastrophic crack growth. In Chapter 7 we showed that if the material behaves elastically, the stress field around cracks is singular near the crack tip. In practice, stresses cannot be infinite since the material will yield at large stress, creating a region near the tip where the material behaves in an inelastic manner. However, it has been found that as long as this region is small, the assumption that the material remains elastic everywhere allows one to analyze the fracture process with a purely elastic model.

Consider, for example, a thin plate of width $2b$ that is uniformly loaded at its ends and contains a centrally located crack of length $2a$, as shown in Fig. 13.15a. When the material behavior in the plate is purely elastic, the stresses near a crack tip (see Fig. 13.15b) can be written in terms of r and θ as

$$\sigma_{ij} \cong \frac{K_I G_{ij}(\theta)}{\sqrt{2\pi r}} \tag{13.17}$$

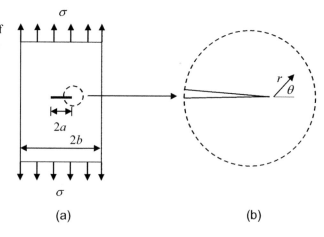

Figure 13.15 (a) A stressed plate containing a central crack of length $2a$. (b) A magnified view of the right end of the crack where the origin of the polar coordinates (r, θ) is at the crack tip.

where K_I is called the *stress intensity factor* for the crack in a mode I opening mode (i.e., when the loading tends to open the crack as shown schematically in Fig. 13.15b, and $G_{ij}(\theta)$ is a non dimensional angular factor that depends on the specific loading and geometry. The dimension of a stress intensity factor is *stress√length*. Stress intensity factors are written in the form

$$K = f(g,a)\sigma\sqrt{\pi a} \tag{13.18}$$

where σ is a reference stress value and a is a reference crack length. The function $f(g,a)$ is a nondimensional *configuration factor* that depends on the reference crack length a and the geometry, g, of the component (g can be a complex function of more than one variable but it is simply designated here by a single symbol). Stress intensity configuration factors have been tabulated for many different geometries. Figures 13.16 and 13.17 show examples of a few cases. Note that in contrast to the central crack of Fig. 13.16a, where a is the half-crack length, for an edge crack (Fig. 13.16b) the length a is taken to be the full length of the crack. For the rectangular beam in Fig. 13.17b, the crack is on the bottom side of the beam because it is the tensile stress on this side that will tend to open the crack and cause it to grow. In contrast, compressive stresses will tend to close a crack and prevent further growth. The opening mode I stress intensity factor, K_I, is the most common stress intensity factor studied since cracks often tend to grow in a tensile stress field with this type of mode. However, a crack can also be subjected to a shearing or tearing type of deformation (see Fig. 13.18) and there are separate stress intensity factors (K_{II}, K_{III}) associated with those cases. Here we will focus only on type I mode opening stress intensity factors.

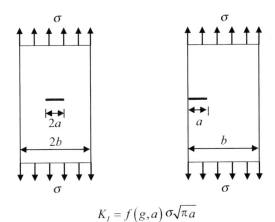

Figure 13.16 Stress intensity factors for uniaxial loading of (a) a centrally cracked plate, and (b) a plate with an edge crack.

$$K_I = f(g,a)\sigma\sqrt{\pi a}$$

(a) $f(g,a) = \sqrt{\sec\left(\dfrac{\pi a}{2b}\right)}$
$\cong 1 \ (b \gg a)$

(b) $f(g,a) = 1.12 - 0.231(a/b) + 10.55(a/b)^2$
$- 21.72(a/b)^3 + 30.39(a/b)^4$

Figure 13.17 Stress intensity factors for (a) uniaxial loading of a plate with cracks on both edges, and (b) for a rectangular beam under a pure bending moment with an edge crack on the tensile stress side of the beam.

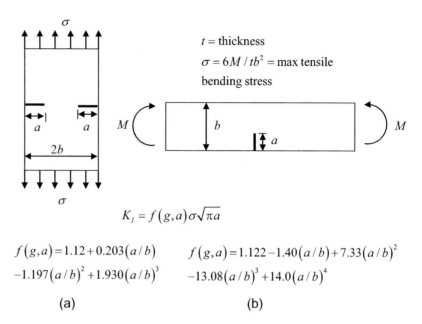

$$K_I = f(g,a)\sigma\sqrt{\pi a}$$

$f(g,a) = 1.12 + 0.203(a/b)$
$-1.197(a/b)^2 + 1.930(a/b)^3$

(a)

$f(g,a) = 1.122 - 1.40(a/b) + 7.33(a/b)^2$
$-13.08(a/b)^3 + 14.0(a/b)^4$

(b)

Figure 13.18 The three types of stress intensity factors.

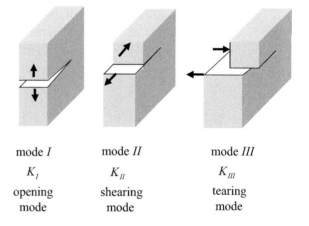

All of the examples shown so far have been for 2-D through-thickness cracks, but one can also examine the stress intensity factor for 3-D cases such as a penny-shaped crack in a block (Fig. 13.19a) or a thumbnail semicircular crack on the edge of a block (Fig. 13.19b). Stress intensity factors for complex geometries and loadings must typically be obtained numerically.

The reason that the stress intensity factors are needed is because it has been found that the rapid growth of a crack to failure occurs when the stress intensity factor exceeds a certain value, called the *fracture toughness*, K_C. The fracture toughness is a function of the specimen thickness, temperature, environmental conditions, etc., so that tests to obtain this toughness

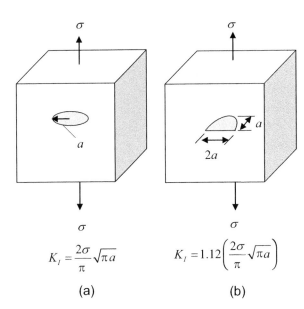

Figure 13.19 Stress intensity factors for (a) a circular crack under a state of uniaxial tension, and for (b) a semicircular crack under a state of uniaxial tension.

must be done under conditions that simulate the actual situations found in practice. As the test specimen thickness increases, the fracture toughness decreases until it reaches a constant value called the plane strain fracture toughness. For a crack in opening mode I, the plane strain fracture toughness is usually written as K_{IC}. The size of the crack at the plane strain toughness is called the *critical flaw size*, a_c, where

$$K_{IC} = f(g, a_c)\sigma\sqrt{\pi a_c} \tag{13.19}$$

Example 13.4 Estimating Critical Flaw Size

If one knows both the yield strength and plane strain fracture toughness of a material, then you can obtain a rough estimate of the critical flaw size. For example, consider a 4340 steel (an airframe material) and a A533B steel (a nuclear pressure vessel steel). For the 4340 steel, the yield stress is $\sigma_Y = 240$ ksi and the fracture toughness is $K_{IC} = 50$ ksi $\sqrt{\text{in.}}$, while for the A533B steel $\sigma_Y = 60$ ksi and $K_{IC} = 140$ ksi$\sqrt{\text{in}}$. Consider the case where we have a small half-penny-shaped crack on the stress-free surface. If the crack is stressed in an opening mode then

$$K_I = 1.12\left(\frac{2\sigma}{\pi}\right)\sqrt{\pi a} \tag{13.20}$$

Fracture occurs when $K_I = K_{IC}$, $a = a_c$, so under these conditions we can solve Eq. (13.20) for the critical flaw size. If we also take the reference stress to be the yield stress (which admittedly is a very large value but here we are just doing a rough comparative estimate), then we find

$$a_c = 0.626\left(\frac{K_{IC}}{\sigma_{yield}}\right)^2 \tag{13.21}$$

For the 4340 steel, Eq. (13.21) gives $a_c = 0.03$ in., while for the A533B steel $a_c = 3.4$ in. Clearly, the critical flaw size can vary significantly with the choice of material. If one inspects a material or structure with a nondestructive evaluation (NDE) test such as ultrasound, for example, in order to ensure that cracks have not grown to near a critical length, the critical flaw size determines the sensitivity required in the inspection.

Fracture toughness is generally obtained experimentally with standard specimens such as a beam or uniaxial tension member with an edge crack at the root of a notch. For plate geometries such as seen in Fig. 13.16 and Fig. 13.17 consistent results can be obtained with such tabulated values as long as the plate thickness, t, and the crack length, a, satisfy [9]:

$$t, a \geq 2.5\left(\frac{K_{IC}}{\sigma_{yield}}\right)^2 \tag{13.22}$$

If Eq. (13.22) is not satisfied, then predicting failure on the criterion $K_I < K_{IC}$ is conservative and an elastic–plastic fracture analysis can lead to a more efficient design.

In order to estimate the remaining safe life in a material undergoing a fatigue environment, it is necessary to know the rate at which cracks are growing in the material. If the minimum and maximum stresses are $(\sigma_{min}, \sigma_{max})$ then the corresponding stress intensity factor will change from K_{min} to K_{max} where

$$K_{max,\,min} = f(g, a)\sigma_{max,\,min}\sqrt{\pi a} \tag{13.23}$$

A commonly used crack growth law is the *Paris law* (also called the Paris–Erdogan law):

$$\frac{da}{dN} = C\Delta K^m, \quad \Delta K = K_{max} - K_{min} \tag{13.24}$$

where C, m are material and environmental dependent constants, and $a = a(N)$ is the crack size as a function of N, the number of cycles of loading. On a log–log plot (see Fig. 13.20), this crack growth law is linear, a behavior that is often found in crack growth studies except at early stages of growth where there can be a threshold value of ΔK below which there is no crack growth and the curve bends down, or at late stages of growth near the critical flaw size, where the crack growth becomes quite rapid and the curve bends up. In terms of stress the Paris law can be written as

$$\frac{da}{dN} = C\left[\sqrt{\pi a}f(g, a)\Delta\sigma\right]^m \quad \Delta\sigma = \sigma_{max} - \sigma_{min} \tag{13.25}$$

which can be rearranged and integrated to yield

$$\Delta N = \int_{N_1}^{N_c} dN = \int_{a_1}^{a_c} \frac{da}{C\left[\sqrt{\pi a}f(g, a)\Delta\sigma\right]^m} \tag{13.26}$$

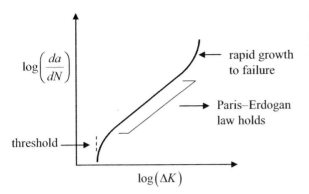

Figure 13.20 Crack growth on a log–log plot, showing a linear region where the Paris–Erdogan law holds.

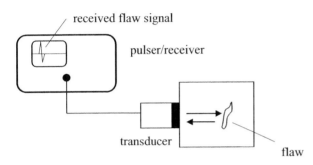

Figure 13.21 An ultrasonic NDE inspection setup for examining components for flaws.

where a_1 is the crack size at N_1 cycles, a_c is the critical flaw size at failure after N_c cycles, and $\Delta N = N_c - N_1$ is the number of cycles left to failure. Thus, if we know (1) the configuration factor from fracture mechanics studies, (2) the critical flaw size from stress analyses and material tests, (3) the range of stresses, $\Delta \sigma$, experienced by the component, and (4) the measured value of the crack size a_1 at N_1 cycles from a nondestructive evaluation (NDE) inspection of the component, then we can estimate the remaining life ΔN.

There are various NDE inspection methods in use. Three of the most common inspection methods are ultrasonics, eddy currents, and X-rays. Other methods include dye penetrants, magnetic particle, thermography, and more. In the case of an ultrasonic inspection, a transducer is driven by short voltage pulses from a pulser/receiver (see Fig. 13.21). The transducer is normally made from a piezoelectric crystal that converts these voltage pulses into motion of the crystal which is then transmitted into the component being inspected as elastic waves. The waves interact with any flaws, such as cracks, present in the component, which reflect those waves back to the transducer. The piezoelectric transducer converts the motion of the crystal back into received voltage pulses, which are sent to the pulser/receiver to be amplified and displayed. In the case of eddy current inspection (Fig. 13.22), an alternating current and voltage are used to drive a coil, which is then placed on or near the surface of an electrically conducting

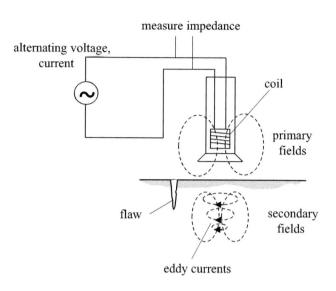

Figure 13.22 An eddy current NDE inspection setup for evaluating surface-breaking flaws.

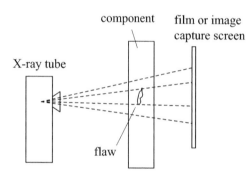

Figure 13.23 An NDE X-ray inspection setup.

component. The time-varying fields surrounding the coil, called the primary fields, induce secondary fields in the component and corresponding "eddies" of currents as shown schematically in Fig. 13.22. If a surface crack is present, it interrupts the flow of these eddy currents and causes a change of the coupled coil-component electrical impedance that can be measured. For X-ray inspections, a stream of electrons is generated in an X-ray tube that then strikes a target, producing X-rays that pass out through a window in the tube. These X-rays interact with a flawed component, producing an X-ray image of the flaw on a piece of film or a screen (Fig. 13.23) due to changes in the absorption of the X-rays caused by the presence of the flaw. Both ultrasonic and eddy current inspections are indirect measurements since the signals they produce must be related to the nature and size of the flaws present. X-ray inspections are more direct since they produce images of the flaws present. There is a radiation hazard associated with X-ray inspections that is not present in either ultrasonic or eddy current inspections, so this can limit the use of this method. Eddy current inspections also are limited since they require the component being inspected to be electrically conducting and only surface or near-surface flaws can be examined due to the limited depth of penetration of the eddy currents below the surface. Ultrasonic inspections have few material constraints and can inspect components for both surface-breaking and interior flaws so that ultrasonic NDE is one of the most commonly used inspection methods. For more details on NDE inspections, see [10], [11], and [12].

13.4 Stability

The previous sections have analyzed how structures fail either due to excessive static loads or due to fracture or fatigue. Another type of failure is when a structure loses stability and can suddenly experience unwanted displacements, and, in some cases, collapse. Most elementary strength of materials texts discuss this type of loss of stability by considering the simply supported member shown in Fig. 13.24a that is placed under a uniaxial load, P. As the load increases, the member compresses without any deflection in the z-direction, but at a certain critical load, P_{cr}, the member develops a sudden z-displacement as illustrated by the dashed line in Fig. 13.24a. The appearance of this sudden lateral deflection is called *buckling*. As can be seen from Fig. 13.24a, when buckling occurs the member acts as a deflected beam where the bending is coupled to the axial load. To examine this type of behavior in general, consider a small element of a deflected beam as shown in Fig. 13.24b. Although the axial load is constant, the shear force and bending moment can vary in the beam, as shown, and we have also allowed for the possibility of a distributed load/unit length, $q_z(x)$. From vertical force equilibrium we find

$$\left(V_z + \frac{dV_z}{dx}dx\right) - V_z + q_z dx = 0$$
$$\rightarrow \frac{dV_z}{dx} = -q_z$$
(13.27)

which is the same relation as found in an ordinary beam problem. From moment equilibrium, however, we have, summing moments about point O, as shown,

$$\left(M_y + \frac{dM_y}{dx}dx\right) - M_y - \left(V_z + \frac{dV_z}{dx}dx\right)dx - P(dw) + q_z dx\left(\frac{dx}{2}\right) = 0$$
$$\rightarrow \frac{dM_y}{dx} = V_z + P\frac{dw}{dx}$$
(13.28)

where we have neglected all the higher-order terms in Eq. (13.28) since they vanish as dx goes to zero. Equation (13.28) show the coupling between the bending moment and the axial load in the buckled beam. Differentiating Eq. (13.28) and combining it with Eq. (13.27) we find

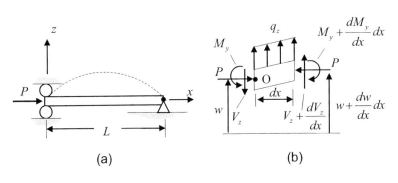

Figure 13.24 (a) A simply supported member under a uniaxial load. (b) A free body diagram of a deflected member.

$$\frac{d^2 M_y}{dx^2} = -q_z + P\frac{d^2 w}{dx^2} \tag{13.29}$$

From the moment–curvature relation

$$M_y = -EI\frac{d^2 w}{dx^2} \tag{13.30}$$

Placing this relation into Eq. (13.29) gives the differential equation

$$\frac{d^4 w}{dx^4} + k^2 \frac{d^2 w}{dx^2} = \frac{q_z}{EI} \tag{13.31}$$

where $k^2 = P/EI$. If we consider cases where the distributed load is absent, Eq. (13.31) is a homogeneous fourth order equation. Its general solution is

$$w(x) = A\cos(kx) + B\sin(kx) + Cx + D \tag{13.32}$$

where the constants (A, B, C, D) are determined from the boundary conditions.

Let's consider the simply supported case of Fig. 13.24a where these boundary conditions are

$$\begin{aligned} M_y(0) &= 0 \rightarrow \left. d^2 w/dx^2 \right|_{x=0} = 0 \\ w(0) &= 0 \\ M_y(L) &= 0 \rightarrow \left. d^2 w/dx^2 \right|_{x=L} = 0 \\ w(L) &= 0 \end{aligned} \tag{13.33}$$

The first two of these boundary conditions give $A = D = 0$ while the second two boundary conditions give $C = 0$ and the condition

$$B\sin(kL) = B\sin\left[\sqrt{\frac{P}{EI}}L\right] = 0 \tag{13.34}$$

There are two ways in which we can satisfy Eq. (13.34). The first solution is $B = 0$, which is certainly a legitimate solution since it says that the beam does not deflect in the z-direction. If B is not zero, we can divide by it and we can only satisfy the boundary condition if

$$\sin\left[\sqrt{\frac{P}{EI}}L\right] = 0 \tag{13.35}$$

which yields a series of buckling loads given by

$$P_n = EI\left(\frac{n\pi}{L}\right)^2 (n = 1, 2, \ldots) \tag{13.36}$$

The smallest ($n = 1$) of these possible loads occurs first, so it is the critical buckling load, P_{cr} (also called the Euler buckling load), given by

$$P_{cr} = EI\left(\frac{\pi}{L}\right)^2 \tag{13.37}$$

and the associated buckling deflection is

$$w(x) = B\sin\left(\frac{\pi x}{L}\right) \tag{13.38}$$

where the amplitude, B, is arbitrary.

From this analysis it is clear that the critical load depends on the boundary conditions so that if one considered other cases the critical load would be different. We will not consider such cases here but they can be explored in the problems at the end of this chapter.

Buckling of long, slender members is an example of a *bifurcation instability* (see Fig. 13.25a) where, on a load path (a plot of generalized force versus generalized displacement), at the critical load there can be more than one load paths possible. Bifurcation stability, however, is only one way in which stability can be lost. One can have a *limit-load instability*, where the load reaches a maximum value (or, rarely, a minimum value) beyond which the system is unstable. Some shell structures and structures that are sensitive to small imperfections of geometry, loading, or boundary conditions can have a limit-load behavior similar to that seen in Fig. 13.25b. Another way in which stability can be lost is via *snap-through buckling* where the system reaches a limit load and then jumps to a new, stable equilibrium position, as seen in Fig.13.25c. Slender arches and shallow shell structures can exhibit such snap-through behavior. Analysis of limit loads and snap-through buckling typically requires a nonlinear stress analysis, because if one assumes a linear behavior (see the dashed lines in Figs. 13.25b, c), these instabilities are not captured by the analysis. Similarly, in bifurcation types of instability the behavior of the system at loads beyond the critical load is nonlinear. Thus, most elementary and advanced strength of materials texts rarely examine limit-load instabilities, snap-through buckling, or postcritical load behavior for buckling. However, if one considers simpler deformable structures composed of systems of rigid bodies and springs, one can easily analyze these types of instabilities, as we will now discuss.

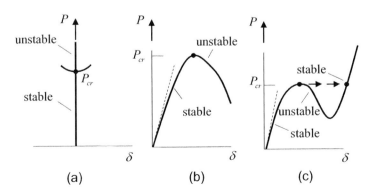

Figure 13.25 (a) A bifurcation type of instability. (b) A limit-load type of instability. (c) A snap-through buckling type of instability. Here P represents a generalized force and δ represents a generalized displacement.

Figure 13.26 A ball at rest on a surface at either an unstable position, A, or a stable position, B.

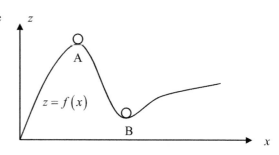

As illustrated by the buckling bifurcation problem just considered, one can obtain the critical load from an examination of equilibrium. Energy methods, however, provide an alternative tool for both obtaining the critical load and for considering nonlinear behavior. Consider, for example, the simple case where the load paths can be described in terms of a single generalized load, P (either a force or a moment), and a single generalized displacement variable, δ (either a displacement or a rotation angle). In these cases the criterion for the stability of the system is that the second derivative of the total potential energy of the system, Π, with respect to the deformation variable, must be positive, i.e., $d^2\Pi/d\delta^2 > 0$, where $\Pi(\delta) = V(\delta) - Pu(\delta)$, where $V(\delta)$ is the potential energy of the deformable bodies present (which is same as the strain energy discussed in Chapter 8) and u is the displacement at the load P in the direction of P. Also, at equilibrium one has $d\Pi/d\delta = 0$, which is an alternate way to guarantee that equilibrium is satisfied. [Note: P is held fixed when these derivatives are taken.] These criteria make sense since they are a generalization of the stability criteria seen, for example, when we consider the stability of a ball on a curved surface, as shown in Fig. 13.26. In that case the surface $z = f(x)$ is proportional to the total potential energy of the ball due to gravity and one has equilibrium at points A and B when $df/dx = 0$, but the equilibrium is unstable at A and stable at B where $d^2f/dx^2 < 0$ and $d^2f/dx^2 > 0$, respectively. The following example will illustrate the use of an energy method to analyze stability.

Example 13.5 Stability Analysis via Equilibrium or Energy Methods

Consider a rigid bar of length L that is allowed to rotate about a smooth pin at O and is attached to a linear torsional spring at O, which supplies a moment, M_0, that is proportional to the angle of rotation, i.e., $M_0 = k_\theta \theta$, where k_θ is the spring constant (see Fig. 13.27a).

(a) From equilibrium of moments about the smooth pin at O, determine all the allowable values for P. Show that there are multiple load paths that the load can follow and that there is a bifurcation point at which load paths cross. Determine the force $P = P_{cr}$ at the bifurcation point. Show that you obtain the same results by setting $d\Pi/d\theta = 0$, where Π is the total potential energy and θ is the angle of rotation of the bar about the pin at O.

(b) By examining the second derivative of the total potential energy, $d^2\Pi/d\theta^2$, determine the stability of all the possible load paths. What is the physical behavior of the system when $P > P_{cr}$?

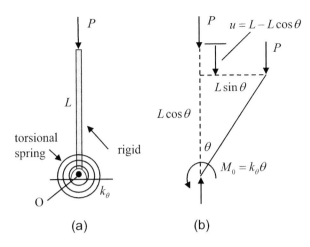

Figure 13.27 (a) A rigid bar of length L that is pinned at O and is attached at O to a linear torsional spring whose spring constant is k_θ. A vertical force, P, acts on the bar. (b) Free body diagram of the deflected bar.

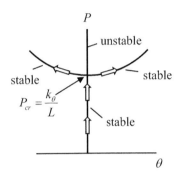

Figure 13.28 The load paths for the bar of Fig. 13.27a, showing a bifurcation behavior.

From a free body diagram of the bar (Fig. 13.27b) and summing moments about O we have

$$\sum M_O = 0$$
$$-PL\sin\theta + k_\theta \theta = 0 \quad (13.39)$$

This equation has two solutions. The first is $\theta = 0$ and the load P is arbitrary. The second solution is $P = k_\theta \theta / L\sin\theta$. These solutions are two intersecting lines so there is a bifurcation behavior as seen in Fig. 13.28. The critical load at the intersection point is $P_{cr} = k_\theta/L$. The total potential energy of the bar is

$$\Pi = \frac{1}{2}k_\theta \theta^2 - PL(1-\cos\theta) \quad (13.40)$$

Setting the derivative equal to zero, we find $d\Pi/d\theta = k_\theta \theta - PL\sin\theta = 0$, which gives us the same result for the critical load as equilibrium. To examine the stability of the under various paths, we must examine the second derivative of the total potential energy, which is

$$\frac{d^2\Pi}{d\theta^2} = -PL\cos\theta + k_\theta \quad (13.41)$$

Along the $\theta = 0$ path $d^2\Pi/d\theta^2 = -PL + k_\theta$, which gives $d^2\Pi/d\theta^2 > 0$ when $P < P_{cr}$ and $d^2\Pi/d\theta^2 < 0$ when $P > P_{cr}$, yielding the stability labels shown in Fig. 13.28 for the vertical load path. For the curved path $P = k_\theta \theta / L\sin\theta = P_{cr}\theta/\sin\theta$ and $d^2\Pi/d\theta^2 = LP_{cr}(1-\theta/\tan\theta)$, which is always positive since $\theta/\tan\theta < 1$. Thus, the curved path is always stable, as shown in Fig. 13.28. As the load P is increased from a value of zero, the stable vertical load path is followed and the bar remains vertical until P reaches the critical load. For $P > P_{cr}$ the vertical load path is no longer stable so the bar must rotate right or left and follow a stable curved path. Note that there is no unwanted or catastrophic behavior for $P > P_{cr}$ so a bifurcation does not always require some unwanted or catastrophic behavior.

There are additional stability problems at the end of this chapter that will allow you to explore bifurcation, limit-load, and snap-through stability cases.

All the cases we have shown are static problems. However, time-varying systems can exhibit dynamic instabilities. These are beyond what we will cover in this book, but you can see some examples in [13].

13.5 PROBLEMS

P13.1 A circular shaft of diameter d is subjected to a combined loading of a bending moment, M, and a torque, T. The material of the shaft has a yield stress σ_y. Determine an expression for the failure surfaces for this case in terms of M, T, and d for the maximum shearing stress theory and the maximum distortional strain energy theory. If the bending moment is $M = 10$ kN-m, the torque is $T = 20$ kN-m, and the yield stress is $\sigma_y = 300$ MPa, determine the minimum allowable shaft diameter to obtain a safety factor of 2 according to these two theories.

P13.2 For a thin plate in a state of plane stress, yielding is observed when $\sigma_{xx} = 90$ MPa, $\sigma_{yy} = 10$ MPa, $\sigma_{xy} = 30$ MPa. (a) What is the corresponding yield stress in uniaxial tension predicted by the maximum shear-stress theory and the maximum distortional strain energy theory? (b) How do your answers change if $\sigma_{xx} = 90$ MPa, $\sigma_{yy} = 30$ MPa, $\sigma_{xy} = 40$ MPa?

P13.3 The maximum distortional strain energy theory written in terms of the principal stresses is given by Eq. (13.5) for a state of plane stress. What is the corresponding theory written in terms of the stresses $(\sigma_{xx}, \sigma_{yy}, \sigma_{xy})$? What is the corresponding theory written in terms of the stresses, as given in a set of Cartesian coordinates, for a general state of stress?

P13.4 A 1.0 m diameter thin wall cylindrical vessel carries an internal pressure $p = 3$ MPa and is made of a material with a uniaxial yield stress $\sigma_y = 400$ MPa. Determine the wall thickness required to prevent yielding with a safety factor of 2.0 according to the maximum shearing stress theory and the maximum distortional strain energy theory.

P13.5 A material with a yield stress $\sigma_y = 650$ MPa, an ultimate strength $\sigma_u = 800$ MPa, and a fatigue strength of 350 MPa at 10^7 cycles is subjected to a mean stress $\sigma_m = 300$ MPa together with an alternating stress $\sigma_a = 200$ MPa. Will the part survive 10^7 cycles according to the Goodman curve? Will it survive based on the more conservative Soderberg curve?

P13.6 A thin-wall cylindrical pressure vessel has a radius $r = 100$ mm and a thickness $t = 5$ mm. The pressure in the vessel varies from zero to p. Using the effective (von Mises) stress to calculate mean and alternating stresses, determine the pressure p so that the vessel will fail after 10^8 cycles based on a Goodman curve. The ultimate strength of the material is $\sigma_u = 800$ MPa and the fatigue strength at 10^8 cycles is $\sigma_{fs} = 250$ MPa.

P13.7 A bolt is subjected to an alternating axial force with maximum and minimum values of (F_{max}, F_{min}), respectively. The ultimate static tensile stress of the material is σ_u and the fatigue strength for an alternating stress at the desired life is σ_{fs}. Based on the Goodman curve, show that the required cross-sectional area of the bolt, A, is

$$A = \frac{1}{\sigma_{fs}}\left[F_{max} - \frac{1}{2}(F_{max} + F_{min})\left(1 - \frac{\sigma_{fs}}{\sigma_u}\right)\right]$$

P13.8 A small leaf spring of rectangular cross-section that has a width $b = 10$ mm wide and a length $L = 150$ mm can be modeled as a simply supported beam subjected to a center concentrated force, P, that varies from zero to 20 N. Using the Goodman curve, determine the value of the height, h, needed for the spring if the fatigue strength at the desired life is $\sigma_{fs} = 350$ MPa and the ultimate static tensile strength is $\sigma_u = 1200$ MPa.

P13.9 Consider a material whose S–N curve is given by

$$S = 800N^{-0.12}$$

where S is the alternating stress in MPa and N is the number of cycles to failure. If this material is placed in an operating environment where it operates for 250 000 cycles at an alternating stress level of 175 MPa and for 250 000 cycles at 150 MPa, will the material fail according to the Palmgren–Miner rule?

P13.10 A 100 mm wide, 10 mm thick plate made with a material having a fracture toughness $K_{IC} = 80$ MPa $\sqrt{m}$ is subjected to a tensile load, P (Fig. P13.1a)), which

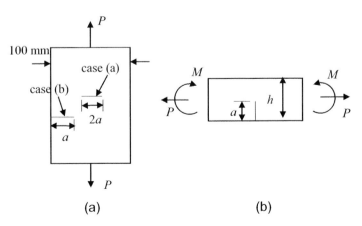

Figure P13.1 (a) An axially loaded cracked plate. (b) A cracked beam under combined loading.

varies from zero to $P = 140$ kN. Use the stress intensity configuration factors in Fig. 13.16.
(a) Determine the size that a through-thickness centrally located fatigue crack must grow to before failure by fast fracture occurs.
(b) How does your answer to (a) change if the crack is located on the edge of the plate instead?

P13.11 A 30 mm wide plate that has a plane strain fracture toughness $K_{IC} = 70$ MPa $\sqrt{m}$ is subjected to an axial tensile stress of 150 MPa. If an ultrasonic NDE method is capable of finding a 3 mm long crack, is this testing method able to detect a surface breaking flaw in the plate before it becomes critical? What is the safety factor for this inspection? Note, however, that in a fatigue environment the frequency of inspection is also important to ensure that a crack does not grow to a critical length between inspections.

P13.12 A rectangular beam of height $h = 50$ mm and width $t = 20$ mm carries a combined loading of $P = 10$ kN and $M = 200$ N-m (Fig. P13.1b). If the beam is made of a material with a fracture toughness $K_{IC} = 40$ MPa $\sqrt{m}$, what is the critical crack length at which the beam fractures due to an edge crack, as shown. Use the stress intensity configuration factors in Fig. 13.16 and Fig. 13.17b.

P13.13 A rolling container is supported by two parallel beams, each of length $L = 40$ ft and having rectangular cross sections of height $h = 10$ in. and width $t = 1.0$ in. The weight of the container and its contents is $W = 15\,000$ lb. A crack of length a exists at the bottom of one of the beams at its center as shown in Fig. P13.2.
(a) If the fracture toughness of the beam material is $K_{IC} = 60$ ksi $\sqrt{in.}$, determine the largest crack that could be tolerated at the configuration shown in Fig. P13.2 before unstable crack growth would occur.
(b) If a one-half inch long crack already exists in the beam at the location shown in Fig. P13.2, how many times could the container be pushed across the beams before the cracked beam fails? Assume the bending moment varies from zero to a maximum value given in part (a) and assume the crack growth law is

Figure P13.2 A container traveling over a cracked beam.

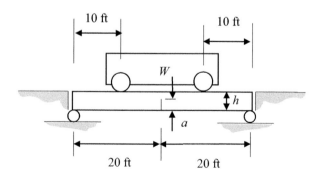

$$\frac{da}{dN} = A(\Delta K)^m$$

where $A = 1.0 \times 10^{-18}$ in./cycle and $m = 3$. Also, assume the configuration factor in the stress intensity expression is equal to 1.122, independent of crack size.

(c) If an NDE inspection is performed every 10 000 crossings, what size of crack must be reliably detected if one wants to guarantee that the beam will not fail before the next inspection? Use the same crack growth law and configuration factor in (b) and the critical flaw size from (a).

P13.14 A rigid bar of length L is attached to a linear spring whose other end rides in a vertical smooth slot so that the spring always remains horizontal. The spring constant is k. A vertical load, P, is applied to the bar, as shown in Fig. P13.3a.

(a) From equilibrium of moments about the smooth pin at O, determine all the allowable values for P. Show that there are multiple load paths that the load can follow and that there a bifurcation point at which load paths cross. Determine the force $P = P_{cr}$ at the bifurcation point. Sketch your results. Show that you obtain the same results by setting $d\Pi/d\theta = 0$, where Π is the total potential energy and θ is the angle of rotation of the bar about the pin at O.

(b) By examining the second derivative of the total potential energy, $d^2\Pi/d\theta^2$, determine the stability of all the possible load paths. What is the physical behavior of the system when $P > P_{cr}$?

P13.15 A rigid bar of length L is attached to a linear spring whose other end rides in a vertical smooth slot so that the spring always remains horizontal. The spring constant is k. A vertical load, P, is applied to the bar eccentrically with a small eccentricity, e, as shown in Fig. P13.3b.

(a) From equilibrium of moments about O, determine the equilibrium values of P as a function of the angle of rotation, θ, about O. Show that you obtain the same

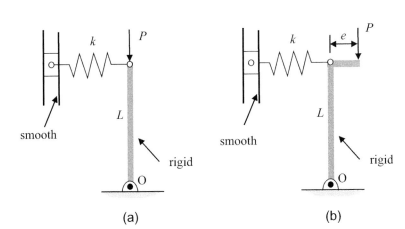

Figure P13.3 (a) A vertical force P acts on rigid bar of length L, which is attached to a linear spring. (b) The same configuration but where the load is applied eccentrically to the bar, where e is the small eccentricity.

results by setting $d\Pi/d\theta = 0$, where Π is the total potential energy and θ is the angle of rotation of the bar about the pin at O. Plot the normalized load, P/kL, versus θ for the normalized eccentricity $e/L = 0.1$ and for $-1.5 < \theta < 1.5$ radians (because the slope of a portion the load curve is very large, also limit the plot range to $0 < P/kL < 2$).

(b) For $P/kL = 0.5$, $e/L = 0.1$ plot the normalized total potential energy, Π/kL^2, over the same angles used in part (a) and determine the equilibrium positions and if they are stable or unstable. Use that information to describe physically what happens in this problem as the load P increases and justify why this type of instability is called a limit-load stability. By making the eccentricity very small, can you see the relationship between this behavior and the bifurcation behavior of this system when $e = 0$?

P13.16 Consider a shallow truss, as shown in Fig. P13.4a, which is composed of two uniaxially loaded members that are pinned at their ends. Because of symmetry, the pin connecting the two members experiences a displacement, u, as shown in Fig. P13.4b and the lengths of the members changes from L_0 to L. If we assume linear elastic behavior for the members, then they behave like linear springs, where $F = -k(L - L_0)$. [Note: the minus sign is present because $(L - L_0)$ is the elongation and we have shown the members in Fig. P13.4c to be compressive.] The spring constant $k = EA/L_0$, where A is the cross-sectional area of each member and E is Young's modulus.

(a) From equilibrium, obtain an expression for the nondimensional load, P/P_0 in terms of the nondimensional displacement, $\delta = u/d$, and a/d, where $P_0 = ka^2/d$. Show that you obtain the same result by setting $d\Pi(\delta)/d\delta = 0$, where Π is the total potential energy. Plot this nondimensional load versus nondimensional displacement from $u/d = 0$ to $u/d = 2.5$ for a shallow truss where $d/a = 0.1$. As the load increases from zero, what is the first maximum load, $P = P_{cr}$, reached?

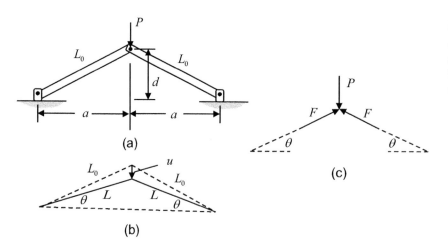

Figure P13.4 (a) A shallow truss. (b) The deformation. (c) A free body diagram of the forces.

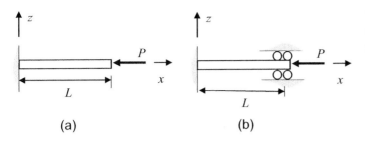

Figure P13.5 (a) An axially loaded cantilever beam. (b) An axially loaded fixed–fixed beam.

(b) Plot the total potential energy, $\Pi(\delta)$, for the same values of displacement in part (a) for $P = 0.25P_0$. Identify the equilibrium positions and state if they are stable or unstable. If these stability results are characteristic of the portions of the $P(\delta)/P_0$ curve on which they occur, describe physically what happens when the truss is loaded from $P = 0$ to $P > P_{cr}$. This type of loss of stability behavior is called snap-through buckling. It can occur, for example, if the loading on a shallow shell roof structure becomes too large. If after snap-through buckling occurs the load is removed, does the truss return to its original unloaded position?

P13.17 Consider an axially loaded cantilever beam, as shown in Fig. P 13.5a. Determine the buckling loads and the corresponding buckled beam shapes. Show that, for the lowest-order mode, we can use the Euler buckling load expression for a simply supported column with the length replaced by an effective length for the cantilever. What is this effective length? If the cantilever has a rectangular cross section of base b and height h, where $b < h$, what area moment should be used to determine the buckling load?

P13.18 Repeat problem P13.17 for the fixed-fixed beam of Fig. P13.5b, solving for the lowest buckling load, corresponding mode, and effective length.

References

1. E. Zahavi, *Fatigue Design – Life Expectancy of Machine Parts* (Boca Raton, FL: CRC Press, 1996)
2. A. F. Liu, *Structural Life Assessment Methods* (ASM International, CRC Press, 1998)
3. B. Farahmoud, *Fatigue and Fracture of High-Risk Parts* (London, UK: Chapman and Hall, 1997)
4. H. Tada, P. C. Paris, and G. R. Irwin, *The Stress Analysis of Cracks Handbook* (ASME Press, 2000)
5. D. P. Rook and D. J. Cartwright, *Compendium of Stress Intensity Factors*, (H.M.S.O, 1976)
6. G. Shih, Ed., *Experimental Evaluation of Stress Concentration and Stress Intensity Factors*, (Switzerland: Springer, 1981)
7. J. F. Knott, *Fundamentals of Fracture Mechanics*, (Oxford, UK: Butterworths, 1973)
8. J. M. Barsom and S. T. Rolfe, *Fracture and Fatigue Control in Structures*, 3rd edn (ASTM, 1999)

9. T. L Anderson, *Fracture Mechanics*, 4th edn (Boca Raton, FL: CRC Press, 2017)
10. L. W. Schmerr, Jr., *Fundamentals of Ultrasonic Nondestructive Evaluation – A Modeling Approach*, 2nd edn (Switzerland: Springer, 2016)
11. N. Bowler, *Eddy-Current Nondestructive Evaluation* (Switzerland: Springer, 2019)
12. R. Halmshaw, *Industrial Radiology: Theory and Practice* (Switzerland: Springer, 2012)
13. L. W. Schmerr, *Engineering Dynamics 2.0 – Fundamentals and Numerical Solutions* (Switzerland: Springer, 2019)

APPENDIX A

Cross-Section Properties

Descriptions of bending and torsional deformations involve numerous properties of the member cross-section. As seen in Chapter 10, the bending cross-sectional properties included the location (Y_c, Z_c) of the centroid and the second area moments and the mixed second area moment (I_{yy}, I_{zz}, I_{yz}), as calculated with respect to centroidal coordinates (y, z). To describe the shear stresses and the shear center location (e_Y, e_Z) for the bending of thin sections the principal sectorial area function, ω_P (see Appendix C), and the sectorial products of areas $(I_{y\omega}, I_{z\omega})$ also appeared, as well as the first area moments (Q_y, Q_z), which are closely related to the first area moments (Q_{yA}, Q_{zA}) of the entire cross-sectional area, A (see Appendix C). For torsion, in Chapter 11 we saw that the effective polar area moment, J_{eff}, and the torsional warping constant, J_ω, were two important quantities. The location of the center of twist/shear center, (e_Y, e_Z), again was needed as was the principal sectorial area function, ω_P, the sectorial products of areas, $(I_{y\omega}, I_{z\omega})$, as well as the first area moments (Q_{yA}, Q_{zA}). All of the quantities that are associated with the sectorial area function are discussed extensively in Appendix C. Figure A.1 lists some of these important torsional properties for common thin cross-sectional areas. For bending, we have not listed approximate expressions for thin sections. Instead, we have written a MATLAB® function cross_section that calculates the location of the centroid and the area moments and mixed area moments of a cross-section composed of rectangles lying parallel to a (Y, Z) coordinate system. This allows one to calculate these properties for all of the cross-sections shown in Fig. A.1 and Fig. A.1 (continued) whether or not they are thin. The calling sequence for the MATLAB® function cross_section is:

```
[ Yct, Zct, Iy, Iy, Iyz] = cross_section( Ly, Lz, Yc, Zc);
```

where (Yc, Zc) are vectors containing the location of the centroids of the rectangular parts in a (Y, Z) coordinate system, and (Ly, Lz) are the lengths of the rectangular parts along the (Y, Z) axes. The outputs of the function are the coordinates of the centroid of the total cross-section, (Yct, Zct), and the area moments and mixed area moment, relative to a set of centroidal axes, (y, z). [Note that, for a thin section, all the dimensions are typically given in terms of distances along the centerlines of the sections, while the function cross_section uses lengths of the sides of individual rectangular components. Thus, to compare the exact results from cross_section with thin cross-section values, one must relate the two different descriptions of the geometry. An example of this process is given in Section A.3.] Figure A.2 shows a number of common cross-sections for which a MATLAB® script (named sections) has

Figure A.1 Torsional properties of thin sections. S is the location of the shear center.

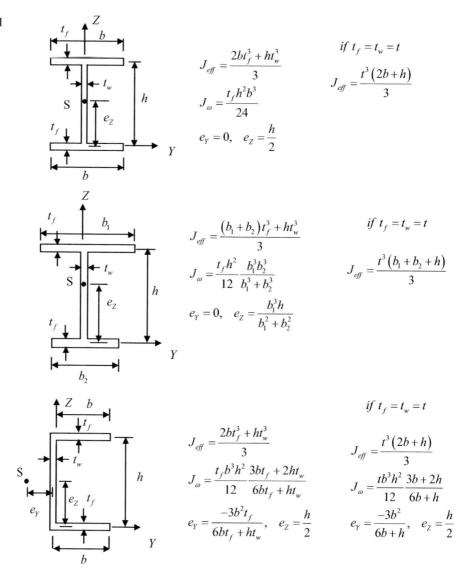

been written that uses the function cross_section and allows one to calculate the centroids and area moments directly from the given dimensions (which again are measured, as shown in Fig. A.2, from the lengths of the sides, not along the centerlines). A description of the script and a sample calculation are given in Section A.3.

In the bending of unsymmetrical sections, several relationships for transforming second area moments and mixed second area moments were discussed in Chapter 10. These included the parallel axis theorem and the transformation to principal coordinates of the cross-section. In this appendix we will briefly outline those relations.

Figure A.1 (cont.)
Torsional properties of thin sections. S is the location of the shear center.

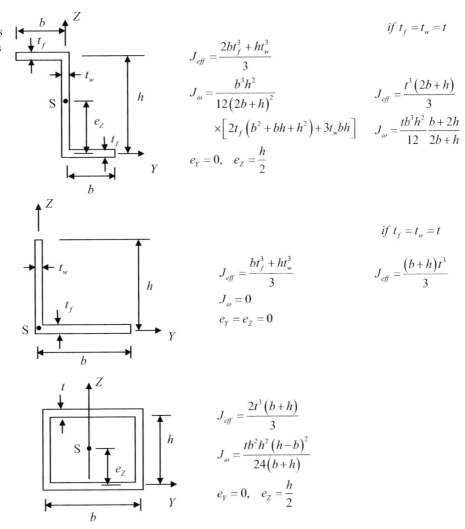

$$J_{eff} = \frac{2bt_f^3 + ht_w^3}{3}$$

$$J_\omega = \frac{b^3 h^2}{12(2b+h)^2} \times \left[2t_f \left(b^2 + bh + h^2 \right) + 3t_w bh \right]$$

$$e_Y = 0, \quad e_Z = \frac{h}{2}$$

if $t_f = t_w = t$

$$J_{eff} = \frac{t^3 (2b+h)}{3}$$

$$J_\omega = \frac{tb^3 h^2}{12} \frac{b+2h}{2b+h}$$

$$J_{eff} = \frac{bt_f^3 + ht_w^3}{3}$$

$$J_\omega = 0$$

$$e_Y = e_Z = 0$$

if $t_f = t_w = t$

$$J_{eff} = \frac{(b+h)t^3}{3}$$

$$J_{eff} = \frac{2t^3(b+h)}{3}$$

$$J_\omega = \frac{tb^2 h^2 (h-b)^2}{24(b+h)}$$

$$e_Y = 0, \quad e_Z = \frac{h}{2}$$

A.1 Parallel Axis Theorem

In calculating the second area moments and mixed second area moments for composite cross-sections, i.e., those that are composed of combinations of simple shapes, it is often convenient to determine how the values of those area moments change when we use coordinates relative to different axes that are translated (and not rotated) with respect to each other. Thus, consider a cross-sectional area A, as shown in Fig. A.3, and two sets of axes (y', z') and (y, z) that are parallel to each other and where the (y', z') axes have their origin at the centroid, C, of A. If we let (y_c, z_c) be the coordinate locations of C as measured in the (y, z) set of axes then from the definition of the area moment, I_{zz}, we have

Figure A.2 Various cross-sections. (a) An L-section. (b) A J-section (which includes a C-section as a special case). (c) A T-section. (d) A Z-section. (e) An I-section. All dimensions are measured along the section sides as shown. Note that in (d) the Z-axis is not along the centerline of the vertical part of the section as it is in the symmetrical sections of (c) and (e).

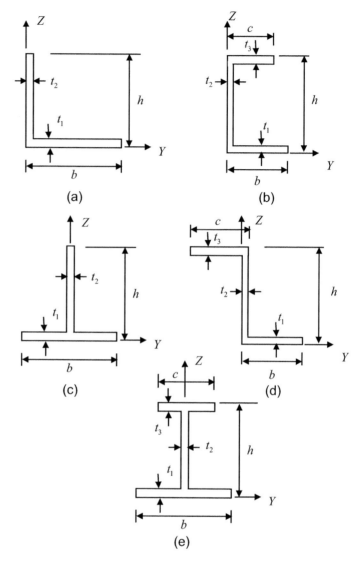

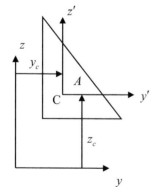

Figure A.3 Two parallel coordinate systems, where the (y', z') axes have their origin at the centroid of the area A.

$$I_{zz} = \int y^2 dA = \int (y_c + y')^2 dA$$
$$= \int \left(y_c^2 + 2y_c y' + (y')^2 \right) dA \quad \text{(A.1)}$$
$$= (I_{z'z'})_c + A y_c^2$$

where the middle integral vanishes since y' is measured from the centroid. In similar fashion one can show

$$I_{yy} = (I_{y'y'})_c + Az_c^2$$
$$I_{yz} = (I_{y'z'})_c + Ay_c z_c \qquad (A.2)$$

Equations (A.1) and (A.2) represent the *parallel axis theorem* for the area moments.

A.2 Area Moments in Rotated Coordinates and Principal Axes

The values of the area moments also change when we use a new rotated set of coordinates to describe their values. Figure A.4 shows a cross-sectional area as measured in a set of (y, z) axes and a set of rotated (y', z') axes. One can again use the definitions of these area moments to relate their values. We will not give the details here but simply show the end result, namely,

$$I_{y'y'} = \frac{I_{yy} + I_{zz}}{2} + \frac{I_{yy} - I_{zz}}{2} \cos(2\theta) - I_{yz} \sin(2\theta)$$

$$I_{z'z'} = \frac{I_{yy} + I_{zz}}{2} - \frac{I_{yy} - I_{zz}}{2} \cos(2\theta) + I_{yz} \sin(2\theta) \qquad (A.3)$$

$$I_{y'z'} = \frac{I_{yy} - I_{zz}}{2} \sin(2\theta) + I_{yz} \cos(2\theta)$$

These area moments transform like the stresses do for a case of plane stress where we have only the nonzero stresses $(\sigma_{yy}, \sigma_{zz}, \sigma_{yz})$. From the general stress transformation, Eq. (2.38), one can show that we find

$$\sigma_{y'y'} = \frac{\sigma_{yy} + \sigma_{zz}}{2} + \frac{\sigma_{yy} - \sigma_{zz}}{2} \cos(2\theta) + \sigma_{yz} \sin(2\theta)$$

$$\sigma_{z'z'} = \frac{\sigma_{yy} + \sigma_{zz}}{2} - \frac{\sigma_{yy} - \sigma_{zz}}{2} \cos(2\theta) - \sigma_{yz} \sin(2\theta) \qquad (A.4)$$

$$\sigma_{y'z'} = -\frac{\sigma_{yy} - \sigma_{zz}}{2} \sin(2\theta) + \sigma_{yz} \cos(2\theta)$$

so the transformations are identical if we use the correspondences $(\sigma_{yy}, \sigma_{zz}, \sigma_{yz}) \to (I_{yy}, I_{zz}, -I_{yz})$ and $(\sigma_{y'y'}, \sigma_{z'z'}, \sigma_{y'z'}) \to (I_{y'y'}, I_{z'z'}, -I_{y'z'})$. This correspondence also means that in a principal set of coordinates, the mixed area moment will be zero and the area moments will have their largest and smallest values (I_{p1}, I_{p2}). As in the plane-stress case (see Eq. (2.67)), we have these principal area moments given by

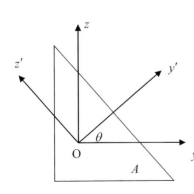

Figure A.4 An area as measured in a set of (y, z) axes and a rotated set of (y', z') axes.

Appendix A

$$I_{p1}, I_{p2} = \frac{I_{yy} + I_{zz}}{2} \pm \sqrt{\left(\frac{I_{yy} - I_{zz}}{2}\right)^2 + I_{yz}^2} \qquad (A.5)$$

This correspondence between stresses and area moments also means that we can use the MATLAB® function eig to calculate the principal area moments and their principal directions as we will see in the following example.

Example A.1 Principal Area Moments and Principal Directions

Consider the thin L-shaped section shown in Fig. A.5a, which has a constant thickness $t = 0.01$ in. We have the centroid C located at

$$Y_c = \frac{Y_1 A_1 + Y_2 A_2}{A_1 + A_2} = \frac{(0)(0.04) + (1)(0.02)}{0.04 + 0.02} = \frac{1}{3} \text{ in.}$$

$$Z_c = \frac{Z_1 A_1 + Z_2 A_2}{A_1 + A_2} = \frac{(0)(0.02) + (2)(0.04)}{0.04 + 0.02} = \frac{4}{3} \text{ in.} \qquad (A.6)$$

and for the area moments and mixed area moments about the centroid

$$I_{yy} = \left[0 + (0.02)\left(-\frac{4}{3}\right)^2\right] + \left[\frac{(0.01)(4)^3}{12} + (0.04)\left(2 - \frac{4}{3}\right)^2\right] = 0.1067 \text{ in.}^4$$

$$I_{zz} = \left[\frac{(0.01)(2)^3}{12} + (0.02)\left(1 - \frac{1}{3}\right)^2\right] + \left[0 + (0.04)\left(-\frac{1}{3}\right)^2\right] = 0.0200 \text{ in.}^4 \qquad (A.7)$$

$$I_{yz} = \left[0 + (0.02)\left(1 - \frac{1}{3}\right)\left(-\frac{4}{3}\right)\right] + \left[0 + (0.04)\left(-\frac{1}{3}\right)\left(2 - \frac{4}{3}\right)\right] = -0.0267 \text{ in.}^4$$

Figure A.5 (a) A thin L-shaped cross-section. The dimensions shown are measured along the centerlines of the sections. (b) The principal coordinate axes (y_p, z_p) at the centroid C.

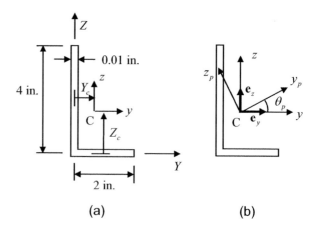

(a) (b)

If we calculate the values in Eq. (A.7) and place them in a 2 × 2 matrix with the values

$$\begin{bmatrix} I_{yy} & -I_{yz} \\ -I_{yz} & I_{zz} \end{bmatrix}$$

we can then use the eig function to calculate the principal values and principal directions. We have in MATLAB®

```
Iyy = .02*(4/3)^2 +(.01)*4^3/12 +.04*(2-4/3)^2
Iyy = 0.1067

Izz = .01*8/12 +.02*(1-1/3)^2 +.04*(1/3)^2
Izz = 0.0200

Iyz = .02*(1-1/3)*(-4/3) +.04*(-1/3)*(2-4/3)
Iyz = -0.0267

I = [ Iyy  -Iyz; -Iyz  Izz]
I =    0.1067    0.0267
       0.0267    0.0200

[pdirs, pvals] = eig(I)
pdirs =   0.2723   -0.9622
         -0.9622   -0.2723

pvals =   0.0125        0
               0   0.1142
```

We see that the two principal values along the principal (y_p, z_p) axes are $I_{y_p} = 0.1142$ in.^{4} and $I_{z_p} = 0.0125$ in.^{4}, respectively, since if you examine the unit vectors in the columns of the pdirs matrix you will see that the corresponding columns of this matrix are in the negative y_p and z_p directions. Thus, the components of unit vectors along the positive y_p and z_p axes are $\mathbf{e}_{y_p} = 0.9622\mathbf{e}_y + 0.2723\mathbf{e}_z$ and $\mathbf{e}_{z_p} = -0.2723\mathbf{e}_y + 0.9622\mathbf{e}_z$, respectively, where $(\mathbf{e}_y, \mathbf{e}_z)$ are unit vectors along the (y, z) axes. We can find the angle θ_p shown in Fig. A.5b via

```
atand(.2723/.9622)
ans = 15.8014
```

so that $\theta_p = 15.8°$, approximately.

A.3 Calculating Centroids and Area Moments in MATLAB®

Consider first the L-shaped cross-section shown in Fig. A.6a. We will decompose the section into the two rectangles shown in Fig. A.6b. The lengths of these rectangles and the locations of their centroids in the (Y, Z) coordinates are as shown in Table A.1.

Table A.1 The properties of the rectangular parts of the L-shaped section shown in Fig. A.6b

Rectangle	Ly	Lz	Yc	Zc
(1)	70	10	45	5
(2)	10	120	5	60

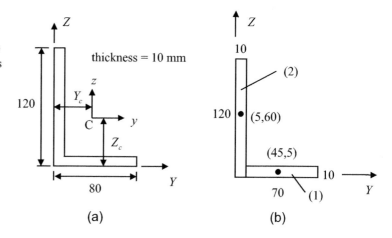

Fig. A.6 (a) An L-shaped cross-section whose centroid is located at C. (b) The cross-section divided into two rectangles, where the dimensions and centroids of the rectangles are given in the (Y, Z) coordinates.

We can place these values into vectors and calculate the cross-sectional area properties in MATLAB®, giving:

```
Ly = [ 70 10];
Lz = [ 10 120];
Yc = [ 45 5];
Zc = [ 5 60];
[ yc, zc, Iy, Iz, Iyz] = cross_section(Ly, Lz, Yc, Zc)
yc = 19.7368
zc = 39.7368
Iy = 2.7832e+06
Iz = 1.0032e+06
Iyz = -9.7263e+05
```

so that the location of the centroid of the entire cross-section is at $Y_c = 19.74$ mm, $Z_c = 39.74$ mm and the area moments are $I_{yy} = 2.78 \times 10^6$ mm^4, $I_{zz} = 1.00 \times 10^6$ mm^4. The mixed area moment is $I_{yz} = -9.73 \times 10^5$ mm^4.

We can also use the cross_section function to check on the accuracy of using the thin wall approximations that are customarily used when dealing with thin sections. Consider, for example, the thin L-shaped section considered in Fig. A.5a. The dimensions in that figure were based on distances as measured along the centerline of the cross-section and the distances to the centroid were also calculated from a set of axes along the centerlines, which

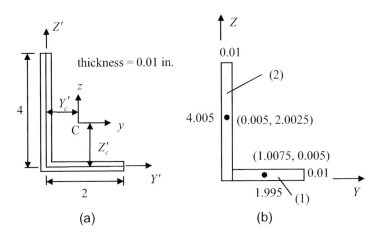

Fig. A.7 (a) Centerline dimensions used for the L-section of Fig. A.5a as well as the distances to the centroid C as measured from those centerlines. (b) The exact dimensions of the cross-section when it is treated as two rectangles as well as the centroid locations of those rectangles as measured from the (Y, Z) axes shown.

we will call (Y', Z'). The exact dimensions of two rectangles making up the cross-section as well as the (Y, Z) axes used for locating the centroid are shown in Fig. A.7b. Using those exact dimensions, we find in MATLAB®:

```
Ly = [ 1.995   0.01];
Lz = [ 0.01    4.005];
Yc = [ 1.0075  0.005];
Zc = [ 0.005   2.0025];

[Yct, Zct, Iy, Iz, Iyz] = cross_section(Ly, Lz, Yc, Zc)
% centroid locations from the (Y, Z) axes
Yct = 0.3383
Zct = 1.3383
% area moments for centroidal coordinates
Iy = 0.1067
Iz = 0.0200
Iyz = -0.0267
% calculate centroid locations from the centerlines to compare
% with thin section results
Ycp = Yct -.005
Ycp = 0.3333
Zcp = Zct - .005
Zcp = 1.3333
```

where $Ycp = Y'_c$, $Zcp = Z'_c$. Thus, we see that as measured from the centerlines the centroid coordinates are $Y'_c = 0.3333$ in., $Z'_c = 1.3333$ in., and $I_{yy} = 0.1067$ in.^{4}, $I_{zz} = 0.0200$ in.^{4}, $I_{yz} = -0.0267$ in.^{4}. These are in agreement with the thin cross-section results of the previous section (see Eq. (A.6) and Eq. (A.7)), showing the validity of the use the thin-section approximations for this cross-section. Generally, the thin-section approximations are used to make the calculations, when done by hand, more manageable and the expressions for the area moments less complex. With the availability of functions like cross_section, however, there is less need to use the thin-section approximations. We have also, as mentioned previously, developed a MATLAB® script, sections, that makes these calculations even easier. Five cross-sections are shown in Fig. A.2 that can be described in terms of up to three lengths (b, h, c) and three thicknesses (t_1, t_2, t_3). The cross-sections are called L-, J-, T-, Z-, and I-sections. If one enters a string 'L', 'J', 'T', 'Z', or 'I' in a variable called type to identify the type of section and then enters the necessary length and parameters for the chosen section into the command window, we can simply run the script sections. That script uses the function cross_section and returns the location of the centroid of the section in the (Y, Z) coordinates shown in Fig. A.2 as well as the area moments (I_{yy}, I_{zz}, I_{yz}) with respect to centroidal axes. For example, consider the L-section of Fig. A.6. In MATLAB® we have

```
clear
type = ' L';
b = 80;
h = 120;
t1 = 10;
t2 = t1;
sections
yc = 19.7368
zc = 39.7368
Iy = 2.7832e+06
Iz = 1.0032e+06
Iyz = -9.7263e+05
```

These values agree with the ones calculated previously.

APPENDIX B
The Beltrami–Michell Compatibility Equations

B.1 Compatibility Equations for Stresses

The compatibility relations discussed in Chapter 4 are relations between the strains in a body. However, if we use the stress–strain relations and the equilibrium equations, we can instead write those compatibility relations in terms of the stresses. For a homogeneous, isotropic elastic solid these compatibility equations are called the Beltrami–Michell relations. From Eq. (4.50) the 81 compatibility equations are

$$\frac{\partial^2 e_{ik}}{\partial x_j \partial x_l} + \frac{\partial^2 e_{jl}}{\partial x_i \partial x_k} - \frac{\partial^2 e_{jk}}{\partial x_i \partial x_l} - \frac{\partial^2 e_{il}}{\partial x_j \partial x_k} = 0 \quad (i,j,k,l = 1,2,3) \tag{B.1}$$

If we use the stress–strain relations (see Eq. (5.34)) in these equations, we have the corresponding 81 equations

$$\frac{\partial^2 \sigma_{ik}}{\partial x_j \partial x_l} + \frac{\partial^2 \sigma_{jl}}{\partial x_i \partial x_k} - \frac{\partial^2 \sigma_{jk}}{\partial x_i \partial x_l} - \frac{\partial^2 \sigma_{il}}{\partial x_j \partial x_k}$$
$$= \frac{v}{1+v}\left[\delta_{ik}\frac{\partial^2 \Theta}{\partial x_j \partial x_l} - \delta_{jk}\frac{\partial^2 \Theta}{\partial x_i \partial x_l} + \delta_{jl}\frac{\partial^2 \Theta}{\partial x_i \partial x_k} - \delta_{il}\frac{\partial^2 \Theta}{\partial x_j \partial x_k}\right] \quad (i,j,k,l = 1,2,3) \tag{B.2}$$

where δ_{ij} is the Kronecker delta (see Eq. (2.34)) and

$$\Theta = \sum_{n=1}^{3} \sigma_{nn} \tag{B.3}$$

If we set $k = j$ and sum over j the resulting equations from 1 to 3, we have nine equations which actually represent six equations because of the symmetry of the stresses:

$$\sum_{j=1}^{3}\left(\frac{\partial^2 \sigma_{ij}}{\partial x_j \partial x_l} + \frac{\partial^2 \sigma_{jl}}{\partial x_j \partial x_i} - \frac{\partial^2 \sigma_{il}}{\partial x_j \partial x_j}\right) - \frac{\partial^2 \Theta}{\partial x_i \partial x_l}$$
$$= \frac{v}{1+v}\sum_{j=1}^{3}\left[\delta_{ij}\frac{\partial^2 \Theta}{\partial x_j \partial x_l} - \delta_{jj}\frac{\partial^2 \Theta}{\partial x_i \partial x_l} + \delta_{jl}\frac{\partial^2 \Theta}{\partial x_i \partial x_j} - \delta_{il}\frac{\partial^2 \Theta}{\partial x_j \partial x_j}\right] \quad (i,l = 1,2,3) \tag{B.4}$$

These equations are linear combinations of the original 81 equations so they can serve as the reduced system to represent compatibility. Again, these equations are not all independent, so that in fact there are only really three independent compatibility relations. If we use the relations

$$\sum_{j=1}^{3} \frac{\partial^2 \sigma_{il}}{\partial x_j \partial x_j} = \nabla^2 \sigma_{il}$$

$$\sum_{j=1}^{3} \frac{\partial^2 \Theta}{\partial x_j \partial x_j} = \nabla^2 \Theta \quad \text{(B.5)}$$

$$\sum_{j=1}^{3} \delta_{jj} = 3$$

Eq. (B.4) can be rewritten as

$$\sum_{j=1}^{3} \left\{ \frac{\partial^2 \sigma_{ij}}{\partial x_j \partial x_l} + \frac{\partial^2 \sigma_{jl}}{\partial x_i \partial x_j} - \frac{\partial^2 \Theta}{\partial x_i \partial x_l} - \nabla^2 \sigma_{il} \right\} = \frac{-\nu}{1+\nu} \left[\delta_{il} \nabla^2 \Theta + \frac{\partial^2 \Theta}{\partial x_i \partial x_l} \right] \quad (i, l = 1, 2, 3) \quad \text{(B.6)}$$

We can express Eq. (B.6) in a more compact form by using the equations of equilibrium (see Eq. (3.10)) written as

$$\sum_{j=1}^{3} \frac{\partial \sigma_{ij}}{\partial x_j} + f_i = 0 \quad (i = 1, 2, 3) \quad \text{(B.7)}$$

where f_i are the body force components. Then we have

$$\sum_{j=1}^{3} \frac{\partial^2 \sigma_{ij}}{\partial x_j \partial x_l} = -\frac{\partial f_i}{\partial x_l}$$

$$\sum_{j=1}^{3} \frac{\partial^2 \sigma_{jl}}{\partial x_i \partial x_j} = \sum_{j=1}^{3} \frac{\partial^2 \sigma_{lj}}{\partial x_j \partial x_i} = -\frac{\partial f_l}{\partial x_i} \quad (i, l = 1, 2, 3) \quad \text{(B.8)}$$

so that Eq. (B.6) becomes

$$\nabla^2 \sigma_{il} + \frac{1}{1+\nu} \frac{\partial^2 \Theta}{\partial x_i \partial x_l} - \frac{\nu}{1+\nu} \delta_{il} \nabla^2 \Theta = -\left(\frac{\partial f_i}{\partial x_l} + \frac{\partial f_l}{\partial x_i} \right) \quad (i, l = 1, 2, 3) \quad \text{(B.9)}$$

To simplify Eq. (B.9) even more, let $i = l$ in Eq. (B.6) and sum over i which, using the symmetry of the stresses, gives

$$2 \sum_{i=1}^{3} \sum_{j=1}^{3} \frac{\partial^2 \sigma_{ij}}{\partial x_i \partial x_j} - 2\nabla^2 \Theta = -\frac{\nu}{1+\nu} (4\nabla^2 \Theta) \quad \text{(B.10)}$$

which gives

$$\sum_{i=1}^{3} \sum_{j=1}^{3} \frac{\partial^2 \sigma_{ij}}{\partial x_i \partial x_j} = \frac{1-\nu}{1+\nu} \nabla^2 \Theta \quad \text{(B.11)}$$

But by differentiating the equations of equilibrium again we have

$$\sum_{i=1}^{3}\sum_{j=1}^{3} \frac{\partial^2 \sigma_{ij}}{\partial x_i \partial x_j} = -\sum_{i=1}^{3} \frac{\partial f_i}{\partial x_i} \tag{B.12}$$

so using this result in Eq. (B.11) we find

$$\nabla^2 \Theta = -\left(\frac{1+\nu}{1-\nu}\right) \sum_{i=1}^{3} \frac{\partial f_i}{\partial x_i} \tag{B.13}$$

Using Eq. (B.13) in Eq. (B.9) gives, finally, the compatibility equations written in terms of the stresses as

$$\nabla^2 \sigma_{il} + \frac{1}{1+\nu}\frac{\partial^2 \Theta}{\partial x_i \partial x_l} = -\frac{\nu}{1-\nu}\delta_{il}\sum_{j=1}^{3}\frac{\partial f_j}{\partial x_j} - \left(\frac{\partial f_i}{\partial x_l} + \frac{\partial f_l}{\partial x_i}\right) \quad (i,l=1,2,3) \tag{B.14}$$

Equations (B.14) are the *Beltrami–Michell compatibility equations*. For the case of no body forces they are simply

$$\nabla^2 \sigma_{il} + \frac{1}{1+\nu}\frac{\partial^2 \Theta}{\partial x_i \partial x_l} = 0 \quad (i,l=1,2,3) \tag{B.15}$$

APPENDIX C

The Sectorial Area Function

In elementary strength of materials problems involving bending of symmetrical sections and torsion of circular shafts, certain geometrical properties of the cross-section appear. These are as follows.

Location $(\bar{y}, \bar{z})$ of the centroid of the cross-section and the associated first area moments (Q_{yA}, Q_{zA}):

$$\bar{y}A = Q_{zA} = \int_A y\, dA$$
$$\bar{z}A = Q_{yA} = \int_A z\, dA \tag{C.1}$$

[Note: these first area moments are based on the entire cross-sectional area. Similar quantities (Q_y, Q_z) were defined in Chapter 10 to describe the shear flow in bending. Those quantities are for a portion, A_q, of the cross-section as defined by a distance, s, in the cross-section. In this appendix (Q_{yA}, Q_{zA}) will always refer to quantities calculated for the entire cross-section.]

Second area moments and mixed second area moment:

$$I_{zz} = \int_A y^2\, dA$$
$$I_{yy} = \int_A z^2\, dA \tag{C.2}$$
$$I_{yz} = \int_A yz\, dA$$

Polar area moment:

$$J = \int_A (y^2 + z^2)\, dA = \int_A r^2\, dA \tag{C.3}$$

In the bending and torsion of unsymmetrical thin sections, additional cross-sectional properties appear that are related to a quantity called the *sectorial area function*. To define this function, consider a thin cross-section, as shown in Fig. C.1a. If we let $\mathbf{r}_A$ be a position vector from a point A in the cross-section to an arbitrary point, Q, along the centerline of the

The Sectorial Area Function 541

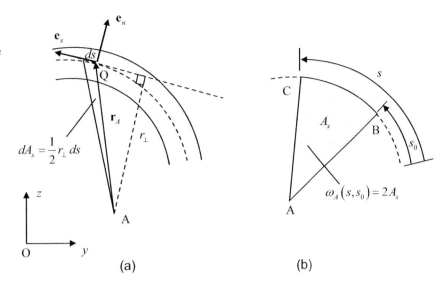

Figure C.1 (a) Geometry of a thin section with the parameters used to define the sectorial area function. (b) The sectorial area function, ω_A, is twice the sectorial area $A_s = ABC$ shown.

section and let e_s be a unit vector tangent to the centerline, then we can define a unit normal to centerline curve, e_n, as $e_n = e_s \times e_x$ (shown in Fig C.1a), where e_x is a unit vector along the x-axis. When the point on the centerline moves a small distance, ds, the line from A to Q sweeps out a small area $dA_s = \mathbf{r}_A \cdot e_n ds/2 = r_\perp ds/2$, where $r_\perp$ is the perpendicular distance to a line tangent to the centerline at Q. The differential of the sectorial area function, $d\omega_A$, is defined as $d\omega_A \equiv 2dA_s$ or, equivalently, $d\omega_A = \mathbf{r}_A \cdot e_s ds = r_\perp ds$ so it is just twice this swept area. The sectorial area function itself is

$$\omega_A(s, s_0) = \int_{s_0}^{s} \mathbf{r}_A \cdot e_n ds = \int_{s_0}^{s} r_\perp ds \qquad (C.4)$$

where s_0 is a position along the centerline at which the sectorial area is zero called the *sectorial area origin* and s is a varying position in the cross-section. There may be more than one place in the cross-section where the sectorial area is zero so in that case there are multiple origins. The sectorial area function depends on both the point A (called the pole of the sectorial area function) as well as the origin, s_0, so when it is important to be explicit about these points the sectorial area function will be written as $\omega_A(s, s_0)$. Otherwise, it will simply be written as $\omega(s)$. From a geometrical standpoint, this sectorial area function represents the twice the total area ABC swept out along the centerline as we move from s_0 to s, as shown in Fig. C.1b. The sectorial area also has a physical meaning since in the torsion of thin cross-sections the out-of-plane warping of the cross-section (along the x-direction) is defined by this function. Note that we can write the perpendicular distance $r_\perp$ as $r_\perp = \mathbf{r}_A \cdot e_n$ or, equivalently, $r_\perp = \mathbf{r}_A \cdot (e_s \times e_x) = (\mathbf{r}_A \times e_s) \cdot e_x$, which shows that ω_A is positive when $\mathbf{r}_A \times e_s$ is in the $+e_x$ direction, corresponding to the area being swept out in the counterclockwise direction. Similarly ω_A is negative when $\mathbf{r}_A \times e_s$ is in the $-e_x$ direction,

Figure C.2 (a, b, c) Sectorial area functions (shaded plots) for different choices of the pole, A, and initial point (origin), I. All sides of the C-section are of length a.

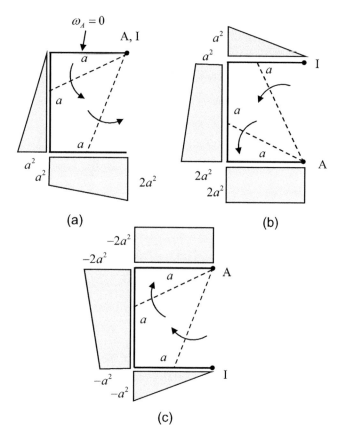

which corresponds to the area being swept out in a clockwise sense. The areas shown in Fig. C.1a, b are both being swept out in the counterclockwise sense, so they are both positive. Figure C.2 shows the sectorial area functions for a C-shaped cross-section with different choices for the pole A and initial point, I, for the integration (which is also an origin). The sense in which the area is being swept out in these cases is shown by dashed lines and arrows that determine if the sectorial area is positive or negative. Note that since the pole A was chosen along one of the legs of the cross-section, in all three cases there is no additional sectorial area swept out for that leg, so the sectorial area function is a constant in that leg. When the constant is zero, any point in that leg is an origin. For a given sectorial area function, $\omega(s)$, we can define a number of cross-sectional properties. The *first sectorial moment*, Q_ω, for a thin section of variable thickness, $t = t(s)$ is defined as

$$Q_\omega = \int_A \omega(s) dA \tag{C.5}$$

where $dA = t(s)ds$, and A is the total area of the cross-section. The *sectorial products of areas* $(I_{y\omega}, I_{z\omega})$ are given by

$$I_{y\omega} = \int_A y(s)\omega(s)dA$$

$$I_{z\omega} = \int_A z(s)\omega(s)dA \tag{C.6}$$

where (y,z) are distances along the y- and z-axes as measured from some origin, O to a general point on the centerline such as point Q (Fig. C.1a). A third quantity, called the *sectorial second moment* (or *warping constant*), $J_{\omega P}$, is defined as

$$J_{\omega P} = \int_A \omega_P^2(s,s_p)dA \tag{C.7}$$

where $\omega_P(s,s_P)$ is a warping function with a specific pole, P, and origin(s), s_P, called the *principal sectorial area function*, which will be defined shortly. Comparing Eqs. (C.5)–(C.7) with Eqs. (C.1)–(C.3) we see that these properties based on the sectorial area function are analogous to the properties that appear in symmetric cross-section bending problems.

Since the sectorial area function depends on both the pole used as well as the sectorial area origin chosen on the centerline, it is instructive to examine how the choice of pole and origin affect that function. Figure C.3 shows a geometry where there are two different poles (A, B), and two different sectorial area origins (s_0, s_1). Let us use those points to define the sectorial area functions $\omega_A(s,s_0)$ and $\omega_B(s,s_1)$. Then from the geometry we have

$$\begin{aligned}\omega_A(s,s_0) &= \int_{s_0}^s \mathbf{r}_A \cdot \mathbf{e}_n ds \\ &= \int_{s_0}^s (\mathbf{r}_B + \mathbf{r}_{B/A}) \cdot \mathbf{e}_n ds \\ &= \int_{s_0}^{s_1} \mathbf{r}_B \cdot \mathbf{e}_n ds + \int_{s_1}^s \mathbf{r}_B \cdot \mathbf{e}_n ds + \int_{s_0}^s \mathbf{r}_{B/A} \cdot \mathbf{e}_n ds \\ &= -\omega_B(s_0,s_1) + \omega_B(s,s_1) + \int_{s_0}^s \mathbf{r}_{B/A} \cdot \mathbf{e}_n ds\end{aligned} \tag{C.8}$$

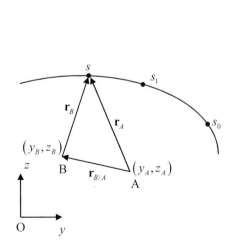

Figure C.3 Geometry for examining the effects of using different poles and initial points to define the sectorial area function.

Now, let us examine the last integral in Eq. (C.8). We have

$$\mathbf{e}_n ds = (\mathbf{e}_s \times \mathbf{e}_x) ds$$
$$= \left[\left(\frac{dy}{ds}\mathbf{e}_y + \frac{dz}{ds}\mathbf{e}_z\right) \times \mathbf{e}_x\right] ds \qquad (C.9)$$
$$= -dy\,\mathbf{e}_z + dz\,\mathbf{e}_y$$

and

$$\mathbf{r}_{B/A} = (y_B - y_A)\mathbf{e}_y + (z_B - z_A)\mathbf{e}_z \qquad (C.10)$$

so that

$$\int_{s_0}^{s} \mathbf{r}_{B/A} \cdot \mathbf{e}_n \, ds = (y_B - y_A)\int_{s_0}^{s} dz - (z_B - z_A)\int_{s_0}^{s} dy$$
$$= (y_B - y_A)(z(s) - z(s_0)) - (z_B - z_A)(y(s) - y(s_0)) \qquad (C.11)$$
$$= (y_B - y_A)(z(s) - z_0) - (z_B - z_A)(y(s) - y_0)$$

where we have let $z(s_0) = z_0$, $y(s_0) = y_0$. Placing Eq. (C.11) into Eq. (C.8) gives

$$\omega_A(s, s_0) = \omega_B(s, s_1) - \omega_B(s_0, s_1)$$
$$+ (y_B - y_A)(z(s) - z_0) - (z_B - z_A)(y(s) - y_0) \qquad (C.12)$$

Equation (C.12) gives us a general relationship between sectorial area functions when we change both the pole and sectorial area origin. Now, consider some special cases. First, consider the case where we do not change the pole (A = B) but have different origins ($s_0 \neq s_1$). Then

$$\omega_A(s, s_0) = \omega_A(s, s_1) - \omega_A(s_0, s_1) \qquad (C.13)$$

But $\omega_A(s_0, s_1)$ is just a constant, so changing the origin simply changes the sectorial area function by a constant amount. Now consider the case where the origins are the same ($s_1 = s_0$) but the pole locations are different (A $\neq$ B). Then we have

$$\omega_A(s, s_0) = \omega_B(s, s_0) + (y_B - y_A)(z(s) - z_0) - (z_B - z_A)(y(s) - y_0) \qquad (C.14)$$

Principal Pole

In the beam problems of Chapter 10 it is shown that, in order to produce bending only (no torsion), the shear forces in the cross-section must be located at a specific point S called the shear center. For thin sections, the shear center point can be located by a physical approach where the shear flow expression (which assumes bending only) is used to evaluate where the shear forces must be located. Alternatively, it is shown that if the pole P is chosen such that sectorial products of area $(I_{y\omega p}, I_{z\omega p})$ are both zero, then P will be the shear center S. In Chapter 10 that result was obtained with the use of the expression for the shear flow in bending but here we will see that, if we calculate the location of a specific pole called the *principal pole*, which is a function only of the geometry, this principal pole is also the shear center.

If we take the general relationship of Eq. (C.12) and multiply it by the distance $y(s)$ to a general point s on the centerline of a thin section, as measured from the coordinate system origin O (Fig. C.3), and integrate both sides over the entire cross-section we find

$$I_{y\omega_A} = I_{y\omega_B} - \omega_B(s_0, s_1)Q_{zA} + (z_A - z_B)(I_{zz} - y_0 Q_{zA}) \\ - (y_A - y_B)(I_{yz} - z_0 Q_{zA}) \tag{C.15}$$

If, instead, we multiply Eq. (C.12) by z and integrate, we find similarly

$$I_{z\omega_A} = I_{z\omega_B} - \omega_B(s_0, s_1)Q_{yA} + (z_A - z_B)\left(I_{yz} - y_0 Q_{yA}\right) \\ - (y_A - y_B)\left(I_{yy} - z_0 Q_{yA}\right) \tag{C.16}$$

Now, take the coordinate system origin O (from which the coordinates (y, z) are measured) to be located at the centroid of the cross-section. Then $Q_{yA} = Q_{zA} = 0$ and Eq. (C.15) and Eq. (C.16) simplify to

$$\begin{aligned} I_{y\omega_A} &= I_{y\omega_B} + (z_A - z_B)I_{zz} - (y_A - y_B)I_{yz} \\ I_{z\omega_A} &= I_{z\omega_B} + (z_A - z_B)I_{yz} - (y_A - y_B)I_{yy} \end{aligned} \tag{C.17}$$

Next, we will require pole A to be located at a specific pole, P, called the principal pole, where

$$\begin{aligned} I_{y\omega P} &= \int_A y(s)\omega_P(s) dA = 0 \\ I_{z\omega P} &= \int_A z(s)\omega_P(s) dA = 0 \end{aligned} \tag{C.18}$$

In this case, Eq. (C.17) becomes a set of simultaneous equations

$$\begin{aligned} I_{y\omega_B} + (z_P - z_B)I_{zz} - (y_P - y_B)I_{yz} &= 0 \\ I_{z\omega_B} + (z_P - z_B)I_{yz} - (y_P - y_B)I_{yy} &= 0 \end{aligned} \tag{C.19}$$

which can be solved to obtain the two coordinates of the principal pole (y_P, z_P), both measured from the centroid of the cross-section, as

$$\begin{aligned} y_P &= y_B + \frac{I_{z\omega_B} I_{zz} - I_{y\omega_B} I_{yz}}{I_{yy} I_{zz} - I_{yz}^2} \\ z_P &= z_B + \frac{I_{z\omega_B} I_{yz} - I_{y\omega_B} I_{yy}}{I_{yy} I_{zz} - I_{yz}^2} \end{aligned} \tag{C.20}$$

which can also be written in terms of the relative distances of the principal pole P from pole B as

$$e_y = y_P - y_B = \frac{I_{z\omega_B}I_{zz} - I_{y\omega_B}I_{yz}}{I_{yy}I_{zz} - I_{yz}^2}$$

$$e_z = z_P - z_B = \frac{I_{z\omega_B}I_{yz} - I_{y\omega_B}I_{yy}}{I_{yy}I_{zz} - I_{yz}^2}$$

(C.21)

which is identical to Eq. (10.52) for the location of the shear center S (relative to pole B). Thus, the principal pole P is indeed the same point as the shear center S. We are free to choose whatever sectorial area origin we want in calculating the sectorial area terms on the right-hand side of Eq. (C.21) since if we change the sectorial area origin used in calculating ω_B, this sectorial area only changes by a constant amount, as previously shown in Eq. (C.13). Thus, if we let $\bar{\omega}_B = \omega_B + \omega_0$ where ω_0 is a constant, we have

$$I_{y\bar{\omega}_B} = \int_A y(\omega_B + \omega_0)dA = I_{y\omega_B} + \omega_0 Q_{zA} = I_{y\omega_B}$$

$$I_{z\bar{\omega}_B} = \int_A z(\omega_B + \omega_0)dA = I_{z\omega_B} + \omega_0 Q_{yA} = I_{z\omega_B}$$

(C.22)

since $Q_{zA} = Q_{yA} = 0$. Because the sectorial products of area remain unchanged when the sectorial area function origin is changed, the coordinates of the principal pole found from Eq. (C.20) or Eq. (C.21) are independent of this sectorial area origin. We also placed no restriction on the pole B used to determine the principal pole, so in practice both the pole and the sectorial area origin used to calculate ω_B can be chosen to make the calculations as easy as possible. The principal pole location does not depend on the particular choice of the pole B since if we use another pole A location instead and assume Eq. (C.21) produces a different principal pole location (y'_P, z'_P) from the (y_P, z_P) values obtained with pole B, then using the relations of Eq. (C.17) we can easily show that

$$y'_P - y_A = \frac{I_{z\omega_A}I_{zz} - I_{y\omega_A}I_{yz}}{I_{yy}I_{zz} - I_{yz}^2} = \frac{I_{z\omega_B}I_{zz} - I_{y\omega_B}I_{yz}}{I_{yy}I_{zz} - I_{yz}^2} + (y_B - y_A)$$

$$= y_P - y_B + (y_B - y_A)$$

$$\rightarrow y'_P = y_P$$

and a similar calculation shows $z'_P = z_P$ also. Thus, the principal pole (shear center) location is a function of the geometry only and is independent of both the pole and sectorial area function origin used in its location.

An alternative to using Eq. (C.20) to determine the principal pole is write a sectorial area function ω_B in terms of a set of unknown pole coordinates (y_B, z_B) *relative to the centroid* and then determine those pole coordinates that satisfy Eq. (C.18), i.e., solve

$$\int_A y(s)\omega_B(s)dA = 0$$

$$\int_A z(s)\omega_B(s)dA = 0 \qquad (C.23)$$

since those coordinates will be the coordinates, relative to the centroid, of the principal pole (shear center). Again, the choice of origin used in obtaining ω_B in Eq. (C.23) is arbitrary.

If the y-axis of the cross-section is an axis of symmetry, then the centroid will lie on that axis of symmetry so that if we place the pole used in calculating ω_B also on that symmetry axis, we will have $z_B = 0$ and one can easily show that $I_{y\omega_B} = 0$. Since we also have $I_{yz} = 0$, Eq. (C.20) shows that $z_P = 0$; i.e., the principal pole also lies on the axis of symmetry. Similarly, if the z-axis is an axis of symmetry, the principal pole will lie on that axis. If both the y- and z-axis are axes of symmetry, then the principal pole will be at the centroid.

Example C.1 Shear Center Location

For a simple example of calculating the shear center, consider the thin unsymmetrical I-beam cross-section shown in Fig. C.4a where the thickness $t \ll a, b, h$. This cross-section does have an axis of symmetry so the centroid and shear center lie on the y-axis and $I_{yz} = 0$ so that we have, from Eq. (C.21)

$$e_y = \frac{I_{z\omega_B}}{I_{yy}} \qquad (C.24)$$
$$e_z = 0$$

As in Chapter 10, when calculating parameters such as centroids, area moments, etc., for thin cross-sections it is common to use the dimensions as measured along the centerline, as shown in Fig C.4a, to calculate the cross-sectional properties and to neglect terms that involve the thickness raised to higher powers. Thus, for example we calculate I_{yy} as

$$I_{yy} = \frac{ta^3}{12} + \frac{ht^3}{12} + \frac{tb^3}{12} \cong \frac{t}{12}(a^3 + b^3) \qquad (C.25)$$

where we have neglected the middle term. We will use point B (Fig. C.4b) as both the pole and the initial point I (origin) to calculate ω_B. The sectorial area with this choice is zero except for the right-hand vertical section where $\omega_B = hs$, with the distance s taken along the section as shown in Fig. C.4b. The sectorial area function is shown as the shaded area in Fig. C.4b. To calculate $I_{z\omega_B}$ we have

$$I_{z\omega_B} = \int_{-b/2}^{b/2} z(s)\omega_B(s)tds = \int_{-b/2}^{b/2} s(hs)tds \qquad (C.26)$$
$$= thb^3/12$$

Figure C.4 (a) An unsymmetrical I-beam cross-section. (b) Geometrical parameters for determining the centroid, shear center, and sectorial area parameters.

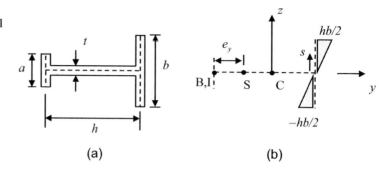

Figure C.5 Parameters for calculating the principal pole (shear center) for the unsymmetrical I-beam in Fig. C.4a using Eq. (C.23) and the resulting sectorial area functions shown as the shaded areas.

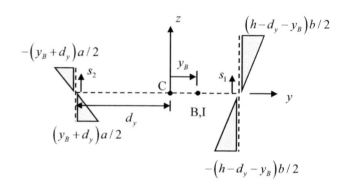

which gives

$$e_y = \frac{hb^3}{(a^3 + b^3)} \quad (C.27)$$

For $a = b$ the I-beam has two axes of symmetry and $e_y = h/2$ so the shear center is at the centroid, as stated earlier. As an alternative way to calculate the shear center, let us use Eq. (C.23) directly. Figure C.5 shows the choice of a pole B and initial point I (origin) that is at an arbitrary distance y_B relative to the centroid. We will see we do not have to know explicitly the centroid location in this example which we have described by the distance, d_y, as shown. Also shown is the sectorial area function ω_B. The first equation in Eq. (C.23) is satisfied automatically by the symmetry present so consider the second equation. We have

$$\int_A z(s)\omega(s)dA = \int_{-b/2}^{b/2} s_1\left[(h - d_y - y_B)s_1\right]tds_1 + \int_{-a/2}^{a/2} s_2\left[-(y_B + d_y)s_2\right]tds_2 = 0 \quad (C.28)$$

which, letting $y_B + d_y = e_y$ gives

$$\frac{(h - e_y)tb^3}{12} - \frac{e_y ta^3}{12} = 0 \quad (C.29)$$

which again gives Eq. (C.27). We see that indeed we did not have to know the distance d_y to the centroid in this example to locate the shear center. Also note that we could have replaced the distance z, from the centroid by $z + d_z$ in Eq. (C.28) and not changed this final result. This is because the sectorial area function ω_B satisfied

$$Q_{\omega_B} = \int_A \omega_B dA = 0 \tag{C.30}$$

When this is the case, the coordinates y and z appearing in Eq. (C.23) can be measured from any fixed location we choose in the cross-section since if we let $y' = y + d_y$, $z' = z + d_z$, where (d_y, d_z) are any two constants, then if Eq. (C.30) is satisfied we have

$$\int_A y' \omega_B dA = \int_A y \omega_B dA + d_y \int_A \omega_B dA = \int_A y \omega_B dA = 0$$

$$\int_A z' \omega_B dA = \int_A z \omega_B dA + d_z \int_A \omega_B dA = \int_A z \omega_B dA = 0 \tag{C.31}$$

A sectorial area function, $\omega_P(s, s_0)$, which satisfies

$$\int_A \omega_P dA = 0$$

$$\int_A y' \omega_P dA = 0 \tag{C.32}$$

$$\int_A z' \omega_P dA = 0$$

for coordinates (y', z') measured with respect to any fixed point in the cross-section is called a *principal sectorial area function*, where s_0 is a *principal origin* of the principal sectorial area function. The three equations in Eq. (C.32) fix the location of the principal pole P and the location of the principal origin(s) (there may be more than one origin location in the cross-section) in the principal sectorial area function.

Principal Origin

Now, consider further the equation for determining the principal origin(s). By definition, if for a sectorial area function computed with a given pole B there is a sectorial area function origin, s_0, such that

$$Q_{\omega_B} = \int_A \omega_B(s, s_0) dA = 0 \tag{C.33}$$

the point s_0 is called a principal origin. To determine the origin(s) consider another sectorial area function $\omega_B(s, s_1)$ with a known origin, s_1. Then from Eq. (C.13)

$$\omega_B(s, s_0) = \omega_B(s, s_1) - \omega_B(s_0, s_1) \tag{C.34}$$

so that if s_0 is a principal origin we must have

$$Q_{\omega_B} = \int_A \omega(s, s_0) dA = \int_A \omega_B(s, s_1) dA - \omega_B(s_0, s_1) A = 0 \tag{C.35}$$

This is an equation that we can use to determine the principal origin(s) in the following way. The quantity $\omega(s_0, s_1)$ is just a constant that we will call ω_0. From Eq. (C.35) we can solve for ω_0 as

$$\omega_0 = \frac{1}{A} \int_A \omega_B(s, s_1) dA \tag{C.36}$$

The right-hand side is known since both the pole B and the center s_1 are specified. The sectorial area function ω_B with the principal origin(s) can then be obtained from Eq. (C.34) as

$$\omega_B(s, s_0) = \omega_B(s, s_1) - \omega_0 \tag{C.37}$$

If we evaluate and plot $\omega_B(s, s_0)$ the zero(s) of this sectorial area function are the location(s) of the principal origin(s). [Note: generally, we do not have to solve explicitly for the s_0 values at the origins because the ω_0 value and $\omega_B(s, s_1)$ are all that we need to obtain the sectorial area function with those principal origin(s).] The principal origin(s) do not depend on the choice of s_1 used in Eq. (C.36) since if we had used a sectorial area function $\omega_B(s, s_2)$ instead to determine the principal origin we would have

$$\int_A \omega_B(s, s_2) dA - \omega_B(s_0, s_2) A = 0 \tag{C.38}$$

But from Eq. (C.13)

$$\omega_B(s, s_2) = \omega_B(s, s_1) - \omega_B(s_2, s_1) \tag{C.39}$$

so Eq. (C.38) becomes

$$\begin{aligned} \int_A \omega_B(s, s_1) dA &- [\omega_B(s_0, s_2) + \omega(s_2, s_1)] A \\ &= \int_A \omega_B(s, s_1) dA - \omega_B(s_0, s_1) A = 0 \end{aligned} \tag{C.40}$$

which leads to the same ω_0 value and hence the same principal origin(s). Thus, a principal origin, like the principal pole, is a cross-sectional property.

If we let the pole B in the above discussion be the principal pole P (or, equivalently the shear center, S) then the sectorial area $\omega_P(s, s_0) = \omega_P(s, s_1) - \omega_0$ is the principal sectorial area function with principal pole P and principal origin s_0.

The Sectorial Second Moment (Warping Constant)

The sectorial second moment defined in Eq. (C.7) uses the principal sectorial area function. We can also define this quantity in terms of a sectorial area function, ω_B, that is defined with respect to an arbitrary pole and origin. To see this let us use Eq. (C.12) to relate the principal sectorial area function to

$$\omega_P(s, s_P) = \omega_B(s, s_1) - \omega_B(s_P, s_1) \qquad \text{(C.41)}$$
$$+ (y_B - y_P)(z(s) - z_0) - (z_B - z_P)(y(s) - y_0)$$

If we integrate Eq. (C.41) and assume that $y(s)$ and $z(s)$ are measured from the centroid of the cross-section, then we find

$$Q_{\omega_P} = 0 = Q_{\omega_B} - \omega_B(s_P, s_1)A - y_0(z_P - z_B)A + z_0(y_P - y_B)A \qquad \text{(C.42)}$$

Placing Eq. (C.42) into Eq. (C.41), we can write that equation as

$$\omega_P(s, s_P) = \omega_B(s, s_1) - \frac{Q_{\omega_B}}{A} + (z_P - z_B)y - (y_P - y_B)z \qquad \text{(C.43)}$$

Using Eq. (C.43), sectorial second moment is

$$J_{\omega_P} = \int_A \omega_P^2 dA = J_{\omega_B} + 2(z_P - z_B)I_{y\omega_B} - 2(y_P - y_B)I_{z\omega_B} - \frac{Q_{\omega_B}^2}{A} \qquad \text{(C.44)}$$
$$+ (y_P - y_B)^2 I_{yy} + (z_P - z_B)^2 I_{zz} - 2(y_P - y_B)(z_P - z_B)I_{yz}$$

However, because P is a principal pole we have from Eq. (C.19)

$$I_{y\omega_B} + (z_P - z_B)I_{zz} - (y_P - y_B)I_{yz} = 0$$
$$I_{z\omega_B} + (z_P - z_B)I_{yz} - (y_P - y_B)I_{yy} = 0 \qquad \text{(C.45)}$$

which can be used to write Eq. (C.44) as

$$J_{\omega_P} = J_{\omega_B} - \frac{Q_{\omega_B}^2}{A} - (y_P - y_B)^2 I_{yy} \qquad \text{(C.46)}$$
$$+ 2(y_P - y_B)(z_P - z_B)I_{yz} - (z_P - z_B)^2 I_{zz}$$

which gives us the final relationship we sought that allows us to compute the sectorial second moment from a knowledge of the relative location of the principal pole with respect to pole B, the second area moments of the cross-section, and ω_B.

Example C.2 Sectorial Second Moment

To illustrate a calculation of the sectorial second moment, consider the unsymmetrical I-beam of Fig. C.4a. The principal sectorial area function is given in the two flanges (see Fig. C.5) as

$$\omega_P = \begin{cases} (h - e_y)s_1 & -b/2 \leq s_1 \leq b/2 \\ -e_y s_2 & -a/2 \leq s_2 \leq a/2 \end{cases} \tag{C.47}$$

where e_y is given by Eq. (C.27). Then

$$J_{\omega_P} = \int_A \omega_P^2 dA = t \int_{-b/2}^{b/2} (h - e_y)^2 s_1^2 ds_1 + t \int_{-a/2}^{a/2} e_y^2 s_2^2 ds_2$$

$$= \frac{tb^3(h - e_y)^2}{12} + \frac{ta^3 e_y^2}{12} = \frac{th^2 a^3 b^3}{12(a^3 + b^3)} \tag{C.48}$$

where the final result comes after using the expression for e_y and some algebra. Now, let us use instead Eq. (C.46) to calculate the sectorial second moment. In this case we have $Q_{\omega_B} = 0$ as well as $I_{yz} = 0$ and also $z_P = z_B = 0$ so we find

$$J_{\omega_P} = J_{\omega_B} - (y_S - y_B)^2 I_{yy}$$

$$= \int_{-b/2}^{b/2} (hs)^2 t ds - e_y^2 \left[\frac{t(a^3 + b^3)}{12} \right] \tag{C.49}$$

$$= \frac{th^2 b^3}{12} - \frac{th^2 b^6}{12(a^3 + b^3)} = \frac{th^2 a^3 b^3}{12(a^3 + b^3)}$$

which gives the same value.

Now, consider these sectorial area calculations in a more general unsymmetrical section example.

Example C.3 Determining Sectorial Properties of an Unsymmetrical Section

All the examples shown so far have been rather simple sections with some symmetry so that the calculations could be easily illustrated. Now, consider a more general thin unsymmetrical section such as the one shown in Fig. C.6a. We will calculate the principal pole (shear center) for the cross-section and the principal sectorial area function. The thickness, t, is the same for all sections. Using the sections (1), (2) (3) shown in Fig. C.6b the centroid is given by

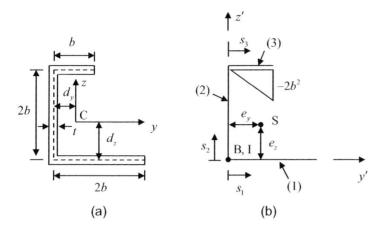

Figure C.6 (a) The geometry of an unsymmetrical section, showing the location, C, of the centroid. (b) The parameters needed to define the location of the shear center and the principal sectorial area function.

$$\bar{y} = \frac{\bar{y}_1 A_1 + \bar{y}_2 A_2 + \bar{y}_3 A_3}{A_1 + A_2 + A_3}$$

$$\bar{z} = \frac{\bar{z}_1 A_1 + \bar{z}_2 A_2 + \bar{z}_3 A_3}{A_1 + A_2 + A_3}$$

(C.50)

where A_m is the area of section (m) for $m = 1, 2, 3$ and where $\bar{y}_m$ is the location of the centroid of A_m in the (y', z') coordinates. Note that all distances are measured along the centerline of the cross-section and areas are computed with these distances also, as is customary when dealing with thin sections. These expressions give

$$\bar{y} = \frac{(b)(2bt) + (0)(2bt) + (b/2)(bt)}{2bt + 2bt + bt} = \frac{b}{2}$$

$$\bar{z} = \frac{(0)(2bt) + (b)(2bt) + (2b)(bt)}{2bt + 2bt + bt} = \frac{4b}{5}$$

(C.51)

where $(\bar{y}, \bar{z})$ here are the coordinates of C in the (y', z') coordinates. For the second area and mixed area moments we have from the parallel axis theorem

$$I_{yy} = I_{\bar{y}1} + A_1 d_{y1}^2 + I_{\bar{y}2} + A_2 d_{y2}^2 + I_{\bar{y}3} + A_3 d_{y3}^2$$
$$I_{zz} = I_{\bar{z}1} + A_1 d_{z1}^2 + I_{\bar{z}2} + A_2 d_{z2}^2 + I_{\bar{z}3} + A_3 d_{z3}^2$$
$$I_{yz} = I_{\bar{y}\bar{z}1} + A_1 d_{y1} d_{z1} + I_{\bar{y}\bar{z}2} + A_2 d_{y2} d_{z2} + I_{\bar{y}\bar{z}3} + A_3 d_{y3} d_{z3}$$

(C.52)

where $I_{\bar{y}m}$ ($m = 1, 2, 3$) are second area moments about a y-axis going through the centroid of each area A_m and $I_{\bar{z}m}$ ($m = 1, 2, 3$) are second area moments about a z-axis going through the centroid of each A_m. The mixed area moments $I_{\bar{y}\bar{z}m}$ ($m = 1, 2, 3$) are similarly measured with to the centroidal axes going through the centroid of each A_m. The distances (d_{ym}, d_{zm}) ($m = 1, 2, 3$) are distances from the (y, z) centroidal axes, respectively, to the individual centroids of the areas A_m. For our specific geometry we have

$$I_{yy} = \frac{2bt^3}{12} + 2bt\left(-\frac{4b}{5}\right)^2 + \frac{t(2b)^3}{12} + 2bt\left(b - \frac{4b}{5}\right)^2 + \frac{bt^3}{12} + bt\left(2b - \frac{4b}{5}\right)^2$$

$$= \frac{52b^3 t}{15}$$

$$I_{zz} = \frac{t(2b)^3}{12} + 2bt\left(\frac{b}{2}\right)^2 + \frac{2bt^3}{12} + 2bt\left(\frac{-b}{2}\right)^2 + \frac{tb^3}{12} + bt(0)^2 \qquad (C.53)$$

$$= \frac{7b^3 t}{4}$$

$$I_{yz} = 0 + 2bt\left(\frac{b}{2}\right)\left(\frac{-4b}{5}\right) + 0 + 2bt\left(\frac{-b}{2}\right)\left(\frac{b}{5}\right) + 0 + bt\left(\frac{6b}{5}\right)(0)$$

$$= -b^3 t$$

where we have neglected all the terms involving t^3 (which was also done in Chapter 10) since these terms are much smaller than the remaining terms for a thin section. To calculate the location of the shear center, we can use point B as both the pole and initial point (origin) to calculate ω_B. These are convenient choices since in the three sections we only have a nonzero function in one section as shown in Fig. C.6b. Specifically, $\omega_B = -2bs_3$ in section (3). Then we can calculate the functions $(I_{y\omega_B}, I_{z\omega_B})$ as

$$I_{y\omega_B} = \int_A y\omega_B dA = \int_0^b -(b/2 - s_3)(-2bs_3)t ds_3 = -\frac{b^4 t}{6}$$

$$I_{z\omega_B} = \int_A z\omega_B dA = \int_0^b (2b - 4b/5)(-2bs_3)t ds_3 = -\frac{6b^4 t}{5} \qquad (C.54)$$

which gives, from Eq. (C.21)

$$e_y = \frac{I_{z\omega_B} I_{zz} - I_{y\omega_B} I_{yz}}{I_{yy} I_{zz} - I_{yz}^2} = \frac{(-6b^4 t/5)(7b^3 t/4) - (-b^4 t/6)(-b^3 t)}{(76b^6 t^2/15)} = \frac{-17b}{38}$$

$$e_z = \frac{I_{z\omega_B} I_{yz} - I_{y\omega_B} I_{yy}}{I_{yy} I_{zz} - I_{yz}^2} = \frac{(-6b^4 t/5)(-b^3 t) - (-b^4 t/6)(52b^3 t/15)}{(76b^6 t^2/15)} = \frac{20b}{57} \qquad (C.55)$$

Now, let us compute the sectorial area function with respect to the principal pole P (the same as the shear center S) and having the same sectorial origin $s_I = I$ as before (Fig. C.7). Then this sectorial area function from the geometry of Fig. C.7 is:

$$\omega_P(s, s_I) = \begin{cases} 20bs_1/57 & 0 \le s_1 \le 2b \\ 17bs_2/38 & 0 \le s_2 \le 2b \\ 17b^2/19 - 94bs_3/57 & 0 \le s_3 \le b \end{cases} \qquad (C.56)$$

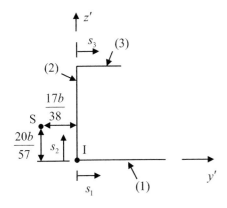

Figure C.7 Calculation of the sectorial area about the shear center with I as the origin.

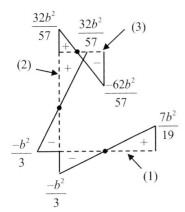

Figure C.8 The principal sectorial area function for the unsymmetrical section of Fig. C.6a.

The first sectorial moment of this function is

$$\int_A \omega_P(s, s_I)dA = \int_0^{2b} \frac{20bs_1}{57} t\,ds_1 + \int_0^{2b} \frac{17bs_2}{38} t\,ds_2 + \int_0^b \left(\frac{17b^2}{19} - \frac{94bs_3}{57}\right) t\,ds_3$$

$$= \left[\frac{40}{57}b^3 t + \frac{34}{38}b^3 t + \frac{17}{19}b^3 t - \frac{47}{57}b^3 t\right] = \left[\frac{80}{114} + \frac{102}{114} + \frac{102}{114} - \frac{94}{114}\right]b^3 t$$

$$= \frac{5}{3}b^3 t$$

(C.57)

Using Eq. (C.36), let us compute

$$\omega_0 = \frac{1}{A}\int_A \omega_P(s, s_I)dA \qquad (C.58)$$

Then

$$\omega_0 = \frac{1}{5bt}\frac{5b^3 t}{3}$$
$$= \frac{b^2}{3}$$
(C.59)

and the principal sectorial area function is $\omega_P(s, s_0) = \omega_P(s, s_I) - \omega_0$ where

$$\omega_P(s, s_0) = \begin{cases} 20bs_1/57 - b^2/3 & 0 \le s_1 \le 2b \\ 17bs_2/38 - b^2/3 & 0 \le s_2 \le 2b \\ 32b^2/57 - 94bs_3/57 & 0 \le s_3 \le b \end{cases}$$
(C.60)

This principal sectorial area function is shown in Fig. C.8 where it can be seen that there are actually three principal origins (shown as black dots) in the principal sectorial area function.

APPENDIX D
MATLAB® Files

At various points in this book, MATLAB® functions or scripts are used to aid in the solution process. The listings of the m-files for those functions or scripts are given below. [Note: because of formatting changes, there may be some minor differences between these listings and the actual m-files.] The m-files can be obtained from Cambridge University Press at www.cambridge.org/schmerr or by sending an e-mail with the subject title "Advanced Mechanics Codes" to the author at lschmerr@iastate.edu

stress_invs

```
function [ I1, I2, I3] = stress_invs(s)
% [ I1, I2, I3] =stress_invs(s) uses the stress matrix s as its
% input argument and returns the three stress invariants (I1,
% I2, I3)

I1 = trace(s);
I2 = s(1,1)*s(2,2) - s(1,2)^2 + s(1,1)*s(3,3) - s(1,3)^2 ...
    + s(2,2)*s(3,3) - s(2,3)^2;
I3 = det(s);
```

cantilever_d

```
%script cantilever_d
% This script plots the 2-d displacement field of a cantilever
% beam from engineering beam theory

% make an 8x4 grid on a beam with normalized length and width
% of 1 and 0.2, respectively.
xn = linspace(0,1,8);
yn = linspace (-0.1, 0.1, 4);
[x,y] = meshgrid(xn, yn);

% choose values for Poisson's ratio and normalized height hn =
% h/L
nu = 1/3;
hn = 0.1;
```

```
% calculate normalized beam displacements

ux = x.^2.*y/2 + (1+nu)*(hn*y - y.^3/3) + nu*y.^3/6 -y/2;
uy = -x.^3/6 - nu*x.*y.^2/2 + x/2 -1/3;

% make a quiver plot of the displacements

quiver(x, y, ux, uy)
axis equal
```

stress_strain

```
function  stress = stress_strain(E, nu, strain)
% stress = stress_strain(E, nu, strain) has as inputs Young's
% modulus, E, (in GPa), Poisson's ratio, nu, and the tensor
% strain matrix, strain, (in micro strain). The function
% calculates the stress matrix (in MPa) for an isotropic
% elastic solid

E = E*10^3; % put E in MPa
e = strain*10^(-6); % put micro factor in strain
C = E/((1+nu)*(1-2*nu));
G = E/(2*(1+nu));
stress = zeros(3,3);
stress(1,1) = C*((1-nu)*e(1,1) +nu*(e(2,2) +e(3,3)));
stress(2,2) = C*((1-nu)*e(2,2) +nu*(e(1,1) +e(3,3)));
stress(3,3) = C*((1-nu)*e(3,3) +nu*(e(1,1) +e(2,2)));
stress(1,2) = 2*G*e(1,2);
stress(1,3) = 2*G*e(1,3);
stress(2,3) = 2*G*e(2,3);
% make stresses symmetric
stress(2,1) = stress(1,2);
stress(3,1) = stress(1,3);
stress(3,2) = stress(2,3);
```

strain_stress

```
function strain = strain_stress(E, nu, stress)
% strain = strain_stress( E, nu ,stress) returns the strain
% matrix strain, measured in microstrain, for a state of
% stress, stress, with the stresses measured in MPa, and where
% E is given in GPa. nu is Poisson's ratio.

% To get strains in microstrain, multiply E by 10^-3
```

```
E = E*10^-3;

G = E/(2*(1+nu));
strain(1,1) = (1/E)*(stress(1,1) -nu*(stress(2,2)... +stress(3,3)));
strain(2,2) = (1/E)*(stress(2,2) -nu*(stress(1,1)... +stress(3,3)));
strain(3,3) = (1/E)*(stress(3,3) -nu*(stress(1,1)... +stress(2,2)));
strain(1,2) = stress(1,2)/(2*G);   % note: these are tensor
% strains
strain(1,3) = stress(1,3)/(2*G);
strain(2,3) = stress(2,3)/(2*G);
strain(2,1) = strain(1,2);
strain(3,1) = strain(1,3);
strain(3,2) = strain(2,3);
```

M_Matrix

```
function M = M_Matrix(l)
% M = M_Matrix(l) takes  as its input argument the direction
% cosine matrix l whose transpose, lt, transforms the x-
% coordinates to the x'-coordinates through the relation {x'}
% = [lt]{x}. The function returns the 6x6 matrix, M, which
% transforms the 6x6 matrix of elastic constants,C, as
% measured in the x-coordinates, to the matrix of elastic
% constants,C', as measured in the x'-coordinates, through the
% relation [C'] = [M][C][Mt], where Mt is the transpose of M.

M = zeros(6,6);
M(1,1) = l(1,1)^2;
M(1,2) = l(2,1)^2;
M(1,3) = l(3,1)^2;
M(1,4) = 2*l(2,1)*l(3,1);
M(1,5) = 2*l(3,1)*l(1,1);
M(1,6) = 2*l(2,1)*l(1,1);
M(2,1) = l(1,2)^2;
M(2,2) = l(2,2)^2;
M(2,3) = l(3,2)^2;
M(2,4) = 2*l(2,2)*l(3,2);
M(2,5) = 2*l(1,2)*l(3,2);
M(2,6) = 2*l(1,2)*l(2,2);
M(3,1) = l(1,3)^2;
M(3,2) = l(2,3)^2;
M(3,3) = l(3,3)^2;
M(3,4) = 2*l(2,3)*l(3,3);
```

```
M(3,5) = 2*l(1,3)*l(3,3);
M(3,6) = 2*l(1,3)*l(2,3);
M(4,1) = l(1,2)*l(1,3);
M(4,2) = l(2,2)*l(2,3);
M(4,3) = l(3,2)*l(3,3);
M(4,4) = l(2,2)*l(3,3) + l(3,2)*l(2,3);
M(4,5) = l(1,2)*l(3,3) + l(3,2)*l(1,3);
M(4,6) = l(1,2)*l(2,3) + l(2,2)*l(1,3);
M(5,1) = l(1,1)*l(1,3);
M(5,2) = l(2,1)*l(2,3);
M(5,3) = l(3,1)*l(3,3);
M(5,4) = l(2,1)*l(3,3) + l(3,1)*l(2,3);
M(5,5) = l(1,1)*l(3,3) + l(3,1)*l(1,3);
M(5,6) = l(1,1)*l(2,3) + l(2,1)*l(1,3);
M(6,1) = l(1,1)*l(1,2);
M(6,2) = l(2,1)*l(2,2);
M(6,3) = l(3,1)*l(3,2);
M(6,4) = l(2,1)*l(3,2) + l(3,1)*l(2,2);
M(6,5) = l(1,1)*l(3,2) + l(3,1)*l(1,2);
M(6,6) = l(1,1)*l(2,2) + l(2,1)*l(1,2);
```

---

m_to_v

```
function y = m_to_v(input, type)
% m_to_v converts a 3x3 matrix of strains or stresses to a 6x1
% vector
% first input argument is the 3x3 matrix
% the type argument is either 'strain' or 'stress'
% if type = 'strain' then the matrix shear strains are doubled
% to give engineering shear strains
if strcmp(type, 'stress')
    y(1) = input(1,1);
    y(2) = input(2,2);
    y(3) = input(3,3);
    y(4) = input(2,3);
    y(5) = input(1,3);
    y(6) = input(1,2);
elseif strcmp(type, 'strain')
    y(1) = input(1,1);
    y(2) = input(2,2);
    y(3) = input(3,3);
```

```
    y(4) = 2*input(2,3);
    y(5) = 2*input(1,3);
    y(6) = 2*input(1,2);
else
    error('wrong type')
end
% put in column vector
y = y';
```

v_to_m

```
function y = v_to_m(input, type)
% v_to_m converts a 6x1 vector of strains or stresses to a 3x3
% matrix
% first input argument is the 6x1 vector
% the type argument is either 'strain' or 'stress'
% if type = 'strain' then the vector engineering shear strains
% are halved to generate the tensor shear strains in the 3x3
% matrix
if strcmp(type, 'stress')
    y(1,1) = input(1);
    y(2,2) = input(2);
    y(3,3) = input(3);
    y(2,3) = input(4);
    y(1,3) = input(5);
    y(1,2) = input(6);
elseif strcmp(type, 'strain')
    y(1,1) = input(1);
    y(2,2) = input(2);
    y(3,3) = input(3);
    y(2,3) = input(4)/2;
    y(1,3) = input(5)/2;
    y(1,2) = input(6)/2;
else
    error('wrong type')
end
y(2,1) = y(1,2);
y(3,1) = y(1,3);
y(3,2) = y(2,3);
```

rosette

```
function [exx, eyy, exy] = rosette(anga, angb, angc, ea, eb,... ec)
% [exx, eyy, exy] = rosette(anga, angb, angc, ea, eb, ec)
% takes the three angles (anga, angb, angc) of the three gages
% in a rosette gage (as measured in degrees from the x-
% axis)and the three strains (ea, eb, ec)measured by those
% gages, and returns the tensor strains(exx, eyy, exy)
nax = cos(anga*pi/180);
nay = sin(anga*pi/180);
nbx = cos(angb*pi/180);
nby = sin(angb*pi/180);
ncx = cos(angc*pi/180);
ncy = sin(angc*pi/180);
Coeff = [ nax^2   nay^2   2*nax*nay;...
    nbx^2   nby^2   2*nbx*nby; ...
    ncx^2   ncy^2   2*ncx*ncy] ;
vals = [ ea eb ec]';
strains = Coeff\vals;
exx = strains(1);
eyy = strains(2);
exy = strains(3);
```

stress_singularity

```
function stress_singularity()
% stress_singularity determines the singular behavior of the
% stresses near a sharp notch and plots the behavior of the
% parameter that controls the order of the singularity as a
% function of the notch angle.
warning off
a = linspace(0, pi, 100); %angle alpha
out = zeros (1,100);
for k =1:100
    val = 0.5 +0.5*a(k)/pi;  % choose initial guesses
    out(k) = fzero(@singularity, val, [ ],a(k)); % find roots
end
plot((180/pi)*a(1:100), out(1:100))
axis([ 0 180 0 1.0])
xlabel('\alpha , degrees')
```

```
ylabel(' smallest root, m')
warning on

function y = singularity(x, a)
y = sin (x*(2*pi-a)) - x*sin(a);
end
end
```

cross_section

```
function [Yct, Zct, Iy, Iz, Iyz] = cross_section( Ly, Lz,...
 Yc, Zc)
% [Yct, Zct, Iy, Iz, Iyz] = cross_section(Ly, Lz, Yc, Zc)
% determines the coordinates of the centroid of a cross-
% section and the area moments relative to that centroid for a
% cross-section composed of rectangular parts. Ly and Lz are
% vectors containing the lengths of the rectangles in the Y-
% and Z-directions, and Yc, Zc are vectors containing the
% locations of the Y- and Z-coordinates of the rectangles. The
% outputs are the Y- and Z-coordinates of the cross-section
% centroid (Yct, Zct) and the area moments (Iy, Iz, Iyz)
% relative a set of centroidal (y,z) axes. Note that the
% rectangles must lie parallel to the (Y,Z) axes.

A = sum(Ly.*Lz);    % total area of cross-section

% calculate location of centroid for the total cross-section
Yct = sum((Yc.*Ly.*Lz)/A); % Y-coordinate of centroid in (Y,Z)
Zct = sum((Zc.*Ly.*Lz)/A); % Z-coordinate of centroid in (Y,Z)

% calculate area moments about the (Y,Z) axes
IY = sum(Ly.*(Lz.^3)/12 +Ly.*Lz.*(Zc.^2)); % IYY in (Y,Z)
% coordinates
IZ = sum(Lz.*(Ly.^3)/12 +Ly.*Lz.*(Yc.^2)); % IZZ in (Y,Z)
% coordinates
IYZ = sum(Ly.*Lz.*Yc.*Zc); % IYZ in (Y,Z) coordinates

% use parallel axis theorem to calculate area moments about
% centroidal coordinates (y,z)
Iy = IY - A*(Zct^2);   % Iyy for centroidal (y,z) axes
Iz = IZ -A*(Yct^2);    % Izz for centroidal (y,z) axes
Iyz =IYZ - A*Yct*Zct;  % Iyz for centroidal (y,z) axes
```

---
sections

```
% script sections
% This script computes the location of the centroid and area
% moments for a L-section, a T-section, a J-section, a Z-
% section, and a I-section. In the command window one must
% enter a string 'L', 'T', 'J', 'Z', or 'L' in a variable
% named type and then enter the lengths b, h, c and the
% thicknesses t1, t2, t3 (or a subset of these values) in the
% command window and then run this script. The script uses the
% function cross_section to return the location of the
% centroid and the area moments
switch type
    case{'L'}
        Ly(1) = b- t2; Ly(2) = t2;
        Lz(1) = t1; Lz(2) = h;
        Yc(1) = t2 + (b-t2)/2; Yc(2) =t2/2;
        Zc(1) = t1/2; Zc(2) = h/2;
    case{'T'}
        Ly(1) = b; Ly(2) = t2;
        Lz(1) = t1; Lz(2) = h-t1;
        Yc(1) = 0; Yc(2) = 0;
        Zc(1) = t1/2; Zc(2) = t1 + (h- t1)/2;
    case{'J'}
        Ly(1) = b -t2; Ly(2) = t2; Ly(3) = c-t2;
        Lz(1) = t1; Lz(2) = h; Lz(3) = t3;
        Yc(1) = t2 + (b-t2)/2; Yc(2) = t2/2; Yc(3) = t2 ...
+(c-t2)/2;
        Zc(1) = t1/2; Zc(2) = h/2; Zc(3) = h - t3/2;
    case{'Z'}
        Ly(1) = b - t2; Ly(2) = t2; Ly(3) = c -t2;
        Lz(1) = t1; Lz(2) = h; Lz(3) = t3;
        Yc(1) = t2 + (b-t2)/2; Yc(2) = t2/2; Yc(3) = -(c-t2)/2;
        Zc(1) = t1/2; Zc(2) = h/2; Zc(3) = h -t3/2;
    case{'I'}
        Ly(1) = b; Ly(2) = t2; Ly(3) = c;
        Lz(1) = t1; Lz(2) = h - t1 -t3; Lz(3)  = t3;
        Yc(1) = 0; Yc(2) = 0; Yc(3) = 0;
        Zc(1) = t1/2; Zc(2) = t1 + (h -t1 -t3)/2;Zc(3) = ...
h -t3/2;
    otherwise
        disp(['Unknown type'])
end

[yc, zc, Iy, Iz, Iyz] = cross_section(Ly, Lz, Yc, Zc)
```

Index

Airy stress function, 149, 153, 190
Airy stress function, Cartesian coordinates, 191
Airy stress function, polar coordinates, 197
alternating stress, 502

Beltrami–Michell compatibility equations, 149, 539
bending moment, 6
Betti–Rayleigh theorem, *see* reciprocal theorem of Betti–Rayleigh
biaxial state of stress, 51
bifurcation instability, 517
biharmonic equation, 158
biharmonic equation, Cartesian coordinates, 191
biharmonic equation, polar coordinates, 197
biharmonic operator, 153
bimoment, 490
boundary conditions, 158
boundary element method, 148, 361
buckling, 515

Castigliano's first theorem, 256
Castigliano's second theorem, 260
center of twist, 433
centroid, 527, 540
Clapeyron's theorem, 229
combined bending and torsion, thin, closed sections, 477
combined bending and torsion, thin, open sections, 475
compatibility equations, 101, 303
compatibility matrix, 304, 312
complementary strain energy density, 233
compliance matrix, 124
concentrated force on a planar surface, 209
concentrated force on a wedge, 208
configuration factor, 509
constant life curves, 503
critical buckling load, 516
critical flaw size, 511
cubic material, 127
curved beam, 202

deviatoric stress, 67
dilatation, 98

direction cosine matrix, 35
distortional strain energy density, 232
dummy load, 261

eddy current NDE, 513
effective maximum radius, 414
effective polar area moment, 414, 527
effective stress, 61, 500
eigenvalue problem, 44
eigenvalues, 44
eigenvectors, 44
elastic bodies in contact, *see* Hertz contact theory
elastic constants, 122
Engesser's first theorem, 260
Engesser's second theorem, 264
engineering beam theory, 9
engineering shear strain, 14, 90
equilibrium equations, Cartesian coordinates, 71
equilibrium equations, cylindrical coordinates, 82
equilibrium equations, plane strain, 154
equilibrium equations, plane stress, 84
equilibrium equations, spherical coordinates, 83
equivalent stress, 61
Euler buckling load, *see* critical buckling load

fatigue failure, 502
fatigue strength, 504
finite differences method, 148
first area moments, 527, 540
first sectorial moment, 542
flexibility matrix, 299
flexure stress, unsymmetrical bending, 375
force-based finite element method, axial loads, 295
force-based finite element method, beam bending, 329
force-based finite element method, general body, 358
force-based finite element method, specified displacements, 347
four pillars, 18
fracture mechanics, 508
fracture toughness, 510
fundamental solutions, 362

Gauss's theorem, 70
Gauss's theorem -2D, 76
generalized displacements, 174
generalized forces, 174
generalized Hooke's law, 119
generalized strains, 174
Gerber curve, 504
Goodman curve, 504
governing equations, 146

Hertz contact theory, 212
Hooke's law, 5

interference, 188
isotropic material, 119, 127

Kronecker delta, 35

Lamé constants, 148
Laplace's equation, 412
limit-load instability, 517
local average rotation, 94

maximum distortional strain energy theory, 499
maximum normal stress theory, 496
maximum shearing stress theory, 497
Maxwell stress function, 149
mean stress, 502
membrane analogy, 423
Michell solutions, 197
Miner rule, *see* Palmgren-Miner rule
mixed second area moment, 8, 527, 540
modified sectorial area function, 459
modified warping constant, 462
Mohr's circle, 53
moment–curvature relationship, 8
Morera stress function, 149
Morrow curve, 504

Navier's equations, 147, 310
Navier's table problem, 166
Neumann problem, 435
neutral axis, 7
nonuniform torsion, 427
normal strain, 3, 88
normal stress, 4, 32
normalized Prandtl stress function, 419
null space, 314

565

octahedral plane, 60
orthogonal matrix, 37
orthotropic material, 124

Palmgren–Miner rule, 507
parallel axis theorem, 531
Paris law, 512
plane strain, 154
plane stress, 51, 73
planes of extreme shear stress, 56
Poisson's equation, 418
Poisson's ratio, 3
polar area moment, 15, 540
Prandtl stress function, 149, 417
primary twisting moment, 436
primary warping function, 428
principal area moments, 531
principal directions, 42
principal origin, 549
principal planes, 42
principal pole, 545
principal sectorial area function, 527, 543
principal strains, 95
principal stresses, 42
principle of complementary virtual work, 250
principle of complementary virtual work - axial loads, 280
principle of complementary virtual work - beam bending, 319
principle of least work, 264
principle of minimum complementary potential energy, 251
principle of virtual work, 240
principle of virtual work, axial loads, 278
principle of virtual work, beam bending, 316
proper orthogonal matrix, 37
pseudostiffness matrix, 311, 333
pure bending of a curved beam, 202

Rayleigh–Ritz method, 257
reciprocal theorem of Betti–Rayleigh, 268
reciprocity, 267
redundants, 263

rosette strain gage, *see* strain gage rosette
row-reduced echelon form, 313

Saint-Venant torsional theory, *see* uniform torsion
Saint-Venant's principle, 161
second area moments, 8, 527, 540
secondary twisting moment, 436
secondary warping, 428
sectorial area function, 391, 540
sectorial area origin, 541
sectorial area pole, 541
sectorial products of areas, 527, 542
sectorial second moment, *see* warping constant
shape functions, 281
shear stress, 30
shear center, 389
shear flow, 17, 385
shear flow expression, unsymmetrical bending, 386
shear force, 9
shear modulus, 14
shear strain, 89
shear stress, 9
shrink-fit, 187
single plane of loading, 376
single-valued displacement, 103
singularity functions, 20
S–N curve, 502
snap-through buckling, 517
Soderberg curve, 504
Somigliana's identity, 364
state of strain, 93
state of stress, 29
stiffness matrix, 148, 283
stiffness-based finite element method, 148
stiffness-based finite element method, axial loads, 281
stiffness-based finite element method, beam bending, 320
stiffness-based finite element method, general body, 356
strain energy, 226
strain energy density, 226
strain gage rosette, 141

strain invariants, 95
strain transformation equations, 95
strain–displacement relations, 91
strains, cylindrical coordinates, 112
stress concentration at a circular hole, 198
stress intensity factor, 509
stress invariants, 43
stress range, 502
stress singularities, 214
stress transformation equations, 34
stress vector, 28

tensor shear strain, 90
theorem of minimum potential energy, 242
thick-wall pressure vessel, 183
total shear stress, 32
traction vector, *see* stress vector
transformation of elastic constants, 132
transversely isotropic material, 126
Tresca theory, *see* maximum shearing stress theory
true stress, 504

ultimate stress, 504
ultrasonic NDE, 513
uniform torsion, 411
uniqueness, 243
unit load method, 262

virtual displacements, 239
virtual work, 240
Voigt notation, 124
von Mises failure theory, 499
von Mises stress, 61, 500

warping constant, 432, 527
warping constant, thin sections, 447
warping function, 411
warping normal stress, 428
work, 224

X-ray NDE, 514

yield normal stress, 498
yield shear stress, 497
Young's modulus, 5